MW01356358

Industrial and Applied Mathematics

Industrial and Applied Mathematics, a Scopus-indexed book series, publishes high-quality research-level monographs, lecture notes and contributed volumes focusing on areas where mathematics is used in a fundamental way, such as industrial mathematics, bio-mathematics, financial mathematics, applied statistics, operations research and computer science.

Abdon Atangana · İlknur Koca

Fractional Differential and Integral Operators with Respect to a Function

Theory Methods and Applications

Abdon Atangana
University of the Free State
Bloemfontein, South Africa

İlknur Koca
Muğla Sıtkı Koçman University
Kötekli/Muğla, Türkiye

ISSN 2364-6837 ISSN 2364-6845 (electronic)
Industrial and Applied Mathematics
ISBN 978-981-97-9950-3 ISBN 978-981-97-9951-0 (eBook)
https://doi.org/10.1007/978-981-97-9951-0

This Springer imprint is published by the registered company Springer Nature Singapore Pte Ltd.
The registered company address is: 152 Beach Road, #21-01/04 Gateway East, Singapore 189721, Singapore

Preface

A remark regarding a future occurrence or piece of data is known as a prediction or forecast. They are frequently, but not always, based on knowledge or experience. There isn't a consensus across all disciplines or authors regarding the precise distinction between estimation and actuality. Because future occurrences are inevitably unpredictable, it is impossible to know the future with certainty. Predictions can be helpful when preparing for potential outcomes. In general, people use this to influence their surroundings. It is important to note that there are various methods for making forecasts. Regression and its several subcategories, such as linear regression, generalized linear models, including logistic regression, Poisson regression, and Probit regression, among others, are statistical techniques used for prediction. For instance, according to theories of gravity, if an apple fell from a tree, it would be seen to move toward the center of the earth with a specified and constant acceleration. A prediction is a rigorous, frequently quantitative, statement that foretells what would be observed under specific conditions. Testing claims that follow logically from scientific ideas is the foundation of the scientific method. Replicable experiments or observational studies are used to accomplish this. The past and future behavior of a process within the parameters of a model is frequently described using mathematical equations, models, and computer simulations. In some circumstances, such as in much of quantum physics, it is possible to forecast the probability of an occurrence rather than a particular outcome. We point out that Bernhard Riemann's invention of the Riemann integral was the first accurate formulation of the integral of a function on an interval in real analysis. Although it was presented to the University of Göttingen faculty in 1854, it wasn't until 1868 that it was published in a journal. The Riemann integral can be calculated using the fundamental theorem of calculus for a variety of functions and real-world applications, or it can be roughly estimated using numerical integration or simulated using Monte Carlo integration. We point out that Bernhard Riemann's invention of the Riemann integral was the first accurate formulation of the integral of a function on an interval in real analysis. Although it was presented to the University of Göttingen faculty in 1854, it wasn't until 1868 that it was published in a journal. The Riemann integral can be calculated using the fundamental theorem of calculus for a variety of functions and real-world applications, or it can be roughly estimated using numerical integration or simulated using Monte Carlo integration. For many theoretical uses, the Riemann integral is inappropriate. The Riemann–Stieltjes integral can fix some of the technical issues with Riemann integration, and the Lebesgue integral eliminates most of them; however, it does not adequately handle improper integrals. There are now more avenues for theoretical and practical research thanks to the extension to the Riemann–Stieltjes integral. In recent decades, various authors separately obtained a derivative about another function to create a differential calculus connected to this integral. On the other hand, throughout the past few decades, there has been a lot of discussion and research into fractional differential and integral operators. Fractional derivatives and integral with respect to another function were proposed as an extension of these operators. The Riemann–Stieltjes integral and derivative can alternatively be seen as an extension of the fractional calculus framework. In recent years, there hasn't been much theoretical study done. This book's goal is to present theoretically sound conclusions that have already been discovered in the fields of inequalities, nonlinear ordinary differential equations, numerical approximations, and certain applications. Basic definitions and characteristics of these differential and integral equations are presented at the beginning of the book. Following this, various inequalities that have applications in many areas of science, technology, and engineering are derived. The theory of the existence and uniqueness of nonlinear ordinary differential equations with these derivatives was thoroughly discussed in the following sections. The theoretical derivation of the numerical technique for nonlinear differential equations related to these derivatives is then covered in detail in the sections that follow. The final section includes some applications using ordinary and partial differential equations.

Bloemfontein, South Africa — Abdon Atangana
Kötekli/Muğla, Türkiye — İlknur Koca

Acknowledgments The LORD is my shepherd; I shall not want. He maketh me to lie down in green pastures: he leadeth me beside the still waters. He restoreth my soul: he leadeth me in the paths of righteousness for his name's sake. Yea, though I walk through the valley of the shadow of death, I will fear no evil: for thou art with me; thy rod and thy staff they comfort me. Thou preparest a table before me in the presence of mine enemies: thou anointest my head with oil; my cup runneth over. Surely goodness and mercy shall follow me all the days of my life: and I will dwell in the house of the LORD forever. Both authors would like to thank God the Almighty who gave them strength, good health, and knowledge to put this book together.

Contents

Symbols

$\mathbb{R}$ is set of real numbers
$\Gamma(.)$ is Gamma Function
$\beta(x, y)$ is Beta Function
$E_{\alpha,\beta}(.)$ is Mittag-Leffler function
$\binom{n}{k}$ is binomial coefficient
$f * g$ is convolution of f and g
$\|\cdot\|_p$ is L_p norm $(1 \leq p < \infty)$
$\sum_{j=1}^{n} a_i$ is sum of a_i from 1 to n
$C[a, b]$ is set of all continuous real-valued functions on the interval $[a, b]$
L^p is p integrable functions
$\overline{B}(b, y_0)$ is a closed ball of center b and ratio y_0
$B(b, y_0)$ is a open ball of center b and ratio y_0
$E(x)$ is expectation of x
$M(\alpha)$ is normalization function for Caputo–Fabrizio derivative
$AB(\alpha)$ is normalization function for Atangana–Baleanu derivative
$f'(t)$ is the derivative of $f(t)$
${}_0D_t^{\alpha}f(t)$ is the fractional derivative of $f(t)$
$D_g f(t)$ is the derivative of $f(t)$ with respect to the function $g(t)$
${}_0^{RL}D_g^{\alpha}f(t)$ is the Riemann–Liouville derivative of $f(t)$ with respect to $g(t)$
${}_0^{C}D_g^{\alpha}f(t)$ is the Caputo derivative of $f(t)$ with respect to $g(t)$
${}_0^{RCF}D_g^{\alpha}f(t)$ is the Caputo–Fabrizio derivative of $f(t)$ with respect to the function $g(t)$ in RL sense
${}_0^{CCF}D_g^{\alpha}f(t)$ is the Caputo–Fabrizio derivative of $f(t)$ with respect to the function $g(t)$ in Caputo sense
${}_0^{ABR}D_g^{\alpha}f(t)$ is the Atangana–Baleanu derivative of $f(t)$ with respect to the function $g(t)$ in RL sense
${}_0^{ABC}D_g^{\alpha}f(t)$is the Atangana–Baleanu derivative of $f(t)$ with respect to the function $g(t)$ in Caputo sense
${}_0J_t^{\alpha}f(t)$ is the fractional integral of $f(t)$

$\int_a^b f(t)dg(t)$ is the integral of f with respect to $g(t)$ within $[a, b]$
${}_0J_g^\alpha f(t)$ is the fractional integral of $f(t)$ with respect to the function $g(t)$
${}_0^{RL}J_g^\alpha f(t)$ is the Riemann–Liouville integral of $f(t)$ with respect to $g(t)$
${}_0^{RCF}J_g^\alpha f(t)$ is the Caputo–Fabrizio integral of $f(t)$ with respect to the function $g(t)$ in RL sense
${}_0^{ABR}J_g^\alpha f(t)$ is the Atangana–Baleanu integral of $f(t)$ with respect to the function $g(t)$ in RL sense

Chapter 1
History of Differential and Integral Calculus

People in all social classes in each society are curious about what will happen tomorrow depending on a particular circumstance. For instance, in a pandemic crisis, policymakers, citizens, and members of other classes are interested in knowing when the curve will flatten and what steps can be taken to ensure that the curve flattens soon. When there is a conflict, everyone citizens, combatants, and governments wants to know when things will return to normal and who will win. They then implemented several systems to foresee how things would turn out. In general, when faced with difficult circumstances, people want to know when things will get better. When something pleasant happens, people often wonder how long it will last and how they might make it better. This uncertainty and doubt prompted our ancestors to explore mathematical ideas, such as the concept of a function with variables t or x, and t, which denotes that such a function depends just on time or on time and space. Infinitesimal was an earlier attempt. Perhaps Nicolaus Mercator or Gottfried Wilhelm Leibniz proposed the idea of infinitesimals for the first time in the year 1670. In his book The Method of Mechanical Theorems, Archimedes utilized what is now known as the method of indivisibles to calculate the volumes and surface areas of solids. Archimedes used the method of exhaustion to resolve the issue in his formal, written treatises. Nicholas of Cusa's work from the fifteenth century was expanded upon by Johannes Kepler in the seventeenth, particularly his method of calculating the surface area of a circle by modeling it as an infinite-sided polygon. The real continuum was made possible by Simon Stevin's work in the sixteenth century on the decimal representation of all numbers. The indivisibles approach of Bonaventura Cavalieri allowed the findings of the ancient authors to be expanded. The concept of indivisibles was applied to geometrical figures as being made up of codimension 1 elements. In contrast to indivisibles, John Wallis' infinitesimals would break down geometrical forms into infinitely thin units of the same dimension, laying the foundation for more widespread integral calculus techniques. The law of continuity,

A. Atangana and İ. Koca, *Fractional Differential and Integral Operators with Respect to a Function*, Industrial and Applied Mathematics,
https://doi.org/10.1007/978-981-97-9951-0_1

which states that what works for finite numbers also works for infinite numbers and vice versa, and the transcendental law of homogeneity, which outlines procedures for replacing expressions involving unassignable quantities with expressions involving only assignable ones, were heuristic principles on which Leibniz's use of infinitesimals was based. Numerous mathematicians of the eighteenth century, including Joseph–Louis Lagrange and Leonhard Euler, used infinitesimals frequently. It is significant to remember that Leibniz's development of calculus includes infinitesimals, along with the laws of continuity and homogeneity. An object that is infinitesimally small but not zero in size or so small that it cannot be distinguished from zero by any means is referred to as an infinitesimal object. Thus, the word infinitesimal in mathematics refers to something that is infinitely little, smaller than any regular real number. While studying the derivative of a function, infinitesimals are frequently compared to other infinitesimals of comparable size. To calculate an integral, an unlimited number of infinitesimals are added together.

In the same way that geometry is the study of shape and algebra is the study of generalizations of arithmetic operations, calculus is the study of continuous change from a mathematical perspective. The two main branches of it are differential calculus and integral calculus: the former deals with instantaneous rates of change and curve slopes, while the latter deals with accumulation of quantities and areas under or between curves. The fundamental theorem of calculus connects these two branches, and they both utilize the fundamental ideas of convergence of infinite sequences and infinite series to a well-defined limit number of infinitesimals that are added to calculate an integral. Issac Newton and Gottfried Wilhelm Leibniz independently created infinitesimal calculus in the late seventeenth century. Subsequent works, such as codifying the concept of limits, gave these advancements a conceptual foundation that was firmer. Calculus is now widely used in social science, engineering, and science. Yet, it is important to highlight that while Isaac Newton and Gottfried Wilhelm Leibniz separately invented modern calculus in seventeenth-century Europe and published their work at roughly the same time, its foundations may be found in ancient Greece, China, the Middle East, medieval Europe, and India. It is worth noting that this concept gave rise to a new field of mathematics and its applications. Indeed, differential calculus is a branch of mathematics that analyzes the rates at which quantities change. It is one of calculus' two traditional divisions, the other being integral calculus, which studies the area beneath a curve. The derivative of a function, related concepts like the differential, and its applications are the core topics of study in differential calculus. A function's derivative at a certain input value describes the function's rate of change near that input value. Differentiation is the process of determining a derivative. The slope of the tangent line to the graph of the function at that position is the derivative at that location if the derivative exists and is defined at that point. The derivative of a function at a place determines the best linear approximation to the function at that position for a real-valued function with a single real variable. It is vital to note that the fundamental theorem of calculus, which asserts that differentiation is the opposite process to integration, connects differential calculus and integral calculus. In the last few decades, differentiation has found use in almost every discipline of science, technology, and engineering. For instance, in some areas of applied research, such as physics, the body's velocity and acceleration are the derivatives of the displacement of a moving body with respect to time and time, respectively. By putting all in order, we can observe that this derivative statement results in the famous $F = ma$ equation connected to Newton's second rule of motion. The derivative of a body's momentum with respect to time equals the force applied to the body. Another important example of the application of this mathematical concept is the derivative of a chemical reaction which is the reaction rate. Derivatives are used in operations research to build factories and find the most effective means of material transportation. They are used to analyze a few significant behaviors of functions; for example, derivatives are typically employed to locate a function's maxima and minima. Differential equations, often known as equations containing derivatives, are essential for describing natural events. Numerous areas of mathematics, including complex analysis, functional analysis, differential geometry, measure theory, and abstract algebra, contain derivatives and their generalizations. In addition to the use of the derivative notion already described, it is important to realize that derivatives are how differential equations are created. A differential equation can be defined mathematically as the relationship between a group of functions and their derivatives. The relationship between functions of one variable and their derivatives with respect to that variable is the basis of an ordinary differential equation. Any differential equation that connects functions of more than one variable to their partial derivatives is referred to as a partial differential equation.

Let us revert to the mathematical concept of derivative. An important theorem known as the Mean Value theorem helps us to have a relation between $(a, f(a))$ and $(b, f(b))$ of a function $f(t)$ within the interval $a < b$. This theorem tells us that the slope between $(a, f(a))$ and $(b, f(b))$ is the same as the slope of the convert line to f a some point $\overline{c}$ found between $a < b$. Mathematically speaking, the theorem tells us that

$$f'(\overline{c}) = \frac{f(b) - f(a)}{b - a} \tag{1.1}$$

where $f'(t)$ can be denoted as

$$\frac{d}{dt} f(t) = \lim_{\Delta t \to 0} \frac{f(t + \Delta t) - f(t)}{\Delta t}. \tag{1.2}$$

We can have

$$f^{(2)}(t) = \left(f'(t)\right)' = \lim_{\Delta t \to 0} \frac{f'(t + \Delta t) - f'(t)}{\Delta t}. \tag{1.3}$$

From the above formula some important equations have been introduced. We can list Newton's second law that explains the relation between a force, mass, and acceleration. Newton's second law of motion states that the force acting on an object is equal to the mass of that object multiplied by its acceleration. Mathematically, it can be expressed as

$$F(t) = ma \tag{1.4}$$

where F is the force applied to the object, m is the mass of the object, and a is the acceleration of the object.

In the case of partial differential equation, we can list the heat equation. The heat equation is a partial differential equation that describes how the distribution of heat evolves over time in a given region. It's commonly used in physics and engineering to model the behavior of temperature in various systems, such as solids, liquids, and gases. The formula of heat equation is given as

$$\frac{\partial u(x,t)}{\partial t} = \alpha \frac{\partial^2 u(x,t)}{\partial x^2}. \tag{1.5}$$

where $u(x,t)$ represents the temperature distribution as a function of space (x) and time (t). $\frac{\partial u(x,t)}{\partial t}$ is the rate of change of temperature with respect to time, $\frac{\partial^2 u(x,t)}{\partial x^2}$ is the second derivative of temperature with respect to space, and α is the thermal diffusivity, a material property that relates how quickly heat spreads through the medium.

The groundwater flow equation is a fundamental equation used to describe the movement of groundwater through porous media such as soil and rock. It is a partial differential equation that governs the flow of water in an aquifer. It is the groundwater flow equation

$$\frac{S}{T}\frac{\partial h(r,t)}{\partial t} = \frac{\partial h(r,t)}{\partial r} + \frac{\partial^2 h(r,t)}{\partial r^2}. \tag{1.6}$$

The above equation is the modified groundwater equation of unsteady flow toward the well. In this equation, h is hydraulic head, r is radial distance from the well, S is storage coefficient, T is transmissivity, and t is the time since the beginning of pumping.

The advection–dispersion equation is a partial differential equation commonly used to describe the transport of solutes (such as pollutants or contaminants) in groundwater and surface water systems. It's a fundamental equation in hydrogeology and contaminant transport modeling. The advection dispersion equation is given as

$$\frac{\partial u(x,t)}{\partial t} = -v\frac{\partial u(x,t)}{\partial x} + \Delta \frac{\partial^2 u(x,t)}{\partial x^2}. \tag{1.7}$$

Here u is the concentration of the solute, t is time, x is spatial position, Δ is the dispersion coefficient (which includes both molecular diffusion and hydrodynamic dispersion), and v is the velocity of groundwater flow.

The SIRD model is a mathematical framework used in epidemiology to understand and predict the spread of infectious diseases within a population. It stands for Susceptible, Infected, Recovered, and Deceased. The model divides the population into these four compartments and tracks the flow of individuals among them over time.

Here's a brief explanation of each compartment:

Susceptible (S): This compartment represents individuals who are susceptible to the disease and can potentially become infected if they come into contact with infectious individuals.

Infected (I): Individuals in this compartment are currently infected with the disease and are capable of spreading it to susceptible individuals through transmission.

Recovered (R): This compartment represents individuals who have recovered from the disease and have developed immunity to it. They are no longer capable of being infected or transmitting the disease.

Deceased (D): This compartment accounts for individuals who have died as a result of the disease.

This model uses the following system of differential equations:

$$\begin{aligned} \frac{dS}{dt} &= -\frac{\beta IS}{N}, \\ \frac{dI}{dt} &= \frac{\beta IS}{N} - \gamma I - \mu I, \\ \frac{dR}{dt} &= \gamma I, \\ \frac{dD}{dt} &= \mu I, \end{aligned} \tag{1.8}$$

where β, γ, μ are the rates of infection, recovery, and mortality parameters, respectively.

The Chen attractor is a type of strange attractor, which is a complex geometric object that arises in the study of chaotic dynamical systems. The Chen attractor is a visual representation of the behavior of a chaotic system. Unlike regular attractors, which draw trajectories toward a specific point or limit cycle, strange attractors like the Chen attractor have a fractal-like structure and exhibit complex, non-repeating patterns.

The equations that govern the Chen system are given as

$$\begin{aligned} \frac{dx}{dt} &= a(y - x), \\ \frac{dy}{dt} &= (c - a)x - xz + cy, \\ \frac{dz}{dt} &= xy - bz \end{aligned} \tag{1.9}$$

where x, y, and z are variables representing the state of the system at any given time, a, b, and c are parameters that determine the behavior of the system, and t represents time. When these equations are solved numerically and plotted in three-dimensional space, they produce a chaotic attractor with intricate, swirling patterns. The shape of the attractor and the behavior of trajectories within it depend on the values of the parameters a, b, and c.

The Lotka–Volterra equations, also known as the predator–prey equations, are a pair of first-order, nonlinear, differential equations frequently used to describe the dynamics of biological systems involving two interacting species. In this case, let's denote the two species as prey and predator. The equations describe how the populations of these species change over time based on certain assumptions. Let x represent the population of the prey species and y represent the population of the predator species. Then the Lotka–Volterra equation is given as

$$\frac{dx}{dt} = \alpha x - \beta xy, \qquad (1.10)$$
$$\frac{dy}{dt} = \delta xy - \gamma y$$

where α represents the prey's intrinsic growth rate, it signifies how fast the prey population would grow in the absence of predators; β represents the predation rate, it shows how often predators consume prey; δ represents the efficiency of turning predated prey into predator population, it accounts for how many new predators are added per prey consumed; and γ represents the predator's death rate, it signifies how fast the predator population declines in the absence of prey. Many others can be listed.

From the fundamental theorem of calculus, we have that the antiderivative of a function f is given as

$$F(t) = \int_0^t f(\tau)\,d\tau. \qquad (1.11)$$

The above quantities like some forces formula can be used to obtain quantities like some force and volumes. The above formula is known as the Riemann integral. It was extended by Stieltjes and the extended definition is given as follows:

$$\int_0^t f(\tau)\,dg(\tau). \qquad (1.12)$$

If $g(t)$ is continuously differentiable, then we have

$$F(t) = \int_0^t f(\tau)\,g'(\tau)d\tau. \qquad (1.13)$$

Taking the derivative on both sides yields

$$F'(t) = f(t)g'(t) \qquad (1.14)$$

which end up with

$$f(t)=\frac{F'(t)}{g'(t)}. \tag{1.15}$$

This therefore leads to the conclusion that a class associate derivative to the Riemann–Stieltjes can be given as

$$D_g f(t)=\lim_{h\to 0}\frac{f(t+h)-f(t)}{g(t+h)-g(t)}. \tag{1.16}$$

If both functions are differentiable then

$$D_g f(t)=\frac{f'(t)}{g'(t)} \tag{1.17}$$

Therefore, assuming the differentiability of g and f with respect to t for a selected t^{α}, we obtain $\frac{df}{dt^{\alpha}}$ representing the fractal derivative. Given the association between fractals and conformability, the retrieval of the conformable derivative ensues.

If both functions are differentiable then

$$D_g f(t_1)=\lim_{t\to t_1}\frac{f(t)-f(t_1)}{g(t)-g(t_1)}=\frac{f'(t_1)}{g'(t_1)}. \tag{1.18}$$

We can give some illustrative examples for fractal derivative as follows:

- $\left(D_t t^2\right)(t_1)=\lim\limits_{t\to t_1}\frac{t^2-t_1^2}{t-t_1}=2t_1.$
- $\left(D_{t^2} t\right)(t_1)=\lim\limits_{t\to t_1}\frac{t-t_1}{t^2-t_1^2}=\frac{1}{2t_1}.$
- $\left(D_{\cos t}\sin t\right)(t_1)=\lim\limits_{t\to t_1}\frac{\sin t-\sin t_1}{\cos t-\cos t_1}=-\csc t_1.$
- $\left(D_{\sin t}\cos t\right)(t_1)=\lim\limits_{t\to t_1}\frac{\cos t-\cos t_1}{\sin t-\sin t_1}=-\tan t_1.$

Several authors have investigated these operators; see [1–10].

References

1. Atangana, A.: Some stochastic chaotic attractors with global derivative and stochastic fractal mapping: existence, uniqueness and applications. Math. Methods Appl. Sci. **46**(7), 7875–7929 (2023)
2. Attia, N., Akgül, A., Atangana, A.: Reproducing kernel method with global derivative. J. Funct. Spaces (2023)
3. Atangana, A., Araz, S.İ: Nonlinear equations with global differential and integral operators: existence, uniqueness with application to epidemiology. Results Phys. **20**, 103593 (2021)
4. Atangana, A., Araz, S.İ: New concept in calculus: piecewise differential and integral operators. Chaos, Solitons & Fractals **145**, 110638 (2021)
5. Atangana, A.: Extension of rate of change concept: from local to nonlocal operators with applications. Results Phys. **19**, 103515 (2020)

6. Kim, Y.J.: Stieltjes derivatives and its applications to integral inequalities of Stieltjes type. Pure Appl. Math. **18**(1), 63–78 (2011)
7. Monteiro, G.A., Satco, B.: Extremal solutions for measure differential inclusions via Stieltjes derivatives. Advances in Difference Equations (2019)
8. Satco, B., Satco, B., Smyrlis, G.: Periodic boundary value problems involving Stieltjes derivatives. J. Fixed Point Theory Appl. **22** (2020)
9. Area, I., Fernandez, F.J., Nieto, J.J., Tojo, F.A.F.: Concept and solution of digital twin based on a Stieltjes differential equation. Math. Methods Appl. Sci. **45**(2), 7451–7465 (2022)
10. Fernandez, F.J., Albes, I.M., Tojo, F.A.F.: On first and second order linear Stieljes differential eqautions. J. Math. Anal. Appl. **511**(1), 123 (2022)

Chapter 2
Derivative with Respect to a Function: Derivatives, Definitions, and Properties

We reviewed the concept of classical differential operator with its history, development, and its use at the end of the first chapter. As such, we looked into how this idea has been essential in calculus and calculus-based applications. At the tail-end of the chapter, we offered an extension of this classical notion-the derivative of one function with respect to another. It was a natural evolution-to deepen the comprehension of differential operators in more intricate situations. Following this framework, the present chapter seeks to widen the field of fractional differential equations, well-known in numerous areas for the ability of modeling behaviors that classical differential equations struggle to reproduce. This covers the fractional derivative concept itself, specifically in relation to these derivatives with respect to another function for many kernels. Through the exploration of different types of these fractional derivatives, we will, at the same time, point out their power to generalize classical derivatives, accommodating a more general framework for tackling more complex equations. Furthermore, in this chapter, we will show how fractional derivatives enable to recover some well-known classical differential operators when conditions are satisfied and the appropriate functions are chosen. This will greatly contribute to the theoretical understanding of the fractional calculus, whilst facilitating modeling of complex systems in physical, biological, and engineering domains exhibiting memory, non-local, and non-linear interactions and efforts.

2.1 Extension Concept of Rate of Change

In the classical way the rate of change of a continuous function $f(t)$ between t_1 and t_2 is given by

$$a = \frac{f(t_2) - f(t_1)}{t_2 - t_1}. \tag{2.1}$$

A. Atangana and İ. Koca, *Fractional Differential and Integral Operators with Respect to a Function*, Industrial and Applied Mathematics,
https://doi.org/10.1007/978-981-97-9951-0_2

The above definition can be viewed as a proportion of a continuous function $f(t)$ between t_1 and t_2 and the function $g(t) = t$, and it can be extended as [1, 2]

$$a = \frac{f(t_2) - f(t_1)}{g(t_2) - g(t_1)}. \tag{2.2}$$

One of the main questions to answer in this work is the following "Can we extend the rate of change into a more general form with $g(t) \neq a \neq t$?".

Definition 2.1 Let f be a continuous function and g be a positive continuous non-constant function. The global rate of change between a and b is given as

$$r = \frac{f(b) - f(a)}{g(b) - g(a)}. \tag{2.3}$$

The function g should be increasing such that always if $a < b$, then $g(a) < g(b)$.

Definition 2.2 Let f be a continuous function and g be a positive continuous necessarily increasing in $[a, b]$ and nonzero for all $t \in [a, b]$. A derivative of f with respect to the function g is defined as

$$D_g f(t_1) = \lim_{t \to t_1} \frac{f(t) - f(t_1)}{g(t) - g(t_1)}. \tag{2.4}$$

Remark 2.1 Assume that $D_g f(t)$ exists without loss of generality, let $t > t_1$, then it holds that

(i) If $D_g f(t) = 0 \Longrightarrow f(t) = f(t_1)$ then the function f is constant,
(ii) If $D_g f(t) > 0 \Longrightarrow f(t) > f(t_1)$ then the function f is increases,
(iii)If $D_g f(t) < 0 \Longrightarrow f(t) < f(t_1)$ then the function f is decreases.

If $g(t) = t$,

$$D_g f(t_1) = \lim_{t \to t_1} \frac{f(t) - f(t_1)}{t - t_1}. \tag{2.5}$$

Therefore, we recover the classical differential operator.

If $g(t) = t^\alpha$, we have

$$D_g f(t_1) = \lim_{t \to t_1} \frac{f(t) - f(t_1)}{t^\alpha - t_1^\alpha} = \frac{df}{dt^\alpha}, \ t > 0, \alpha > 0. \tag{2.6}$$

Hence we recover the fractal derivative.

If $g(t) = \frac{t^{2-\alpha}}{2-\alpha}$, we have

$$D_g f(t_1) = \lim_{t \to t_1} \frac{f(t) - f(t_1)}{t^{2-\alpha} - t_1^{2-\alpha}}(2 - \alpha) = \frac{df}{dt^\alpha}, t > 0, \ 1 \leq \alpha < 2. \tag{2.7}$$

If $g(t) = t^{\beta(t)}$ condition on $\beta(t)$ to guarantee $g(t)$ is increasing, we have

$$D_g f(t_1) = \lim_{t \to t_1} \frac{f(t) - f(t_1)}{t^{\beta(t)} - t_1^{\beta(t_1)}}. \tag{2.8}$$

If f and g are differentiable, then

$$D_g f(t_1) = \lim_{t \to t_1} \frac{f(t) - f(t_1)}{g(t) - g(t_1)} = \frac{f^{'}(t_1)}{g^{'}(t_1)}. \tag{2.9}$$

Thus, under the condition that f and g are differentiable by choosing t^{α}, we recover $\frac{df}{dt^{\alpha}}$ which is the fractal derivative. Since there is a connection of fractal and conformable then conformable is also recovered.

Remark 2.2 The integral associated to the general derivative is nothing more than the Riemann–Stieltjes integral, with a small difference since the Riemann–Stieltjes integral is defined as [3, 4]

$$\int_a^b f(x)dg(x). \tag{2.10}$$

Definition 2.3 Let f and g be continuous functions, then the integral of f with respect to the function g in general concept is given as

$$F(t) = \int f(t)dg(t). \tag{2.11}$$

Thus within the closed interval $[a, b]$, we have

$$\int_a^b f(t)dg(t). \tag{2.12}$$

If $g(t)$ is differentiable, then

$$F(t) = \int f g^{'}(t)dt \tag{2.13}$$

or

$$\int_a^b f(t)g^{'}(t)dt. \tag{2.14}$$

If $g(t) = t$, then we recover the classical integral

$$\int_a^b f(t)dt. \tag{2.15}$$

If $g(t) = t^{\alpha}$, we recover the Atangana–Goufo integral

$$\int_{a}^{b} t^{1-\alpha} f(t)dt. \tag{2.16}$$

In this section, the concept of convolution will be used to help readers that are not used to this concept. We present some important information above it. We start with the definition.

Definition 2.4 Let $f(t)$ and $g(t)$ be two continuous functions in $\mathbb{R}$. The convolution of f and g is denoted by $(f * g)(t)$ and defined as

$$(f * g)(t) = \int_{-\infty}^{\infty} f(\tau) g(t-\tau) d\tau. \tag{2.17}$$

If f and g are defined in $[0, \infty]$ then

$$\begin{aligned}(f * g)(t) &= \int_{0}^{t} g(t-\tau) f(\tau) d\tau = \int_{0}^{t} f(t-\tau) g(\tau) d\tau \\ &= (g * f)(t).\end{aligned} \tag{2.18}$$

We now present some properties of the above operator.

Let $f(t)$ and $g(t)$ be two continuous

- $f * g(t - t_0) = (f * g)(t - t_0)$.
- $f(t - t_0) * g(t - t_0) = (f * g)(t - 2t_0)$.

A very useful relation used in many fields of science is the Laplace transform of a convolution. We recall that the Laplace transform of a continuous function $y(t)$ is defined as

$$L(y(t))(s) = \int_{0}^{\infty} \exp(-st) f(t) dt. \tag{2.19}$$

Now having

$$F(s) = \int_{0}^{\infty} \exp(-st) f(t) dt \tag{2.20}$$

$$G(s) = \int_{0}^{\infty} \exp(-st) g(t) dt$$

$$L\left(f*g\left(t\right)\right)\left(s\right)=F\left(s\right)\cdot G\left(s\right). \tag{2.21}$$

The above is due to the fact that

$$F\left(s\right)\cdot G\left(s\right)=\int_{0}^{\infty}\int_{0}^{\infty}\exp\left(-s\left(t+\bar{t}\right)\right)f\left(t\right)g\left(\bar{t}\right)dtd\bar{t}. \tag{2.22}$$

Letting

$$z=t+\bar{t} \tag{2.23}$$

leads to

$$\begin{aligned}F\left(s\right)\cdot G\left(s\right)&=\int_{0}^{\infty}\int_{0}^{\infty}\exp\left(-sz\right)f\left(t\right)g\left(z-t\right)dtdz \\ &=\int_{0}^{\infty}\exp\left(-sz\right)f*gdz.\end{aligned} \tag{2.24}$$

If $f=\delta\left(t\right)$ then

$$\int_{0}^{t}\delta\left(t-\tau\right)g\left(\tau\right)d\tau=g\left(t\right). \tag{2.25}$$

If the convolution of f and g exists and if $f,g\in L^{1}\left[\mathbb{R}\right]$, then $f*g\in L^{1}\left[\mathbb{R}\right]$. If $f\in L^{1}\left[\mathbb{R}\right]$ and $g\in L^{q}\left[\mathbb{R}\right]$ with $1\leq g<\infty$ then $f*g\in L^{p}\left[\mathbb{R}\right]$ and

$$\|f*g\|_{L^{p}[\mathbb{R}]}\leq\|f\|_{L^{1}[\mathbb{R}]}\|g\|_{L^{p}[\mathbb{R}]}. \tag{2.26}$$

If f and g is differentiable then

$$\frac{d}{dt}\left(f*g\right)=\frac{d}{dt}f*g\text{ or }f*\frac{dg}{dt}. \tag{2.27}$$

We shall present example of convolution.

Example 2.1 Let $g\left(t\right)=\frac{t^{-\alpha}}{\Gamma(1-\alpha)}$, $f\left(t\right)=t^{n}, 0<\alpha\leq 1$.

$$\left(g*f\right)\left(t\right)=\frac{1}{\Gamma\left(1-\alpha\right)}\int_{0}^{t}\tau^{n}\left(t-\tau\right)^{-\alpha}d\tau. \tag{2.28}$$

We let $ty = \tau$, then

$$(g * f)(t) = \frac{1}{\Gamma(1-\alpha)} \int_0^1 t^n y^n t^{-\alpha} (1-y)^{-\alpha} t dy \tag{2.29}$$

$$= \frac{t^{n-\alpha+1}}{\Gamma(1-\alpha)} \int_0^1 y^{n+1-1} (1-y)^{1-\alpha-1} dy$$

$$= \frac{t^{n-\alpha+1}}{\Gamma(n-\alpha)} B(n+1, 1-\alpha) = \frac{t^{n-\alpha+1}\Gamma(n+1)\Gamma(1-\alpha)}{\Gamma(1-\alpha)\Gamma(n+2-\alpha)}$$

$$(f * g)(t) = t^{n-\alpha+1} \frac{\Gamma(n+1)}{\Gamma(n+2-\alpha)} = \frac{t^{n-\alpha+1} n!}{\Gamma(n+2-\alpha)}.$$

Example 2.2 Let $g(t) = \frac{t^{-\alpha}}{\Gamma(1-\alpha)}$ and $f(t) = E_\alpha(-\lambda t^\alpha)$. We note that

$$E_\alpha(-\lambda t^\alpha) = \sum_{j=0}^{\infty} \frac{(-\lambda t^\alpha)^j}{\Gamma(\alpha j + 1)} \tag{2.30}$$

$$(f * g)(t) = \frac{1}{\Gamma(1-\alpha)} \int_0^t (t-\tau)^{-\alpha} \sum_{j=0}^{\infty} \frac{(-\lambda \tau^\alpha)^j}{\Gamma(\alpha j + 1)} d\tau \tag{2.31}$$

$$= \frac{1}{\Gamma(1-\alpha)} \sum_{j=0}^{\infty} \int_0^t (t-\tau)^{-\alpha} \frac{(-\lambda \tau^\alpha)^j}{\Gamma(\alpha j + 1)} d\tau$$

$$= \frac{1}{\Gamma(1-\alpha)} \sum_{j=0}^{\infty} \frac{(-\lambda)^j}{\Gamma(\alpha j + 1)} \int_0^t (t-\tau)^{-\alpha} \tau^{\alpha j} d\tau$$

$$= \frac{1}{\Gamma(1-\alpha)} \sum_{j=0}^{\infty} \frac{(-\lambda)^j}{\Gamma(\alpha j + 1)} t^{\alpha j + 1 - \alpha} B(1-\alpha, \alpha j + 1)$$

$$= \frac{t^{1-\alpha}}{\Gamma(1-\alpha)} \sum_{j=0}^{\infty} \frac{(-\lambda t^\alpha)^j}{\Gamma(\alpha j + 1)} \frac{\Gamma(1-\alpha)\Gamma(\alpha j + 1)}{\Gamma(1-\alpha+\alpha j + 1)}$$

$$= t^{1-\alpha} \sum_{j=0}^{\infty} \frac{(-\lambda t^\alpha)^j}{\Gamma(2-\alpha+\alpha j)} = t^{1-\alpha} E_{\alpha, 2-\alpha}(-\lambda t^\alpha)$$

$$(f * g)(t) = t^{1-\alpha} E_{\alpha, 2-\alpha}(-\lambda t^\alpha).$$

Example 2.3 Let $g(t) = \frac{t^{\alpha-1}}{\Gamma(\alpha)}$ and $f(t) = t^{\beta}, 0 < \alpha \leq 1$ and $\beta > 0$

$$(f * g)(t) = \frac{1}{\Gamma(\alpha)} \int_0^t (t-\tau)^{\alpha-1} \tau^{\beta} d\tau \tag{2.32}$$
$$= \frac{t^{\alpha+\beta}}{\Gamma(\alpha)} \int_0^1 (1-z)^{\alpha-1} z^{\beta} dz$$
$$= \frac{t^{\alpha+\beta}}{\Gamma(\alpha)} B(\alpha, \beta+1) = \frac{t^{\alpha+\beta}\Gamma(\beta+1)}{\Gamma(\alpha+\beta+1)}.$$

2.1.1 Convolution of Global Rate with Kernels

Due to the wider applicability of the concept of nonlocal operators and their abilities to replicate complex real-world problems, we present the case with nonlocal operators [5–9].

Definition 2.5 Let $f(t)$ be continuous and $g(t)$ be a non-constant increasing positive function. Let $K(t)$ be a kernel, singular or non-singular. For $0 < \alpha \leq 1$, a fractional global derivative of Caputo sense is defined by

$${}_0^C D_g^{\alpha} f(t) = D_g f(t) * K(t). \tag{2.33}$$

With Riemann–Liouville type, we have

$${}_0^{RL} D_g^{\alpha} f(t) = D_g (f(t) * K(t)), \tag{2.34}$$

where $(*)$ denotes the convolution operator. For example if $K(t) = \frac{t^{-\alpha}}{\Gamma(1-\alpha)}$, we have the general power law type

$${}_0^C D_g^{\alpha} f(t) = \frac{1}{\Gamma(1-\alpha)} \int_0^t D_g f(\tau)(t-\tau)^{-\alpha} d\tau, \tag{2.35}$$

$${}_0^{RL} D_g^{\alpha} f(t) = \frac{1}{\Gamma(1-\alpha)} D_g \int_0^t f(\tau)(t-\tau)^{-\alpha} d\tau. \tag{2.36}$$

If

$$K(t) = \frac{\exp[-\frac{\alpha}{1-\alpha} t]}{1-\alpha}, \tag{2.37}$$

then we recover the Caputo–Fabrizio case

$$ {}_{0}^{CCF}D_{g}^{\alpha}f(t)=\frac{M(\alpha)}{1-\alpha}\int_{0}^{t}D_{g}f(\tau)\exp\left[-\alpha\frac{(t-\tau)}{1-\alpha}\right]d\tau, \tag{2.38} $$

$$ {}_{0}^{RCF}D_{g}^{\alpha}f(t)=\frac{M(\alpha)}{1-\alpha}D_{g}\int_{0}^{t}f(\tau)\exp\left[-\alpha\frac{(t-\tau)}{1-\alpha}\right]d\tau. \tag{2.39} $$

If the kernel

$$ K(t)=\frac{AB(\alpha)}{1-\alpha}E_{\alpha}[-\frac{\alpha}{1-\alpha}t^{\alpha}], \tag{2.40} $$

we have

$$ {}_{0}^{ABC}D_{g}^{\alpha}f(t)=\frac{AB(\alpha)}{1-\alpha}\int_{0}^{t}D_{g}f(\tau)E_{\alpha}\left[-\alpha\frac{(t-\tau)^{\alpha}}{1-\alpha}\right]d\tau, \tag{2.41} $$

$$ {}_{0}^{ABR}D_{g}^{\alpha}f(t)=\frac{AB(\alpha)}{1-\alpha}D_{g}\int_{0}^{t}f(\tau)E_{\alpha}\left[-\alpha\frac{(t-\tau)^{\alpha}}{1-\alpha}\right]d\tau. \tag{2.42} $$

If f and g are differentiable, then

$$ {}_{0}^{CCF}D_{g}^{\alpha}f(t)=\frac{M(\alpha)}{1-\alpha}\int_{0}^{t}\frac{d}{d\tau}f(\tau)\gamma'(\tau)\exp\left[-\alpha\frac{(t-\tau)}{1-\alpha}\right]d\tau, \tag{2.43} $$

where

$$ \gamma'(t)=\frac{1}{g'(t)}. \tag{2.44} $$

$$ {}_{0}^{RCF}D_{g}^{\alpha}f(t)=\frac{M(\alpha)}{1-\alpha}\frac{d}{dt}\int_{0}^{t}f(\tau)\exp\left[-\alpha\frac{(t-\tau)}{1-\alpha}\right]d\tau\gamma'(t). \tag{2.45} $$

Then

$$ {}_{0}^{C}D_{g}^{\alpha}f(t)=\frac{1}{\Gamma(1-\alpha)}\int_{0}^{t}\frac{d}{d\tau}f(\tau)\gamma'(\tau)(t-\tau)^{-\alpha}d\tau, \tag{2.46} $$

$$ {}_{0}^{RL}D_{g}^{\alpha}f(t)=\frac{1}{\Gamma(1-\alpha)}\frac{d}{dt}\int_{0}^{t}f(\tau)(t-\tau)^{-\alpha}d\tau\gamma'(t). \tag{2.47} $$

Finally,

$$ {}_{0}^{ABC}D_{g}^{\alpha}f(t) = \frac{AB(\alpha)}{1-\alpha}\int_{0}^{t}\frac{d}{d\tau}f(\tau)\gamma'(\tau)E_{\alpha}\left[-\alpha\frac{(t-\tau)^{\alpha}}{1-\alpha}\right]d\tau, \tag{2.48} $$

$$ {}_{0}^{ABR}D_{g}^{\alpha}f(t) = \frac{AB(\alpha)}{1-\alpha}\frac{d}{dt}\int_{0}^{t}f(\tau)E_{\alpha}\left[-\alpha\frac{(t-\tau)^{\alpha}}{1-\alpha}\right]d\tau\gamma'(t). \tag{2.49} $$

Many other kernels can be used to recover other derivatives. To grasp the concept of fractional integral with regard to another function, one must first understand the Riemann–Stieltjes integral, including its characteristics and others. In the following chapter, we will offer a full analysis of the Riemann–Stieltjes integral in series and continuous terms, as well as several essential characteristics.

References

1. Atangana, A.: Extension of rate of change concept: From local to nonlocal operators with applications. Resul. Phys. **19**, Article id. 103515 (2020)
2. Bardi, J.S.: The calculus wars: Newton, Leibniz, and the greatest mathematical clash of all time. Thunder's Mouth Press, New York (2006). ISBN 1-56025-706-7
3. Hildebrandt, T.H.: Definitions of Stieltjes integrals of the Riemann type. Am. Math. Month. **45**(5), 265–78 (1938)
4. Dahlquist, G.: Convergence and stability in the numerical integration of ordinary differential equations. Math. Scand. **4**, 33–53 (1956)
5. Oldham, K.B., Spanier, J.: The fractional calculus; theory and applications of differentiation and integration to arbitrary order. Mathematics in Science and Engineering. V. Academic Press (1974). ISBN 978-0-12-525550-9
6. Miller, K.S. Ross, B. (Eds.): An Introduction to the Fractional Calculus and Fractional Differential Equations. Wiley (1993). ISBN 978-0-471-58884-9
7. Caputo, M.: Linear model of dissipation whose Q is almost frequency independent II. Geophys. J. Int. **13**(5), 529–39 (1967)
8. Atangana, A., Baleanu, D.: New fractional derivatives with nonlocal and non-singular kernel: theory and application to heat transfer model. Therm. Sci. **20**(2), 763–9 (2016)
9. Caputo, M., Fabrizio, M.: Applications of new time and spatial fractional derivatives with exponential kernels. Prog. Fract. Differ. Appl. **2**(1), 1–11 (2016)

Chapter 3
Integral Operators, Definitions, and Properties

In fact, as we already discussed, the Riemann–Stieltjes integral, which can be thought of as an integral of a continuous function with respect to another continuous function, is the fundamental operator used to extend the classical and fractional differential and integral operators that are the subject of discussion in this book. We will notice that when the other function is differentiable or continuous, various important results have been found by authors who have studied this operator. We will offer a selection of recent findings that will be relevant to the other chapters in this book in this chapter. Just to refresh your memory, the Riemann sum is often the region that lies roughly under the function curve [1]. The Riemann sum differs from the real area because the area of the region filled with little forms is not exactly equal to the area to be measured. This inaccuracy can be corrected by subdividing the region into smaller shapes. The sum approaches the Riemann integral as the forms become smaller. The Riemann–Stieltjes integral, an essential generalization of the Riemann integral, will be discussed in this chapter [2]. And numerous previously acquired results will be shown to orient readers who are unfamiliar with this issue. In addition, we will provide several key derivations based on the features of the other function.

Definition 3.1 ([1–7]) Given a function h that is bounded and defined on the closed interval $I = [u, v]$, a function β that is defined and monotonically increasing on I, and a partition $p = \{t_0, t_1, ..., t_{n-1}, t_n\}$ of I with corresponding subdivision Δ, let $M_j = \sup_{t \in I_j} |h(t)|$ and $m_j = \inf_{t \in I_j} |h(t)|$, for $I_j = \left[t_{j-1}, t_j\right]$. Then the upper Riemann–Stieltjes sum of h over β with respect to the partition p, denoted by $U(p, h, \beta)$ or $U(\Delta, h, \beta)$, is defined by

$$U(p, h, \beta) = \sum_{j=1}^{n} M_j \Delta\beta_j \tag{3.1}$$

A. Atangana and İ. Koca, *Fractional Differential and Integral Operators with Respect to a Function*, Industrial and Applied Mathematics,
https://doi.org/10.1007/978-981-97-9951-0_3

and the lower Riemann–Stieltjes sum of h over β with respect to the partition p, denoted by $L(p, h, \beta)$ or $L(\Delta, h, \beta)$, is defined by

$$L(p, h, \beta) = \sum_{j=1}^{n} m_j \Delta\beta_j \tag{3.2}$$

where

$$\Delta\beta_j = \left(\beta(t_j) - \beta(t_{j-1})\right). \tag{3.3}$$

The preceding definition informs us about the possible partition of a function defined in a closed interval in terms of the lower and upper Riemann–Stieltjes sums. This definition can also be found in a number of previously published documents. The above definition's information leads us to the following Lemma. The lemma tells us something about the boundedness of the lower and higher Riemann–Stieltjes integrals. This research will be extremely important for inequality and other theories such as existence and uniqueness.

Lemma 3.1 ([1–7]) *Suppose that h is a bounded function with domain $I = [u, v]$ and β is a function that is defined and monotonically increasing on I. Let p be a partition of I, $M = \sup_{t\in I} |h(t)|$ and $m = \inf_{t\in I} |h(t)|$. Then*

$$m(\beta(v) - \beta(u)) \le L(p, h, \beta) \le U(p, h, \beta) \le M(\beta(v) - \beta(u)) \tag{3.4}$$

and

$$L(p, h, \beta) \le L(p^*, h, \beta) \le U(p^*, h, \beta) \le U(p, h, \beta), \tag{3.5}$$

for any refinement p^* of p. Furthermore, if Δ_γ and Δ_λ are any two subdivisions of I, then

$$L(\Delta_\gamma, h, \beta) \le U(\Delta_\lambda, h, \beta). \tag{3.6}$$

Lemma 3.1's constraints with the greatest lower and least upper bound qualities of the real-valued function lead to the following formulations. The concepts below are now provided in terms of integrals rather than sums, as we previously presented. The integrals, on the other hand, are equal to the lower and upper bounds. It should be noted that the concepts offered here can be extended to the context of fractional differential and integral operators.

Definition 3.2 ([1–7]) Suppose that h is a function on R that is defined and bounded on the interval $I = [u, v]$, $\wp = \wp[u, v]$ is the set of all partitions of $[u, v]$, and β is a function that is defined and monotonically increasing on I. Then the upper Riemann–Stieltjes integral and the lower Riemann–Stieltjes integral are defined by

$$\int_u^{\overline{v}} h(t) d\beta(t) = \inf_{p\in\wp} |U(p, h, \beta)| \tag{3.7}$$

and

$$\underline{\int_{u}^{v}} h(t)d\beta(t) = \sup_{p \in \wp} |L(p, h, \beta)|, \tag{3.8}$$

respectively. If

$$\int_{u}^{\overline{v}} h(t)d\beta(t) = \int_{\underline{a}}^{b} h(t)d\beta(t), \tag{3.9}$$

in this case, one can see that, indeed within the framework of integral calculus, h is Riemann–Stieltjes integrable or integrable with respect to β in the Riemann sense, on I, and the common value of the integral is provided below by

$$\int_{u}^{v} h(t)d\beta(t) \tag{3.10}$$

or

$$\int_{u}^{v} hd\beta. \tag{3.11}$$

As in other circumstances, to build a space, one can collect all the functions that satisfy the following properties and define a norm. In our situation, we can now use the following definition to categorize all of these functions.

Definition 3.3 ([1–7]) Suppose that β is a function that is defined and monotonically increasing on the interval $I = [u, v]$. Then the set of all functions that are integrable with respect to β in the Riemann sense is denoted by $\mathbb{R}(\beta)$. We showed some key inequalities using the Riemann–Stieltjes sum, specifically the boundness of the lower and upper Riemann–Stieltjes sums. We can now give the same inequality when the sum is replaced by an integral and establish the lower and upper bounds in the following theorem using the aforementioned formulation.

Theorem 3.1 ([1–7]) *Suppose that h is a bounded function with domain $I = [u, v]$, β is a function that is defined and monotonically increasing on I and $m \le h(t) \le M$ for all $t \in I$. Then*

$$m(\beta(v) - \beta(u)) \le \underline{\int_{u}^{v}} hd\beta \le \int_{u}^{\overline{v}} hd\beta \le M(\beta(v) - \beta(u)). \tag{3.12}$$

Furthermore, if h is Riemann–Stieltjes integrable on I, then

$$m(\beta(v) - \beta(u)) \leq \int_u^v h(t)d\beta(t) \leq M(\beta(v) - \beta(u)). \tag{3.13}$$

We should remark that the function $\beta(t)$ is considered in general here; we did not limit it to the smooth function class. But one should note that, in elementary calculus, we focused on Riemann integrals of continuous functions. Even there, we either ignored the tight requirement of checking all potential partitions or restricted ourselves to functions where a technique might be utilized. Depending on how rigorous your course was, some examples of obtaining the integral from the definition might have used partitions of equal length and summation formulas, or they might have used a particular bounding lemma that applied to t^n for each $n \in j$. It is not worth our time to go through some tedious methods just to demonstrate that special functions are integrable. Integrability is only useful if it can be verified with a fair amount of effort. To that purpose, we want to look for function qualities that guarantee integrability. We will give a theorem that will enlighten us about the integrability conditions. We will recall that the Riemann integral is inappropriate for many theoretical applications. Some of the technical flaws in Riemann integration can be addressed with the Riemann–Stieltjes integral, and the majority vanish with the Lebesgue integral, but the latter does not provide an acceptable solution of improper integrals. The gauge integral is a generalization of the Lebesgue integral that is closer to the Riemann integral at the same time. We shall now start this discussion with the following theorem.

Corollary 3.1 ([1–7]) *If h is a function that is continuous on the interval $I = [u, v]$, then h is Riemann–Stieltjes integrable on $[u, v]$.*

Proof Let β be monotonically increasing on I and h be continuous on I. Suppose that $\varepsilon > 0$ is given. Then there exists an $\eta > 0$ such that

$$[\beta(v) - \beta(u)]\,\eta < \varepsilon. \tag{3.14}$$

Here we make use of the uniform continuous theorem, h is uniformly continuous in $[u, v]$; therefore, there exists a $\delta > 0$ such that

$$\forall(q, w), [q, w \in I \wedge |q - w| < \delta \Rightarrow |h(q) - h(w)| < \varepsilon]. \tag{3.15}$$

Let $p = \{u = t_0, t_1, ..., t_{n-1}, t_n = v\}$ be a partition of $[u, v]$ for which mesh $p < \delta$ and, for each j, $j = 1, 2, ..., n$, set $M_j = \sup_{t_{j-1} \leq t \leq t_j} |h(t)|$ and $m_j = \inf_{t_{j-1} \leq t \leq t_j} |h(t)|$. Then $M_j - m_j \leq \eta$ and

$$U(p,h,\beta) - L(p,h,\beta) = \sum_{j=1}^{n} \left(M_j - m_j\right) \Delta\beta_j \le \eta \sum_{j=1}^{n} \Delta\beta_j = \eta\,[\beta(v) - \beta(u)] < \varepsilon. \tag{3.16}$$

Since $\varepsilon > 0$ was arbitrary, we have that

$$(\forall\varepsilon, \varepsilon > 0 \Rightarrow \exists p, p \in \wp[u,v] \wedge (U(p,h,\beta) - L(p,h,\beta)) < \varepsilon)\,. \tag{3.17}$$

According to the Integrability Criterion, $h \in \mathbb{R}(\beta)$. Since β was arbitrary, one can indeed conclude that h is Riemann–Stieltjes Integrable (with respect to any monotonically increasing function on $[u, v]$). The following corollary can then be established and its proof will be provided.

Corollary 3.2 ([1–7]) *If h is a function that is monotonic on the interval $I = [u, v]$ and β is continuous and monotonically increasing on I, then $h \in \mathbb{R}(\beta)$. We shall present below the proof of the above corollary.*

Proof Suppose that h is a function that is monotonic on the interval $I = [u, v]$ and β is continuous and monotonically increasing on I. For $\varepsilon > 0$ given, let $n \in j$ be such that

$$(\beta(v) - \beta(u))\,|h(v) - h(u)| < n\varepsilon. \tag{3.18}$$

Because β is continuous and monotonically increasing, we can choose a partition

$p = \{u = t_0, t_1, ..., t_{n-1}, t_n = v\}$ of $[u, v]$ such that

$$\Delta\beta_j = \left(\beta(t_j) - \beta(t_{j-1})\right) = \frac{\beta(v) - \beta(u)}{n}. \tag{3.19}$$

If h is monotonically increasing in I, then, for each $j \in \{1, 2, ..., n\}$,

$$M_j = \sup_{t_{j-1} \le t \le t_j} |h(t)| = h(t_j) \tag{3.20}$$

and

$$m_j = \inf_{t_{j-1} \le t \le t_j} |h(t)| = h(t_{j-1}). \tag{3.21}$$

Also

$$\begin{aligned} U(p,h,\beta) - L(p,h,\beta) &= \sum_{j=1}^{n} \left(M_j - m_j\right) \Delta\beta_j, \\ &= \frac{\beta(v) - \beta(u)}{n} \sum_{j=1}^{n} \left(h(t_j) - h(t_{j-1})\right), \\ &= \frac{\beta(v) - \beta(u)}{n} (h(v) - h(u)) < \varepsilon \end{aligned} \tag{3.22}$$

while h monotonically decreasing yields that

$$M_j = h(t_{j-1}), m_j = h(t_j) \tag{3.23}$$

and

$$\begin{aligned} U(p,h,\beta) - L(p,h,\beta) &= \frac{\beta(v)-\beta(u)}{n} \sum_{j=1}^{n} \left(h(t_{j-1}) - h(t_j)\right), \\ &= \frac{\beta(v)-\beta(u)}{n} (h(u) - h(v)) < \varepsilon. \end{aligned} \tag{3.24}$$

Since $\varepsilon > 0$ was arbitrary, we shall then have the following:

$$(\forall \varepsilon, \varepsilon > 0 \Rightarrow \exists p, p \in \wp[u,v] \wedge (U(p,h,\beta) - L(p,h,\beta)) < \varepsilon). \tag{3.25}$$

In view of the Integrability Criterion, $h \in \mathbb{R}(\beta)$. From the above, one can now present the following corollary. For this corollary the function does not need to be continuous everywhere, thus we are expecting it to have some points of discontinuity.

Corollary 3.3 ([1–7]) *Suppose that h is bounded on* $[u, v]$, *h has only finitely many points of discontinuity in* $I = [u, v]$, *and that the monotonically increasing function* β *is continuous at each point of discontinuity of h. Then* $h \in \mathbb{R}(\beta)$. *We shall present below a detailed proof of the above corollary.*

Proof Let $\varepsilon > 0$ be given. Suppose that h is bounded on $[u, v]$ and continuous on $[u, v] - E$ where $E = \{\zeta_1, \zeta_2, ..., \zeta_k\}$ is the nonempty finite set of points of discontinuity of h in $[u, v]$. Suppose further that β is a monotonically increasing function on $[u, v]$ that is continuous at each element of E. Because E is finite and β is continuous at each $\zeta_j \in E$, we can find k pairwise disjoint intervals $\left[q_j, w_j\right]$, $j = 1, 2, ..., k$, such that

$$E \subset \bigcup_{j=1}^{k} \left[q_j, w_j\right] \subseteq [u, v] \tag{3.26}$$

and

$$\sum_{j=1}^{k} \left(\beta(w_j) - \beta(q_j)\right) < \varepsilon^*, \tag{3.27}$$

For a suitable choiceof $\varepsilon^* > 0$, furthermore, one can choose adequate intervals in such a way that each point $\zeta_m \in E \cap (u, v)$ belongs to the interior of the corresponding interval, $[q_m, w_m,]$. Let

$$K = [u, v] - \bigcup_{j=1}^{k} \left(q_j, w_j\right). \tag{3.28}$$

Here we note that K is compact and h continuous on K implies that h is uniformly continuous there. Thus, corresponding to $\varepsilon^* > 0$, there exists a $\delta > 0$ such that

$$\left(\forall s, t \in K \wedge |s - t| < \delta \Rightarrow |h(s) - h(t)| < \varepsilon^*\right). \tag{3.29}$$

For this case, we shall let $p = \{u = t_0, t_1, ..., t_{n-1}, t_n = v\}$ be partition of $[u, v]$ satisfying the following conditions:

- $\left(\forall j \in \{1, 2, ..., k\} \Rightarrow q_j \in p \wedge w_j \in p\right)$,
- $\left(\forall j \in \{1, 2, ..., k\} \Rightarrow \left(q_j, w_j\right) \cap p = \varnothing\right)$,
- $\left[\left(\forall p, jp \in \{1, 2, ..., n\} \wedge j \in \{1, 2, ..., k\} \wedge t_{p-1} \neq q_j\right) \Rightarrow \Delta t_p < \delta\right]$.

It should be noted that given the established conditions, $t_{q-1} = q_j$ implies that $t_q = w_j$. If

$$M = \sup_{t \in I} |h(t)|, M_p = \sup_{t_{p-1} \leq t \leq t_p} |h(t)| \tag{3.30}$$

and

$$m_p = \inf_{t_{p-1} \leq t \leq t_p} |h(t)|, \tag{3.31}$$

then for each p, $M_p - m_p \leq 2M$. Furthermore, $M_p - m_p < \varepsilon^*$ as long as $t_{p-1} \neq q_j$. Using commutativity to regroup the summation according to the available bounds yields that

$$\begin{aligned} U(p, h, \beta) - L(p, h, \beta) &= \sum_{j=1}^{n} \left(M_j - m_j\right) \Delta\beta_j, \\ &\leq [\beta(v) - \beta(u)]\, \varepsilon^* + 2M\varepsilon^* < \varepsilon \end{aligned} \tag{3.32}$$

whenever

$$\varepsilon^* < \frac{\varepsilon}{2M + [\beta(v) - \beta(u)]}. \tag{3.33}$$

Since $\varepsilon > 0$ was random, we derive from the integrability criteria that $f \in \mathbb{R}(\beta)$. We can now present the following theorem with three major conditions.

Theorem 3.2 ([1–7]) *Suppose that f is bounded on $[u, v]$ and β is monotonically increasing on $[u, v]$.*

(a) If there exist an $\varepsilon > 0$ and a partition p^* of $[u, v]$ such that Eq. (3.14) is satisfied, then Eq. (3.14) is satisfied for every refinement p of p^*.

(b) If Eq. (3.14) is satisfied for the partition $p = \{u = t_0, t_1, ..., t_{n-1}, t_n = v\}$ and for each j, $j = 1, 2, ..., n$, $s_j$and t_j are arbitrary points in $[t_{j-1}, t_j]$, then

$$\sum_{j=1}^{n} \left|h(s_j) - h(t_j)\right| \Delta\beta_j < \varepsilon. \tag{3.34}$$

(c) If $f \in \mathbb{R}(\beta)$, Eq. (3.14) is satisfied for the partition

$$p = \{u = t_0, t_1, ..., t_{n-1}, t_n = v\} \tag{3.35}$$

and for each $j = 1, 2, ..., n$, t_j are arbitrary points in $[t_{j-1}, t_j]$, then

$$\left| \sum_{j=1}^{n} h(t_j)\Delta\beta_j - \int_u^v h(t)d\beta(t) \right| < \varepsilon. \tag{3.36}$$

Remark 3.1 Remember the following definition of Riemann Integrals from elementary calculus: Given a function h defined on an interval, $I = \{t : u \leq t \leq v\}$, the $\mathbb{R}_j$ sum for $\Delta = \{I_1, I_2, ..., I_n\}$ a subdivision of I is given by

$$\sum_{j=1}^{j=n} h\left(\zeta_j\right) K\left(I_j\right) \tag{3.37}$$

where ζ_j is any element of I_j. The point ζ_j is referred to as a sampling point. To get the $\mathbb{R}_j$ we wish to take the limit over such sums as the mesh of the partitions associated with integral. Δ goes to 0. In particular, if the function f is defined on $I = \{t : u \leq t \leq v\}$ and $p[u, v]$ denotes the set of all partitions $\{u = t_0, t_1, ..., t_{n-1}, t_n = v\}$ of the interval I, then f is said to be $\mathbb{R}_j$ integrable over I if and only if

$$\lim_{\text{mesh } p[u,v] \to 0} \sum_{j=1}^{j=n} h\left(\zeta_j\right)\left(t_j - t_{j-1}\right) \tag{3.38}$$

exists for any choices of $\zeta_j \in [t_{j-1} - t_j]$. The limit is called the $\mathbb{R}_j$ integral and is denoted by $\int_u^v h(t)dt$.

Taking $\beta(t) = t$ in Theorem 3.3 justifies that the old concept of an $\mathbb{R}_j$ Integrability is the same as Riemann integrability, which was introduced at the start of this chapter. The following theorem states that the composition of a function with a Riemann–Stieltjes integrable function must satisfy a sufficient condition. This result is especially significant when these operations are located in restricted areas. We will observe that, under certain conditions, this finding can be extended to nonlocal operators; specifically, a specific condition should be imposed on the fractional order.

Theorem 3.3 ([1–7]) *Suppose $h \in \mathbb{R}(\beta)$ on $[u, v]$, $m \leq h \leq M$ on $[u, v]$, and Φ a continuous function on $[m, M]$, let $h(t) = \Phi(h(t))$ for $t \in [u, v]$. Then $h \in \mathbb{R}(\beta)$ on $[u, v]$.*

3.1 Properties of Riemann–Stieltjes Integrals

This section provides a list of the properties of the numerous Riemann–Stieltjes integrals. The first lemma enables us to derive conclusions about the upper and lower Riemann–Stieltjes sums of a constant times a limited function in connection to the function's upper and lower Riemann–Stieltjes sums [3].

Lemma 3.2 ([1–7]) *Suppose that h is a bounded function that is defined on the interval $I = [u, v]$. For k is a nonzero real number and $g = kh$, we have*

$$\inf_{t\in I} |g(t)| = \begin{cases} k.\inf_{t\in I} |h(t)| & \text{if } k > 0 \\ k.\sup_{t\in I} |h(t)| & \text{if } k < 0 \end{cases} \tag{3.39}$$

$$\sup_{t\in I} |g(t)| = \begin{cases} k.\sup_{t\in I} |h(t)| & \text{if } k > 0 \\ k.\inf_{t\in I} |h(t)| & \text{if } k < 0. \end{cases}$$

Proof In this situation, only the proof of two of the four inequalities will be shown because the remainder can be derived similarly. For h is a function that is defined and bounded on the interval $I = [u, v]$ and k a nonzero real number, let $g(t) = k.h(t)$.

Suppose that $k > 0$ and that $M = \sup_{t\in I} |h(t)|$. Then $h(t) \leq M$ for all $t \in I$ and

$$g(t) = k.h(t) \leq kM \text{ for all } t \in I. \tag{3.40}$$

Hence, kM is an upper bound for $g(t)$ on the interval I. If kM is not the least upper bound, then there exists an $\varepsilon > 0$ such that $g(t) \leq kM - \varepsilon$ for all $t \in I$. (Here, ε can be taken to be any positive real that is less than or equal to the distance between kM and $\sup_{t\in I} |g(t)|$.) By substitution, we have $k.h(t) \leq kM - \varepsilon$ for all $t \in I$. Since k is positive, the latter is equivalent to

$$h(t) \leq M - \left(\frac{\varepsilon}{k}\right) \text{ for all } t \in I, \tag{3.41}$$

which contradicts that M is the supremum of f over I. Therefore,

$$\sup_{t\in I} |g(t)| = kM = k\sup_{t\in I} |h(t)|. \tag{3.42}$$

Next, suppose that $k < 0$ and that $M = \sup_{t\in I} |h(t)|$. Now, $h(t) \leq M$ for all $t \in I$ implies that

$$g(t) = k.h(t) \geq kM. \tag{3.43}$$

Hence, kM is a lower bound for $g(t)$ on I. If kM is not a greatest lower bound, then there exists an $\varepsilon > 0$, such that

$$g(t) \geq kM + \varepsilon, \tag{3.44}$$

for all $t \in I$. But from

$$k.h(t) \geq kM + \varepsilon \tag{3.45}$$

and $k < 0$, we conclude that

$$h(t) \leq M + \left(\frac{\varepsilon}{k}\right) \tag{3.46}$$

for all $t \in I$. Since $\frac{\varepsilon}{k}$ is negative $M + \left(\frac{\varepsilon}{k}\right) < M$ which gives us a contradiction to M being the $\sup_{t\in I} |h(t)|$. Therefore,

$$\sup_{t\in I} |g(t)| = kM = k\sup_{t\in I} |h(t)|. \tag{3.47}$$

Theorem 3.4 ([1–7] Properties of Upper and Lower Riemann–Stieltjes Integrals) *Suppose that the functions h, h_1, and h_2 are bounded and defined on the closed interval $I = [u, v]$ and β is a function that is defined and monotonically increasing in I.*

(a) If $g = kh$ for $k \in \mathbb{R}\backslash\{0\}$, then

$$\int_{\underline{u}}^{v} g d\beta = \begin{cases} k \displaystyle\int_{\underline{u}}^{v} h(t)d\beta(t), & \text{if } k > 0 \\ k \displaystyle\int_{u}^{\overline{v}} h(t)d\beta(t), & \text{if } k < 0 \end{cases} \tag{3.48}$$

and

$$\int_{u}^{\overline{v}} g d\beta = \begin{cases} k \displaystyle\int_{u}^{\overline{v}} h(t)d\beta(t), & \text{if } k > 0 \\ k \displaystyle\int_{\underline{u}}^{v} h(t)d\beta(t), & \text{if } k < 0 \end{cases}. \tag{3.49}$$

(b) If $s = h_1 + h_2$, then

$$\int_{\underline{u}}^{v} s(t)d\beta(t) \geq \int_{\underline{u}}^{v} h_1(t)d\beta(t) + \int_{\underline{u}}^{v} h_2(t)d\beta(t) \tag{3.50}$$

and

$$\overline{\int_{u}^{v}} s(t)d\beta(t) \leq \overline{\int_{u}^{v}} h_1(t)d\beta(t) + \overline{\int_{u}^{v}} h_2(t)d\beta(t). \tag{3.51}$$

(c) If $h_1(t) \leq h_2(t)$ *for all* $t \in I$, *then*

$$\underline{\int_{u}^{v}} h_1(t)d\beta(t) \leq \underline{\int_{u}^{v}} h_2(t)d\beta(t) \tag{3.52}$$

and

$$\overline{\int_{u}^{v}} h_1(t)d\beta(t) \leq \overline{\int_{u}^{v}} h_2(t)d\beta(t). \tag{3.53}$$

(d) If $u < v < r$ *and* h *is bounded on* $I^* = \{t : u \leq t \leq r\}$ *and* β *is monotonically increasing on* I^*, *then*

$$\underline{\int_{u}^{r}} h(t)d\beta(t) = \underline{\int_{u}^{v}} h(x)d\beta(t) + \underline{\int_{v}^{r}} h(t)d\beta(t) \tag{3.54}$$

and

$$\overline{\int_{u}^{r}} h(t)d\beta(t) = \overline{\int_{u}^{v}} h(t)d\beta(t) + \overline{\int_{v}^{r}} h(t)d\beta(t). \tag{3.55}$$

The conclusions of Theorem 3.5 are directly applicable to some of the algebraic features given in the following Theorem for Riemann–Stieltjes integrable functions.

Theorem 3.5 (Algebraic Properties of Riemann–Stieltjes Integrals [1–7]) *Suppose that the functions* h, h_1, *and* $h_2 \in \mathbb{R}(\beta)$ *on the closed interval* $I = [u, v]$.

(a) If $g(t) = kh(t)$ *for all* $t \in I$, *then* $g \in \mathbb{R}(\beta)$ *and*

$$\int_{u}^{v} g(t)d\beta(t) = k\int_{u}^{v} h(t)d\beta(t). \tag{3.56}$$

(b) If $s = h_1 + h_2$, *then* $h_1 + h_2 \in \mathbb{R}(\beta)$ *and*

$$\int_{u}^{v} s(t)d\beta(t) = \int_{u}^{v} h_1(t)d\beta(t) + \int_{u}^{v} h_2(t)d\beta(t). \tag{3.57}$$

(c) If $h_1(t) \leq h_2(t)$ *for all* $t \in I$, *then*

$$\int_u^v h_1(t)d\beta(t) \leq \int_u^v h_2(t)d\beta(t). \tag{3.58}$$

(d) If the function $h \in \mathbb{R}(\beta)$ *also on* $I^* = \{t : v \leq t \leq r\}$, *then* h *is Riemann–Stieltjes integrable on* $I \cup I^*$ *and*

$$\int_u^r h(t)d\beta(t) = \int_u^v h(t)d\beta(t) + \int_v^r h(t)d\beta(t). \tag{3.59}$$

(e) If $|h(t)| \leq M$ *for all* $t \in I$, *then*

$$\left|\int_u^v h(t)d\beta(t)\right| \leq M\left[\beta(v) - \beta(u)\right]. \tag{3.60}$$

(f) If $h \in \mathbb{R}(\beta^*)$ *on* I, *then* $h \in \mathbb{R}(\beta + \beta^*)$ *and*

$$\int_u^v hd(\beta + \beta^*) = \int_u^v h(t)d\beta(t) + \int_u^v h(t)d\beta^*(t). \tag{3.61}$$

(g) If r *is any positive real constant, then* $h \in \mathbb{R}(r\alpha)$ *and*

$$\int_u^v hd(r\beta) = r\int_u^v h(t)d\beta(t). \tag{3.62}$$

Remark 3.2 As long as the integrals exist, the formula given in (d) of the Corollary holds regardless of the location of v, i.e., v need not be between u and r.

Remark 3.3 Since a point has no dimension (that is, has length 0), we note that

$$\int_u^u h(t)d\beta(t) = 0, \tag{3.63}$$

for any function h.

Remark 3.4 If we think of the definition of the Riemann–Stieltjes integrals as taking direction into account (for example, with $\int_u^v h(t)d\beta(t)$ we had $u < v$ and took the sums over subdivisions as we were going from u to v), then it makes sense to introduce the convention that

$$\int_u^v h(t)d\beta(t) = -\int_v^u h(t)d\beta(t), \tag{3.64}$$

for Riemann–Stieltjes integrable functions h. The fact that we can deduce a partition of any subinterval and vice versa for any partition of an interval leads straight to the next result.

Theorem 3.6 (Restrictions of Integrable Functions [1–7]) *If the function h is (Riemann) integrable on an interval I, then $h\backslash_{I^*}$ is integrable on I^* for any subinterval I^* of I.*

We produce a set of Riemann–Stieltjes integrable functions by selecting various continuous functions for Φ in Theorem 3.4 and combining them with the fundamental characteristics of Riemann–Stieltjes integrals.

Theorem 3.7 ([1–7]) *If $h \in \mathbb{R}(\beta)$ and $g \in \mathbb{R}(\beta)$ on $[u, v]$, then $hg \in \mathbb{R}(\beta)$.*

Proof Suppose that $h \in \mathbb{R}(\beta)$ and $g \in \mathbb{R}(\beta)$ on $[u, v]$. From the Algebraic Properties of the Riemann–Stieltjes Integral, it follows that $(h + g) \in \mathbb{R}(\beta)$ on $[u, v]$ and $(h - g) \in \mathbb{R}(\beta)$ on $[u, v]$.

Theorem 3.8 ([1–7]) *If $h \in \mathbb{R}(\beta)$ on $[u, v]$, then $|h| \in \mathbb{R}(\beta)$ and*

$$\left|\int_u^v h(t)d\beta(t)\right| \leq \int_u^v |h(t)|\, d\beta(t). \tag{3.65}$$

Proof Suppose that $h \in \mathbb{R}(\beta)$ on $[u, v]$. Taking $\Phi(t) = |t|$ in Theorem 3.4 yields that $|h| \in \mathbb{R}(\beta)$ on $[u, v]$. Choose $\gamma = 1$, if

$$\int h(t)d\beta(t) \geq 0 \tag{3.66}$$

and $\gamma = -1$, if

$$\int h(t)d\beta(t) \leq 0. \tag{3.67}$$

Then

$$\left|\int_u^v h(t)d\beta(t)\right| = \gamma \int_u^v h(t)d\beta(t) \tag{3.68}$$

and

$$\gamma h(t) \leq |h(t)| \text{ for } t \in [u, v]. \tag{3.69}$$

It follows from Algebraic Properties of the Riemann–Stieltjes Integrals (a) and (c) that

$$\left|\int_u^v h(t)d\beta(t)\right| = \gamma\int_u^v h(t)d\beta(t) = \int_u^v \gamma h(t)d\beta(t) \leq \int_u^v |h(t)|\,d\beta(t). \tag{3.70}$$

Specific instances of utilizing the concept to find the integral for integrable functions have conspicuously been missing from our presentation. Consider for a moment what the definition calls for us to discover: The set of all upper Riemann–Stieltjes sums and the set of all lower Riemann–Stieltjes sums must first be identified. Here, the interval's subdivisions, across which we are integrating, cover every conceivable scenario. We need uniformity, a straightforward interpretation for the necessary suprema and infima, and a methodical approach to determine when we have examined enough subdivisions or sums. When we have general criteria that ensure integrability, the uniqueness of the least upper and greatest lower bounds allows us to calculate the value of the integral by studying carefully chosen special subsets of the set of all subdivisions of an interval. The following conclusion provides a sufficient condition for obtaining a Riemann–Stieltjes integral as a point evaluation. It makes use of the distinctive function. Remember that the characteristic function for a set S and A is subset of S is $\chi_A : S \to \{0, 1\}$ is defined by

$$\chi_A(t) = \begin{cases} 1, & \text{if } t \in A \\ 0, & \text{if } t \in S - A \end{cases}. \tag{3.71}$$

In the following $\chi_{(0,\infty)}$ denotes the characteristic function with $S = \mathbb{R}$ and $A = (0, \infty)$, i.e.,

$$\chi_{(0,\infty)}(t) = \begin{cases} 1, & \text{if } t > 0 \\ 0, & \text{if } t \leq 0 \end{cases}. \tag{3.72}$$

Lemma 3.3 ([1–7]) *Suppose that h is bounded on $[u, v]$ and continuous at $s \in (u, v)$. If*

$$\beta(t) = \chi_{(0,\infty)}(t - s), \tag{3.73}$$

then

$$\int_u^v h(t)d\beta(t) = h(s). \tag{3.74}$$

Theorem 3.9 ([1–7]) *Suppose the sequence of nonnegative real numbers $\{c_n\}_{n=1}^{\infty}$ is such that $\sum_{n=1}^{\infty} c_n$ is convergent, $\{s_n\}_{n=1}^{\infty}$ is a sequence of distinct points in (u, v), and f is a function that is continuous on $[u, v]$. If*

$$\beta(t) = \sum_{n=1}^{\infty} c_n \chi_{(0,\infty)}(t - s_n), \tag{3.75}$$

then

$$\int_a^b f(t) d\beta(t) = \sum_{n=1}^{\infty} c_n h(s_n). \tag{3.76}$$

Proof For $q, w \in (u, v)$ such that $q < w$, let

$$S_q = \{n \in J : u < s_n \leq q\} \tag{3.77}$$

and

$$T_w = \{n \in J : u < s_n \leq w\}. \tag{3.78}$$

Then

$$\beta(q) = \sum_{n \in S_q} c_n \leq \sum_{n \in T_w} c_n = \beta(w), \tag{3.79}$$

from which we conclude that β is monotonically increasing. Furthermore, $\beta(u) = 0$ and $\beta(v) = \sum_{n=1}^{\infty} c_n$. Let $\varepsilon > 0$ be given. Since $\sum_{n=1}^{\infty} c_n$ is convergent, there exists a positive integer K such that

$$\sum_{n=K+1}^{\infty} c_n < \frac{\varepsilon}{M}, \tag{3.80}$$

where $M = \sup_{t \in [u,v]} |h(t)|$. Let

$$\beta_1(t) = \sum_{n=1}^{K} c_n \chi_{(0,\infty)}(t - s_n) \tag{3.81}$$

and

$$\beta_2(t) = \sum_{n=K+1}^{\infty} c_n \chi_{(0,\infty)}(t - s_n). \tag{3.82}$$

It follows from Lemma 3.3 that

$$\int_u^v h(t) d\beta_1(t) = \sum_{n=1}^{K} c_n h(s_n), \tag{3.83}$$

while $\beta_2(v) - \beta_2(u) < \frac{\varepsilon}{M}$ yields that

$$\left| \int_u^v h(t) d\beta_2(t) \right| < \varepsilon. \tag{3.84}$$

Because $\beta = \beta_1 + \beta_2$, we conclude that

$$\left| \int_u^v h(t) d\beta_1(t) - \sum_{n=1}^{K} c_n h(s_n) \right| < \varepsilon. \tag{3.85}$$

Since $\varepsilon > 0$ arbitrary,

$$\int_u^v h(t) d\beta(t) = \sum_{n=1}^{\infty} c_n h(s_n). \tag{3.86}$$

Theorem 3.10 ([1–7]) *Suppose that β is a monotonically increasing function such that $\beta' \in \mathbb{R}$ on $[u, v]$ and h is a real function that is bounded on $[u, v]$. Then $h \in \mathbb{R}(\beta)$ if and only if $h\beta' \in \mathbb{R}$. Furthermore,*

$$\int_u^v h(t) d\beta(t) = \int_u^v h(t) \beta'(t) dt. \tag{3.87}$$

Theorem 3.11 (Change of variables [1–7]) *Suppose that Φ is a strictly increasing continuous function that maps an interval $[A, B]$ onto $[u, v]$, β is monotonically increasing on $[u, v]$, and $h \in \mathbb{R}(\beta)$ on $[u, v]$. For $y \in [A, B]$, let $\beta(y) = \beta(\Phi(y))$ and $g(y) = h(\Phi(y))$. Then $g \in \mathbb{R}(\beta)$ and*

$$\int_A^B g(y) d\beta(y) = \int_u^v h(t) d\beta(t). \tag{3.88}$$

3.2 Riemann Integrals and Differentiation

We have some good results that allow us to leverage our understanding of derivatives to compute integrals when we limit ourselves to Riemann integrals. In addition to being of general interest, the first result can be used to demonstrate some of the traits we are looking for. The discussion supporting several significant conclusions of the classical differential and integral calculus will then receive our full focus. We'll discuss some fundamental, well-known theorems that can later be expanded to the context of nonlocal differential and integral operators. We shall start with the Mean Value Theorem for integrals in the following theorem.

Theorem 3.12 (Mean Value Theorem for Integrals [1–7]) *Suppose that h is continuous on $I = [u, v]$. Then there exists a number ζ in I such that*

$$\int_u^v h(t)dt = h(\zeta)(v-u). \tag{3.89}$$

Proof This result follows directly from the bounds on integrals. Since h is continuous on $[u, v]$, it is integrable there and

$$m(v-u) \le \int_u^v h(t)dt \le M(v-u) \tag{3.90}$$

where

$$m = \inf_{t\in I} |h(t)| = \min_{t\in I} |h(t)| = h(t_0) \tag{3.91}$$

for some $t_0 \in I$ and

$$M = \sup_{t\in I} |h(t)| = \max_{t\in I} |h(t)| = h(t_1) \tag{3.92}$$

for some $t_1 \in I$. Now

$$A = \frac{\int_u^v h(t)dt}{(v-u)} \tag{3.93}$$

is a real number such that $m \le A \le M$. By the Intermediate Value Theorem, $h(t_0) \le A \le h(t_1)$ implies that there exists a $\zeta \in I$ such that $h(\zeta) = A$.

Theorem 3.13 (The First Fundamental Theorem of Calculus [1–7]) *Suppose that $h \in \mathbb{R}$ on $I = [u, v]$. Then the function H given by*

$$H(t) = \int_u^t h(s)ds \tag{3.94}$$

is uniformly continuous on $[u, v]$. If h is continuous on I, then H is differentiable in (u, v) and, for each

$$t \in (u, v),\ H'(t) = h(t). \tag{3.95}$$

However, if the function $\beta(t) = t$ have that, the function's integral over a specified interval, or the second fundamental theorem of calculus, on the other hand, is equal to the change of any antiderivative F between the endpoints of the interval. If an antiderivative can be derived via symbolic integration, without numerical integration, then this substantially simplifies the calculation of a definite integral.

Theorem 3.14 (The Second Fundamental Theorem of Calculus [1–7]) *If $h \in \mathbb{R}$ on $I = [u, v]$ and there exists a function H that is differentiable on $[u, v]$ with $H'(t) = h(t)$, then*

$$\int_{u}^{v} h(s)ds = H(b) - H(a). \tag{3.96}$$

Remark 3.5 The first calculus fundamental theorem is stated differently than it was in elementary calculus. If h were not assumed to be integrable, if instead of taking h to be integrable in $I = [u, v]$, we take h to be integrable on an open interval containing $I = [u, v]$, we can claim that $G(t) = \int_{a}^{t} h(s)ds$ is differentiable on $[u, v]$ with $G^{'}(t) = h(t)$ on $[u, v]$. This allows us to provide a somewhat different proof for the second calculus fundamental theorem. In particular, if H is any antiderivative, then for h then $H - G = c$ for some constant c and we have that

$$\begin{aligned} H(v) - H(u) &= [G(v) + c] - [G(u) + c] \\ &= [G(v) - G(u)] = \int_{u}^{v} h(s)ds. \end{aligned} \tag{3.97}$$

Remark 3.6 The fundamental theorem of calculus provides a situation in which finding the integral of a function is identical to finding its primal or antiderivative. We conclude that h has a primitive when it is a continuous function and designate the set of all primitives by $\int h(t)dt$ to find the definite integral $\int_{u}^{v} h(t)d(t)$. We find any primitive of h, H, and we conclude that

$$\int_{u}^{v} h(t)dt = H(t)|_{u}^{v} = H(v) - H(u). \tag{3.98}$$

Theorem 3.15 ([1–7]) *Suppose that the function h is continuous on a segment I, the functions q and $\frac{dq}{dt}$ are continuous on a segment J, and the range of q is contained in I. If $u, v \in J$, then*

$$\int_{u}^{v} h(q(t))q'(t)dt = \int_{q(u)}^{q(v)} h(q)dq. \tag{3.99}$$

References

1. Riemann, B.: On the Hypotheses which Lie at the Bases of Geometry. Birkhäuser (2016)
2. Natanson, I.P.: Theory of Functions of a Real Variable. Boron, Frederick Ungar Pub. Co., New York, Translated from the Russian by Leo F (1964)
3. Thomas, G.B., Finney, R.L.: Calculus and Analytic Geometry, 9th edn. Addison Wesley (1996). ISBN 0-201-53174-7
4. Bullock, G.L.: A geometric interpretation of the Riemann-Stieltjes integral. Amer. Math. Month. **95**(5) (1988). Mathematical Association of America: 448–455
5. Hildebrandt, T.H.: Definitions of Stieltjes integrals of the Riemann type. Amer. Math. Month. **45**(5), 265–278 (1938)
6. Pollard, H.: The Stieltjes integral and its generalizations. Q. J. Pure Appl. Math. **49** (1920)
7. Young, L.C.: An inequality of the Hölder type, connected with Stieltjes integration. Acta Math. **67**(1), 251–282 (1936)

Chapter 4
Inequalities Related to Global Fractional Derivatives

Inequalities are a great tool in mathematics and a core concept in nearly every area of pure and applied mathematics. The relevance of the term lies in its emerging technical adoption for multiple applications, from research to applications. Here, we will derive and establish the new inequalities in which further inequalities will immediately follow from it, and will be essentially related to a fractional derivative and integral of a function with respect to another function. The concept of these operators introduced in this book as global fractional differential and integral operators provides a generalized structure of analysis.

In addition, this section explains the versions of some already studied inequalities, adapted to provide context for the localization of these global fractional differential and integral operators, as they have the ability to easily branch to that new setting. Based on these adaptations, new inequalities will also be obtained that is more in line with the context of fractional calculus and we are interested in presenting new results that extend the mathematical tools there. The goal we attempt to accomplish with these results is to satisfy the audience in different levels and provide them some insight on the topic at hand, in addition to its theoretical significance and applications for readers coming from different areas of knowledge.

Harnack's inequality: Let f be continuous on a closed ball and harmonic on its interior, then $\forall t$ with $|t - t_0| = r < R$

$$\frac{1 - \frac{r}{R}}{1 + \frac{r}{R}} f(t_0) \leq f(t) \leq \frac{1 + \frac{r}{R}}{1 - \frac{r}{R}} f(t_0). \tag{4.1}$$

Minkowski's inequality: If $p > 1$ then

A. Atangana and İ. Koca, *Fractional Differential and Integral Operators with Respect to a Function*, Industrial and Applied Mathematics,
https://doi.org/10.1007/978-981-97-9951-0_4

$$\left|\int_a^b |f(t)+g(t)|^p \, dt\right|^{\frac{1}{p}} \leq \left(\int_a^b |f(t)|^p \, dt\right)^{\frac{1}{p}} + \left(\int_a^b |g(t)|^p \, dt\right)^{\frac{1}{p}}. \tag{4.2}$$

Clarkson's inequality: Let $\left(X, \sum, \mu\right)$ be a measure space; Let $f, g : X \to R$ be measurable functions in $L^p[X]$. Then for $2 \leq p < n$,

$$\left\|\frac{f+g}{2}\right\|_{L^p[R]}^p + \left\|\frac{f-g}{2}\right\|_{L^2[R]}^2 \leq \frac{1}{2}\left(\|f\|_{L^p[R]}^p + \|g\|_{L^p[R]}^p\right). \tag{4.3}$$

For $n < p < 2$

$$\left\|\frac{f+g}{2}\right\|_{L^p[R]}^q + \left\|\frac{f-g}{2}\right\|_{L^p[R]}^q \leq \left(\frac{1}{2}\|f\|_{L^p[R]}^p + \frac{1}{2}\|g\|_{L^p[R]}^p\right)^{\frac{q}{p}} \tag{4.4}$$

where $\frac{1}{q} + \frac{1}{p} = 1$.

Cauchy–Schwarz inequality: Let f and $g \in L^p[R]$, then

$$\left|\int_R f(x)\, g(x)\, dx\right| \leq \int_R |f(x)|^2 \, dx \int_R |g(x)|^2 \, dx. \tag{4.5}$$

Hölder inequality: Let $\left(X, \sum, \mu\right)$ be a measurable space and let $p, q \in [1, \infty)$ with $\frac{1}{q} + \frac{1}{p} = 1$. Then for all $f, g \in X$

$$\|fg\|_{L^{\prime}[\mathbb{R}]} \leq \|f\|_{L^p[\mathbb{R}]} \|q\|_{L^q[\mathbb{R}]}. \tag{4.6}$$

Landau–Kolmogorov inequality: Let $\|f\|$ be the supremum of $|f(t)|$ where f is defined in $(0, \infty)$. If then $f \in C^2[0, \infty]$ and f and f'' are bounded then

$$\|f'\| \leq 2\|f\|^{\frac{1}{2}} \|f''\|^{\frac{1}{2}}. \tag{4.7}$$

But if f is defined in $\mathbb{R}$ then

$$\|f'\| \leq \sqrt{2}\|f\|^{\frac{1}{2}} \|f''\|^{\frac{1}{2}}. \tag{4.8}$$

Gronwall inequality: Let I denote an interval of the real line of the form $[a, \infty)$ or $[a, b]$ or $[a, b)$ with $a < b$. Let β and u be real-valued continuous functions defined on I. If u is differentiable in the interior I° of I and satisfies the differential inequality

$$u'(t) \leq \beta(t)\, u(t), t \in I^{\circ}, \tag{4.9}$$

then u is bounded by the solution of the corresponding differential equation

$$u(t) \leq u(a) \exp\left(\int_a^t \beta(s)\, ds\right) \tag{4.10}$$

for all $t \in I$.

Global fractional derivatives have recently been introduced and have attracted attention of many scholars from different backgrounds in particular in applied mathematics [1–6]. However studies related to inequalities have not been developed. In this chapter, we devote our attention to the discussion underpinning inequalities related to these differential operators.

4.1 Inequalities Related to Power Law Decay-Based Global Derivatives

We recall that the power law-based global fractional derivative is defined as

$${}_{t_0}^{RL}D_g^{\alpha} f(t) = \frac{1}{\Gamma(1-\alpha)} D_g \int_{t_0}^{t} f(\tau)(t-\tau)^{-\alpha} d\tau. \tag{4.11}$$

We assume that $t \in [t_0, T]$ the function $g(t)$ is differentiable $g'(t) > 0$ and it is continuous and bounded. In summary, $\forall t \in [t_0, t]$ and there exists $M_{g'} > 0$.

Lemma 4.1 *Let m and M be two real values such that $\forall t \in [a, b]$,*

$$m < f(t) < M. \tag{4.12}$$

Then

$$\frac{m(t-t_0)^{-\alpha}}{\Gamma(1-\alpha)g'(t)} < {}_{t_0}^{RL}D_g^{\alpha} f(t) < \frac{M(t-t_0)^{-\alpha}}{\Gamma(1-\alpha)g'(t)}. \tag{4.13}$$

Proof From the hypothesis, we have

$$m < f(t) < M, \tag{4.14}$$

$$\frac{1}{\Gamma(1-\alpha)} D_g \int_{t_0}^{t} m(t-\tau)^{-\alpha} d\tau < {}_{t_0}^{RL}D_g^{\alpha} f(t) < \frac{1}{\Gamma(1-\alpha)} D_g \int_{t_0}^{t} M(t-\tau)^{-\alpha} d\tau, \tag{4.15}$$

$$\frac{m}{\Gamma(1-\alpha)}\frac{1}{g'(t)}\frac{d}{dt}\int_{t_0}^{t}(t-\tau)^{-\alpha}d\tau < {}_{t_0}^{RL}D_g^{\alpha}f(t) < \frac{M}{\Gamma(1-\alpha)}\frac{1}{g'(t)}\frac{d}{dt}\int_{t_0}^{t}(t-\tau)^{-\alpha}d\tau, \tag{4.16}$$

$$\frac{m(t-t_0)^{-\alpha}}{\Gamma(1-\alpha)g'(t)} < {}_{t_0}^{RL}D_g^{\alpha}f(t) < \frac{M(t-t_0)^{-\alpha}}{\Gamma(1-\alpha)g'(t)},$$

which completes the proof.

In this section, the relationship between Caputo and Riemann–Liouville derivatives will be used in several proofs. Thus we shall present below the demonstration which is obtained using integration by part.

$${}_{t_0}^{RL}D_t^{\alpha}f(t) = \frac{1}{\Gamma(1-\alpha)}\frac{d}{dt}\int_{t_0}^{t}f(\tau)(t-\tau)^{-\alpha}d\tau. \tag{4.17}$$

The function $f(t)$ needs to be differentiable. By using integration by part, we have

$$v'(\tau) = (t-\tau)^{-\alpha}, u(t) = f(t) \tag{4.18}$$

$$\begin{aligned}
{}_{t_0}^{RL}D_t^{\alpha}f(t) &= \frac{1}{\Gamma(1-\alpha)}\frac{d}{dt}\left[\frac{-(t-\tau)^{1-\alpha}}{1-\alpha}f(t)\Big|_{t_0}^{t} + \int_{t_0}^{t}f'(\tau)\frac{(t-\tau)^{1-\alpha}}{1-\alpha}d\tau\right] \\
&= \frac{1}{\Gamma(1-\alpha)}\left[\left(\frac{f(t_0)(t-t_0)^{1-\alpha}}{1-\alpha}\right)' + \frac{d}{dt}\int_{t_0}^{t}f'(\tau)\frac{(t-\tau)^{1-\alpha}}{1-\alpha}d\tau\right] \\
&= \frac{1}{\Gamma(1-\alpha)}f(t_0)(t-t_0)^{-\alpha} + \frac{1}{\Gamma(1-\alpha)}\int_{t_0}^{t}f'(\tau)(t-\tau)^{-\alpha}d\tau \\
&= \frac{1}{\Gamma(1-\alpha)}f(t_0)(t-t_0)^{-\alpha} + {}_{t_0}^{C}D_t^{\alpha}f(t).
\end{aligned} \tag{4.19}$$

Thus if $f(t_0) = 0$ then

$${}_{t_0}^{RL}D_t^{\alpha}f(t) = {}_{t_0}^{C}D_t^{\alpha}f(t). \tag{4.20}$$

Lemma 4.2 *Let f be continuously differentiable on $[a, b]$. Let $g'(t)$ be bounded the if $f'(t)$ is also bounded and $1-2\alpha > 0$*

$$\left|{}_{t_0}^{RL}D_t^{\alpha}f(t)\right| < \frac{M_1}{\Gamma(1-\alpha)}\left\{\frac{M_2(t-t_0)^{1-\alpha}}{(1-2\alpha)^{\frac{1}{2}}} + (t-t_0)^{-\alpha}\left|f(t_0)\right|\right\}. \tag{4.21}$$

Proof Let $g'(t)$ be bounded in $[a, b]$ then there exists $M_1 > 0$ such that $\left|\frac{1}{g'(t)}\right| < M_1$. $\forall t \in [a, b]$,

$$\left|{}_{t_0}^{RL}D_g^\alpha f(t)\right| = \left|\frac{1}{\Gamma(1-\alpha)} D_g \int_{t_0}^{t} f(\tau)(t-\tau)^{-\alpha} d\tau\right|, \tag{4.22}$$

$$= \frac{1}{\Gamma(1-\alpha)}\left|\frac{1}{g'(t)}\frac{d}{dt}\int_{t_0}^{t} f(\tau)(t-\tau)^{-\alpha} d\tau\right|,$$

$$= \frac{1}{\Gamma(1-\alpha)}\left|\frac{1}{g'(t)}\left\{\int_{t_0}^{t} f'(\tau)(t-\tau)^{-\alpha} d\tau + (t-t_0)^{-\alpha} f(t_0)\right\}\right|,$$

$$\leq \frac{1}{\Gamma(1-\alpha)}\sup_{t\in[a,b]}\left|\frac{1}{g'(t)}\right|\left\{\begin{array}{c}\left|\int_{t_0}^{t} f'(\tau)(t-\tau)^{-\alpha} d\tau\right| \\ +(t-t_0)^{-\alpha}|f(t_0)|\end{array}\right\},$$

$$\leq \frac{M_1}{\Gamma(1-\alpha)}\left\{\begin{array}{c}\left(\int_{t_0}^{t}|f'(\tau)|^2 d\tau\right)^{\frac{1}{2}}\left(\int_{t_0}^{t}(t-\tau)^{-2\alpha} d\tau\right)^{\frac{1}{2}} \\ +(t-t_0)^{-\alpha}|f(t_0)|\end{array}\right\},$$

$$\leq \frac{M_1}{\Gamma(1-\alpha)}\left\{\left(M_2^2(t-t_0)\right)^{\frac{1}{2}}\left(\frac{(t-t_0)^{1-2\alpha}}{1-2\alpha}\right)^{\frac{1}{2}} + (t-t_0)^{-\alpha}|f(t_0)|\right\},$$

$$\leq \frac{M_1}{\Gamma(1-\alpha)}\left\{\frac{M_2(t-t_0)^{1-\alpha}}{(1-2\alpha)^{\frac{1}{2}}} + (t-t_0)^{-\alpha}|f(t_0)|\right\}.$$

Therefore,

$$\left|{}_{t_0}^{RL}D_g^\alpha f(t)\right| < \frac{M_1}{\Gamma(1-\alpha)}\left\{\frac{M_2(t-t_0)^{1-\alpha}}{(1-2\alpha)^{\frac{1}{2}}} + (t-t_0)^{-\alpha}|f(t_0)|\right\} \tag{4.23}$$

which completes the proof. Noting that the proof was achieved using the Hölder inequality. Let f and h be differentiable in $[a, b]$ such that $f'(t)$ and $h'(t)$ are bounded and $f', h' \in [t_0, t_0 + a]$, then, $\beta' = f'(t) + h'(t)$.

$${}_{t_0}^{RL}D_g^\alpha \beta(t) = \frac{1}{\Gamma(1-\alpha)}\left|D_g\int_{t_0}^{t}(f+h)(\tau)(t-\tau)^{-\alpha} d\tau\right|, \tag{4.24}$$

$$= \frac{1}{\Gamma(1-\alpha)} \left| \frac{1}{g'(t)} \frac{d}{dt} \int_{t_0}^{t} (f+h)(\tau)(t-\tau)^{-\alpha} d\tau \right|,$$

$$= \frac{1}{\Gamma(1-\alpha)} \sup_{t\in[a,b]} \left| \frac{1}{g'(t)} \right| \left| \frac{d}{dt} \int_{t_0}^{t} (f+h)(\tau)(t-\tau)^{-\alpha} d\tau \right|,$$

$$= \frac{M_1}{\Gamma(1-\alpha)} \left| \left\{ \int_{t_0}^{t} (f'+h')(\tau)(t-\tau)^{-\alpha} d\tau + (t-t_0)^{-\alpha} (f+h)(t_0) \right\} \right|,$$

$$\leq \frac{M_1}{\Gamma(1-\alpha)} \left| \left\{ \begin{array}{c} \left(\int_{t_0}^{t} \left|(f'+h')(\tau)\right|^2 d\tau \right)^{\frac{1}{2}} \left(\int_{t_0}^{t} |(t-\tau|^{-2\alpha} d\tau \right)^{\frac{1}{2}} \\ +(t-t_0)^{-\alpha} (f+h)(t_0) \end{array} \right\} \right|.$$

With the help of the Minkowski inequality, we obtain

$$\left|{}_{t_0}^{RL} D_g^{\alpha} \beta(t)\right| \leq \frac{M_1}{\Gamma(1-\alpha)} \left\{ \begin{array}{c} \left(\int_{t_0}^{t} \left|f'\right|^2 d\tau \right)^{\frac{1}{2}} + \left(\int_{t_0}^{t} \left|h'\right|^2 d\tau \right)^{\frac{1}{2}} \\ + \left(\frac{(t-t_0)^{1-2\alpha}}{1-2\alpha} \right)^{\frac{1}{2}} + (t-t_0)^{-\alpha} (f+h)(t_0) \end{array} \right\}, \tag{4.25}$$

$$< \frac{M_1}{\Gamma(1-\alpha)} \left\{ \begin{array}{c} M_{f'}(t-t_0)^{\frac{1}{2}} + M_{h'}(t-t_0)^{\frac{1}{2}} \\ + \left(\frac{(t-t_0)^{\frac{1}{2}-\alpha}}{(1-2\alpha)^{\frac{1}{2}}} \right) + (t-t_0)^{-\alpha} (f+h)(t_0) \end{array} \right\},$$

$$< \frac{M_1}{\Gamma(1-\alpha)} \left\{ \frac{\left(M_{f'} + M_{h'}\right)(t-t_0)^{1-\alpha}}{(1-2\alpha)^{\frac{1}{2}}} + (t-t_0)^{-\alpha} (f+h)(t_0) \right\}.$$

Therefore

$$\left|{}_{t_0}^{RL} D_g^{\alpha} \beta(t)\right| \leq \frac{M_1}{\Gamma(1-\alpha)} \left\{ \frac{\left(M_{f'} + M_{h'}\right)(t-t_0)^{1-\alpha}}{(1-2\alpha)^{\frac{1}{2}}} + (t-t_0)^{-\alpha} (f+h)(t_0) \right\}, \tag{4.26}$$

if $(1-2\alpha) > 0$.

Corollary 4.1 *Let f be defined on a closed ball of R with ratio λ with center x_0. It is assumed that f is nonnegative and harmonic on its interior but continuous on the closed ball then*

$$\frac{\left(1-\frac{\lambda_1}{\lambda}\right) f(x_0)(x-x_0)^{-\alpha}}{\Gamma(1-\alpha)g'(x)} \leq \left|{}_{x_0}^{RL} D_g^{\alpha} f(x)\right| \leq \frac{\left(1+\frac{\lambda_1}{\lambda}\right) f(x_0)(x-x_0)^{-\alpha}}{\Gamma(1-\alpha)g'(x)}. \tag{4.27}$$

Proof Under the conditions prescribed above using Harnack's inequality, we have that

$$f(x_0)\left(1-\frac{\lambda_1}{\lambda}\right) \leq f(x) \leq \left(1+\frac{\lambda_1}{\lambda}\right) f(x_0) \tag{4.28}$$

$$\frac{f(x_0)\left(1-\frac{\lambda_1}{\lambda}\right)}{\Gamma(1-\alpha)} \frac{1}{g'(x)} \frac{d}{dx} \int_{x_0}^{x} (x-\tau)^{-\alpha} d\tau$$

$$\leq {}_{x_0}^{RL}D_g^{\alpha} f(x) \leq \frac{f(x_0)\left(1+\frac{\lambda_1}{\lambda}\right)}{\Gamma(1-\alpha)} \frac{1}{g'(x)} \frac{d}{dx} \int_{x_0}^{x} (x-\tau)^{-\alpha} d\tau,$$

$$\frac{f(x_0)\left(1-\frac{\lambda_1}{\lambda}\right)(1-\alpha)(x-x_0)^{-\alpha}}{\Gamma(2-\alpha) g'(x)}$$

$$\leq {}_{x_0}^{RL}D_g^{\alpha} f(x) \leq \frac{f(x_0)\left(1+\frac{\lambda_1}{\lambda}\right)(1-\alpha)(x-x_0)^{-\alpha}}{\Gamma(2-\alpha) g'(x)}$$

$$\frac{f(x_0)\left(1-\frac{\lambda_1}{\lambda}\right)(x-x_0)^{-\alpha}}{\Gamma(1-\alpha) g'(x)}$$

$$\leq \left| {}_{x_0}^{RL}D_g^{\alpha} f(x) \right| \leq \frac{\left(1+\frac{\lambda_1}{\lambda}\right) f(x_0)(x-x_0)^{-\alpha}}{\Gamma(1-\alpha) g'(x)},$$

which completes the proof.

Corollary 4.2 *Let (X, Σ, μ) be a measure space. Let $g'(t)$ be bounded such that $\left|\frac{1}{g'(t)}\right| < M_{g'}$. Let $f: X \to R$ be measurable in $L^p[X]$, then if $p = 2$ and $-2\alpha + 1 > 0$. We have the following inequality:*

$$\left| {}_{t_0}^{RL}D_g^{\alpha} f(t) \right| \leq \frac{M_{g'}}{\Gamma(1-\alpha)} \left((t-t_0)^{-\alpha} |f(t_0)| + \left\| f' \right\|_{L^2[X]} + \left(\frac{(t-t_0)^{1-2\alpha}}{1-2\alpha} \right)^{\frac{1}{2}} \right). \tag{4.29}$$

Proof To prove the above, we shall first use the integration by part of the Riemann–Liouville integral as follows:

$${}_{t_0}^{RL}D_g^{\alpha} f(t) = \frac{1}{\Gamma(1-\alpha)} \frac{d}{dt} \int_{t_0}^{t} f(\tau)(t-\tau)^{-\alpha} d\tau \tag{4.30}$$

$$= \frac{1}{\Gamma(1-\alpha)} \frac{d}{dt} \left[-f(\tau) \left. \frac{(t-\tau)^{1-\alpha}}{1-\alpha} \right|_{t_0}^{t} + \int_{t_0}^{t} f'(\tau) \frac{(t-\tau)^{1-\alpha}}{1-\alpha} d\tau \right]$$

$$= \frac{1}{\Gamma(1-\alpha)} \frac{d}{dt} \left[f(t_0) \frac{(t-t_0)^{1-\alpha}}{1-\alpha} + \int_{t_0}^{t} f'(\tau) \frac{(t-\tau)^{1-\alpha}}{1-\alpha} d\tau \right]$$

$$= \frac{1}{\Gamma(1-\alpha)} \left[f(t_0)(t-t_0)^{-\alpha} + \frac{d}{dt} \int_{t_0}^{t} f'(\tau) \frac{(t-\tau)^{1-\alpha}}{1-\alpha} d\tau \right]$$

$$= \frac{1}{\Gamma(1-\alpha)} \left[f(t_0)(t-t_0)^{-\alpha} + \frac{d}{dt} \int_{t_0}^{t} f'(\tau)(t-\tau)^{-\alpha} d\tau \right].$$

Therefore applying the norm on both sides yields

$$\left| {}_{t_0}^{RL}D_g^{\alpha} f(t) \right| \leq \frac{1}{\Gamma(1-\alpha)} \left| \frac{1}{g'(t)} \right| \left| f(t_0)(t-t_0)^{-\alpha} + \int_{t_0}^{t} f'(\tau)(t-\tau)^{-\alpha} d\tau \right| \tag{4.31}$$

$$\leq \frac{M_{g'}}{\Gamma(1-\alpha)} \left\{ |f(t_0)| \, (t-t_0)^{-\alpha} + \left(\int_{t_0}^{t} |f'|^2 d\tau \right)^{\frac{1}{2}} \left(\int_{t_0}^{t} (t-\tau)^{-2\alpha} d\tau \right)^{\frac{1}{2}} \right\}$$

$$\leq \frac{M_{g'}}{\Gamma(1-\alpha)} \left\{ |f(t_0)| \, (t-t_0)^{-\alpha} + \left\| f' \right\|_{L^2[X]} \left(\frac{(t-t_0)^{1-2\alpha}}{(1-2\alpha)} \right)^{\frac{1}{2}} \right\}.$$

So we get

$$\left| {}_{t_0}^{RL}D_g^{\alpha} f(t) \right| \leq \frac{M_{g'}}{\Gamma(1-\alpha)} \left\{ |f(t_0)| \, (t-t_0)^{-\alpha} + \left\| f' \right\|_{L^2[X]} \left(\frac{(t-t_0)^{1-2\alpha}}{(1-2\alpha)} \right)^{\frac{1}{2}} \right\}$$

which completes the proof.

Corollary 4.3 *Let (X, Σ, μ) be a measure space. Let $g'(t)$ be bounded such that $\left| \frac{1}{g'(t)} \right| < M_{g'}$. Let f' and $h : X \to R$ be measurable in $L^2[X]$, if $1 - 2\alpha > 0$, then we have*

$$\left| {}_{t_0}^{RL}D_g^{\alpha} f(t) \right| \leq \frac{M_{g'}}{\Gamma(1-\alpha)} \left((t-t_0)^{-\alpha} |f(t_0)| + \left\{ \left\| f' \right\|_{L^2[X]} + \|h\|_{L^2[X]} \right\} \left(\frac{(t-t_0)^{1-2\alpha}}{1-2\alpha} \right)^{\frac{1}{2}} \right). \tag{4.32}$$

Proof Using the integration by part of ${}_{t_0}^{RL}D_g^{\alpha} f(t)$, we had

$${}_{t_0}^{RL}D_g^{\alpha} f(t) = \frac{1}{\Gamma(1-\alpha)} \left[f(t_0)(t-t_0)^{-\alpha} + \int_{t_0}^{t} f'(\tau)(t-\tau)^{-\alpha} d\tau \right]. \tag{4.33}$$

Therefore

$$\left|{}_{t_0}^{RL}D_g^{\alpha}f(t)\right| \tag{4.34}$$
$$\leq \frac{M_{g'}}{\Gamma(1-\alpha)}\left\{|f(t_0)|(t-t_0)^{-\alpha}+\left(\int_{t_0}^{t}|f'(\tau)|^2d\tau\right)^{\frac{1}{2}}\left(\int_{t_0}^{t}(t-\tau)^{-2\alpha}d\tau\right)^{\frac{1}{2}}\right\}$$
$$\leq \frac{M_{g'}}{\Gamma(1-\alpha)}\left\{|f(t_0)|(t-t_0)^{-\alpha}+\left(\frac{(t-t_0)^{1-2\alpha}}{1-2\alpha}\right)^{\frac{1}{2}}\left(\int_{t_0}^{t}\left|\frac{f'-h}{2}+\frac{f'+h}{2}\right|^2d\tau\right)^{\frac{1}{2}}\right\}$$
$$\leq \frac{M_{g'}}{\Gamma(1-\alpha)}\left\{|f(t_0)|(t-t_0)^{-\alpha}+\left(\frac{(t-t_0)^{1-2\alpha}}{1-2\alpha}\right)^{\frac{1}{2}}\left(\frac{\|f'-h\|_{L^2[X]}^2}{2}+\frac{\|f'+h\|_{L^2[X]}^2}{2}\right)\right\}$$
$$\leq \frac{M_{g'}}{\Gamma(1-\alpha)}\left((t-t_0)^{-\alpha}|f(t_0)|+\left(\frac{\left\|f'\right\|_{L^2[X]}^2+\|h\|_{L^2[X]}^2}{2}\right)^{\frac{1}{2}}\left(\frac{(t-t_0)^{1-2\alpha}}{1-2\alpha}\right)^{\frac{1}{2}}\right).$$

The above is obtained due to Minkowski and Clakson's inequalities. So finally we have the following inequality:

$$\left|{}_{t_0}^{RL}D_g^{\alpha}f(t)\right| \leq \frac{M_{g'}}{\Gamma(1-\alpha)}\left(\begin{array}{c}(t-t_0)^{-\alpha}|f(t_0)| \\ +\left\{\left\|f'\right\|_{L^2[X]}+\|h\|_{L^2[X]}\right\}\left(\frac{(t-t_0)^{1-2\alpha}}{1-2\alpha}\right)^{\frac{1}{2}}\end{array}\right).$$

Corollary 4.4 *Let f be twice differentiable on $[a,b]$ such that $f \in L^p[a,b]$ and $f'' \in L^q[a,b]$ where $\frac{1}{q}+\frac{1}{p}=1$ and $f' \in L^2[a,b]$ then*

$$\left|{}_{t_0}^{RL}D_g^{\alpha}f(t)\right| \leq \frac{M_1}{\Gamma(1-\alpha)}\left\{\frac{(t-t_0)^{\frac{1}{2}-\alpha}}{(1-2\alpha)^{\frac{1}{2}}}K\|f\|_{L^p[a,b]}^{\beta}\left\|f''\right\|_{L^q[a,b]}^{1-\beta}+(t-t_0)^{-\alpha}|f(t_0)|\right\}, \tag{4.35}$$

β and K are constants and p, q are positive integer numbers.

Proof By definition of ${}_{t_0}^{RL}D_g^{\alpha}f(t)$, we have

$$\left|{}_{t_0}^{RL}D_g^{\alpha}f(t)\right| = \left|\frac{1}{\Gamma(1-\alpha)}D_g\int_{t_0}^{t}f(\tau)(t-\tau)^{-\alpha}d\tau\right|, \tag{4.36}$$
$$= \frac{1}{\Gamma(1-\alpha)}\left|\frac{1}{g'(t)}\frac{d}{dt}\int_{t_0}^{t}f(\tau)(t-\tau)^{-\alpha}d\tau\right|,$$
$$= \frac{1}{\Gamma(1-\alpha)}\left|\frac{1}{g'(t)}\left\{\int_{t_0}^{t}f'(\tau)(t-\tau)^{-\alpha}d\tau+(t-t_0)^{-\alpha}f(t_0)\right\}\right|,$$
$$\leq \frac{1}{\Gamma(1-\alpha)}\sup_{t\in[a,b]}\left|\frac{1}{g'(t)}\right|\left\{\left|\int_{t_0}^{t}f'(\tau)(t-\tau)^{-\alpha}d\tau\right|+(t-t_0)^{-\alpha}|f(t_0)|\right\},$$

$$\leq \frac{M_1}{\Gamma(1-\alpha)} \left\{ \left(\int_{t_0}^{t} \left| f'(\tau) \right|^2 d\tau \right)^{\frac{1}{2}} \left(\int_{t_0}^{t} (t-\tau)^{-2\alpha} d\tau \right)^{\frac{1}{2}} + (t-t_0)^{-\alpha} \left| f(t_0) \right| \right\},$$

$$\leq \frac{M_1}{\Gamma(1-\alpha)} \left\{ \frac{(t-t_0)^{\frac{1}{2}-\alpha}}{(1-2\alpha)^{\frac{1}{2}}} \left\| f' \right\|_{L^2[a,b]} + (t-\tau)^{-\alpha} \left| f(t_0) \right| \right\}.$$

By the Landau–Kolmogorov inequality we have that

$$\left| {}_{t_0}^{RL} D_g^{\alpha} f(t) \right| \leq \frac{M_1}{\Gamma(1-\alpha)} \left\{ \frac{(t-t_0)^{\frac{1}{2}-\alpha}}{(1-2\alpha)^{\frac{1}{2}}} K \left\| f \right\|_{L^p[a,b]}^{\beta} \left\| f'' \right\|_{L^q[a,b]}^{q-\beta} + (t-t_0)^{-\alpha} \left| f(t_0) \right| \right\} \tag{4.37}$$

which completes the proof.

Lemma 4.3 *Let f be continuously differentiable on $[a, b]$. Let $g'(t)$ be continuously bounded then if $f'(t)$ is bounded and $(1-2\alpha) > 0$ then*

$$\left| {}_{t_0}^{RL} D_g^{\alpha} f(t) \right|^2 \leq \frac{\left(M_{g'}\right)^2}{\Gamma^2(1-\alpha)} \left\{ \begin{array}{c} M_{f'}^2 (t-t_0) \left(\frac{(t-t_0)^{1-2\alpha}}{(1-2\alpha)} \right) \\ +2(t-t_0)^{-\alpha} \left| f(t_0) \right| M_{f'} (t-t_0)^{\frac{1}{2}} \left(\frac{(t-t_0)^{\frac{1}{2}-\alpha}}{(1-2\alpha)^{\frac{1}{2}}} \right) \\ +(t-t_0)^{-2\alpha} \left| f(t_0) \right|^2 \end{array} \right\}. \tag{4.38}$$

Proof We now evaluate the square of this derivative.

$$\left| {}_{t_0}^{RL} D_g^{\alpha} f(t) \right|^2 = \left| \frac{1}{\Gamma(1-\alpha)} D_g \int_{t_0}^{t} f(\tau)(t-\tau)^{-\alpha} d\tau \right|^2, \tag{4.39}$$

$$= \frac{1}{\Gamma^2(1-\alpha)} \left| \frac{1}{g'(t)} \frac{d}{dt} \int_{t_0}^{t} f(\tau)(t-\tau)^{-\alpha} d\tau \right|^2,$$

$$\leq \frac{1}{\Gamma^2(1-\alpha)} \left| \frac{1}{g'(t)} \right|^2 \left| \int_{t_0}^{t} f'(\tau)(t-\tau)^{-\alpha} d\tau + (t-t_0)^{-\alpha} f(t_0) \right|^2,$$

$$\leq \frac{1}{\Gamma^2(1-\alpha)} \sup_{t\in[a,b]} \left| \frac{1}{g'(t)} \right|^2 \left\{ \begin{array}{c} \left| \int_{t_0}^{t} f'(\tau)(t-\tau)^{-\alpha} d\tau \right|^2 \\ +2(t-t_0)^{-\alpha} \left| f(t_0) \right| \left| \int_{t_0}^{t} f'(\tau)(t-\tau)^{-\alpha} d\tau \right| \\ +(t-t_0)^{-2\alpha} \left| f(t_0) \right|^2 \end{array} \right\},$$

$$\leq \frac{\left(M_{g'}\right)^2}{\Gamma^2(1-\alpha)} \left\{ \begin{array}{c} \int_{t_0}^{t} \left| f'(\tau) \right|^2 d\tau \int_{t_0}^{t} (t-\tau)^{-2\alpha} d\tau \\ +2(t-t_0)^{-\alpha} \left| f(t_0) \right| \left(\int_{t_0}^{t} \left| f'(\tau) \right|^2 d\tau \right)^{\frac{1}{2}} \left(\int_{t_0}^{t} (t-\tau)^{-2\alpha} d\tau \right)^{\frac{1}{2}} \\ +(t-t_0)^{-2\alpha} \left| f(t_0) \right|^2 \end{array} \right\},$$

$$\leq \frac{(M_{g'})^2}{\Gamma^2(1-\alpha)} \left\{ \begin{array}{c} M_{f'}^2(t-t_0)\left(\frac{(t-t_0)^{1-2\alpha}}{(1-2\alpha)}\right) \\ +2(t-t_0)^{-\alpha}|f(t_0)|\, M_{f'}(t-t_0)^{\frac{1}{2}}\left(\frac{(t-t_0)^{\frac{1}{2}-\alpha}}{(1-2\alpha)^{\frac{1}{2}}}\right) \\ +(t-t_0)^{-2\alpha}|f(t_0)|^2 \end{array} \right\},$$

which completes the proof. Noting that the proof was achieved with the help of Cauchy–Schwarz and Hölder inequalities.

Corollary 4.5 *If the function f is only continuous on $[a,b]$ but $|g'(t)|^{-1} < M_1$ is continuous in $[a,b]$. Then if f is twice differentiable with $f''(t)$, $f'(t)$ bounded then*

$$\left|{}_{t_0}^{RL}D_g^{\alpha} f(t)\right|^2 \leq \frac{\sqrt{2}M_1}{\Gamma(1-\alpha)} \sup_{a\leq t\leq b}\left(M_f \frac{(t-t_0)^{1-\alpha}}{(1-2\alpha)^{\frac{1}{2}}}\right)^{\frac{1}{2}} \times \left(\sup_{a\leq t\leq b} \begin{array}{c} \alpha(t-t_0)^{-\alpha-1}|f(t_0)| \\ +|f'(t_0)|\,(t-t_0)^{-\alpha} \\ +M_{f''}\frac{(t-t_0)^{1-\alpha}}{(1-2\alpha)^{\frac{1}{2}}} \end{array}\right), \tag{4.40}$$

if $(1-2\alpha) > 0$.

Proof By definition of ${}_{t_0}^{RL}D_g^{\alpha} f(t)$, we have

$$\left|{}_{t_0}^{RL}D_g^{\alpha} f(t)\right| = \frac{1}{\Gamma(1-\alpha)}\left|\frac{1}{g'(t)}\frac{d}{dt}\int_{t_0}^{t} f(\tau)(t-\tau)^{-\alpha}d\tau\right|. \tag{4.41}$$

Then by the Landau–Kolmogorov inequality we have that

$$\left|{}_{t_0}^{RL}D_g^{\alpha} f(t)\right|^2 \leq \frac{\sqrt{2}M_1}{\Gamma(1-\alpha)}\|F\|^{\frac{1}{2}}\|F''\|^{\frac{1}{2}}, \tag{4.42}$$

$$\leq \frac{\sqrt{2}M_1}{\Gamma(1-\alpha)}\left\|\int_{t_0}^{t} f(\tau)(t-\tau)^{-\alpha}d\tau\right\|^{\frac{1}{2}}\left\|\frac{d^2}{dt^2}\int_{t_0}^{t} f(\tau)(t-\tau)^{-\alpha}d\tau\right\|^{\frac{1}{2}},$$

$$\leq \frac{\sqrt{2}M_1}{\Gamma(1-\alpha)}\left\|\int_{t_0}^{t} f(\tau)(t-\tau)^{-\alpha}d\tau\right\|^{\frac{1}{2}}$$

$$\times\left\|\frac{d}{dt}\left((t-t_0)^{-\alpha}f(t_0)+\int_{t_0}^{t} f'(\tau)(t-\tau)^{-\alpha}d\tau\right)\right\|^{\frac{1}{2}},$$

$$\leq \frac{\sqrt{2}M_1}{\Gamma(1-\alpha)}\left\|\int_{t_0}^{t} f(\tau)(t-\tau)^{-\alpha}d\tau\right\|^{\frac{1}{2}}$$

$$\times\left\|-\alpha(t-t_0)^{-\alpha-1}|f(t_0)|+\frac{d}{dt}\int_{t_0}^{t} f'(\tau)(t-\tau)^{-\alpha}d\tau\right\|^{\frac{1}{2}},$$

$$\leq \frac{\sqrt{2}M_1}{\Gamma(1-\alpha)}\left(\sup_{a\leq t\leq b}\left|\int_{t_0}^{t} f(\tau)(t-\tau)^{-\alpha}d\tau\right|\right)^{\frac{1}{2}}$$

$$\times\left(\sup_{a\leq t\leq b}\left|\begin{array}{c}-\alpha(t-t_0)^{-\alpha-1}|f(t_0)|\\ +f''(t_0)(t-t_0)^{-\alpha}+\int_{t_0}^{t} f''(\tau)(t-\tau)^{-\alpha}d\tau\end{array}\right|\right)^{\frac{1}{2}},$$

$$\leq \frac{\sqrt{2}M_1}{\Gamma(1-\alpha)}\left[\begin{array}{c}\sup_{a\leq t\leq b}\left\{\left(\int_{t_0}^{t}\left|f'(\tau)\right|^2 d\tau\right)^{\frac{1}{2}}\left(\int_{t_0}^{t}(t-\tau)^{-2\alpha}d\tau\right)^{\frac{1}{2}}\right\}^{\frac{1}{2}}\\ \sup_{a\leq t\leq b}\left\{\begin{array}{c}\alpha(t-t_0)^{-\alpha-1}|f(t_0)|+\left|f'(t_0)\right|(t-t_0)^{-\alpha}\\ +\left(\int_{t_0}^{t}\left|f''(\tau)\right|^2\right)^{\frac{1}{2}}\left(\int_{t_0}^{t}(t-\tau)^{-2\alpha}d\tau\right)^{\frac{1}{2}}\end{array}\right\}^{\frac{1}{2}}\end{array}\right],$$

$$\leq \frac{\sqrt{2}M_1}{\Gamma(1-\alpha)}\left[\begin{array}{c}\left(\sup_{a\leq t\leq b} M_{f'}(t-t_0)^{\frac{1}{2}}\frac{(t-t_0)^{\frac{1}{2}-\alpha}}{(1-2\alpha)^{\frac{1}{2}}}\right)^{\frac{1}{2}}\\ \times\sup_{a\leq t\leq b}\left\{\begin{array}{c}\alpha(t-t_0)^{-\alpha-1}|f(t_0)|\\ +\left|f'(t_0)\right|(t-t_0)^{-\alpha}+M_{f''}\frac{(t-t_0)^{1-\alpha}}{(1-2\alpha)^{\frac{1}{2}}}\end{array}\right\}^{\frac{1}{2}}\end{array}\right],$$

$$\leq \frac{\sqrt{2}M_1}{\Gamma(1-\alpha)}\left[\begin{array}{c}\sup_{a\leq t\leq b}\left(M_{f'}\frac{(t-t_0)^{1-\alpha}}{(1-2\alpha)^{\frac{1}{2}}}\right)^{\frac{1}{2}}\\ \times\sup_{a\leq t\leq b}\left\{\begin{array}{c}\alpha(t-t_0)^{-\alpha-1}|f(t_0)|\\ +\left|f'(t_0)\right|(t-t_0)^{-\alpha}+M_{f''}\frac{(t-t_0)^{1-\alpha}}{(1-2\alpha)^{\frac{1}{2}}}\end{array}\right\}^{\frac{1}{2}}\end{array}\right],$$

which completes the proof.

4.2 Inequalities Related to Exponential Law Decay-Based Global Derivative

We shall recall that the exponential-based global derivative is given by

$$ {}_{t_0}^{RE}D_g^{\alpha} f(t)=\frac{1}{1-\alpha}D_g\int_{t_0}^{t} f(\tau)\exp\left(-\frac{\alpha}{1-\alpha}(t-\tau)\right)d\tau. \tag{4.43}$$

Lemma 4.4 *Let m and M be two real values such that $\forall t\in[a,b]$,*

$$m<f(t)<M. \tag{4.44}$$

Then

$$\frac{m}{g^{'}(t)(1-\alpha)}\exp\left(-\frac{\alpha}{1-\alpha}(t-t_0)\right) <{}_{t_0}^{RE}D_g^{\alpha}f(t) < \frac{M}{g^{'}(t)(1-\alpha)}\exp\left(-\frac{\alpha}{1-\alpha}(t-t_0)\right). \tag{4.45}$$

Proof From the hypothesis, we have

$$m < f(t) < M, \tag{4.46}$$

then

$${}_{t_0}^{RE}D_g^{\alpha}m <{}_{t_0}^{RE}D_g^{\alpha}f(t) <{}_{t_0}^{RE}D_g^{\alpha}M. \tag{4.47}$$

$$\begin{aligned}&\frac{m}{g^{'}(t)(1-\alpha)}\frac{d}{dt}\int_{t_0}^{t}\exp\left(-\frac{\alpha}{1-\alpha}(t-\tau)\right)d\tau \\ &<{}_{t_0}^{RE}D_g^{\alpha}f(t) \\ &<\frac{M}{g^{'}(t)(1-\alpha)}\frac{d}{dt}\int_{t_0}^{t}\exp\left(-\frac{\alpha}{1-\alpha}(t-\tau)\right)d\tau,\end{aligned} \tag{4.48}$$

$$\begin{aligned}&\frac{m}{g^{'}(t)(1-\alpha)}\frac{d}{dt}\left\{\frac{1-\alpha}{\alpha}\left(1-\exp\left(-\frac{\alpha}{1-\alpha}(t-t_0)\right)\right)\right\} \\ &<{}_{t_0}^{RE}D_g^{\alpha}f(t) \\ &<\frac{M}{g^{'}(t)(1-\alpha)}\frac{d}{dt}\left\{\frac{1-\alpha}{\alpha}\left(1-\exp\left(-\frac{\alpha}{1-\alpha}(t-t_0)\right)\right)\right\},\end{aligned} \tag{4.49}$$

$$\frac{m}{g^{'}(t)(1-\alpha)}\exp\left(-\frac{\alpha}{1-\alpha}(t-t_0)\right) <{}_{t_0}^{RE}D_g^{\alpha}f(t) < \frac{M\alpha}{\alpha g^{'}(t)(1-\alpha)}\exp\left(-\frac{\alpha}{1-\alpha}(t-t_0)\right),$$

$$\frac{m}{g^{'}(t)(1-\alpha)}\exp\left(-\frac{\alpha}{1-\alpha}(t-t_0)\right) <{}_{t_0}^{RE}D_g^{\alpha}f(t) < \frac{M}{g^{'}(t)(1-\alpha)}\exp\left(-\frac{\alpha}{1-\alpha}(t-t_0)\right), \tag{4.50}$$

which completes the proof.

Lemma 4.5 *Let f be continuously differentiable with a bounded derivative $f^{'}(t)$ on $[a,b]$. Let $g^{'}(t)$ be bounded then*

$$\left|{}_{t_0}^{RE}D_g^{\alpha}f(t)\right| < \frac{M_{g'}}{(1-\alpha)\,g^{'}(t)}\left\{|f(t_0)| + M_{f'}(t-t_0)^{\frac{1}{2}}\right\}. \tag{4.51}$$

Proof Let $g'(t)$ be bounded in $[a, b]$ there exists M_g such that $\left|\frac{1}{g'(t)}\right| < M_g$. Therefore

$$\left|{}_{t_0}^{RE}D_g^{\alpha}f(t)\right| = \left|\frac{1}{1-\alpha}\frac{1}{g'(t)}\frac{d}{dt}\int_{t_0}^{t} f(\tau)\exp\left(-\frac{\alpha}{1-\alpha}(t-\tau)\right)d\tau\right|, \tag{4.52}$$

$$\leq \frac{M_g}{(1-\alpha)g'(t)}\left|\begin{array}{c} f(t_0)\exp\left(-\frac{\alpha}{1-\alpha}(t-t_0)\right) \\ +\int_{t_0}^{t} f'(\tau)\exp\left(-\frac{\alpha}{1-\alpha}(t-\tau)\right)d\tau \end{array}\right|,$$

$$\leq \frac{M_g}{(1-\alpha)g'(t)}\left\{\begin{array}{c} |f(t_0)|\exp\left(-\frac{\alpha}{1-\alpha}(t-t_0)\right) \\ +\left(\int_{t_0}^{t}|f'(\tau)|^2 d\tau\right)^{\frac{1}{2}}\left(\int_{t_0}^{t}\exp\left(-\frac{2\alpha}{1-\alpha}(t-\tau)d\tau\right)\right)^{\frac{1}{2}} \end{array}\right\},$$

$$\leq \frac{M_g}{(1-\alpha)g'(t)}\left\{|f(t_0)| + \left(M_{f'}^2(t-t_0)\right)^{\frac{1}{2}}\right\},$$

$$\leq \frac{M_g}{(1-\alpha)g'(t)}\left\{|f(t_0)| + M_{f'}(t-t_0)^{\frac{1}{2}}\right\},$$

which completes the proof.

Corollary 4.6 *Let f be a nonnegative function defined on a closed ball of R with ratio λ with center x_0. It is assumed that f is harmonic on the interior on the ball and continuous on the closed ball then*

$$\frac{m\left(1-\frac{\lambda_1}{\lambda}\right)f(x_0)}{(1-\alpha)g'(x)}\exp\left(-\frac{\alpha}{1-\alpha}(x_0-x)\right) \leq {}_{x_0}^{RE}D_g^{\alpha}f(x)$$
$$\leq \frac{M\left(1+\frac{\lambda_1}{\lambda}\right)f(x_0)}{(1-\alpha)g'(x)}\exp\left(-\frac{\alpha}{1-\alpha}(x_0-x)\right). \tag{4.53}$$

Proof Using Harnacks's inequality we have that $\forall x$, $|x - x_0| < \lambda_1 < \lambda$.

$$f(x_0)\left(1-\frac{\lambda_1}{\lambda}\right) \leq f(x) \leq f(x_0)\left(1+\frac{\lambda_1}{\lambda}\right), \tag{4.54}$$

$$\frac{m\left(1-\frac{\lambda_1}{\lambda}\right)f(x_0)}{(1-\alpha)g'(x)}\exp\left(-\frac{\alpha}{1-\alpha}(x_0-x)\right) \leq {}_{x_0}^{RE}D_g^{\alpha}f(x)$$
$$\leq \frac{M\left(1+\frac{\lambda_1}{\lambda}\right)f(x_0)}{(1-\alpha)g'(x)}\exp\left(-\frac{\alpha}{1-\alpha}(x_0-x)\right), \tag{4.55}$$

which completes the proof.

Corollary 4.7 *Let (X, Σ, μ) be a measure space. Let f' and g' : $X \to R$ measurable in $L^p[a, b]$, then if $p = 2$, we have*

$$\left|{}_{t_0}^{RE}D_g^{\alpha} f(t)\right| \leq \frac{M_1}{(1-\alpha)}\left\{|f(t_0)| + \frac{\|f'\|_{L^2[X]}^2}{2} + \frac{\|g'\|_{L^2[X]}^2}{2}\right\}. \tag{4.56}$$

Proof By definition, we have

$$\begin{aligned}\left|{}_{t_0}^{RE}D_g^{\alpha} f(t)\right| &= \left|\frac{1}{1-\alpha}\frac{1}{g'(t)}\frac{d}{dt}\int_{t_0}^{t} f(\tau)\exp\left(-\frac{\alpha}{1-\alpha}(t-\tau)\right)d\tau\right|, \\ &\leq \frac{M_1}{(1-\alpha)}\left|f(t_0)\exp\left(-\frac{\alpha}{1-\alpha}(t-t_0)\right)\right. \\ &\quad \left.+\int_{t_0}^{t} f'(\tau)\exp\left(-\frac{\alpha}{1-\alpha}(t-\tau)\right)d\tau\right|, \\ &\leq \frac{M_1}{(1-\alpha)}\left\{|f(t_0)|\exp\left(-\frac{\alpha}{1-\alpha}(t-t_0)\right)\right. \\ &\quad \left.+\int_{t_0}^{t} |f'(\tau)|\exp\left(-\frac{\alpha}{1-\alpha}(t-\tau)\right)d\tau\right\}.\end{aligned} \tag{4.57}$$

Noting that $\forall t \in X$ and $\alpha \in (0, 1)$ the function $\exp\left(-\frac{\alpha}{1-\alpha}t\right)$ is decreasing and is bounded by 1; therefore,

$$\begin{aligned}\left|{}_{t_0}^{RE}D_g^{\alpha} f(t)\right| &\leq \frac{M_1}{(1-\alpha)}\left\{|f(t_0)| + \int_{t_0}^{t} |f'(\tau)|\,d\tau\right\}, \\ &\leq \frac{M_1}{(1-\alpha)}\left\{|f(t_0)| + \left(\int_{t_0}^{t} |f'(\tau)|^2 d\tau\right)^{\frac{1}{2}}\right\}.\end{aligned} \tag{4.58}$$

Since $f' \in L^2[a, b]$, then

$$\begin{aligned}&\left|{}_{t_0}^{RE}D_g^{\alpha} f(t)\right| \\ &\leq \frac{M_1}{(1-\alpha)}\left\{|f(t_0)| + \left(\int_{t_0}^{t}\left|\frac{f'(\tau)-g_1(\tau)}{2} + \frac{f'(\tau)+g_1(\tau)}{2}\right|^2 d\tau\right)^{\frac{1}{2}}\right\}, \\ &\leq \frac{M_1}{(1-\alpha)}\left(|f(t_0)| + \left(\int_{t_0}^{t}\frac{2|f'(\tau)-g_1(\tau)|^2}{4} + \frac{2|f'(\tau)+g_1(\tau)|^2}{4}\right)^{\frac{1}{2}} d\tau\right),\end{aligned} \tag{4.59}$$

$$\leq \frac{M_1}{(1-\alpha)}\left(|f(t_0)| + \left(\int_{t_0}^{t} \frac{|f'(\tau)-g_1(\tau)|^2}{2} d\tau\right)^{\frac{1}{2}} + \left(\int_{t_0}^{t} \frac{|f'(\tau)+g_1(\tau)|^2}{2} d\tau\right)^{\frac{1}{2}}\right).$$

By the Minkovski inequality, thus

$$\left|{}_{t_0}^{RE}D_g^{\alpha} f(t)\right| \leq \frac{M_1}{(1-\alpha)}\left(|f(t_0)| + \left\|\frac{f' - g_1}{2}\right\|_{L^2[X]}^2 + \left\|\frac{f' + g_1}{2}\right\|_{L^2[X]}^2\right). \quad (4.60)$$

By Clarkson's inequality we have

$$\left|{}_{t_0}^{RE}D_g^{\alpha} f(t)\right| \leq \frac{M_1}{(1-\alpha)}\left\{|f(t_0)| + \frac{\|f'\|_{L^2[X]}^2}{2} + \frac{\|g'\|_{L^2[X]}^2}{2}\right\}, \quad (4.61)$$

which completes the proof.

Corollary 4.8 *Let f be continuously twice differentiable such that $f \in L^p[a,b]$, $f'' \in L^q[a,b]$ then*
if $p > 2$

$$\left|{}_{t_0}^{RE}D_g^{\alpha} f(t)\right| \leq \frac{KM_1}{(1-\alpha)}\left\{\begin{array}{l}\left(|f(t_0)|(t-t_0)^{\frac{1}{p}}\right)^{\beta} + \left(K\|f\|_{L^p[a,b]}^{\beta}\|f''\|_{L^q[a,b]}^{1-\beta}\right)^{\beta} \\ +\left(|f(t_0)|(t-t_0)^{\frac{1}{q}}\right)^{1-\beta} + \left(\|f''\|_{L^2[a,b]}^2 (t-t_0)^{\frac{1}{q}}\right)^{1-\beta}\end{array}\right\}. \quad (4.62)$$

Proof By definition, we have that

$$\begin{aligned}{}_{t_0}^{RE}D_g^{\alpha} f(t) &= \left|\frac{1}{1-\alpha} D_g \int_{t_0}^{t} f(\tau)\exp\left(-\frac{\alpha}{1-\alpha}(t-\tau)\right) d\tau\right|, \quad (4.63)\\ &= \left|\frac{1}{1-\alpha}\frac{1}{g'(t)}\frac{d}{dt}\int_{t_0}^{t} f(\tau)\exp\left(-\frac{\alpha}{1-\alpha}(t-\tau)\right) d\tau\right|.\end{aligned}$$

We put

$$F(t)=\int_{t_0}^{t} f(\tau)\exp\left(-\frac{\alpha}{1-\alpha}(t-\tau)\right)d\tau, \tag{4.64}$$

$$ {}_{t_0}^{RE}D_g^{\alpha} f(t)\le \frac{M}{1-\alpha}\left|F'(t)\right|\le \frac{M}{1-\alpha}\left\|F'\right\|_{L^2[a,b]}.$$

By the Landau–Kolmogorov inequality we have that

$$\begin{aligned}\left|{}_{t_0}^{RE}D_g^{\alpha} f(t)\right| &\le \frac{M_1}{(1-\alpha)}\left(K\left\|F\right\|_{L^p[a,b]}^{\beta}\left\|F''\right\|_{L^q[a,b]}^{1-\beta}\right), \\ &\le \frac{M_1K}{(1-\alpha)}\left\|\int_{t_0}^{t} f(\tau)\exp\left(-\frac{\alpha}{1-\alpha}(t-\tau)\right)d\tau\right\|_{L^p[a,b]}^{\beta} \\ &\quad\times\left\|\frac{d^2}{dt^2}\int_{t_0}^{t} f(\tau)\exp\left(-\frac{\alpha}{1-\alpha}(t-\tau)\right)d\tau\right\|_{L^q[a,b]}^{1-\beta}.\end{aligned} \tag{4.65}$$

Let us consider first integral as

$$\int_{t_0}^{t} f(\tau)\exp\left(-\frac{\alpha}{1-\alpha}(t-\tau)\right)d\tau=\int_{t_0}^{t} f(\tau)\exp\left(\frac{\alpha}{1-\alpha}(\tau-t)\right)d\tau \tag{4.66}$$

$$\begin{aligned}\int_{t_0}^{t} f(\tau)\exp\left(\frac{\alpha}{1-\alpha}(\tau-t)\right)d\tau &= \frac{1-\alpha}{\alpha}f(\tau)\exp\left(\frac{\alpha}{1-\alpha}(\tau-t)\right)\Big|_{t_o}^{t} \\ &\quad-\frac{1-\alpha}{\alpha}\int_{t_0}^{t}\exp\left(\frac{\alpha}{1-\alpha}(\tau-t)\right)f'(\tau)d\tau \\ &=\frac{1-\alpha}{\alpha}\left[f(t)-f(t_0)\exp\left(\frac{\alpha}{1-\alpha}(t_0-t)\right)\right] \\ &\quad-\frac{1-\alpha}{\alpha}\int_{t_0}^{t}\exp\left(\frac{\alpha}{1-\alpha}(\tau-t)\right)f'(\tau)d\tau.\end{aligned}$$

$$\frac{d^2}{dt^2}\int_{t_0}^{t} f(\tau)\exp\left(-\frac{\alpha}{1-\alpha}(t-\tau)\right)d\tau = \frac{d}{dt}\left[\begin{array}{c} \frac{1-\alpha}{\alpha}\left\{f(t)-f(t_0)\exp\left(-\frac{\alpha}{1-\alpha}(t-t_0)\right)\right\} \\ -\frac{1-\alpha}{\alpha}\int_{t_0}^{t} f'(\tau)\exp\left(-\frac{\alpha}{1-\alpha}(t-\tau)\right)d\tau \end{array}\right]$$

$$= \frac{1-\alpha}{\alpha}f'(t) - f(t_0)\exp\left(-\frac{\alpha}{1-\alpha}(t-t_0)\right)$$

$$-\frac{d}{dt}\left(\frac{1-\alpha}{\alpha}\int_{t_0}^{t} f'(\tau)\exp\left(-\frac{\alpha}{1-\alpha}(t-\tau)\,d\tau\right)\right)$$

$$= \frac{1-\alpha}{\alpha}f'(t) - f(t_0)\exp\left(-\frac{\alpha}{1-\alpha}(t-t_0)\right)$$

$$-\left(\frac{1-\alpha}{\alpha}\right)^2 f''(t) + \frac{\alpha}{1-\alpha}\exp(t-t_0)\,f(t_0)$$

$$+\left(\frac{1-\alpha}{\alpha}\right)^2\int_{t_0}^{t} f''(\tau)\exp\left(-\frac{\alpha}{1-\alpha}(t-\tau)\,d\tau\right).$$

Applying the norm on both sides yields

$$\left\|\int_{t_0}^{t} f(\tau)\exp\left(\frac{-\alpha}{1-\alpha}(t-\tau)\right)d\tau\right\|_{L^p[a,b]}^{\beta} \tag{4.67}$$

$$\leq \frac{1-\alpha}{\alpha}\left\{\begin{array}{c} \|f\|_{L^p[a,b]} \\ +\|f(t_0)\|_{L^p[a,b]}\left\|\exp\left(\frac{\alpha}{1-\alpha}(t_0-t)\right)\right\|_{L^p[a,b]} \\ +\left\|\int_{t_0}^{t}\exp\left(\frac{\alpha}{1-\alpha}(\tau-t)\right)f'(\tau)d\tau\right\|_{L^p[a,b]} \end{array}\right\}^{\beta}$$

$$\leq \left\{\left(\frac{1-\alpha}{\alpha}\right)\left(\begin{array}{c} \|f\|_{L^p[a,b]} \\ +|f(t_0)|_{L^p[a,b]}(t-t_0)^{\frac{1}{p}} \\ +\|f(t)-f(t_0)\|_{L^p[a,b]} \end{array}\right)\right\}^{\beta}$$

$$\leq \left\{\left(\frac{1-\alpha}{\alpha}\right)\left(\begin{array}{c} \|f\|_{L^p[a,b]} \\ +|f(t_0)|_{L^p[a,b]}(t-t_0)^{\frac{1}{p}} \\ +\|K\|_{L^p[a,b]} \end{array}\right)\right\}^{\beta}.$$

Also we have

$$\left\| \frac{d^2}{dt^2} \int_{t_0}^{t} f(\tau) \exp\left(-\frac{\alpha}{1-\alpha}(t-\tau) \right) d\tau \right\|_{L^q[a,b]}^{1-\beta} \tag{4.68}$$
$$\leq \left(\frac{1-\alpha}{\alpha} \right)^{1-\beta} \|f''\|_{L^q_{[a,b]}}^{1-\beta} + \left(\frac{\alpha}{1-\alpha} \right)^{1-\beta} |f(t_0)|^{1-\beta}$$
$$+ \left(\frac{1-\alpha}{\alpha} \right)^{2-2\beta} \|f''\|_{L^q_{[a,b]}}^{1-\beta}$$
$$+ \left(\frac{\alpha}{1-\alpha} \right)^{1-\beta} |f'(t_0)|^{1-\beta} \left\| \exp\left(-\frac{\alpha}{1-\alpha}(t-t_0) \right) \right\|_{L^q_{[a,b]}}^{1-\beta}$$
$$+ \left(\frac{1-\alpha}{\alpha} \right)^{2-2\beta} \left\| \int_{t_0}^{t} f''(\tau) \exp\left(-\frac{\alpha}{1-\alpha}(t-\tau) \right) d\tau \right\|_{L^q_{[a,b]}}^{1-\beta}$$

replacing all in the main equation yields

$$\left| {}_{t_0}^{RE}D_g^{\alpha} f(t) \right| \leq \frac{KM_1}{(1-\alpha)} \left\{ \begin{array}{c} \left(\frac{1-\alpha}{\alpha}\right)^{1-\beta} \|f''\|_{L^q_{[a,b]}}^{1-\beta} + \left(\frac{\alpha}{1-\alpha}\right)^{1-\beta} |f(t_0)|^{1-\beta} \\ + \left(\frac{1-\alpha}{\alpha}\right)^{2-2\beta} \|f''\|_{L^q_{[a,b]}}^{1-\beta} \\ + \left(\frac{\alpha}{1-\alpha}\right)^{1-\beta} |f'(t_0)|^{1-\beta} \left\| \exp\left(-\frac{\alpha}{1-\alpha}(t-t_0)\right) \right\|_{L^q_{[a,b]}}^{1-\beta} \\ + \left(\frac{1-\alpha}{\alpha}\right)^{2-2\beta} \left\| \int_{t_0}^{t} f''(\tau) \exp\left(-\frac{\alpha}{1-\alpha}(t-\tau) \right) d\tau \right\|_{L^q_{[a,b]}}^{1-\beta} \end{array} \right\}$$
$$\times \left\{ \left(\frac{1-\alpha}{\alpha} \right) \left(\begin{array}{c} \|f\|_{L^p_{[a,b]}} + |f(t_0)| \left\| \exp\left(-\frac{\alpha}{1-\alpha}(t-\tau)\right) \right\|_{L^p_{[a,b]}} \\ + \left\| \int_{t_0}^{t} f'(\tau) \exp\left(-\frac{\alpha}{1-\alpha}(t-\tau) \right) d\tau \right\|_{L^p_{[a,b]}}^{\beta} \end{array} \right) \right\} \tag{4.69}$$

where K, β, q, p are constants from the Landau–Kolmogorov inequality which completes the proof.

Lemma 4.6 *Let f be continuously differentiable on $[a, b]$. If $f'(t)$ and $g'(t)$ are bounded then*

$$\left| {}_{t_0}^{RE}D_g^{\alpha} f(t) \right|^2 \leq \frac{3M_{g'}^2}{\alpha^2} \left\{ \begin{array}{c} \|f\|_{\infty}^2 + f^2(t_0) \exp\left(\frac{-2\alpha}{1-\alpha}(t-t_0)\right) \\ + \|f\|_{L^2_{[a,b]}}^2 \left\| \exp\left(-\frac{\alpha}{1-\alpha}(t-t_0)\right) \right\|_{L^2_{[a,b]}}^2 \end{array} \right\}. \tag{4.70}$$

Proof If $f'(t)$ is bounded in $[a,b]$ $\forall t \in [a,b]$, $\left|f'(t)\right| \leq M_{f'}$ for $M_{f'} > 0$ also $\left|\frac{1}{g'(t)}\right| \leq M_{g'} < \infty$.

$$\left|{}_{t_0}^{RE}D_g^\alpha f(t)\right|^2 = \left|\frac{1}{1-\alpha}\frac{1}{g'(t)}\frac{d}{dt}\int_{t_0}^{t} f(\tau)\exp\left(-\frac{\alpha}{1-\alpha}(t-\tau)\right)d\tau\right|^2, \tag{4.71}$$

$$\leq \frac{M_{g'}^2}{(1-\alpha)^2}\left|\frac{d}{dt}\int_{t_0}^{t} f(\tau)\exp\left(-\frac{\alpha}{1-\alpha}(t-\tau)\right)d\tau\right|^2,$$

$$\leq \frac{M_{g'}^2}{(1-\alpha)^2}\left|\begin{array}{c}\frac{1-\alpha}{\alpha}f(t) - \frac{1-\alpha}{\alpha}f(t_0)\exp\left(-\frac{\alpha}{1-\alpha}(t-t_0)\right)\\ -\frac{1-\alpha}{\alpha}\int_{t_0}^{t} f'(\tau)\exp\left(-\frac{\alpha}{1-\alpha}(t-\tau)\right)d\tau\end{array}\right|^2,$$

$$\leq \frac{3M_{g'}^2}{\alpha^2}\left\{\begin{array}{c}|f(t)|^2 + f^2(t_0)\exp\left(\frac{-2\alpha}{1-\alpha}(t-t_0)\right)\\ +\left|\int_{t_0}^{t} f'(\tau)\exp\left(-\frac{\alpha}{1-\alpha}(t-\tau)\right)d\tau\right|^2\end{array}\right\}$$

$$\leq \frac{3M_{g'}^2}{\alpha^2}\left\{\begin{array}{c}\|f\|_\infty^2 + f^2(t_0)\exp\left(\frac{-2\alpha}{1-\alpha}(t-t_0)\right)\\ +\|f\|_{L^2_{[a,b]}}^2\left\|\exp\left(-\frac{\alpha}{1-\alpha}(t-t_0)\right)\right\|_{L^2_{[a,b]}}^2\end{array}\right\}.$$

Noting that $\forall\alpha \in (0,1)$, $\forall t \in [a,b]$,

$$\exp\left(-\frac{\alpha}{1-\alpha}(t-\tau)\right) \leq 1, \tag{4.72}$$

therefore

$$\left|{}_{t_0}^{RE}D_g^\alpha f(t)\right|^2 \leq \frac{3M_{g'}^2}{\alpha^2}\left\{\|f\|_\infty^2 + f^2(t_0) + \|f\|_{L^2_{[a,b]}}^2\right\}, \tag{4.73}$$

which completes the proof.

Corollary 4.9 *Let f be twice continuous differentiable on $[a,b]$ and $f''(t)$, $f'(t)$ bounded if $\left|g'(t)\right|^{-1} < M_{g'}$ then*

$$\left|{}_{t_0}^{RE}D_g^\alpha f(t)\right| \leq \frac{M_{g'}\sqrt{2}}{(1-\alpha)}\left\{\frac{1-\alpha}{\alpha}\|f\|_\infty + \frac{1-\alpha}{\alpha}|f(t_0)|\left\|\exp\left(-\frac{\alpha}{1-\alpha}(t-t_0)\right)\right\|_\infty \right. \tag{4.74}$$

$$\left. +\frac{1-\alpha}{\alpha}\|f - f(t_0)\|_\infty\right\}^{\frac{1}{2}} \times \left\{\begin{array}{c}\frac{1-\alpha}{\alpha}\|f'\|_\infty + |f(t_0)| + \left(\frac{1-\alpha}{\alpha}\right)^2\|f'\|_\infty\\ +\left(\frac{1-\alpha}{\alpha}\right)^2|f'(t_0)| + \left(\frac{1-\alpha}{\alpha}\right)^2\|f' - f'(t_0)\|_\infty\end{array}\right\}^{\frac{1}{2}}.$$

Proof By definition of ${}_{t_0}^{RE}D_g^\alpha f(t)$, we have

$$\left|{}_{t_0}^{RE}D_g^{\alpha}f(t)\right| = \left|\frac{1}{1-\alpha}\frac{1}{g'(t)}\frac{d}{dt}\int_{t_0}^{t} f(\tau)\exp\left(-\frac{\alpha}{1-\alpha}(t-\tau)\right)d\tau\right|. \tag{4.75}$$

If we put

$$F(t) = \int_{t_0}^{t} f(\tau)\exp\left(-\frac{\alpha}{1-\alpha}(t-\tau)\right)d\tau, \tag{4.76}$$

then we get

$$\begin{aligned}\left|{}_{t_0}^{RE}D_g^{\alpha}f(t)\right| &= \left|\frac{1}{1-\alpha}\frac{1}{g'(t)}\frac{d}{dt}\int_{t_0}^{t} f(\tau)\exp\left(-\frac{\alpha}{1-\alpha}(t-\tau)\right)d\tau\right|, \\ &\leq \frac{M_{g'}}{(1-\alpha)}\left|\frac{dF}{dt}\right|^2.\end{aligned} \tag{4.77}$$

By the Landau–Kolmogorov inequality we have that

$$\begin{aligned}\left|{}_{t_0}^{RE}D_g^{\alpha}f(t)\right| &\leq \frac{M_{g'}\sqrt{2}}{(1-\alpha)}\|F\|^{\frac{1}{2}}\|F''\|^{\frac{1}{2}}, \\ &\leq \frac{M_{g'}\sqrt{2}}{(1-\alpha)}\left\|\int_{t_0}^{t} f(\tau)\exp\left(-\frac{\alpha}{1-\alpha}(t-\tau)\right)d\tau\right\|_{\infty}^{\frac{1}{2}} \\ &\quad\times\left\|\frac{d^2}{dt^2}\int_{t_0}^{t} f(\tau)\exp\left(-\frac{\alpha}{1-\alpha}(t-\tau)\right)d\tau\right\|_{\infty}^{\frac{1}{2}}, \\ &\leq \frac{M_{g'}\sqrt{2}}{(1-\alpha)}\left\|\begin{array}{c}\frac{1-\alpha}{\alpha}f(t)-\frac{1-\alpha}{\alpha}f(t_0)\exp\left(-\frac{\alpha}{1-\alpha}(t-t_0)\right) \\ -\frac{1-\alpha}{\alpha}\int_{t_0}^{t} f'(\tau)\exp\left(-\frac{\alpha}{1-\alpha}(t-\tau)\right)d\tau\end{array}\right\|_{\infty}^{\frac{1}{2}} \\ &\quad\times\left\|\begin{array}{c}\frac{1-\alpha}{\alpha}f'(t)+f(t_0)\exp\left(-\frac{\alpha}{1-\alpha}(t-t_0)\right)+\left(\frac{1-\alpha}{\alpha}\right)^2 f'(t) \\ +\left(\frac{1-\alpha}{\alpha}\right)^2 f'(t_0)\exp\left(-\frac{\alpha}{1-\alpha}(t-t_0)\right) \\ +\left(\frac{1-\alpha}{\alpha}\right)^2\int_{t_0}^{t} f''(\tau)\exp\left(-\frac{\alpha}{1-\alpha}(t-\tau)\right)d\tau\end{array}\right\|_{\infty}^{\frac{1}{2}}, \\ &\leq \frac{M_{g'}\sqrt{2}}{(1-\alpha)}\left\{\frac{1-\alpha}{\alpha}\|f\|_{\infty}+\frac{1-\alpha}{\alpha}|f(t_0)|\left\|\exp\left(-\frac{\alpha}{1-\alpha}(t-t_0)\right)\right\|_{\infty}\right. \\ &\quad\left.+\frac{1-\alpha}{\alpha}\|f-f(t_0)\|_{\infty}\right\}^{\frac{1}{2}}\times\left\{\begin{array}{c}\frac{1-\alpha}{\alpha}\|f'\|_{\infty}+|f(t_0)|+\left(\frac{1-\alpha}{\alpha}\right)^2\|f'\|_{\infty} \\ +\left(\frac{1-\alpha}{\alpha}\right)^2|f'(t_0)|+\left(\frac{1-\alpha}{\alpha}\right)^2\|f'-f'(t_0)\|_{\infty}\end{array}\right\}^{\frac{1}{2}},\end{aligned} \tag{4.78}$$

which completes the proof.

4.3 Inequalities Related to Mittag–Leffler Law Decay-Based Global Derivative

We shall recall that the Mittag–Leffler-based global derivative is given by

$$ {}_{t_0}^{RM}D_g^{\alpha} f(t) = \frac{1}{1-\alpha} D_g \int_{t_0}^{t} f(\tau) E_{\alpha}\left(-\frac{\alpha}{1-\alpha}(t-\tau)^{\alpha}\right) d\tau. \tag{4.79} $$

Lemma 4.7 *Let m and M be two real values such that $\forall t \in [a, b]$,*

$$ m < f(t) < M. \tag{4.80} $$

Then

$$ \begin{aligned} & \frac{m}{g'(t)(1-\alpha)}\left[E_{\alpha,2}\left(-\frac{\alpha}{1-\alpha}(t-t_0)^{\alpha}\right) + (t-t_0)\, E_{\alpha}\left(-\frac{\alpha}{1-\alpha}(t-t_0)^{\alpha}\right)\right] \\ & < {}_{t_0}^{RM}D_g^{\alpha} f(t) \\ & < \frac{M}{g'(t)(1-\alpha)}\left[E_{\alpha,2}\left(-\frac{\alpha}{1-\alpha}(t-t_0)^{\alpha}\right) - (t-t_0)\, E_{\alpha}\left(-\frac{\alpha}{1-\alpha}(t-t_0)^{\alpha}\right)\right]. \end{aligned} \tag{4.81} $$

Proof From the hypothesis, we have

$$ m < f(t) < M, \tag{4.82} $$

applying the derivative we get

$$ {}_{t_0}^{RM}D_g^{\alpha} m < {}_{t_0}^{RM}D_g^{\alpha} f(t) < {}_{t_0}^{RM}D_g^{\alpha} M, \tag{4.83} $$

$$ \begin{aligned} & \frac{1}{1-\alpha} D_g \int_{t_0}^{t} m E_{\alpha}\left(-\frac{\alpha}{1-\alpha}(t-\tau)^{\alpha}\right) d\tau < {}_{t_0}^{RM}D_g^{\alpha} f(t) \\ & \qquad < \frac{1}{1-\alpha} D_g \int_{t_0}^{t} M E_{\alpha}\left(-\frac{\alpha}{1-\alpha}(t-\tau)^{\alpha}\right) d\tau, \end{aligned} \tag{4.84} $$

$$ \begin{aligned} & \frac{m}{(1-\alpha)}\frac{1}{g'(t)}\frac{d}{dt}\int_{t_0}^{t} E_{\alpha}\left(-\frac{\alpha}{1-\alpha}(t-\tau)^{\alpha}\right) d\tau < {}_{t_0}^{RM}D_g^{\alpha} f(t) \\ & < \frac{M}{(1-\alpha)}\frac{1}{g'(t)}\frac{d}{dt}\int_{t_0}^{t} E_{\alpha}\left(-\frac{\alpha}{1-\alpha}(t-\tau)^{\alpha}\right) d\tau. \end{aligned} \tag{4.85} $$

Noting that

$$\begin{aligned}\int_{t_0}^{t} E_\alpha\left(-\frac{\alpha}{1-\alpha}(t-\tau)^\alpha\right)d\tau &= \int_{t_0}^{t}\sum_{j=0}^{\infty}\left(-\frac{\alpha}{1-\alpha}\right)^j \frac{(t-\tau)^{\alpha j}}{\Gamma(\alpha j+1)}d\tau, \\ &= \sum_{j=0}^{\infty}\frac{\left(-\frac{\alpha}{1-\alpha}\right)^j}{\Gamma(\alpha j+1)}\int_{t_0}^{t}(t-\tau)^{\alpha j}d\tau, \\ &= \sum_{j=0}^{\infty}\frac{\left(-\frac{\alpha}{1-\alpha}\right)^j}{\Gamma(\alpha j+1)}\frac{(t-t_0)^{\alpha j+1}}{\alpha j+1}, \\ &= \sum_{j=0}^{\infty}\frac{\left(-\frac{\alpha}{1-\alpha}(t-t_0)^\alpha\right)^j}{\Gamma(\alpha j+2)}(t-t_0), \\ &= (t-t_0)E_{\alpha,2}\left(-\frac{\alpha}{1-\alpha}(t-t_0)^\alpha\right).\end{aligned} \tag{4.86}$$

Thus

$$\begin{aligned}&\frac{d}{dt}\left[(t-t_0)E_{\alpha,2}\left(-\frac{\alpha}{1-\alpha}(t-t_0)^\alpha\right)\right] \\ &= E_{\alpha,2}\left(-\frac{\alpha}{1-\alpha}(t-t_0)^\alpha\right) + (t-t_0)E_\alpha\left(-\frac{\alpha}{1-\alpha}(t-t_0)^\alpha\right).\end{aligned} \tag{4.87}$$

Therefore

$$\begin{aligned}&\frac{m}{g'(t)(1-\alpha)}\left[E_{\alpha,2}\left(-\frac{\alpha}{1-\alpha}(t-t_0)^\alpha\right) + (t-t_0)E_\alpha\left(-\frac{\alpha}{1-\alpha}(t-t_0)^\alpha\right)\right] \\ &< {}_{t_0}^{RM}D_g^\alpha f(t) \\ &< \frac{M}{g'(t)(1-\alpha)}\left[E_{\alpha,2}\left(-\frac{\alpha}{1-\alpha}(t-t_0)^\alpha\right) + (t-t_0)E_\alpha\left(-\frac{\alpha}{1-\alpha}(t-t_0)^\alpha\right)\right]\end{aligned} \tag{4.88}$$

which completes the proof.

It is also important to note that

$$\begin{aligned}\frac{d}{dt}\int_{t_0}^{t} E_\alpha\left(-\frac{\alpha}{1-\alpha}(t-\tau)^\alpha\right)d\tau &= \frac{d}{dt}\int_{t_0}^{t}\sum_{j=0}^{\infty}\left(-\frac{\alpha}{1-\alpha}\right)^j \frac{(t-\tau)^{\alpha j}}{\Gamma(\alpha j+1)}d\tau, \\ &= \frac{d}{dt}\sum_{j=0}^{\infty}\frac{\left(-\frac{\alpha}{1-\alpha}\right)^j}{\Gamma(\alpha j+1)}\int_{t_0}^{t}(t-\tau)^{\alpha j}d\tau,\end{aligned} \tag{4.89}$$

$$= \frac{d}{dt}\sum_{j=0}^{\infty}\frac{\left(-\frac{\alpha}{1-\alpha}\right)^{j}}{\Gamma(\alpha j+1)}\frac{(t-t_0)^{\alpha j+1}}{\alpha j+1}$$

$$= \sum_{j=0}^{\infty}\frac{\left(-\frac{\alpha}{1-\alpha}\right)^{j}}{\Gamma(\alpha j+1)}(t-t_0)^{\alpha j}$$

$$= E_{\alpha}\left(-\frac{\alpha}{1-\alpha}(t-t_0)^{\alpha}\right).$$

Therefore, we will also have

$$\frac{m}{g'(t)(1-\alpha)}E_{\alpha}\left(-\frac{\alpha}{1-\alpha}(t-t_0)^{\alpha}\right) \leq {}_{t_0}^{RM}D_g^{\alpha}f(t) \leq \frac{M}{g'(t)(1-\alpha)}E_{\alpha}\left(-\frac{\alpha}{1-\alpha}(t-t_0)^{\alpha}\right). \tag{4.90}$$

Lemma 4.8 *Let f be continuous differentiable and f' bounded on $[a,b]$. If $g'(t)$ is bounded, then*

$$\left|{}_{t_0}^{RM}D_g^{\alpha}f(t)\right| \leq \frac{1}{1-\alpha}\frac{1}{g'(t)}\left\{\begin{array}{c} |f(t_0)|\left|E_{\alpha,2}\left(-\frac{\alpha}{1-\alpha}(t-t_0)^{\alpha}\right)\right| + (t-t_0)\left|E_{\alpha}\left(-\frac{\alpha}{1-\alpha}(t-t_0)^{\alpha}\right)\right| \\ +\left|f'(t_0)\right|(t-t_0)\left|E_{\alpha,2}\left(-\frac{\alpha}{1-\alpha}(t-t_0)^{\alpha}\right)\right| \\ +\left\|f''\right\|_{\infty}(t-t_0)^2\left|E_{\alpha,3}\left(-\frac{\alpha}{1-\alpha}(t-t_0)^{\alpha}\right)\right| \end{array}\right\}. \tag{4.91}$$

Proof To show the proof, we shall use the integration by part as

$$\left|{}_{t_0}^{RM}D_g^{\alpha}f(t)\right| = \left|\frac{1}{1-\alpha}\frac{1}{g'(t)}\frac{d}{dt}\int_{t_0}^{t}f(\tau)E_{\alpha}\left(-\frac{\alpha}{1-\alpha}(t-\tau)^{\alpha}\right)d\tau\right|, \tag{4.92}$$

$$\leq \frac{1}{1-\alpha}\left|\frac{1}{g'(t)}\right|\left|\frac{d}{dt}\int_{t_0}^{t}f(\tau)E_{\alpha}\left(-\frac{\alpha}{1-\alpha}(t-\tau)^{\alpha}\right)d\tau\right|.$$

We are interested in

$$\frac{d}{dt}\int_{t_0}^{t}f(\tau)E_{\alpha}\left(-\frac{\alpha}{1-\alpha}(t-\tau)^{\alpha}\right)d\tau \tag{4.93}$$

$$= \frac{d}{dt}\left[\begin{array}{c} -f(\tau)(t-\tau)\,E_{\alpha,2}\left(-\frac{\alpha}{1-\alpha}(t-\tau)^{\alpha}\right)\Big|_{t_0}^{t} \\ +\int_{t_0}^{t}f'(\tau)(t-\tau)\,E_{\alpha,2}\left(-\frac{\alpha}{1-\alpha}(t-\tau)^{\alpha}\right)d\tau \end{array}\right]$$

$$= \frac{d}{dt}\left[\begin{array}{c} f(t_0)(t-t_0)E_{\alpha,2}\left(-\frac{\alpha}{1-\alpha}(t-t_0)^{\alpha}\right) \\ +\int\limits_{t_0}^{t} f'(\tau)(t-\tau)E_{\alpha,2}\left(-\frac{\alpha}{1-\alpha}(t-\tau)^{\alpha}\right)d\tau \end{array}\right]$$

$$= \left[\begin{array}{c} E_{\alpha,2}\left(-\frac{\alpha}{1-\alpha}(t-t_0)^{\alpha}\right)f(t_0)+(t-t_0)E_{\alpha,2}\left(-\frac{\alpha}{1-\alpha}(t-t_0)^{\alpha}\right) \\ +\frac{d}{dt}\int\limits_{t_0}^{t} f'(\tau)(t-\tau)E_{\alpha,2}\left(-\frac{\alpha}{1-\alpha}(t-\tau)^{\alpha}\right)d\tau \end{array}\right]$$

$$= E_{\alpha,2}\left(-\frac{\alpha}{1-\alpha}(t-t_0)^{\alpha}\right)f(t_0)+(t-t_0)E_{\alpha}\left(-\frac{\alpha}{1-\alpha}(t-t_0)^{\alpha}\right)$$

$$+ f'(t_0)(t-t_0)E_{\alpha,2}\left(-\frac{\alpha}{1-\alpha}(t-t_0)^{\alpha}\right)$$

$$+\int\limits_{t_0}^{t} f''(\tau)(t-\tau)E_{\alpha,2}\left(-\frac{\alpha}{1-\alpha}(t-\tau)^{\alpha}\right)d\tau$$

replacing yields

$$\left|{}_{t_0}^{RM}D_g^{\alpha}f(t)\right| \le \frac{1}{1-\alpha}\left|\frac{1}{g'(t)}\right|\left|\begin{array}{c} E_{\alpha,2}\left(-\frac{\alpha}{1-\alpha}(t-t_0)^{\alpha}\right)f(t_0)+(t-t_0)E_{\alpha}\left(-\frac{\alpha}{1-\alpha}(t-t_0)^{\alpha}\right) \\ +f'(t_0)(t-t_0)E_{\alpha,2}\left(-\frac{\alpha}{1-\alpha}(t-t_0)^{\alpha}\right) \\ +\int\limits_{t_0}^{t} f''(\tau)(t-\tau)E_{\alpha,2}\left(-\frac{\alpha}{1-\alpha}(t-\tau)^{\alpha}\right)d\tau \end{array}\right| \tag{4.94}$$

$$\le \frac{1}{1-\alpha}\left|\frac{1}{g'(t)}\right|\left\{\begin{array}{c} |f(t_0)|\left|E_{\alpha,2}\left(-\frac{\alpha}{1-\alpha}(t-t_0)^{\alpha}\right)\right|+(t-t_0)\left|E_{\alpha}\left(-\frac{\alpha}{1-\alpha}(t-t_0)^{\alpha}\right)\right| \\ +\left|f'(t_0)\right|(t-t_0)\left|E_{\alpha,2}\left(-\frac{\alpha}{1-\alpha}(t-t_0)^{\alpha}\right)\right| \\ +\left\|f''\right\|_{\infty}(t-t_0)^2\left|E_{\alpha,3}\left(-\frac{\alpha}{1-\alpha}(t-t_0)^{\alpha}\right)\right| \end{array}\right\}.$$

Therefore

$$\left|{}_{t_0}^{RM}D_g^{\alpha}f(t)\right| \le \frac{1}{1-\alpha}\frac{1}{g'(t)}\left\{\begin{array}{c} |f(t_0)|\left|E_{\alpha,2}\left(-\frac{\alpha}{1-\alpha}(t-t_0)^{\alpha}\right)\right|+(t-t_0)\left|E_{\alpha}\left(-\frac{\alpha}{1-\alpha}(t-t_0)^{\alpha}\right)\right| \\ +\left|f'(t_0)\right|(t-t_0)\left|E_{\alpha,2}\left(-\frac{\alpha}{1-\alpha}(t-t_0)^{\alpha}\right)\right| \\ +\left\|f''\right\|_{\infty}(t-t_0)^2\left|E_{\alpha,3}\left(-\frac{\alpha}{1-\alpha}(t-t_0)^{\alpha}\right)\right| \end{array}\right\} \tag{4.95}$$

which completes the proof.

Corollary 4.10 *Let f be a nonnegative function defined on a closed ball of R with ratio λ with center x_0. It is assumed that f is harmonic on the interior on the ball and continous on the closed ball then if $\lambda_1 < \lambda$, we have*

$$\frac{f(x_0)\left(1-\frac{\lambda_1}{\lambda}\right)m}{(1-\alpha)g'(x)}\left\{E_{\alpha,2}\left(\begin{array}{c}-\frac{\alpha}{1-\alpha}(t-t_0)^{\alpha}\\+(t-t_0)E_{\alpha}\left(-\frac{\alpha}{1-\alpha}(t-t_0)^{\alpha}\right)\end{array}\right)\right\} \tag{4.96}$$
$$\leq {}_{t_0}^{RM}D_g^{\alpha}f(t)$$
$$\leq \frac{f(x_0)\left(1+\frac{\lambda_1}{\lambda}\right)M}{(1-\alpha)g'(x)}\left\{E_{\alpha,2}\left(\begin{array}{c}-\frac{\alpha}{1-\alpha}(t-t_0)^{\alpha}\\+(t-t_0)E_{\alpha}\left(-\frac{\alpha}{1-\alpha}(t-t_0)^{\alpha}\right)\end{array}\right)\right\}.$$

Proof Using Harnack's inequality we have that
$\forall x,\ |x-x_0|<\lambda_1<\lambda,$

$$f(x_0)\left(1-\frac{\lambda_1}{\lambda}\right)\leq f(x)\leq f(x_0)\left(1+\frac{\lambda_1}{\lambda}\right). \tag{4.97}$$

Applying the derivative we have

$$\frac{f(x_0)\left(1-\frac{\lambda_1}{\lambda}\right)m}{(1-\alpha)g'(x)}\left\{E_{\alpha,2}\left(-\frac{\alpha}{1-\alpha}(t-t_0)^{\alpha}+(t-t_0)E_{\alpha}\left(-\frac{\alpha}{1-\alpha}(t-t_0)^{\alpha}\right)\right)\right\} \tag{4.98}$$
$$\leq {}_{t_0}^{RM}D_g^{\alpha}f(t)$$
$$\leq \frac{f(x_0)\left(1+\frac{\lambda_1}{\lambda}\right)M}{(1-\alpha)g'(x)}\left\{E_{\alpha,2}\left(-\frac{\alpha}{1-\alpha}(t-t_0)^{\alpha}+(t-t_0)E_{\alpha}\left(-\frac{\alpha}{1-\alpha}(t-t_0)^{\alpha}\right)\right)\right\},$$

which completes the proof.

Corollary 4.11 *Let (X,Σ,μ) be a measure space. Let $f'', h: X\to R$ be measurable and in $L^p[a,b]$, then if $p=2$, we have*

$$\left|{}_{t_0}^{RM}D_g^{\alpha}f(t)\right|\leq\frac{M_1}{(1-\alpha)}\left\{\begin{array}{c}|f(t_0)|\left|E_{\alpha,2}\left(-\frac{\alpha}{1-\alpha}(t-t_0)^{\alpha}\right)\right|+(t-t_0)\left|E_{\alpha}\left(-\frac{\alpha}{1-\alpha}(t-t_0)^{\alpha}\right)\right|\\+\left|f'(t_0)\right|\left|E_{\alpha,2}\left(-\frac{\alpha}{1-\alpha}(t-t_0)^{\alpha}\right)\right|(t-t_0)\\+\frac{(t-t_0)^{\frac{3}{2}}}{\sqrt{6}}\left\{\|f''\|_{L^2[a,b]}+\|h\|_{L^2[a,b]}\right\}\end{array}\right\}. \tag{4.99}$$

Proof By definition, we have

$$\left|{}_{t_0}^{RM}D_g^{\alpha}f(t)\right|\leq\frac{1}{1-\alpha}\left|\frac{1}{g'(t)}\right|\left|\frac{d}{dt}\int_{t_0}^{t}f(\tau)E_{\alpha}\left(-\frac{\alpha}{1-\alpha}(t-\tau)^{\alpha}\right)d\tau\right| \tag{4.100}$$
$$\leq\frac{1}{(1-\alpha)}\sup_{t\in[a,b]}\left|\frac{1}{g'(t)}\right|\left|\frac{d}{dt}\int_{t_0}^{t}f(\tau)E_{\alpha}\left(-\frac{\alpha}{1-\alpha}(t-\tau)^{\alpha}\right)d\tau\right|$$
$$\leq\frac{M_1}{(1-\alpha)}\left|\frac{d}{dt}\int_{t_0}^{t}f(\tau)E_{\alpha}\left(-\frac{\alpha}{1-\alpha}(t-\tau)^{\alpha}\right)d\tau\right|$$

$$\leq \frac{M_1}{(1-\alpha)} \left| \begin{array}{c} E_{\alpha,2}\left(-\frac{\alpha}{1-\alpha}(t-t_0)^{\alpha}\right) f(t_0) + (t-t_0) E_{\alpha}\left(-\frac{\alpha}{1-\alpha}(t-t_0)^{\alpha}\right) \\ + f'(t_0)(t-t_0) E_{\alpha,2}\left(-\frac{\alpha}{1-\alpha}(t-t_0)^{\alpha}\right) \\ + \int_{t_0}^{t} f''(\tau) E_{\alpha}\left(-\frac{\alpha}{1-\alpha}(t-\tau)^{\alpha}\right) d\tau \end{array} \right|$$

$$\leq \frac{M_1}{(1-\alpha)} \left\{ \begin{array}{c} |f(t_0)| \left|E_{\alpha,2}\left(-\frac{\alpha}{1-\alpha}(t-t_0)^{\alpha}\right)\right| + (t-t_0) \left|E_{\alpha}\left(-\frac{\alpha}{1-\alpha}(t-t_0)^{\alpha}\right)\right| \\ + |f'(t_0)| \left|E_{\alpha,2}\left(-\frac{\alpha}{1-\alpha}(t-t_0)^{\alpha}\right)\right| (t-t_0) \\ + \left(\int_{t_0}^{t} |f''|^2 d\tau\right)^{\frac{1}{2}} \left(\int_{t_0}^{t} (t-\tau)^2 d\tau\right)^{\frac{1}{2}} \end{array} \right\}$$

$$\leq \frac{M_1}{(1-\alpha)} \left\{ \begin{array}{c} |f(t_0)| \left|E_{\alpha,2}\left(-\frac{\alpha}{1-\alpha}(t-t_0)^{\alpha}\right)\right| + (t-t_0) \left|E_{\alpha}\left(-\frac{\alpha}{1-\alpha}(t-t_0)^{\alpha}\right)\right| \\ + |f'(t_0)| \left|E_{\alpha,2}\left(-\frac{\alpha}{1-\alpha}(t-t_0)^{\alpha}\right)\right| (t-t_0) \\ + \frac{(t-t_0)^{\frac{3}{2}}}{\sqrt{3}} \left(\int_{t_0}^{t} \left|\frac{f''-h}{2} + \frac{h+f''}{2}\right|^2 d\tau\right)^{\frac{1}{2}} \end{array} \right\}$$

$$\leq \frac{M_1}{(1-\alpha)} \left\{ \begin{array}{c} |f(t_0)| \left|E_{\alpha,2}\left(-\frac{\alpha}{1-\alpha}(t-t_0)^{\alpha}\right)\right| + (t-t_0) \left|E_{\alpha}\left(-\frac{\alpha}{1-\alpha}(t-t_0)^{\alpha}\right)\right| \\ + |f'(t_0)| \left|E_{\alpha,2}\left(-\frac{\alpha}{1-\alpha}(t-t_0)^{\alpha}\right)\right| (t-t_0) \\ + \frac{(t-t_0)^{\frac{3}{2}}}{\sqrt{3}} \left\{ \frac{\|f''-h\|^2_{L^2[a,b]}}{2} + \frac{\|h+f''\|^2_{L^2[a,b]}}{2} \right\}^{\frac{1}{2}} \end{array} \right\}$$

Now due to Minkowski and Clarkson's inequality we obtain

$$\left|{}_{t_0}^{RM} D_g^{\alpha} f(t)\right| \leq \frac{M_1}{(1-\alpha)} \left\{ \begin{array}{c} |f(t_0)| \left|E_{\alpha,2}\left(-\frac{\alpha}{1-\alpha}(t-t_0)^{\alpha}\right)\right| + (t-t_0) \left|E_{\alpha}\left(-\frac{\alpha}{1-\alpha}(t-t_0)^{\alpha}\right)\right| \\ + |f'(t_0)| \left|E_{\alpha,2}\left(-\frac{\alpha}{1-\alpha}(t-t_0)^{\alpha}\right)\right| (t-t_0) \\ + \frac{(t-t_0)^{\frac{3}{2}}}{\sqrt{6}} \left\{ \|f''\|_{L^2[a,b]} + \|h\|_{L^2[a,b]} \right\} \end{array} \right\} \tag{4.101}$$

which completes the proof.

Corollary 4.12 *Let f be twice differentiable such that f, f'' are bounded on $[a, b]$ then the following inequality is obtained:*

$$\left|{}_{t_0}^{RM} D_g^{\alpha} f(t)\right| \leq \frac{M_1\sqrt{2}}{(1-\alpha)} \left\{ \|f\|_{\infty} (t-t_0) E_{\alpha,2}\left(-\frac{\alpha}{1-\alpha}(t-t_0)^{\alpha}\right) \right\}^{\frac{1}{2}} \tag{4.102}$$

$$\times \left\{ \begin{array}{c} \|f'\|_{\infty} (t-t_0) E_{\alpha,2}\left(-\frac{\alpha}{1-\alpha}(t-t_0)^{\alpha}\right) \\ + |f'(t_0)| \left|E_{\alpha,2}\left(-\frac{\alpha}{1-\alpha}(t-t_0)^{\alpha}\right)\right| (t-t_0) \\ + \frac{(t-t_0)^3}{3} \|f''\|_{\infty} \left|E_{\alpha,3}\left(-\frac{\alpha}{1-\alpha}(t-t_0)^{\alpha}\right)\right| \end{array} \right\}^{\frac{1}{2}}.$$

Proof By definition, we have that

$$
\begin{aligned}
{}_{t_0}^{RM}D_g^{\alpha} f(t) &= \left| \frac{1}{(1-\alpha)} D_g \int_{t_0}^{t} f(\tau) E_{\alpha}\left(-\frac{\alpha}{1-\alpha}(t-\tau)^{\alpha}\right) d\tau \right|, \\
&= \left| \frac{1}{(1-\alpha)} \frac{1}{g'(t)} \frac{d}{dt} \int_{t_0}^{t} f(\tau) E_{\alpha}\left(-\frac{\alpha}{1-\alpha}(t-\tau)^{\alpha}\right) d\tau \right|.
\end{aligned}
\tag{4.103}
$$

Let us take

$$
F(t) = \int_{t_0}^{t} f(\tau) E_{\alpha}\left(-\frac{\alpha}{1-\alpha}(t-\tau)^{\alpha}\right) d\tau, \tag{4.104}
$$

then

$$
\begin{aligned}
\left|{}_{t_0}^{RM}D_g^{\alpha} f(t)\right| &= \left| \frac{1}{(1-\alpha)} \frac{1}{g'(t)} \frac{d}{dt} \int_{t_0}^{t} f(\tau) E_{\alpha}\left(-\frac{\alpha}{1-\alpha}(t-\tau)^{\alpha}\right) d\tau \right| \\
&= \left| \frac{1}{(1-\alpha) g'(t)} \frac{d}{dt} F(t) \right|, \\
&\leq \frac{M_1}{(1-\alpha)} \left| F' \right|, \\
&\leq \frac{M_1}{(1-\alpha)} \left\| F' \right\|_{\infty}.
\end{aligned}
\tag{4.105}
$$

By the Landau–Kolmogorov inequality we obtain

$$
\left|{}_{t_0}^{RM}D_g^{\alpha} f(t)\right| \leq \frac{M_1\sqrt{2}}{(1-\alpha)} \left\| F \right\|_{\infty}^{\frac{1}{2}} \left\| F'' \right\|_{\infty}^{\frac{1}{2}}. \tag{4.106}
$$

$$
\begin{aligned}
|F(t)| &= \left| \int_{t_0}^{t} f(\tau) E_{\alpha}\left(-\frac{\alpha}{1-\alpha}(t-\tau)^{\alpha}\right) d\tau \right| \\
&\leq \|f\|_{\infty} (t-t_0) E_{\alpha,2}\left(-\frac{\alpha}{1-\alpha}(t-t_0)^{\alpha}\right)
\end{aligned}
\tag{4.107}
$$

$$\left|\frac{d^2}{dt^2}\int_{t_0}^{t} f(\tau)E_\alpha\left(-\frac{\alpha}{1-\alpha}(t-\tau)^\alpha\right)d\tau\right| \leq \left\{\begin{array}{c} \|f\|_\infty (t-t_0)\, E_{\alpha,2}\left(-\frac{\alpha}{1-\alpha}(t-t_0)^\alpha\right) \\ +f(t_0)E_{\alpha,2}\left(-\frac{\alpha}{1-\alpha}(t-t_0)^\alpha\right) \\ +(t-t_0)\, E_\alpha\left(-\frac{\alpha}{1-\alpha}(t-t_0)^\alpha\right) \\ +f'(t_0)E_{\alpha,2}\left(-\frac{\alpha}{1-\alpha}(t-t_0)^\alpha\right)(t-t_0) \\ +\left\|f''\right\|_\infty E_{\alpha,3}\left(-\frac{\alpha}{1-\alpha}(t-t_0)^\alpha\right)(t-t_0)^2 \end{array}\right\} \tag{4.108}$$

Replacing yields

$$\left|{}_{t_0}^{RM}D_g^\alpha f(t)\right| \leq \frac{M_1\sqrt{2}}{(1-\alpha)}\left\{\|f\|_\infty (t-t_0)\, E_{\alpha,2}\left(-\frac{\alpha}{1-\alpha}(t-t_0)^\alpha\right)\right\}^{\frac{1}{2}}$$

$$\times\left\{\begin{array}{c} \left\|f'\right\|_\infty (t-t_0)\, E_{\alpha,2}\left(-\frac{\alpha}{1-\alpha}(t-t_0)^\alpha\right) \\ +\left|f'(t_0)\right|\left|E_{\alpha,2}\left(-\frac{\alpha}{1-\alpha}(t-t_0)^\alpha\right)\right|(t-t_0) \\ +\frac{(t-t_0)^3}{3}\left\|f''\right\|_\infty \left|E_{\alpha,3}\left(-\frac{\alpha}{1-\alpha}(t-t_0)^\alpha\right)\right| \end{array}\right\}^{\frac{1}{2}}. \tag{4.109}$$

4.4 Inequalities with Caputo-Type Fractional Global Derivative

We shall recall that the Caputo-type fractional global derivatives are given as

$${}_{t_0}^{CP}D_g^\alpha f(t) = \frac{1}{\Gamma(1-\alpha)}\int_{t_0}^{t} D_g f(\tau)(t-\tau)^{-\alpha}d\tau, \tag{4.110}$$

$${}_{t_0}^{CE}D_g^\alpha f(t) = \frac{1}{1-\alpha}\int_{t_0}^{t} D_g f(\tau)\exp\left(-\frac{\alpha}{1-\alpha}(t-\tau)\right)d\tau \tag{4.111}$$

and

$${}_{t_0}^{CM}D_g^\alpha f(t) = \frac{1}{1-\alpha}\int_{t_0}^{t} D_g f(\tau)E_\alpha\left(-\frac{\alpha}{1-\alpha}(t-\tau)^\alpha\right)d\tau. \tag{4.112}$$

Lemma 4.9 *Assuming that* $\forall t \in [a,b]$, $\left|\frac{f'(t)}{g'(t)}\right|$ *is bounded then*

$$\left|{}_{t_0}^{CP}D_g^\alpha f(t)\right| \leq \frac{1}{\Gamma(2-\alpha)}\left\|\frac{f'}{g'}\right\|_\infty (t-t_0)^{1-\alpha}. \tag{4.113}$$

Proof By the definition of ${}_{t_0}^{CP}D_g^{\alpha} f(t)$ we have

$$\begin{aligned}
\left|{}_{t_0}^{CP}D_g^{\alpha} f(t)\right| &= \left|\frac{1}{\Gamma(1-\alpha)}\int_{t_0}^{t} D_g f(\tau)(t-\tau)^{-\alpha}d\tau\right|, \\
&\leq \frac{1}{\Gamma(1-\alpha)}\int_{t_0}^{t} \left|D_g f(\tau)\right|(t-\tau)^{-\alpha}d\tau, \\
&\leq \frac{1}{\Gamma(1-\alpha)}\int_{t_0}^{t} \left|\frac{f'(\tau)}{g'(\tau)}\right|(t-\tau)^{-\alpha}d\tau, \\
&\leq \frac{1}{\Gamma(1-\alpha)}\int_{t_0}^{t} \sup_{l\in[t_0,\tau]}\left|\frac{f'(l)}{g'(l)}\right|(t-\tau)^{-\alpha}d\tau, \\
&\leq \frac{1}{\Gamma(1-\alpha)}\left\|\frac{f'}{g'}\right\|_{\infty}\frac{(t-t_0)^{1-\alpha}}{1-\alpha}, \\
&\leq \frac{1}{\Gamma(2-\alpha)}\left\|\frac{f'}{g'}\right\|_{\infty}(t-t_0)^{1-\alpha}.
\end{aligned} \tag{4.114}$$

Since $\left\|\frac{f'}{g'}\right\|_{\infty} < M$, then we have

$$\left|{}_{t_0}^{CP}D_g^{\alpha} f(t)\right| \leq \frac{M}{\Gamma(2-\alpha)}(t-t_0)^{1-\alpha}, \tag{4.115}$$

which completes the proof.

Lemma 4.10 *Assuming that* $\forall t \in [a,b]$, $\left|\frac{f'(t)}{g'(t)}\right|$ *is bounded then*

$$\left|{}_{t_0}^{CE}D_g^{\alpha} f(t)\right| = \frac{1}{\alpha}\left\|\frac{f'}{g'}\right\|_{\infty}\left(1-\exp\left(-\frac{\alpha}{1-\alpha}(t-t_0)\right)\right). \tag{4.116}$$

Proof By the definitions of ${}_{t_0}^{CE}D_g^{\alpha} f(t)$,

$$\begin{aligned}
\left|{}_{t_0}^{CE}D_g^{\alpha} f(t)\right| &= \left|\frac{1}{1-\alpha}\int_{t_0}^{t} D_g f(\tau)\exp\left(-\frac{\alpha}{1-\alpha}(t-\tau)\right)d\tau\right|, \\
&\leq \frac{1}{1-\alpha}\int_{t_0}^{t} \left|D_g f(\tau)\right|\exp\left(-\frac{\alpha}{1-\alpha}(t-\tau)\right)d\tau,
\end{aligned} \tag{4.117}$$

$$\leq \frac{1}{1-\alpha}\int_{t_0}^{t}\left|\frac{f'(\tau)}{g'(\tau)}\right|\exp\left(-\frac{\alpha}{1-\alpha}(t-\tau)\right)d\tau,$$

$$\leq \frac{1}{1-\alpha}\int_{t_0}^{t}\sup_{l\in[t_0,\tau]}\left|\frac{f'(l)}{g'(l)}\right|\exp\left(-\frac{\alpha}{1-\alpha}(t-\tau)\right)d\tau,$$

$$\leq \frac{1}{1-\alpha}\left\|\frac{f'}{g'}\right\|_{\infty}\int_{t_0}^{t}\exp\left(-\frac{\alpha}{1-\alpha}(t-\tau)\right)d\tau,$$

$$\leq \frac{1}{1-\alpha}\left\|\frac{f'}{g'}\right\|_{\infty}\exp\left(-\frac{\alpha}{1-\alpha}t\right)\int_{t_0}^{t}\exp\left(\frac{\alpha}{1-\alpha}\tau\right)d\tau,$$

$$\leq \frac{1}{1-\alpha}\left\|\frac{f'}{g'}\right\|_{\infty}\exp\left(-\frac{\alpha}{1-\alpha}t\right)\left(\frac{1-\alpha}{\alpha}\right)\exp\left[\frac{\alpha}{1-\alpha}\tau\right]_{t_0}^{t},$$

$$\leq \frac{1}{1-\alpha}\left\|\frac{f'}{g'}\right\|_{\infty}\left(\frac{1-\alpha}{\alpha}\right)\left[\begin{array}{c}\exp(\frac{\alpha}{1-\alpha}t)\\ -\exp(\frac{\alpha}{1-\alpha}t_0)\end{array}\right]\exp\left(-\frac{\alpha}{1-\alpha}t\right),$$

$$\leq \frac{1}{\alpha}\left\|\frac{f'}{g'}\right\|_{\infty}\left(1-\exp\left(-\frac{\alpha}{1-\alpha}(t-t_0)\right)\right),$$

which completes the proof.

Lemma 4.11 *Assuming that* $\forall t\in[a,b]$, $\left|\frac{f'(t)}{g'(t)}\right|$ *is bounded then*

$$\left|{}_{t_0}^{CM}D_g^{\alpha}f(t)\right| = \frac{1}{1-\alpha}\left\|\frac{f'}{g'}\right\|_{\infty}(t-t_0)E_{\alpha,2}\left(-\frac{\alpha}{1-\alpha}(t-t_0)^{\alpha}\right). \tag{4.118}$$

Proof By definition of ${}_{t_0}^{CM}D_g^{\alpha}f(t)$, we have

$$\left|{}_{t_0}^{CM}D_g^{\alpha}f(t)\right| = \left|\frac{1}{1-\alpha}\int_{t_0}^{t}D_gf(\tau)E_{\alpha}\left(-\frac{\alpha}{1-\alpha}(t-\tau)^{\alpha}\right)d\tau\right|, \tag{4.119}$$

$$\leq \frac{1}{1-\alpha}\int_{t_0}^{t}\left|D_gf(\tau)\right|E_{\alpha}\left(-\frac{\alpha}{1-\alpha}(t-\tau)^{\alpha}\right)d\tau,$$

$$\leq \frac{1}{1-\alpha}\int_{t_0}^{t}\left|\frac{f'(\tau)}{g'(\tau)}\right|E_{\alpha}\left(-\frac{\alpha}{1-\alpha}(t-\tau)^{\alpha}\right)d\tau,$$

$$\leq \frac{1}{1-\alpha}\int_{t_0}^{t} \sup_{l\in[t_0,\tau]}\left|\frac{f^{'}(l)}{g^{'}(l)}\right| E_\alpha\left(-\frac{\alpha}{1-\alpha}(t-\tau)^\alpha\right)d\tau,$$

$$\leq \frac{1}{1-\alpha}\left\|\frac{f^{'}}{g^{'}}\right\|_\infty \int_{t_0}^{t} E_\alpha\left(-\frac{\alpha}{1-\alpha}(t-\tau)^\alpha\right)d\tau,$$

$$\leq \frac{1}{1-\alpha}\left\|\frac{f^{'}}{g^{'}}\right\|_\infty \int_{t_0}^{t}\sum_{j=0}^{\infty}\left(-\frac{\alpha}{1-\alpha}\right)^j \frac{(t-\tau)^{\alpha j}}{\Gamma(\alpha j+1)}d\tau,$$

$$\leq \frac{1}{1-\alpha}\left\|\frac{f^{'}}{g^{'}}\right\|_\infty \sum_{j=0}^{\infty}\frac{\left(-\frac{\alpha}{1-\alpha}\right)^j}{\Gamma(\alpha j+1)}\int_{t_0}^{t}(t-\tau)^{\alpha j}d\tau,$$

$$\leq \frac{1}{1-\alpha}\left\|\frac{f^{'}}{g^{'}}\right\|_\infty \sum_{j=0}^{\infty}\frac{\left(-\frac{\alpha}{1-\alpha}\right)^j}{\Gamma(\alpha j+1)}\frac{(t-t_0)^{\alpha j+1}}{\alpha j+1},$$

$$\leq \frac{1}{1-\alpha}\left\|\frac{f^{'}}{g^{'}}\right\|_\infty (t-t_0)E_{\alpha,2}\left(-\frac{\alpha}{1-\alpha}(t-t_0)^\alpha\right),$$

$$\leq \frac{M}{1-\alpha}(t-t_0)E_{\alpha,2}\left(-\frac{\alpha}{1-\alpha}(t-t_0)^\alpha\right),$$

which completes the proof.

4.5 Inequalities with Power Law Decay-Based Caputo Global Derivative

Corollary 4.13 *Let f be differentiable monotonically increasing on $[a, b]$ such that there exist*

$$m \leq f'(t) \leq M, \tag{4.120}$$

there exists a function $\overline{g}(t) = t$ such that

$$m\,{}_{t_0}^{CP}D_g^\alpha \overline{g}(t) \leq {}_{t_0}^{CP}D_g^\alpha f(\tau) \leq M\,{}_{t_0}^{CP}D_g^\alpha \overline{g}(t). \tag{4.121}$$

Proof Let f' be monotonically increasing on $[a, b]$ such that there exist m and M verifying $\forall t \in [a, b]$,

$$m \leq f'(t) \leq M. \tag{4.122}$$

Then since $g'(t) > 0$ we have that

$$\frac{m}{g'(t)} \leq \frac{f'(t)}{g'(t)} \leq \frac{M}{g'(t)}, \tag{4.123}$$

$$\frac{m}{g'(t)} \leq D_g^\alpha f(t) \leq \frac{M}{g'(t)},$$

$$\frac{m}{\Gamma(1-\alpha)} \int_{t_0}^{t} \frac{1}{g'(\tau)}(t-\tau)^{-\alpha} d\tau <_{t_0}^{CP} D_g^\alpha f(t) < \frac{M}{\Gamma(1-\alpha)} \int_{t_0}^{t} \frac{1}{g'(\tau)}(t-\tau)^{-\alpha} d\tau,$$

$$m_{t_0}^{CP} D_g^\alpha \overline{g}(t) \leq_{t_0}^{CP} D_g^\alpha f(\tau) \leq M_{t_0}^{CP} D_g^\alpha \overline{g}(t).$$

Lemma 4.12 *Let f be continuously differentiable on $[a, b]$. Let $g'(t)$ be bounded, then if f is increasing monotonically, we have if $\left|\frac{f'(t)}{g'(t)^2}\right| < M$, then*

$$\left|{}_{t_0}^{CP} D_g^\alpha f(t)\right| = \left|\frac{1}{\Gamma(1-\alpha)} \int_{t_0}^{t} D_g f(\tau)(t-\tau)^{-\alpha} d\tau\right|, \tag{4.124}$$

$$\leq \frac{1}{\Gamma(1-\alpha)} \left(\int_{t_0}^{t} \left|D_g f(\tau)\right|^2 d\tau\right)^{\frac{1}{2}} \left(\int_{t_0}^{t} (t-\tau)^{-2\alpha} d\tau\right)^{\frac{1}{2}},$$

$$\leq \frac{1}{\Gamma(1-\alpha)} \left(\int_{t_0}^{t} \left|\frac{f'(\tau)}{g'(\tau)}\right|^2 d\tau\right)^{\frac{1}{2}} \left(\frac{(t-t_0)^{1-2\alpha}}{1-2\alpha}\right)^{\frac{1}{2}},$$

$$\leq \frac{1}{\Gamma(1-\alpha)} \left(\int_{t_0}^{t} \sup_{l \in [t_0, \tau]} \left|\frac{f'(l)}{g'(l)^2}\right| \left|f'(\tau)\right| d\tau\right)^{\frac{1}{2}} \left(\frac{(t-t_0)^{1-2\alpha}}{1-2\alpha}\right)^{\frac{1}{2}},$$

$$\leq \frac{\sqrt{M}}{\Gamma(1-\alpha)} (f(t) - f(t_0))^{\frac{1}{2}} \left(\frac{(t-t_0)^{1-2\alpha}}{1-2\alpha}\right)^{\frac{1}{2}},$$

$$\leq \frac{\sqrt{M}}{\Gamma(1-\alpha)} (f(b) - f(t_0))^{\frac{1}{2}} \left(\frac{(b-t_0)^{1-2\alpha}}{1-2\alpha}\right)^{\frac{1}{2}},$$

providing that $2\alpha - 1 > 0$ which completes the proof.

Corollary 4.14 *Let f be a nonnegative function defined on a closed ball of R with ratio λ with center x_0. It is assumed that f is harmonic on the interior on the ball and continous on the closed ball then if $\forall x;\ |x - x_0| < \lambda_1 < \lambda$ then we have*

$$f(x_0)\left(1 - \frac{\lambda_1}{\lambda}\right) {}_{x_0}^{CP} D_g^\alpha(x) \leq {}_{x_0}^{CP} D_g^\alpha f(x) \leq f(x_0)\left(1 + \frac{\lambda_1}{\lambda}\right) {}_{x_0}^{CP} D_g^\alpha(x). \tag{4.125}$$

Proof By the hypothesis conditions and using Harnacks's inequality we have that $\forall x,\ |x - x_0| < \lambda_1 < \lambda$,

$$f(x_0)\left(1 - \frac{\lambda_1}{\lambda}\right) \leq f'(x) \leq f(x_0)\left(1 + \frac{\lambda_1}{\lambda}\right). \tag{4.126}$$

Dividing by $g'(x)$ we get

$$\frac{f(x_0)}{g'(x)}\left(1 - \frac{\lambda_1}{\lambda}\right) \leq \frac{f'(x)}{g'(x)} \leq \frac{f(x_0)}{g'(x)}\left(1 + \frac{\lambda_1}{\lambda}\right), \tag{4.127}$$

then integrate on both sides

$$\begin{aligned}\frac{1}{\Gamma(1-\alpha)}\int_{x_0}^{x}\frac{f(x_0)}{g'(x)}\left(1 - \frac{\lambda_1}{\lambda}\right)(x-\tau)^{-\alpha}d\tau &\leq \frac{1}{\Gamma(1-\alpha)}\int_{x_0}^{x}\frac{f'(x)}{g'(x)}(x-\tau)^{-\alpha}d\tau \\ &\leq \frac{1}{\Gamma(1-\alpha)}\int_{x_0}^{x}\frac{f(x_0)}{g'(x)}\left(1 + \frac{\lambda_1}{\lambda}\right)(x-\tau)^{-\alpha}d\tau,\end{aligned} \tag{4.128}$$

$$\begin{aligned}\frac{f(x_0)\left(1 - \frac{\lambda_1}{\lambda}\right)}{\Gamma(1-\alpha)}\int_{x_0}^{x} D_g(\tau)(x-\tau)^{-\alpha}d\tau &\leq \frac{1}{\Gamma(1-\alpha)}\int_{x_0}^{x} D_g f(\tau)(x-\tau)^{-\alpha}d\tau \\ &\leq \frac{f(x_0)\left(1 + \frac{\lambda_1}{\lambda}\right)}{\Gamma(1-\alpha)}\int_{x_0}^{x} D_g(\tau)(x-\tau)^{-\alpha}d\tau,\end{aligned}$$

$$f(x_0)\left(1 - \frac{\lambda_1}{\lambda}\right){}_{x_0}^{CP}D_g^{\alpha}(x) \leq {}_{x_0}^{CP}D_g^{\alpha}f(x) \leq f(x_0)\left(1 + \frac{\lambda_1}{\lambda}\right){}_{x_0}^{CP}D_g^{\alpha}(x),$$

which completes the proof.

Corollary 4.15 *Let (x, Σ, μ) be a measure space. Let f' and $h : x \to R$ be measurable in L^p, then if $p = 2$, we have*

$$\left|{}_{t_0}^{CP}D_g^{\alpha}f(t)\right| < \frac{\sqrt{2}}{M_1\Gamma(1-\alpha)}\left(\frac{(t-t_0)^{1-2\alpha}}{1-2\alpha}\right)^{\frac{1}{2}}\left\{\frac{\|f'\|_{L^2[a,b]}}{2} + \frac{\|h\|_{L^2[a,b]}}{2}\right\}. \tag{4.129}$$

Proof Let us write the definition of ${}_{t_0}^{CP}D_g^{\alpha}f(t)$ below

$$\begin{aligned}\left|{}_{t_0}^{CP}D_g^{\alpha}f(t)\right| &= \frac{1}{\Gamma(1-\alpha)}\left|\int_{t_0}^{t} D_g f(\tau)(t-\tau)^{-\alpha}d\tau\right|, \\ &= \frac{1}{\Gamma(1-\alpha)}\left|\int_{t_0}^{t}\frac{f'(\tau)}{g'(\tau)}(t-\tau)^{-\alpha}d\tau\right|,\end{aligned} \tag{4.130}$$

$$\leq \frac{1}{\Gamma(1-\alpha)} \left| \int_{t_0}^{t} \frac{f'(\tau)}{\min\limits_{l\in[t_0,\tau]} g'(l)} (t-\tau)^{-\alpha} d\tau \right|,$$

$$\leq \frac{1}{\Gamma(1-\alpha) \min\limits_{t\in[a,b]} g'(t)} \left| \int_{t_0}^{t} f'(\tau)(t-\tau)^{-\alpha} d\tau \right|,$$

$$\leq \frac{1}{\Gamma(1-\alpha)M_1} \left(\int_{t_0}^{t} \left|f'(\tau)\right|^2 d\tau \right)^{\frac{1}{2}} \left(\int_{t_0}^{t} (t-\tau)^{-2\alpha} d\tau \right)^{\frac{1}{2}}, \tag{4.131}$$

$$\leq \frac{1}{\Gamma(1-\alpha)M_1} \left[\int_{t_0}^{t} \left| \frac{f'(\tau)+h(\tau)}{2} + \frac{f'(\tau)-h(\tau)}{2} \right|^2 \right]^{\frac{1}{2}} \left(\frac{(t-t_0)^{1-2\alpha}}{1-2\alpha} \right)^{\frac{1}{2}},$$

$$\leq \frac{1}{\Gamma(1-\alpha)M_1} \left(\frac{(t-t_0)^{1-2\alpha}}{1-2\alpha} \right)^{\frac{1}{2}} \left\{ \left\| \frac{f'+h}{2} \right\|_{L^2[a,b]} + \left\| \frac{f'-h}{2} \right\|_{L^2[a,b]} \right\}.$$

By Clarkson's inequality we have

$$\left| {}_{t_0}^{CP} D_g^{\alpha} f(t) \right| < \frac{\sqrt{2}}{M_1 \Gamma(1-\alpha)} \left(\frac{(t-t_0)^{1-2\alpha}}{1-2\alpha} \right)^{\frac{1}{2}} \left\{ \frac{\left\| f' \right\|_{L^2[a,b]}}{2} + \frac{\|h\|_{L^2[a,b]}}{2} \right\}, \tag{4.132}$$

which completes the proof.

Corollary 4.16 *Let f be two differentiables on $[a, b]$ such that f'' and $f \in L^2[a, b]$ then*

$$\left| {}_{t_0}^{CP} D_g^{\alpha} f(t) \right| < \frac{1}{M_1 \Gamma(1-\alpha)} \left(\frac{(t-t_0)^{1-2\alpha}}{1-2\alpha} \right)^{\frac{1}{2}} K \left\| f \right\|_{L^p[a,b]}^{\beta} \left\| f'' \right\|_{L^r[a,b]}^{1-\beta}, \tag{4.133}$$

where K, α, β are positive constants.

Proof

$$\left| {}_{t_0}^{CP} D_g^{\alpha} f(t) \right| = \frac{1}{\Gamma(1-\alpha)} \left| \int_{t_0}^{t} D_g f(\tau)(t-\tau)^{-\alpha} d\tau \right|, \tag{4.134}$$

$$= \frac{1}{\Gamma(1-\alpha)} \left| \int_{t_0}^{t} \frac{f'(\tau)}{g'(\tau)} (t-\tau)^{-\alpha} d\tau \right|,$$

$$\leq \frac{1}{\Gamma(1-\alpha)} \left| \int_{t_0}^{t} \frac{f'(\tau)}{\inf\limits_{l \in [t_0,\tau]} g'(l)} (t-\tau)^{-\alpha} d\tau \right|,$$

$$\leq \frac{1}{\Gamma(1-\alpha) \min\limits_{t \in [a,b]} g'(t)} \left| \int_{t_0}^{t} f'(\tau)(t-\tau)^{-\alpha} d\tau \right|,$$

$$\leq \frac{1}{\Gamma(1-\alpha) M_1} \left(\int_{t_0}^{t} \left| f'(\tau) \right|^2 d\tau \right)^{\frac{1}{2}} \left(\int_{t_0}^{t} (t-\tau)^{-2\alpha} d\tau \right)^{\frac{1}{2}},$$

$$\leq \frac{1}{\Gamma(1-\alpha) M_1} \left(\frac{(t-t_0)^{1-2\alpha}}{1-2\alpha} \right)^{\frac{1}{2}} \left(\int_{t_0}^{t} \left| f'(\tau) \right|^2 d\tau \right)^{\frac{1}{2}}.$$

By the Landau–Kolmogorov inequality we have

$$\left\| f' \right\|_{L^q[a,b]} \leq K \left\| f \right\|_{L^p[a,b]}^{\beta} \left\| f'' \right\|_{L^r[a,b]}^{1-\beta}. \tag{4.135}$$

Therefore

$$\left| {}_{t_0}^{CP} D_g^{\alpha} f(t) \right| < \frac{1}{\Gamma(1-\alpha) M_1} \left(\frac{(t-t_0)^{1-2\alpha}}{1-2\alpha} \right)^{\frac{1}{2}} K \left\| f \right\|_{L^p[a,b]}^{\beta} \left\| f'' \right\|_{L^r[a,b]}^{1-\beta}, \tag{4.136}$$

which completes the proof with the condition that $1 - 2\alpha > 0$.

Lemma 4.13 *Let f be continuously differentiable on $[a, b]$. Let $g'(t)$ be continuously bounded then if $f'(t)$ is bounded and $(1 - 2\alpha) > 0$ then*

$$\left| {}_{t_0}^{CP} D_g^{\alpha} f(t) \right|^2 = \left| \frac{1}{\Gamma(1-\alpha)} \int_{t_0}^{t} D_g f(\tau)(t-\tau)^{-\alpha} d\tau \right|^2, \tag{4.137}$$

$$\leq \frac{1}{\Gamma^2(1-\alpha)} \left(\int_{t_0}^{t} \left| D_g f(\tau) \right|^2 d\tau \right) \left(\int_{t_0}^{t} (t-\tau)^{-2\alpha} d\tau \right),$$

$$\leq \frac{1}{\Gamma^2(1-\alpha)} \left(\int_{t_0}^{t} \left| \frac{f'(\tau)}{g'(\tau)} \right|^2 d\tau \right) \left(\int_{t_0}^{t} (t-\tau)^{-2\alpha} d\tau \right),$$

$$\leq \frac{1}{\Gamma^2(1-\alpha)} \left(\int_{t_0}^{t} \max_{l \in [t_0,\tau]} \left| \frac{f'(l)}{g'(l)^2} \right| \left| f'(\tau) \right| d\tau \right) \frac{(t-t_0)^{1-2\alpha}}{1-2\alpha},$$

$$\leq \frac{\overline{M}}{\Gamma^2(1-\alpha)} \int_{t_0}^{t} f'(\tau)d\tau \frac{(t-t_0)^{1-2\alpha}}{1-2\alpha},$$
$$\leq \frac{\overline{M}}{\Gamma^2(1-\alpha)} \frac{(t-t_0)^{1-2\alpha}}{1-2\alpha} \left(f(t) - f(t_0)\right),$$
$$\leq \frac{\overline{M}}{\Gamma^2(1-\alpha)} \frac{(b-t_0)^{1-2\alpha}}{1-2\alpha} \left(f(b) - f(t_0)\right),$$

which completes the proof.

4.6 Inequalities with Exponential Law Decay-Based Caputo Global Derivative

Corollary 4.17 *Using the procedure presented before, we have that if f is monotonically increasing function on $[a, b]$. We assume that there exist m and M such that*

$$m \leq f'(t) \leq M, \tag{4.138}$$

then

$$m_{t_0}^{CF} D_t^{\alpha} \overline{g}(t) \leq_{t_0}^{CE} D_g^{\alpha} f(t) \leq M_{t_0}^{CF} D_t^{\alpha} \overline{g}(t). \tag{4.139}$$

Proof Let m and M be two real value numbers such that

$$m \leq f'(t) \leq M. \tag{4.140}$$

Then since

$$\frac{1}{1-\alpha} \exp\left(-\frac{\alpha}{1-\alpha}(t-\tau)\right) > 0. \tag{4.141}$$

$\forall t \in [a, b]$, then multiplying on above yields

$$\frac{m}{1-\alpha} \exp\left(-\frac{\alpha}{1-\alpha}(t-\tau)\right) \leq \frac{f'(t)}{1-\alpha} \exp\left(-\frac{\alpha}{1-\alpha}(t-\tau)\right), \tag{4.142}$$
$$\leq \frac{M}{1-\alpha} \exp\left(-\frac{\alpha}{1-\alpha}(t-\tau)\right).$$

Now multiplying by $\frac{1}{g'(t)}$ and since $\frac{1}{g'(t)} > 0 \; \forall t \in [a, b]$, then

$$\frac{m}{1-\alpha} \exp\left(-\frac{\alpha}{1-\alpha}(t-\tau)\right) \frac{1}{g'(t)} \leq \frac{f'(t)}{1-\alpha} \frac{1}{g'(t)} \exp\left(-\frac{\alpha}{1-\alpha}(t-\tau)\right), \tag{4.143}$$
$$\leq \frac{M}{1-\alpha} \frac{1}{g'(t)} \exp\left(-\frac{\alpha}{1-\alpha}(t-\tau)\right).$$

Integrating on both sides yields

$$\frac{m}{1-\alpha}\int_{t_0}^{t}\frac{1}{g'(t)}\exp\left(-\frac{\alpha}{1-\alpha}(t-\tau)\right)d\tau \leq {}_{t_0}^{CE}D_g^{\alpha}f(t) \tag{4.144}$$

$$\leq \frac{M}{1-\alpha}\int_{t_0}^{t}\frac{1}{g'(t)}\exp\left(-\frac{\alpha}{1-\alpha}(t-\tau)\right)d\tau,$$

$$\frac{m}{1-\alpha}\int_{t_0}^{t}\overline{g}'(t)\exp\left(-\frac{\alpha}{1-\alpha}(t-\tau)\right)d\tau \leq {}_{t_0}^{CE}D_g^{\alpha}f(t)$$

$$\leq \frac{M}{1-\alpha}\int_{t_0}^{t}\overline{g}'(t)\exp\left(-\frac{\alpha}{1-\alpha}(t-\tau)\right)d\tau,$$

$$m\,{}_{t_0}^{CF}D_t^{\alpha}\overline{g}(t) \leq {}_{t_0}^{CE}D_g^{\alpha}f(t) \leq M\,{}_{t_0}^{CF}D_t^{\alpha}\overline{g}(t),$$

which completes the proof.

Lemma 4.14 *Let f be increasing monotonically, if $\frac{f'}{g'^2}$ is bounded for $\forall t \in [a, b]$, then*

$$\left|{}_{t_0}^{CE}D_g^{\alpha}f(t)\right| < \sqrt{\frac{\overline{M}}{2\alpha(1-\alpha)}}\left(f(t)-f(t_0)\right)^{\frac{1}{2}}. \tag{4.145}$$

Proof If $\frac{f'}{g'^2}$ is bounded in $[a, b]$ then $\forall t \in [a, b]$ there exists $\overline{M} > 0$ such that $\forall t \in [a, b]$ if $f'(t) > 0$ then

$$\left|\frac{f'(t)}{g'^2(t)}\right| < \overline{M}. \tag{4.146}$$

But by definition we have that

$$\left|{}_{t_0}^{CE}D_g^{\alpha}f(t)\right| = \frac{1}{1-\alpha}\left|\int_{t_0}^{t}D_g f(\tau)\exp\left(-\frac{\alpha}{1-\alpha}(t-\tau)\right)d\tau\right|, \tag{4.147}$$

$$\leq \frac{1}{1-\alpha}\left(\int_{t_0}^{t}\left|D_g f(\tau)\right|^2 d\tau\right)^{\frac{1}{2}}\left(\int_{t_0}^{t}\exp\left(-\frac{2\alpha}{1-\alpha}(t-\tau)\right)d\tau\right)^{\frac{1}{2}},$$

$$\leq \frac{1}{1-\alpha}\left(\int_{t_0}^{t}\left|\frac{f'(\tau)}{g'(\tau)}\right|^2 d\tau\right)^{\frac{1}{2}}\left(\int_{t_0}^{t}\exp\left(-\frac{2\alpha}{1-\alpha}(t-\tau)\right)d\tau\right)^{\frac{1}{2}},$$

$$\leq \frac{1}{1-\alpha}\left(\int_{t_0}^{t}\left|\frac{f'(\tau)}{g'(\tau)^2}\right|\left|f'(\tau)\right|d\tau\right)^{\frac{1}{2}}\left(\int_{t_0}^{t}\exp\left(-\frac{2\alpha}{1-\alpha}(t-\tau)\right)d\tau\right)^{\frac{1}{2}},$$

$$\leq \frac{1}{1-\alpha}\left(\int_{t_0}^{t} \max_{l\in[t_0,\tau]}\left|\frac{f'(l)}{g'(l)^2}\right| \left|f'(\tau)\right| d\tau\right)^{\frac{1}{2}} \left(\int_{t_0}^{t} \exp\left(-\frac{2\alpha}{1-\alpha}(t-\tau)\right) d\tau\right)^{\frac{1}{2}},$$

$$\leq \frac{\sqrt{\overline{M}}}{1-\alpha}\left(f(t)-f(t_0)\right)^{\frac{1}{2}}\left(\left(\frac{1-\alpha}{2\alpha}\right)\left(1-\exp\left(-\frac{2\alpha}{1-\alpha}(t-t_0)\right)\right)\right)^{\frac{1}{2}},$$

$$\leq \frac{\sqrt{\overline{M}}}{1-\alpha}\left(f(t)-f(t_0)\right)^{\frac{1}{2}}\sqrt{\frac{1-\alpha}{2\alpha}},$$

$$\leq \sqrt{\frac{\overline{M}}{2\alpha(1-\alpha)}}\left(f(t)-f(t_0)\right)^{\frac{1}{2}},$$

which completes the proof.

Corollary 4.18 *Let f be a nonnegative function defined on a closed ball of R with ratio λ with center x_0. It is assumed that f is harmonic on the interior on the ball and continous on the closed ball then if $\forall x;\ |x-x_0|<\lambda_1<\lambda$ then we have*

$$f(x_0)\left(1-\frac{\lambda_1}{\lambda}\right){}_{x_0}^{CE}D_g^{\alpha} f(x) \leq {}_{x_0}^{CE}D_g^{\alpha} f(x) \leq f(x_0)\left(1+\frac{\lambda_1}{\lambda}\right){}_{x_0}^{CE}D_g^{\alpha} f(x). \tag{4.148}$$

Proof By the hypothesis conditions and using Harnack's inequality we have that $\forall x,\ |x-x_0|<\lambda_1<\lambda$.

$$f(x_0)\left(1-\frac{\lambda_1}{\lambda}\right) \leq f'(x) \leq f(x_0)\left(1+\frac{\lambda_1}{\lambda}\right). \tag{4.149}$$

Dividing by $g'(x)$ we get

$$\frac{f(x_0)}{g'(x)}\left(1-\frac{\lambda_1}{\lambda}\right) \leq \frac{f'(x)}{g'(x)} \leq \frac{f(x_0)}{g'(x)}\left(1+\frac{\lambda_1}{\lambda}\right), \tag{4.150}$$

then integrate on both sides

$$\frac{1}{1-\alpha}\int_{x_0}^{x} \frac{f(x_0)}{g'(x)}\left(1-\frac{\lambda_1}{\lambda}\right)\exp\left(-\frac{\alpha}{1-\alpha}(x-\tau)\right) d\tau, \tag{4.151}$$

$$\leq \frac{1}{1-\alpha}\int_{x_0}^{x} \frac{f'(x)}{g'(x)}\exp\left(-\frac{\alpha}{1-\alpha}(x-\tau)\right) d\tau,$$

$$\leq \frac{1}{1-\alpha}\int_{x_0}^{x} \frac{f(x_0)}{g'(x)}\left(1+\frac{\lambda_1}{\lambda}\right)\exp\left(-\frac{\alpha}{1-\alpha}(x-\tau)\right) d\tau,$$

$$\frac{f(x_0)\left(1-\frac{\lambda_1}{\lambda}\right)}{1-\alpha}\int_{x_0}^{x} D_g(\tau)(x-\tau)^{-\alpha}d\tau \leq \frac{1}{1-\alpha}\int_{x_0}^{x} D_g f(\tau)(x-\tau)^{-\alpha}d\tau,$$
$$\leq \frac{f(x_0)\left(1+\frac{\lambda_1}{\lambda}\right)}{1-\alpha}\int_{x_0}^{x} D_g(\tau)(x-\tau)^{-\alpha}d\tau, \tag{4.152}$$

$$f(x_0)\left(1-\frac{\lambda_1}{\lambda}\right){}_{x_0}^{CE}D_g^{\alpha}(x) \leq_{x_0}^{CE} D_g^{\alpha} f(x) \leq f(x_0)\left(1+\frac{\lambda_1}{\lambda}\right)_{x_0}^{CE} D_g^{\alpha}(x), \tag{4.153}$$

which completes the proof.

Corollary 4.19 *Let (x, Σ, μ) be a measure space. Let $f^{'}$ and $h : x \to R$ be measurable in L^p, then if $p = 2$, we have*

$$\left|{}_{t_0}^{CE}D_g^{\alpha} f(t)\right| < \frac{1}{\sqrt{\alpha(1-\alpha)}M_1}\left\{\frac{\|f^{'}\|_{L^2[a,b]}}{2}+\frac{\|h\|_{L^2[a,b]}}{2}\right\}. \tag{4.154}$$

Proof

$$\left|{}_{t_0}^{CE}D_g^{\alpha} f(t)\right| = \frac{1}{1-\alpha}\left|\int_{t_0}^{t} D_g f(\tau)\exp\left(-\frac{\alpha}{1-\alpha}(t-\tau)\right)d\tau\right|, \tag{4.155}$$
$$= \frac{1}{1-\alpha}\left|\int_{t_0}^{t} \left|\frac{f'(\tau)}{g'(\tau)}\right|\exp\left(-\frac{\alpha}{1-\alpha}(t-\tau)\right)d\tau\right|,$$
$$\leq \frac{1}{1-\alpha}\left|\int_{t_0}^{t} \frac{|f'(\tau)|}{\inf\limits_{l\in[t_0,\tau]}|g'(l)|}\exp\left(-\frac{\alpha}{1-\alpha}(t-\tau)\right)d\tau\right|,$$
$$\leq \frac{1}{(1-\alpha)g'(t)}\left|\int_{t_0}^{t} f'(\tau)\exp\left(-\frac{\alpha}{1-\alpha}(t-\tau)\right)d\tau\right|,$$
$$\leq \frac{1}{(1-\alpha)M_1}\left(\int_{t_0}^{t}\left|\frac{f'(\tau)+h(\tau)}{2}+\frac{f'(\tau)-h(\tau)}{2}\right|^2 d\tau\right)^{\frac{1}{2}}$$
$$\times\left(\int_{t_0}^{t}\exp\left(-\frac{2\alpha}{1-\alpha}(t-\tau)\right)\right)^{\frac{1}{2}} d\tau,$$

$$\leq \frac{1}{(1-\alpha)M_1}\left(\int_{t_0}^{t}\left|\frac{f'(\tau)+h(\tau)}{2}+\frac{f'(\tau)-h(\tau)}{2}\right|^2 d\tau\right)^{\frac{1}{2}}$$

$$\times\left(\left(\frac{1-\alpha}{2\alpha}\right)\exp\left(-\frac{2\alpha}{1-\alpha}(t-\tau)\right)\right)^{\frac{1}{2}},$$

$$\leq \frac{1}{(1-\alpha)M_1}\left\{\frac{\left\|f'+h\right\|_{L^2[a,b]}}{2}+\frac{\left\|f'-h\right\|_{L^2[a,b]}}{2}\right\}\sqrt{\frac{1-\alpha}{2\alpha}}.$$

By Clarkson's inequality

$$\left|{}_{t_0}^{CE}D_g^{\alpha}f(t)\right| < \frac{1}{\sqrt{\alpha(1-\alpha)}M_1}\left\{\frac{\left\|f'\right\|_{L^2[a,b]}}{2}+\frac{\|h\|_{L^2[a,b]}}{2}\right\}, \tag{4.156}$$

which completes the proof.

Corollary 4.20 *Let f be twice differentiable on $[a,b]$ such f'' and $f \in L^2[a,b]$. Then if $\frac{1}{g'^2}$ is bounded then*

$$\left|{}_{t_0}^{CE}D_g^{\alpha}f(t)\right| < \frac{1}{(1-\alpha)Mg'}\left(\frac{1-\alpha}{2\alpha}\right)^{\frac{1}{2}} K\,\|f\|_{L^2[a,b]}^{\beta}\left\|f''\right\|_{L^2[a,b]}^{1-\beta}. \tag{4.157}$$

Proof Let f'' and f be in $L^2[a,b]$

$$\left|{}_{t_0}^{CE}D_g^{\alpha}f(t)\right| = \frac{1}{(1-\alpha)}\left|\int_{t_0}^{t}D_g f(\tau)\exp\left(-\frac{\alpha}{1-\alpha}(t-\tau)\right)d\tau\right|, \tag{4.158}$$

$$= \frac{1}{(1-\alpha)}\left|\int_{t_0}^{t}\frac{f'(\tau)}{g'(\tau)}\exp\left(-\frac{\alpha}{1-\alpha}(t-\tau)\right)d\tau\right|,$$

$$\leq \frac{1}{(1-\alpha)}\left(\int_{t_0}^{t}\left|\frac{f'(\tau)}{g'(\tau)}\right|^2 d\tau\right)^{\frac{1}{2}}\left(\int_{t_0}^{t}\exp\left(-\frac{2\alpha}{1-\alpha}(t-\tau)\right)d\tau\right)^{\frac{1}{2}},$$

$$\leq \frac{1}{(1-\alpha)}\left(\int_{t_0}^{t}\frac{1}{\inf\limits_{l\in[t_0,\tau]}|g'(l)|^2}\left|f'(\tau)\right|^2 d\tau\right)^{\frac{1}{2}}\left(\int_{t_0}^{t}\exp\left(-\frac{2\alpha}{1-\alpha}(t-\tau)\right)d\tau\right)^{\frac{1}{2}},$$

$$\leq \frac{1}{(1-\alpha)M_{g'}}\left(\int_{t_0}^{t}\left|f'\right|^2 d\tau\right)^{\frac{1}{2}}\left(\frac{1-\alpha}{2\alpha}\right)^{\frac{1}{2}}\left(1-\exp\left(-\frac{2\alpha}{1-\alpha}(t-t_0)\right)\right)^{\frac{1}{2}},$$

$$\leq \frac{1}{(1-\alpha)M_{g'}}\left\|f'\right\|_{L^2[a,b]}\left(\frac{1-\alpha}{2\alpha}\right)^{\frac{1}{2}}.$$

Using the Landau–Kolmogorov inequality we have

$$\left|{}_{t_0}^{CE} D_g^{\alpha} f(t)\right| < \frac{1}{(1-\alpha) M g'} \left(\frac{1-\alpha}{2\alpha}\right)^{\frac{1}{2}} K \left\| f \right\|_{L^2[a,b]}^{\beta} \left\| f'' \right\|_{L^2[a,b]}^{1-\beta}, \tag{4.159}$$

which completes the proof.

4.7 Inequalities with Mittag–Leffler Law Decay-Based Caputo Global Derivative

Corollary 4.21 *Using the procedure presented before, we have that if f is monotonically increasing on $[a, b]$ such that there exist m and M such that then*

$$m \leq f'(t) \leq M, \tag{4.160}$$

there exists a function $\overline{g}$ such that

$$m_{t_0}^{CM} D_t^{\alpha} \overline{g}(t) \leq_{t_0}^{CM} D_g^{\alpha} f(t) \leq M_{t_0}^{CM} D_t^{\alpha} \overline{g}(t). \tag{4.161}$$

Proof Let f' be monotonically increasing on $[a, b]$ such that there exist m and M verifying $\forall t \in [a, b]$,

$$m \leq f'(t) \leq M. \tag{4.162}$$

Then since

$$\frac{1}{1-\alpha} E_{\alpha}\left(-\frac{\alpha}{1-\alpha}(t-\tau)^{\alpha}\right) > 0. \tag{4.163}$$

$\forall t \in [a, b]$, then multiplying on above yields

$$\begin{aligned} \frac{m}{1-\alpha} E_{\alpha}\left(-\frac{\alpha}{1-\alpha}(t-\tau)^{\alpha}\right) &\leq \frac{f'(t)}{1-\alpha} E_{\alpha}\left(-\frac{\alpha}{1-\alpha}(t-\tau)^{\alpha}\right), \\ &\leq \frac{M}{1-\alpha} E_{\alpha}\left(-\frac{\alpha}{1-\alpha}(t-\tau)^{\alpha}\right). \end{aligned} \tag{4.164}$$

Now multiplying by $\frac{1}{g'(t)} > 0$, $\forall t \in [a, b]$, then we have

$$\begin{aligned} \frac{m}{(1-\alpha) g'(t)} E_{\alpha}\left(-\frac{\alpha}{1-\alpha}(t-\tau)^{\alpha}\right) &\leq \frac{f'(t)}{(1-\alpha) g'(t)} E_{\alpha}\left(-\frac{\alpha}{1-\alpha}(t-\tau)^{\alpha}\right), \\ &\leq \frac{M}{(1-\alpha) g'(t)} E_{\alpha}\left(-\frac{\alpha}{1-\alpha}(t-\tau)^{\alpha}\right). \end{aligned} \tag{4.165}$$

Integrating on both sides

$$\frac{m}{(1-\alpha)}\int_{t_0}^{t}\frac{1}{g'(t)}E_\alpha\left(-\frac{\alpha}{1-\alpha}(t-\tau)^\alpha\right)d\tau \tag{4.166}$$

$$\leq\frac{1}{(1-\alpha)}\int_{t_0}^{t}\frac{f'(t)}{g'(t)}E_\alpha\left(-\frac{\alpha}{1-\alpha}(t-\tau)^\alpha\right)d\tau,$$

$$\leq\frac{M}{(1-\alpha)}\int_{t_0}^{t}\frac{1}{g'(t)}E_\alpha\left(-\frac{\alpha}{1-\alpha}(t-\tau)^\alpha\right)d\tau,$$

$$\frac{m}{(1-\alpha)}\int_{t_0}^{t}\overline{g}'(t)E_\alpha\left(-\frac{\alpha}{1-\alpha}(t-\tau)^\alpha\right)d\tau \tag{4.167}$$

$$\leq\frac{1}{(1-\alpha)}\int_{t_0}^{t}\frac{f'(t)}{g'(t)}E_\alpha\left(-\frac{\alpha}{1-\alpha}(t-\tau)^\alpha\right)d\tau,$$

$$\leq\frac{M}{(1-\alpha)}\int_{t_0}^{t}\overline{g}'(t)E_\alpha\left(-\frac{\alpha}{1-\alpha}(t-\tau)^\alpha\right)d\tau.$$

Finally we have

$$m_{t_0}^{CM}D_t^\alpha\overline{g}(t)\leq_{t_0}^{CM}D_g^\alpha f(t)\leq M_{t_0}^{CM}D_t^\alpha\overline{g}(t), \tag{4.168}$$

which completes the proof.

Lemma 4.15 *Let f be increasing monotonically, if $\frac{f'}{g'^2}$ is bounded for $\forall t\in[a,b]$, then*

$$\left|{}_{t_0}^{CM}D_g^\alpha f(t)\right|\leq\frac{\sqrt{M}}{1-\alpha}(f(t)-f(t_0))^{\frac{1}{2}}\left\|E_\alpha\left(-\frac{\alpha}{1-\alpha}(t-t_0)^\alpha\right)\right\|_{L^2[a,b]}. \tag{4.169}$$

Proof If $\frac{f'}{g'^2}$ is bounded in $[a,b]$ then $\forall t\in[a,b]$ there exists $M>0$ such that $\forall t\in[a,b]$ if $f'(t)>0$ then

$$\left|\frac{f'(t)}{g'^2(t)}\right|<M. \tag{4.170}$$

But by definition we have that

$$
\begin{aligned}
\left|{}_{t_0}^{CM}D_g^{\alpha}f(t)\right| &= \frac{1}{1-\alpha}\left|\int_{t_0}^{t} D_g f(\tau)E_\alpha\left(-\frac{\alpha}{1-\alpha}(t-\tau)^\alpha\right)d\tau\right|, \qquad (4.171)\\
&\leq \frac{1}{1-\alpha}\left(\int_{t_0}^{t}\left|D_g f(\tau)\right|^2 d\tau\right)^{\frac{1}{2}}\left(\int_{t_0}^{t}\left(E_\alpha\left(-\frac{\alpha}{1-\alpha}(t-\tau)^\alpha\right)\right)^2 d\tau\right)^{\frac{1}{2}},\\
&\leq \frac{1}{1-\alpha}\left(\int_{t_0}^{t}\left|\frac{f'(\tau)}{g'(\tau)}\right|^2 d\tau\right)^{\frac{1}{2}}\left\|E_\alpha\left(-\frac{\alpha}{1-\alpha}(t-t_0)^\alpha\right)\right\|_{L^2[a,b]},\\
&\leq \frac{1}{1-\alpha}\left(\int_{t_0}^{t}\left|\frac{f'(\tau)}{g'(\tau)^2}\right|\left|f'(\tau)\right|d\tau\right)^{\frac{1}{2}}\left\|E_\alpha\left(-\frac{\alpha}{1-\alpha}(t-t_0)^\alpha\right)\right\|_{L^2[a,b]},\\
&\leq \frac{1}{1-\alpha}\left(\int_{t_0}^{t}\sup_{l\in[t_0,\tau]}\left|\frac{f'(l)}{g'(l)^2}\right|\left|f'(\tau)\right|d\tau\right)^{\frac{1}{2}}\left\|E_\alpha\left(-\frac{\alpha}{1-\alpha}(t-t_0)^\alpha\right)\right\|_{L^2[a,b]},\\
&\leq \frac{\sqrt{M}}{1-\alpha}\left(f(t)-f(t_0)\right)^{\frac{1}{2}}\left\|E_\alpha\left(-\frac{\alpha}{1-\alpha}(t-t_0)^\alpha\right)\right\|_{L^2[a,b]},\\
&\leq \frac{\sqrt{M}}{1-\alpha}\left(f(t)-f(t_0)\right)^{\frac{1}{2}}\left\|E_\alpha\left(-\frac{\alpha}{1-\alpha}(t-t_0)^\alpha\right)\right\|_{L^2[a,b]},\\
&\leq \frac{\sqrt{M}}{1-\alpha}\left(f(t)-f(t_0)\right)^{\frac{1}{2}}\left\|E_\alpha\left(-\frac{\alpha}{1-\alpha}(t-t_0)^\alpha\right)\right\|_{L^2[a,b]},
\end{aligned}
$$

which completes the proof.

Corollary 4.22 *Let f be a nonnegative function defined on a closed ball of R with ratio λ with center x_0. It is assumed that f is harmonic on the interior on the ball and continous on the closed ball then if $\forall x;\ |x-x_0|<\lambda_1<\lambda$ then we have*

$$
f(x_0)\left(1-\frac{\lambda_1}{\lambda}\right){}_{x_0}^{CM}D_g^{\alpha}f(x) \leq {}_{x_0}^{CM}D_g^{\alpha}f(x) \leq f(x_0)\left(1+\frac{\lambda_1}{\lambda}\right){}_{x_0}^{CM}D_g^{\alpha}f(x). \tag{4.172}
$$

Proof By the hypothesis conditions and using Harnack's inequality we have that $\forall x,\ |x-x_0|<\lambda_1<\lambda$,

$$
f(x_0)\left(1-\frac{\lambda_1}{\lambda}\right) \leq f'(x) \leq f(x_0)\left(1+\frac{\lambda_1}{\lambda}\right). \tag{4.173}
$$

Dividing by $g'(x)$ we get

$$
\frac{f(x_0)}{g'(x)}\left(1-\frac{\lambda_1}{\lambda}\right) \leq \frac{f'(x)}{g'(x)} \leq \frac{f(x_0)}{g'(x)}\left(1+\frac{\lambda_1}{\lambda}\right), \tag{4.174}
$$

then integrate on both sides

$$\frac{1}{1-\alpha}\int_{x_0}^{x}\frac{f(x_0)}{g'(x)}\left(1-\frac{\lambda_1}{\lambda}\right)E_\alpha\left(-\frac{\alpha}{1-\alpha}(t-\tau)^\alpha\right)d\tau \tag{4.175}$$

$$\leq\frac{1}{1-\alpha}\int_{x_0}^{x}\frac{f'(x)}{g'(x)}E_\alpha\left(-\frac{\alpha}{1-\alpha}(t-\tau)^\alpha\right)d\tau,$$

$$\leq\frac{1}{1-\alpha}\int_{x_0}^{x}\frac{f(x_0)}{g'(x)}\left(1+\frac{\lambda_1}{\lambda}\right)E_\alpha\left(-\frac{\alpha}{1-\alpha}(t-\tau)^\alpha\right)d\tau,$$

$$\frac{f(x_0)\left(1-\frac{\lambda_1}{\lambda}\right)}{1-\alpha}\int_{x_0}^{x}D_g(\tau)\,E_\alpha\left(-\frac{\alpha}{1-\alpha}(t-\tau)^\alpha\right)d\tau \tag{4.176}$$

$$\leq\frac{1}{1-\alpha}\int_{x_0}^{x}D_gf(\tau)\,E_\alpha\left(-\frac{\alpha}{1-\alpha}(t-\tau)^\alpha\right)d\tau,$$

$$\leq\frac{f(x_0)\left(1+\frac{\lambda_1}{\lambda}\right)}{1-\alpha}\int_{x_0}^{x}D_g(\tau)\,E_\alpha\left(-\frac{\alpha}{1-\alpha}(t-\tau)^\alpha\right)d\tau,$$

$$f(x_0)\left(1-\frac{\lambda_1}{\lambda}\right){}_{x_0}^{CM}D_g^\alpha(x)\leq{}_{x_0}^{CM}D_g^\alpha f(x)\leq f(x_0)\left(1+\frac{\lambda_1}{\lambda}\right){}_{x_0}^{CM}D_g^\alpha(x), \tag{4.177}$$

which completes the proof.

Corollary 4.23 *Let* (x,Σ,μ) *be a measure space. Let* f' *and* $h:x\to R$ *be measurable in* L^p*, then if* $p=2$*, we have*

$$\left|{}_{t_0}^{CM}D_g^\alpha f(t)\right|<\frac{1}{(1-\alpha)M_1}\left\{\frac{\left\|f'\right\|_{L^2[a,b]}}{2}+\frac{\|h\|_{L^2[a,b]}}{2}\right\}\left\|E_\alpha\left(-\frac{\alpha}{1-\alpha}(t-t_0)^\alpha\right)\right\|_{L^2[a,b]}. \tag{4.178}$$

Proof

$$\left|{}_{t_0}^{CM}D_g^\alpha f(t)\right|=\frac{1}{1-\alpha}\left|\int_{t_0}^{t}D_gf(\tau)E_\alpha\left(-\frac{\alpha}{1-\alpha}(t-\tau)^\alpha\right)d\tau\right|, \tag{4.179}$$

$$=\frac{1}{1-\alpha}\left|\int_{t_0}^{t}\left|\frac{f'(\tau)}{g'(\tau)}\right|E_\alpha\left(-\frac{\alpha}{1-\alpha}(t-\tau)^\alpha\right)d\tau\right|,$$

$$
\begin{aligned}
&\leq \frac{1}{1-\alpha}\left|\int_{t_0}^{t} \frac{f'(\tau)}{\min\limits_{l\in[t_0,\tau]}|g'(l)|} E_\alpha\left(-\frac{\alpha}{1-\alpha}(t-\tau)^\alpha\right) d\tau\right|, \\
&\leq \frac{1}{(1-\alpha)\, g'(t)}\left|\int_{t_0}^{t} f'(\tau) E_\alpha\left(-\frac{\alpha}{1-\alpha}(t-\tau)^\alpha\right) d\tau\right|, \\
&\leq \frac{1}{(1-\alpha)\, M_1}\left(\int_{t_0}^{t} \left|\frac{f'(\tau)+h(\tau)}{2}+\frac{f'(\tau)-h(\tau)}{2}\right|^2 d\tau\right)^{\frac{1}{2}} \\
&\times \left(\int_{t_0}^{t} \left(E_\alpha\left(-\frac{\alpha}{1-\alpha}(t-\tau)^\alpha\right)\right)^2 d\tau\right)^{\frac{1}{2}}, \\
&\leq \frac{1}{(1-\alpha)\, M_1}\left(\int_{t_0}^{t} \left|\frac{f'(\tau)+h(\tau)}{2}+\frac{f'(\tau)-h(\tau)}{2}\right|^2 d\tau\right)^{\frac{1}{2}} \\
&\times \left\| E_\alpha\left(-\frac{\alpha}{1-\alpha}(t-t_0)^\alpha\right)\right\|_{L^2[a,b]}, \\
&\leq \frac{1}{(1-\alpha)\, M_1}\left\{\frac{\left\|f'+h\right\|_{L^2[a,b]}}{2}+\frac{\left\|f'-h\right\|_{L^2[a,b]}}{2}\right\} \\
&\times \left\| E_\alpha\left(-\frac{\alpha}{1-\alpha}(t-t_0)^\alpha\right)\right\|_{L^2[a,b]},
\end{aligned}
$$

By Clarkson's inequality

$$
\left|{}_{t_0}^{CM}D_g^\alpha f(t)\right| < \frac{1}{(1-\alpha)M_1}\left\{\frac{\left\|f'\right\|_{L^2[a,b]}}{2}+\frac{\|h\|_{L^2[a,b]}}{2}\right\}\left\| E_\alpha\left(-\frac{\alpha}{1-\alpha}(t-t_0)^\alpha\right)\right\|_{L^2[a,b]}, \tag{4.180}
$$

which completes the proof.

Corollary 4.24 *Let f be twice differentiable on $[a,b]$ such f'' and $f \in L^2[a,b]$. Then if $\frac{1}{g'^2}$ is bounded then*

$$
\left|{}_{t_0}^{CM}D_g^\alpha f(t)\right| < \frac{1}{(1-\alpha)\, Mg'}\left\| E_\alpha\left(-\frac{\alpha}{1-\alpha}(t-t_0)^\alpha\right)\right\|_{L^2[a,b]} K\,\|f\|_{L^2[a,b]}^{\frac{1}{2}}\,\|f''\|_{L^2[a,b]}^{\frac{1}{2}}. \tag{4.181}
$$

Proof Let f'' and f be in $L^2[a,b]$

$$\left|{}^{CM}_{t_0}D^{\alpha}_{g}f(t)\right| = \frac{1}{(1-\alpha)}\left|\int_{t_0}^{t} D_g f(\tau)E_\alpha\left(-\frac{\alpha}{1-\alpha}(t-\tau)^\alpha\right)d\tau\right|, \tag{4.182}$$

$$= \frac{1}{(1-\alpha)}\left|\int_{t_0}^{t} \frac{f'(\tau)}{g'(\tau)}E_\alpha\left(-\frac{\alpha}{1-\alpha}(t-\tau)^\alpha\right)d\tau\right|,$$

$$\leq \frac{1}{(1-\alpha)}\left(\int_{t_0}^{t}\left|\frac{f'(\tau)}{g'(\tau)}\right|^2 d\tau\right)^{\frac{1}{2}}\left(\int_{t_0}^{t}\left(E_\alpha\left(-\frac{\alpha}{1-\alpha}(t-\tau)^\alpha\right)\right)^2 d\tau\right)^{\frac{1}{2}},$$

$$\leq \frac{1}{(1-\alpha)}\left(\int_{t_0}^{t}\frac{1}{\inf\limits_{l\in[t_0,\tau]}|g'(l)|^2}\left|f'(\tau)\right|^2 d\tau\right)^{\frac{1}{2}}\left\|E_\alpha\left(-\frac{\alpha}{1-\alpha}(t-t_0)^\alpha\right)\right\|_{L^2[a,b]},$$

$$\leq \frac{1}{(1-\alpha)M_{g'}}\left(\int_{t_0}^{t}\left|f'\right|^2 d\tau\right)^{\frac{1}{2}}\left\|E_\alpha\left(-\frac{\alpha}{1-\alpha}(t-t_0)^\alpha\right)\right\|_{L^2[a,b]},$$

$$\leq \frac{1}{(1-\alpha)M_{g'}}\left\|f'\right\|_{L^2[a,b]}\left\|E_\alpha\left(-\frac{\alpha}{1-\alpha}(t-t_0)^\alpha\right)\right\|_{L^2[a,b]}.$$

Using the Landau–Kolmogorov inequality we have

$$\left|{}^{CM}_{t_0}D^{\alpha}_{g}f(t)\right| < \frac{1}{(1-\alpha)Mg'}\left\|E_\alpha\left(-\frac{\alpha}{1-\alpha}(t-t_0)^\alpha\right)\right\|_{L^2[a,b]} K\left\|f\right\|^{\beta}_{L^2[a,b]}\left\|f''\right\|^{1-\beta}_{L^2[a,b]}, \tag{4.183}$$

which completes the proof.

References

1. Atangana, A., Goufo, E.F.D.: Modern and generalized analysis of exogenous growth models. Chaos, Solitons & Fractals **163**, 112605 (2022)
2. Scalas, E., Gorenflo, R., Mainardi, F.: Global fractional differentiation for non-Markovian random walks. Phys. A: Stat. Mech. Appl. **284**(1–4), 376–384 (2000)
3. Petras, Ivo, Dziatkiewicz, Jolanta: Global fractional derivatives of electrochemical processes. J. Electrochem. Soc. **158**(11), E147–E150 (2011)
4. Podlubny, I., Trinh, H.X., Petras, O.S.: Numerical solution of fractional differential equations with global fractional derivative. Fract. Calcul. Appl. Anal. **17**(1), 10–29 (2014)
5. Ahmad, B., Nieto, J.J.: On global fractional differential equations. Comput. & Math. Appl. **62**(3), 1351–1356 (2011)
6. Li, F., Duan, J.: Global fractional derivative and controllability of partial neutral functional integrodifferential systems with impulsive conditions. Abstr. Appl. Anal. 2014, Article ID 493647 (2014)

Chapter 5
Inequalities Associated to Integrals

Integral equations are extremely important in the modelling of real-world problems. These equations are at the heart of the understanding of the dynamics of many systems and play a key role in the development of theoretical tools for the study of them. Integral equations, in particular, are central to the theory of existence and uniqueness of solutions, i.e., the problem is well posed and the solution can be meaningfully defined. Additionally, their theory is at the heart of numerical analysis, being central to the design and theoretical analysis of computational methods for approximating solutions.

Inequalities related to integral equations are of the foremost importance for both of these theories-existence and uniqueness of solutions (in other words, any solution which satisfies the integral expression must equal the actual solution) and numerical analysis together. Inequalities play crucial role giving bounds and estimates which can be used to perform rigorous analysis on problems and systems and prove the viability of solutions and algorithms. This chapter will hence focus on some of the important inequalities related to integral equations found in the literature [1–8] and raise new developments in this domain, contributing to the reasonable attention that this topic deserves, as a step forward to addressing the above inequalities.

We note that we will start with inequalities that are established in a power-law setting from which many fractional and generalized integral operators are constructed. This juxtaposition will serve not only to clarify the classical theory in an introductory sequence of text, but also, in sections to come, a preamble to the extension of these concepts to more general fractional integral operators. We intend to use this systematic study to highlight the importance of inequalities in the development of integral equations theory and its applications.

A. Atangana and İ. Koca, *Fractional Differential and Integral Operators with Respect to a Function*, Industrial and Applied Mathematics,
https://doi.org/10.1007/978-981-97-9951-0_5

5.1 Inequalities with Caputo-Type Global Integral

We shall recall that if $g(t)$ is differentiable on $[a, b]$ then

$$ {}_{t_0}^{P}J_g^{\alpha} f(t) = \frac{1}{\Gamma(\alpha)} \int_{t_0}^{t} g^{'}(\tau) f(\tau)(t-\tau)^{\alpha-1} d\tau. \tag{5.1} $$

Corollary 5.1 *Let f be monotonically increasing on $[a, b]$ such that there exist two real numbers m and M such that then*

$$ m \leq f(t) \leq M, \tag{5.2} $$

then following is satisfied

$$ m_{t_0}^{P}J_g^{\alpha}(1) \leq_{t_0}^{P} J_g^{\alpha} f(t) \leq M_{t_0}^{P}J_g^{\alpha}(1). \tag{5.3} $$

Proof Let f be monotically increasing on $[a, b]$ such that there exist m and M verifying $\forall t \in [a, b]$,

$$ m \leq f(t) \leq M. \tag{5.4} $$

Since $g'(t)$ is positive

$$ g'(t)m \leq g'(t)f(t) \leq g'(t)M. \tag{5.5} $$

Integrating on both sides

$$ \begin{aligned} \frac{m}{\Gamma(\alpha)} \int_{t_0}^{t} g^{'}(\tau)(t-\tau)^{\alpha-1} d\tau \quad &\leq \quad \frac{1}{\Gamma(\alpha)} \int_{t_0}^{t} g^{'}(\tau) f(\tau)(t-\tau)^{\alpha-1} d\tau, \\ &\leq \quad \frac{M}{\Gamma(\alpha)} \int_{t_0}^{t} g^{'}(\tau) f(\tau)(t-\tau)^{\alpha-1} d\tau. \end{aligned} \tag{5.6} $$

So we have

$$ m_{t_0}^{P}J_g^{\alpha}(1) \leq_{t_0}^{P} J_g^{\alpha} f(t) \leq M_{t_0}^{P}J_g^{\alpha}(1). \tag{5.7} $$

If in addition if m and M are positive $2\alpha - 1 > 0$ and $g' \in L^2[a, b]$, then

$$ \frac{m}{\Gamma(\alpha)} \left(\int_{t_0}^{t} \left(g^{'}(\tau) \right)^2 d\tau \right)^{\frac{1}{2}} \left(\int_{t_0}^{t} (t-\tau)^{2\alpha-2} d\tau \right)^{\frac{1}{2}} \quad \leq \quad {}_{t_0}^{P}J_g^{\alpha} f(t) \tag{5.8} $$

$$\leq \frac{M}{\Gamma(\alpha)}\left(\int_{t_0}^{t}\left(g'(\tau)\right)^2 d\tau\right)^{\frac{1}{2}} \times \left(\int_{t_0}^{t}(t-\tau)^{2\alpha-2}d\tau\right)^{\frac{1}{2}},$$

$$m\left\|g'\right\|_{L^2[a,b]}\left(\frac{(t-t_0)^{2\alpha-1}}{2\alpha-1}\right)^{\frac{1}{2}} \leq {}_{t_0}^{P}J_g^{\alpha}f(t) \leq M\left\|g'\right\|_{L^2[a,b]} \times \left(\frac{(t-t_0)^{2\alpha-1}}{2\alpha-1}\right)^{\frac{1}{2}}.$$

So we have

$$m \leq \frac{1}{\|g'\|_{L^2[a,b]}\left(\frac{(t-t_0)^{2\alpha-1}}{2\alpha-1}\right)^{\frac{1}{2}}}\, {}_{t_0}^{P}J_g^{\alpha}f(t) \leq M, \tag{5.9}$$

which completes the proof.

Corollary 5.2 *Let f and $g' \in L^2[a,b]$ then if $2\alpha-1>0$, we will have*

$$\left|{}_{t_0}^{P}J_g^{\alpha}f(t)\right| \leq \frac{1}{\Gamma(\alpha)}\left(\frac{(t-t_0)^{2\alpha-1}}{2\alpha-1}\right)^{\frac{1}{2}}\|f\|_{L^2[a,b]}\left\|g'\right\|_{L^2[a,b]}. \tag{5.10}$$

Proof We put definition of ${}_{t_0}^{P}J_g^{\alpha}f(t)$ below

$$\left|{}_{t_0}^{P}J_g^{\alpha}f(t)\right| = \left|\frac{1}{\Gamma(\alpha)}\int_{t_0}^{t}f(\tau)g'(\tau)(t-\tau)^{\alpha-1}d\tau\right|, \tag{5.11}$$

$$\leq \frac{1}{\Gamma(\alpha)}\left(\int_{t_0}^{t}\left|f(\tau)g'(\tau)\right|^2 d\tau\right)^{\frac{1}{2}}\left(\int_{t_0}^{t}(t-\tau)^{2\alpha-2}d\tau\right)^{\frac{1}{2}},$$

$$\leq \frac{1}{\Gamma(\alpha)}\left(\frac{(t-t_0)^{2\alpha-1}}{2\alpha-1}\right)^{\frac{1}{2}}\left(\int_{t_0}^{t}\left|f(\tau)g'(\tau)\right|^2 d\tau\right)^{\frac{1}{2}},$$

$$\leq \frac{1}{\Gamma(\alpha)}\left(\frac{(t-t_0)^{2\alpha-1}}{2\alpha-1}\right)^{\frac{1}{2}}\left(\int_{t_0}^{t}|f(\tau)|^2 d\tau\int_{t_0}^{t}\left|g'(\tau)\right|^2 d\tau\right)^{\frac{1}{2}},$$

$$\left|{}_{t_0}^{P}J_g^{\alpha}f(t)\right| \leq \frac{1}{\Gamma(\alpha)}\left(\frac{(t-t_0)^{2\alpha-1}}{2\alpha-1}\right)^{\frac{1}{2}}\|f\|_{L^2[a,b]}\left\|g'\right\|_{L^2[a,b]}$$

which completes the proof. Noting that this was achieved using Hölder inequality and Cauchy–Schwarz inequality.

Corollary 5.3 *Let f, g', and h be L^p, then assuming that $2 \leq p \leq \infty$, if $([a,b], \Sigma, \mu)$ in measurable space then*

$$\begin{aligned} \left|{}_{t_0}^{P}J_g^{\alpha} f(t)\right| &\leq \frac{1}{2\Gamma(\alpha)}\left(\frac{(t-t_0)^{2\alpha-1}}{2\alpha-1}\right)^{\frac{1}{2}} \\ &\quad \times \left\{\|f\|^2_{L^2[a,b]} + \|h\|^2_{L^2[a,b]}\right\}\left\{\left\|g'\right\|^2_{L^2[a,b]} + \|h\|^2_{L^2[a,b]}\right\}. \end{aligned} \tag{5.12}$$

Proof Let f, g', and h be L^2 and we put definition of ${}_{t_0}^{P}J_g^{\alpha} f(t)$ below

$$\begin{aligned} \left|{}_{t_0}^{P}J_g^{\alpha} f(t)\right| &= \left|\frac{1}{\Gamma(\alpha)}\int_{t_0}^{t} f(\tau)g'(\tau)(t-\tau)^{\alpha-1}d\tau\right|, \\ &\leq \frac{1}{\Gamma(\alpha)}\left(\int_{t_0}^{t}\left|f(\tau)g'(\tau)\right|^2 d\tau\right)^{\frac{1}{2}}\left(\int_{t_0}^{t}(t-\tau)^{2\alpha-2}d\tau\right)^{\frac{1}{2}}, \\ &\leq \frac{1}{\Gamma(\alpha)}\left(\frac{(t-t_0)^{2\alpha-1}}{2\alpha-1}\right)^{\frac{1}{2}}\left(\int_{t_0}^{t}\left|f(\tau)g'(\tau)\right|^2 d\tau\right)^{\frac{1}{2}}, \\ &\leq \frac{1}{\Gamma(\alpha)}\left(\frac{(t-t_0)^{2\alpha-1}}{2\alpha-1}\right)^{\frac{1}{2}}\left(\int_{t_0}^{t}|f(\tau)|^2 d\tau\int_{t_0}^{t}\left|g'(\tau)\right|^2 d\tau\right)^{\frac{1}{2}}, \\ &\leq \frac{1}{\Gamma(\alpha)}\left(\frac{(t-t_0)^{2\alpha-1}}{2\alpha-1}\right)^{\frac{1}{2}}\left\{\begin{array}{l}\left(\int_{t_0}^{t}\left|\frac{f(\tau)+h(\tau)}{2}+\frac{f(\tau)-h(\tau)}{2}\right|^2 d\tau\right) \\ \times\left(\int_{t_0}^{t}\left|\frac{g'(\tau)+h(\tau)}{2}+\frac{g'(\tau)-h(\tau)}{2}\right|^2 d\tau\right)\end{array}\right\}^{\frac{1}{2}}, \\ &\leq \frac{1}{\Gamma(\alpha)}\left(\frac{(t-t_0)^{2\alpha-1}}{2\alpha-1}\right)^{\frac{1}{2}}\left\{\left\|\frac{f-h}{2}\right\|^2_{L^2[a,b]} + \left\|\frac{f+h}{2}\right\|^2_{L^2[a,b]}\right\}^{\frac{1}{2}} \\ &\quad \times\left\{\left\|\frac{g'-h}{2}\right\|^2_{L^2[a,b]} + \left\|\frac{g'+h}{2}\right\|^2_{L^2[a,b]}\right\}^{\frac{1}{2}}. \end{aligned} \tag{5.13}$$

Using Clarkson's inequality we obtain

$$\begin{aligned} \left|{}_{t_0}^{P}J_g^{\alpha} f(t)\right| &\leq \frac{1}{2\Gamma(\alpha)}\left(\frac{(t-t_0)^{2\alpha-1}}{2\alpha-1}\right)^{\frac{1}{2}} \\ &\quad \times \left\{\|f\|^2_{L^2[a,b]} + \|h\|^2_{L^2[a,b]}\right\}^{\frac{1}{2}}\left\{\left\|g'\right\|^2_{L^2[a,b]} + \|h\|^2_{L^2[a,b]}\right\}^{\frac{1}{2}}, \end{aligned}$$

which completes the proof.

Corollary 5.4 *Let f be a nonnegative function defined on a closed ball of $\mathbb{R}$ with ratio R with center x_0. It is assumed that f is harmonic on the interior on the ball and continous on the closed ball then if $\forall x;\ |x - x_0| < r < R$ then we have*

$$\left(1 - \frac{r}{R}\right) \leq \frac{{}^{P}_{x_0}J^{\alpha}_{g} f(x)}{f(x_0) {}^{P}_{x_0}J^{\alpha}_{g}(1)} \leq \left(1 + \frac{r}{R}\right). \tag{5.14}$$

Proof Let us assume the hypothesis on the corollary holds, then by Harnack's inequality we have $\forall x;\ |x - x_0| < r < R$

$$\left(1 - \frac{r}{R}\right) f(x_0) \leq f(x) \leq \left(1 + \frac{r}{R}\right) f(x_0). \tag{5.15}$$

Since $\forall x$ in the ball, $g'(x) > 0$ then we multiply on both sides with $g'(x)$

$$g'(x)\left(1 - \frac{r}{R}\right) f(x_0) \leq g'(x) f(x) \leq \left(1 + \frac{r}{R}\right) f(x_0) g'(x), \tag{5.16}$$

then integrate on both sides

$$\begin{aligned} \frac{1}{\Gamma(\alpha)} \int_{x_0}^{x} (x-\tau)^{\alpha-1} g'(\tau)\left(1 - \frac{r}{R}\right) f(x_0) d\tau &\leq {}^{P}_{x_0}J^{\alpha}_{g} f(x) \\ &\leq \frac{1}{\Gamma(\alpha)} \int_{x_0}^{x} (x-\tau)^{\alpha-1} g'(\tau)\left(1 + \frac{r}{R}\right) f(x_0) d\tau, \end{aligned} \tag{5.17}$$

$$\begin{aligned} f(x_0)\left(1 - \frac{r}{R}\right) \frac{1}{\Gamma(\alpha)} \int_{x_0}^{x} (x-\tau)^{\alpha-1} g'(\tau) d\tau &\leq {}^{P}_{x_0}J^{\alpha}_{g} f(x) \\ &\leq f(x_0)\left(1 + \frac{r}{R}\right) \frac{1}{\Gamma(\alpha)} \int_{x_0}^{x} (x-\tau)^{\alpha-1} g'(\tau) d\tau, \\ f(x_0)\left(1 - \frac{r}{R}\right) {}^{P}_{x_0}J^{\alpha}_{g}(1) &\leq {}^{P}_{x_0}J^{\alpha}_{g} f(x) \leq f(x_0)\left(1 + \frac{r}{R}\right) {}^{P}_{x_0}J^{\alpha}_{g}(1), \end{aligned} \tag{5.18}$$

$$\left(1 - \frac{r}{R}\right) \leq \frac{{}^{P}_{x_0}J^{\alpha}_{g} f(x)}{f(x_0) {}^{P}_{x_0}J^{\alpha}_{g}(1)} \leq \left(1 + \frac{r}{R}\right), \tag{5.19}$$

which completes the proof.

Corollary 5.5 *Let $g, g'' \in L^p[a, b]$ and f be bounded in $[a, b]$, then*

$$\left| {}^{P}_{t_0}J^{\alpha}_{g} f(t) \right| \leq \frac{MK}{\Gamma(\alpha)} \|g\|^{\beta}_{L^2[a,b]} \left\|g''\right\|^{1-\beta}_{L^2[a,b]}. \tag{5.20}$$

Proof By the definition of ${}_{t_0}^{P}J_g^{\alpha} f(t)$

$$\begin{aligned}
\left|{}_{t_0}^{P}J_g^{\alpha} f(t)\right| &= \left|\frac{1}{\Gamma(\alpha)}\int_{t_0}^{t} f(\tau)g^{'}(\tau)(t-\tau)^{\alpha-1}d\tau\right|, \qquad (5.21)\\
&\leq \frac{1}{\Gamma(\alpha)}\int_{t_0}^{t} |f(\tau)|\left|g^{'}(\tau)\right|(t-\tau)^{\alpha-1}d\tau,\\
&\leq \frac{1}{\Gamma(\alpha)}\int_{t_0}^{t} \sup_{l\in[t_0,\tau]}|f(l)|\left|g^{'}(\tau)\right|(t-\tau)^{\alpha-1}d\tau,\\
&\leq \frac{M}{\Gamma(\alpha)}\int_{t_0}^{t}\left|g^{'}(\tau)\right|(t-\tau)^{\alpha-1}d\tau,\\
&\leq \frac{M}{\Gamma(\alpha)}\left(\int_{t_0}^{t}\left|g^{'}(\tau)\right|^2 d\tau\right)^{\frac{1}{2}}\left(\int_{t_0}^{t}(t-\tau)^{2\alpha-2}d\tau\right)^{\frac{1}{2}},\\
&\leq \frac{M}{\Gamma(\alpha)}\left\|g^{'}\right\|_{L^2[a,b]}\left(\frac{(t-t_0)^{2\alpha-1}}{2\alpha-1}\right)^{\frac{1}{2}}.
\end{aligned}$$

By the Landau–Kolmogorov inequality there exist K, β positive constants such that

$$\left|{}_{t_0}^{P}J_g^{\alpha} f(t)\right| \leq \frac{MK}{\Gamma(\alpha)}\|g\|_{L^2[a,b]}^{\beta}\left\|g''\right\|_{L^2[a,b]}^{1-\beta} \qquad (5.22)$$

which completes the proof.

Lemma 5.1 *Let f be continuously differentiable on $[a,b]$. Let $g'(t)$ be continuously bounded then if $f'(t)$ is bounded and $(2\alpha-1)>0$ then*

$$\left|{}_{t_0}^{P}J_g^{\alpha} f(t)\right|^2 \leq \frac{1}{\Gamma^2(\alpha)}\left(\frac{(t-t_0)^{2\alpha-1}}{2\alpha-1}\right)\|f\|_{L^2[a,b]}^2\left\|g^{'}\right\|_{L^2[a,b]}^2. \qquad (5.23)$$

Proof From the definition

$$\begin{aligned}
\left|{}_{t_0}^{P}J_g^{\alpha} f(t)\right|^2 &= \left|\frac{1}{\Gamma(\alpha)}\int_{t_0}^{t} f(\tau)g^{'}(\tau)(t-\tau)^{\alpha-1}d\tau\right|^2, \qquad (5.24)\\
&\leq \frac{1}{\Gamma^2(\alpha)}\left(\int_{t_0}^{t}\left|f(\tau)g^{'}(\tau)\right|^2 d\tau\right)\left(\int_{t_0}^{t}(t-\tau)^{2\alpha-2}d\tau\right),
\end{aligned}$$

$$\leq \frac{1}{\Gamma^2(\alpha)}\left(\frac{(t-t_0)^{2\alpha-1}}{2\alpha-1}\right)\left(\int_{t_0}^{t}|f(\tau)|^2\,d\tau\right)\left(\int_{t_0}^{t}\left|g^{'}(\tau)\right|^2 d\tau\right),$$

$$\leq \frac{1}{\Gamma^2(\alpha)}\left(\frac{(t-t_0)^{2\alpha-1}}{2\alpha-1}\right)\|f\|^2_{L^2[a,b]}\left\|g^{'}\right\|^2_{L^2[a,b]},$$

with condition $(2\alpha-1)>0$, which completes the proof.

5.2 Inequalities with Caputo–Fabrizio-Type Global Integral

We shall recall that the Caputo/indexCaputo-type integral is given as

$$ {}^{E}_{t_0}J^{\alpha}_{g}f(t) = (1-\alpha)g^{'}(t)f(t) + \alpha\int_{t_0}^{t} f(\tau)g^{'}(\tau)d\tau. \tag{5.25}$$

Corollary 5.6 *Let f be monotonically increasing on $[a,b]$, such that there exist two real values m and M such that then*

$$m \leq f(t) \leq M. \tag{5.26}$$

Then we have

$$m \leq \frac{{}^{E}_{t_0}J^{\alpha}_{g}f(t)}{(1-\alpha)g^{'}(t)+\alpha\left(g(t)-g(t_0)\right)} \leq M. \tag{5.27}$$

Proof If $\forall t \in [a,b]$, we have

$$m \leq f(t) \leq M \tag{5.28}$$

then since $g^{'}(t) > 0$ then

$$\begin{aligned} g^{'}(t)m &\leq g^{'}(t)f(t) \leq g^{'}(t)M, \\ \int_{t_0}^{t} g^{'}(\tau)m\,d\tau &\leq \int_{t_0}^{t} g^{'}(\tau)f(\tau)d\tau \leq \int_{t_0}^{t} g^{'}(\tau)M\,d\tau, \\ \alpha m\int_{t_0}^{t} g^{'}(\tau)d\tau &\leq \alpha\int_{t_0}^{t} g^{'}(\tau)f(\tau)d\tau \leq \alpha\int_{t_0}^{t} g^{'}(\tau)M\,d\tau, \end{aligned} \tag{5.29}$$

on the other hand

$$\begin{aligned} g^{'}(t)m &\le g^{'}(t)f(t) \le g^{'}(t)M, \\ (1-\alpha)g^{'}(t)m &\le (1-\alpha)g^{'}(t)f(t) \le (1-\alpha)g^{'}(t)M. \end{aligned} \tag{5.30}$$

Then by simple addition, we have

$$\begin{aligned} (1-\alpha)g^{'}(t)m + \alpha\int_{t_0}^{t} g^{'}(\tau)md\tau &\le {}_{t_0}^{E}J_g^{\alpha} f(t) \le (1-\alpha)g^{'}(t)M + \alpha\int_{t_0}^{t} g^{'}(\tau)Md\tau, \quad (5.31) \\ (1-\alpha)g^{'}(t)m + \alpha(g(t)-g(t_0))m &\le {}_{t_0}^{E}J_g^{\alpha} f(t) \le (1-\alpha)g^{'}(t)M + \alpha(g(t)-g(t_0))M. \end{aligned}$$

Therefore

$$m \le \frac{{}_{t_0}^{E}J_g^{\alpha} f(t)}{(1-\alpha)g^{'}(t) + \alpha\left(g(t)-g(t_0)\right)} \le M, \tag{5.32}$$

which completes the proof.

If in addition if m and M are positive

$$\begin{aligned} &(1-\alpha)g^{'}(t)m + \alpha\int_{t_0}^{t} g^{'}(\tau)md\tau \qquad (5.33) \\ &\le {}_{t_0}^{E}J_g^{\alpha} f(t) \\ &\le (1-\alpha)g^{'}(t)M + \alpha\int_{t_0}^{t} g^{'}(\tau)Md\tau, \end{aligned}$$

$$\begin{aligned} &(1-\alpha)g^{'}(t)m + \alpha m\left(\int_{t_0}^{t}\left(g^{'}(\tau)\right)^2 d\tau\right)^{\frac{1}{2}}\left(\int_{t_0}^{t}(1)^2 d\tau\right)^{\frac{1}{2}} \\ &\le {}_{t_0}^{E}J_g^{\alpha} f(t) \\ &\le (1-\alpha)g^{'}(t)M + \alpha M\left(\int_{t_0}^{t}\left(g^{'}(\tau)\right)^2 d\tau\right)^{\frac{1}{2}}\left(\int_{t_0}^{t}(1)^2 d\tau\right)^{\frac{1}{2}}, \end{aligned}$$

$$\begin{aligned} &(1-\alpha)m\left\|g'\right\|_{\infty} + \alpha m\left\|g'\right\|_{L^2[a,b]}\sqrt{t-t_0} \\ &\le {}_{t_0}^{E}J_g^{\alpha} f(t) \\ &\le (1-\alpha)M\left\|g'\right\|_{\infty} + \alpha M\left\|g'\right\|_{L^2[a,b]}\sqrt{t-t_0}, \end{aligned}$$

$$m \le \frac{{}_{t_0}^{E}J_g^{\alpha} f(t)}{(1-\alpha)\left\|g'\right\|_{\infty} + \alpha\left\|g'\right\|_{L^2[a,b]}\sqrt{t-t_0}} \le M. \tag{5.34}$$

Corollary 5.7 *Let f and $g' \in L^2[a,b]$ and then g' and f are bounded in [a,b] there exist M_g, with $|g'| < M_g$ and M, with $|f| < M$. Then we have*

$$\left|{}_{t_0}^{E}J_g^{\alpha} f(t)\right| \leq (1-\alpha)\left\|g'\right\|_{\infty}\|f\|_{\infty} + \alpha\|f\|_{L^2[a,b]}\left\|g'\right\|_{L^2[a,b]}. \tag{5.35}$$

Proof We put definition of ${}_{t_0}^{E}J_g^{\alpha} f(t)$ below

$$\begin{aligned}
\left|{}_{t_0}^{E}J_g^{\alpha} f(t)\right| &\leq \left|(1-\alpha)g'(t)f(t) + \alpha\int_{t_0}^{t} f(\tau)g'(\tau)d\tau\right|, \qquad (5.36)\\
&\leq \left|(1-\alpha)g'(t)f(t)\right| + \left|\alpha\int_{t_0}^{t} f(\tau)g'(\tau)d\tau\right|,\\
&\leq (1-\alpha)\sup_{t\in[a,b]}\left|g'(t)\right|\sup_{t\in[a,b]}|f(t)|\\
&\quad + \alpha\left(\int_{t_0}^{t}|f(\tau)|^2 d\tau\right)^{\frac{1}{2}}\left(\int_{t_0}^{t}\left|g'(\tau)\right|^2 d\tau\right)^{\frac{1}{2}},\\
&\leq (1-\alpha)\left\|g'\right\|_{\infty}\|f\|_{\infty} + \alpha\|f\|_{L^2[a,b]}\left\|g'\right\|_{L^2[a,b]}
\end{aligned}$$

which completes the proof. Noting that this was achieved using Hölder inequality.

Corollary 5.8 *Let f, g' and h be L^p, then assuming that $2 \leq p \leq \infty$, if $([a,b], \Sigma, \mu)$ in measurable space then*

$$\begin{aligned}
\left|{}_{t_0}^{E}J_g^{\alpha} f(t)\right| &\leq (1-\alpha)\left\|g'\right\|_{\infty}\|f\|_{\infty} \qquad (5.37)\\
&\quad + \frac{\alpha}{2}\left\{\|f\|_{L^2[a,b]}^2 + \|h\|_{L^2[a,b]}^2\right\}^{\frac{1}{2}}\left\{\left\|g'\right\|_{L^2[a,b]}^2 + \|h\|_{L^2[a,b]}^2\right\}^{\frac{1}{2}}.
\end{aligned}$$

Proof Let f, g', and h be L^2 and we put definition of ${}_{t_0}^{E}J_g^{\alpha} f(t)$ below

$$\begin{aligned}
\left|{}_{t_0}^{E}J_g^{\alpha} f(t)\right| &\leq \left|(1-\alpha)g'(t)f(t) + \alpha\int_{t_0}^{t} f(\tau)g'(\tau)d\tau\right|, \qquad (5.38)\\
&\leq (1-\alpha)\left|g'(t)\right||f(t)| + \alpha\left(\int_{t_0}^{t}|f(\tau)|^2 d\tau\right)^{\frac{1}{2}}\left(\int_{t_0}^{t}\left|g'(\tau)\right|^2 d\tau\right)^{\frac{1}{2}},\\
&\leq (1-\alpha)\sup_{t\in[a,b]}\left|g'(t)\right|\sup_{t\in[a,b]}|f(t)|
\end{aligned}$$

$$+\alpha\left\{\begin{array}{l}\left(\int\limits_{t_0}^{t}\left|\frac{f(\tau)+h(\tau)}{2}+\frac{f(\tau)-h(\tau)}{2}\right|^2 d\tau\right)\\ \times\left(\int\limits_{t_0}^{t}\left|\frac{g'(\tau)+h(\tau)}{2}+\frac{g'(\tau)-h(\tau)}{2}\right|^2 d\tau\right)\end{array}\right\}^{\frac{1}{2}},$$

$$\leq (1-\alpha)\left\|g'\right\|_\infty \|f\|_\infty+\alpha\left\{\left\|\frac{f-h}{2}\right\|^2_{L^2[a,b]}+\left\|\frac{f+h}{2}\right\|^2_{L^2[a,b]}\right\}^{\frac{1}{2}}$$

$$\times\left\{\left\|\frac{g'-h}{2}\right\|^2_{L^2[a,b]}+\left\|\frac{g'+h}{2}\right\|^2_{L^2[a,b]}\right\}^{\frac{1}{2}}.$$

Using Clarkson's inequality we obtain

$$\leq (1-\alpha)\left\|g'\right\|_\infty \|f\|_\infty+\frac{\alpha}{2}\left\{\|f\|^2_{L^2[a,b]}+\|h\|^2_{L^2[a,b]}\right\}^{\frac{1}{2}}\left\{\|g'\|^2_{L^2[a,b]}+\|h\|^2_{L^2[a,b]}\right\}^{\frac{1}{2}}, \tag{5.39}$$

which completes the proof.

Corollary 5.9 *Let f be a nonnegative function defined on a closed ball of $\mathbb{R}$ with ratio R with center x_0. It is assumed that f is harmonic on the interior on the ball and continuous on the closed ball then if $\forall x;\ |x-x_0|<r<R$ then we have*

$$\left(1-\frac{r}{R}\right)\leq\frac{{}^{E}_{x_0}J^{\alpha}_{g}f(x)}{f(x_0){}^{P}_{x_0}J^{\alpha}_{g}(x_0)}\leq\left(1+\frac{r}{R}\right). \tag{5.40}$$

Proof Let us assume the hypothesis on the corollary holds, then by Harnack's inequality we have $\forall x;\ |x-x_0|<r<R$

$$\left(1-\frac{r}{R}\right)f(x_0)\leq f(x)\leq\left(1+\frac{r}{R}\right)f(x_0). \tag{5.41}$$

Since $\forall x$ in the ball, $g'(x)>0$ then we multiply on both sides with $(1-\alpha)\,g'(x)$

$$(1-\alpha)\,g'(x)\left(1-\frac{r}{R}\right)f(x_0)\leq(1-\alpha)\,g'(x)f(x)\leq(1-\alpha)\left(1+\frac{r}{R}\right)f(x_0)g'(x), \tag{5.42}$$

also let us write again

$$\left(1-\frac{r}{R}\right)f(x_0)\leq f(x)\leq\left(1+\frac{r}{R}\right)f(x_0), \tag{5.43}$$

then integrate on both sides with $\alpha\int\limits_{x_0}^{x}g'(\tau)d\tau$

$$\alpha \int_{x_0}^{x} \left(1 - \frac{r}{R}\right) f(x_0) g'(\tau) d\tau \leq \alpha \int_{x_0}^{x} f(x) g'(\tau) d\tau \leq \alpha \int_{x_0}^{x} \left(1 + \frac{r}{R}\right) f(x_0) g'(\tau) d\tau. \tag{5.44}$$

If we add two inequalities as below

$$(1-\alpha) g'(x) \left(1 - \frac{r}{R}\right) f(x_0) \leq (1-\alpha) g'(x) f(x) \leq (1-\alpha) \left(1 + \frac{r}{R}\right) f(x_0) g'(x), \tag{5.45}$$

$$\alpha \int_{x_0}^{x} \left(1 - \frac{r}{R}\right) f(x_0) g'(\tau) d\tau \leq \alpha \int_{x_0}^{x} f(x) g'(\tau) d\tau \leq \alpha \int_{x_0}^{x} \left(1 + \frac{r}{R}\right) f(x_0) g'(\tau) d\tau, \tag{5.46}$$

then we have

$$\begin{aligned} \left(1 - \frac{r}{R}\right) \left\{ (1-\alpha) g'(x) f(x_0) + \alpha \int_{x_0}^{x} f(x_0) g'(\tau) d\tau \right\} &\leq {}_{x_0}^{E} J_g^{\alpha} f(x) \qquad (5.47) \\ &\leq \left(1 + \frac{r}{R}\right) \left\{ \begin{array}{c} (1-\alpha) g'(x) f(x_0) \\ +\alpha \int_{x_0}^{x} f(x_0) g'(\tau) d\tau \end{array} \right\}, \\ \left(1 - \frac{r}{R}\right) {}_{x_0}^{E} J_g^{\alpha} f(x_0) &\leq {}_{x_0}^{E} J_g^{\alpha} f(x) \leq \left(1 + \frac{r}{R}\right) {}_{x_0}^{E} J_g^{\alpha} f(x_0), \\ \left(1 - \frac{r}{R}\right)_{x_0}^{E} &\leq \frac{{}_{x_0}^{E} J_g^{\alpha} f(x)}{J_g^{\alpha} f(x_0)} \leq \left(1 + \frac{r}{R}\right)_{x_0}^{E}, \end{aligned}$$

which completes the proof.

Corollary 5.10 *Let $g, g'' \in L^2[a,b]$ and f be bounded in $[a,b]$, then*

$$\left| {}_{t_0}^{E} J_g^{\alpha} f(t) \right| \leq (1-\alpha) \left\| g' \right\|_{\infty} M + \alpha M K \left\| g \right\|_{L^2[a,b]}^{\beta} \left\| g'' \right\|_{L^2[a,b]}^{1-\beta}. \tag{5.48}$$

Proof By the definition of ${}_{t_0}^{E} J_g^{\alpha} f(t)$

$$\begin{aligned} \left| {}_{t_0}^{E} J_g^{\alpha} f(t) \right| &\leq \left| (1-\alpha) g'(t) f(t) + \alpha \int_{t_0}^{t} f(\tau) g'(\tau) d\tau \right|, \qquad (5.49) \\ &\leq (1-\alpha) \left| g'(t) \right| \left| f(t) \right| + \alpha \int_{t_0}^{t} \left| f(\tau) \right| \left| g'(\tau) \right| d\tau, \end{aligned}$$

$$
\begin{aligned}
&\leq (1-\alpha)\sup_{t\in[a,b]}\left|g'(t)\right|\sup_{t\in[a,b]}|f(t)|+\alpha\int_{t_0}^{t}\sup_{l\in[t_0,\tau]}|f(l)|\left|g'(\tau)\right|d\tau,\\
&\leq (1-\alpha)\sup_{t\in[a,b]}\left|g'(t)\right|M+\alpha M\int_{t_0}^{t}\left|g'(\tau)\right|d\tau,\\
&\leq (1-\alpha)\left\|g'\right\|_{\infty}M+\alpha M\left(\int_{t_0}^{t}\left|g'(\tau)\right|^2 d\tau\right)^{\frac{1}{2}}\left(\int_{t_0}^{t}(1)^2 d\tau\right)^{\frac{1}{2}},\\
&\leq (1-\alpha)\left\|g'\right\|_{\infty}M+\alpha M\left\|g'\right\|_{L^2[a,b]}\sqrt{t-t_0}.
\end{aligned}
$$

By the Landau-Kolmogorov inequality there exist K, β positive constants such that

$$
\left|{}_{t_0}^{E}J_g^{\alpha}f(t)\right|\leq(1-\alpha)\left\|g'\right\|_{\infty}M+\alpha MK\left\|g\right\|_{L^2[a,b]}^{\beta}\left\|g''\right\|_{L^2[a,b]}^{1-\beta}\sqrt{t-t_0}, \tag{5.50}
$$

which completes the proof.

5.3 Inequalities with Mittag–Leffler-Type Global Integral

We present in this section inequalities associated to the global derivative with the Atangana–Baleanu integral. We recall that

$$
{}_{t_0}^{M}J_g^{\alpha}f(t)=(1-\alpha)f(t)g'(t)+\frac{\alpha}{\Gamma(\alpha)}\int_{t_0}^{t}f(\tau)g'(\tau)(t-\tau)^{\alpha-1}d\tau. \tag{5.51}
$$

Corollary 5.11 *Let f be monotonically increasing function such that $\forall t\in[a,b]$, there exist two real values m and M such that then*

$$
m\leq f(t)\leq M. \tag{5.52}
$$

Then we have

$$
m\leq\frac{{}_{t_0}^{M}J_g^{\alpha}f(t)}{{}_{t_0}^{M}J_g^{\alpha}(1)}\leq M. \tag{5.53}
$$

Proof Let the condition prescribed in the corollary holds then since $g'(t)>0$, $\forall t\in[a,b]$, we have

$$
g'(t)m\leq g'(t)f(t)\leq g'(t)M. \tag{5.54}
$$

If we take integrate on both sides

$$m\frac{\alpha}{\Gamma(\alpha)}\int_{t_0}^{t} g^{'}(\tau)(t-\tau)^{\alpha-1}d\tau \leq \frac{\alpha}{\Gamma(\alpha)}\int_{t_0}^{t} g^{'}(\tau)f(\tau)(t-\tau)^{\alpha-1}d\tau$$
$$\leq M\frac{\alpha}{\Gamma(\alpha)}\int_{t_0}^{t} g^{'}(\tau)(t-\tau)^{\alpha-1}d\tau. \tag{5.55}$$

Also we have

$$(1-\alpha)g^{'}(t)m \leq (1-\alpha)g^{'}(t)f(t) \leq (1-\alpha)g^{'}(t)M. \tag{5.56}$$

By simple addition, we have

$$\begin{aligned} m(1-\alpha)g^{'}(t)+m\frac{\alpha}{\Gamma(\alpha)}\int_{t_0}^{t} g^{'}(\tau)(t-\tau)^{\alpha-1}d\tau \quad &\leq \quad {}_{t_0}^{M}J_g^{\alpha}f(t) \qquad (5.57)\\ &\leq \quad M(1-\alpha)g^{'}(t)\\ &\quad +M\frac{\alpha}{\Gamma(\alpha)}\int_{t_0}^{t} g^{'}(\tau)(t-\tau)^{\alpha-1}d\tau,\\ m\left\{(1-\alpha)g^{'}(t)+\frac{\alpha}{\Gamma(\alpha)}\int_{t_0}^{t} g^{'}(\tau)(t-\tau)^{\alpha-1}d\tau\right\} \quad &\leq \quad {}_{t_0}^{M}J_g^{\alpha}f(t)\\ &\leq \quad M\Big\{(1-\alpha)g^{'}(t)\\ &\quad +\frac{\alpha}{\Gamma(\alpha)}\int_{t_0}^{t} g^{'}(\tau)(t-\tau)^{\alpha-1}d\tau\Big\} \end{aligned}$$

$$m{}_{t_0}^{M}J_g^{\alpha}(1) \leq {}_{t_0}^{M}J_g^{\alpha}f(t) \leq M{}_{t_0}^{M}J_g^{\alpha}(1). \tag{5.58}$$

Dividing yields

$$m \leq \frac{{}_{t_0}^{M}J_g^{\alpha}f(t)}{{}_{t_0}^{M}J_g^{\alpha}(1)} \leq M, \tag{5.59}$$

which completes the proof.

Corollary 5.12 *Let f be continuously differentiable on $[a, b]$. Let $g'(t)$ be continuously bounded then if $f'(t)$ is bounded and $(2\alpha - 1) > 0$ then*

Proof By the definition of ${}_{t_0}^{M}J_g^{\alpha} f(t)$

$$\begin{aligned}
\left|{}_{t_0}^{M}J_g^{\alpha} f(t)\right| &= \left|(1-\alpha)f(t)g^{'}(t)+\frac{\alpha}{\Gamma(\alpha)}\int_{t_0}^{t} f(\tau)g^{'}(\tau)(t-\tau)^{\alpha-1}d\tau\right|, \qquad (5.60)\\
&\leq (1-\alpha)\left|f(t)g^{'}(t)\right|+\frac{\alpha}{\Gamma(\alpha)}\left(\int_{t_0}^{t}\left|f(\tau)g^{'}(\tau)\right|^2 d\tau\right)^{\frac{1}{2}}\left(\int_{t_0}^{t}(t-\tau)^{2\alpha-2}d\tau\right)^{\frac{1}{2}},\\
&\leq (1-\alpha)\left|f(t)g^{'}(t)\right|+\frac{\alpha}{\Gamma(\alpha)}\left(\int_{t_0}^{t}|f(\tau)|^2 d\tau\right)^{\frac{1}{2}}\\
&\quad\times \|g\|_{\infty}\left(\frac{(t-t_0)^{2\alpha-1}}{2\alpha-1}\right)^{\frac{1}{2}},\\
&\leq (1-\alpha)\sup_{t\in[a,b]}|f(t)|\sup_{t\in[a,b]}\left|g^{'}(t)\right|+\frac{\alpha}{\Gamma(\alpha)}\left(\frac{(t-t_0)^{2\alpha-1}}{2\alpha-1}\right)^{\frac{1}{2}}\|f\|_{L^2[a,b]}\left\|g'\right\|_{\infty},\\
&\leq (1-\alpha)\|f\|_{\infty}\left\|g^{'}\right\|_{\infty}+\frac{\alpha}{\Gamma(\alpha)}\left(\frac{(t-t_0)^{2\alpha-1}}{2\alpha-1}\right)^{\frac{1}{2}}\|f\|_{L^2[a,b]}\left\|g'\right\|_{\infty},
\end{aligned}$$

if in addition g'' and g are in $L^p[a,b]$ then using Clarkson's inequality, we have

$$\left|{}_{t_0}^{M}J_g^{\alpha} f(t)\right| \leq (1-\alpha)\|f\|_{\infty}\left\|g^{'}\right\|_{\infty}+\frac{\alpha}{\Gamma(\alpha)}\|f\|_{L^2[a,b]}K\|g\|_{L^p[a,b]}^{\beta}\left\|g''\right\|_{L^q[a,b]}^{1-\beta}, \qquad (5.61)$$

which completes the proof.

Under the same conditions if

$$\begin{aligned}
\left|{}_{t_0}^{M}J_g^{\alpha} f(t)\right|^2 &= \left|(1-\alpha)f(t)g^{'}(t)+\frac{\alpha}{\Gamma(\alpha)}\int_{t_0}^{t} f(\tau)g^{'}(\tau)(t-\tau)^{\alpha-1}d\tau\right|^2, \qquad (5.62)\\
&\leq \left|(1-\alpha)f(t)g^{'}(t)\right|^2\\
&\quad+2(1-\alpha)\left|f(t)g^{'}(t)\right|\left|\frac{\alpha}{\Gamma(\alpha)}\int_{t_0}^{t} f(\tau)g^{'}(\tau)(t-\tau)^{\alpha-1}d\tau\right|\\
&\quad+\left|\frac{\alpha}{\Gamma(\alpha)}\int_{t_0}^{t} f(\tau)g^{'}(\tau)(t-\tau)^{\alpha-1}d\tau\right|^2,
\end{aligned}$$

$$
\begin{aligned}
&\leq \quad (1-\alpha)^2 \left| f(t) \right|^2 \left| g^{'}(t) \right|^2 \\
&\quad + 2(1-\alpha) \left| f(t) \right| \left| g^{'}(t) \right| \frac{\alpha}{\Gamma(\alpha)} \left(\int_{t_0}^{t} \left| f(\tau) g^{'}(\tau) \right|^2 d\tau \right)^{\frac{1}{2}} \left(\int_{t_0}^{t} (t-\tau)^{2\alpha-2} d\tau \right)^{\frac{1}{2}} \\
&\quad + \frac{\alpha^2}{\Gamma^2(\alpha)} \left(\int_{t_0}^{t} \left| f(\tau) g^{'}(\tau) \right|^2 d\tau \right) \left(\int_{t_0}^{t} (t-\tau)^{2\alpha-2} d\tau \right), \\
&\leq \quad (1-\alpha)^2 \sup_{t\in[a,b]} \left| f(t) \right|^2 \sup_{t\in[a,b]} \left| g^{'}(t) \right|^2 \\
&\quad + 2(1-\alpha) \sup_{t\in[a,b]} \left| f(t) \right| \sup_{t\in[a,b]} \left| g^{'}(t) \right| \frac{\alpha}{\Gamma(\alpha)} \left(\int_{t_0}^{t} \left| f(\tau) g^{'}(\tau) \right|^2 d\tau \right)^{\frac{1}{2}} \left(\int_{t_0}^{t} (t-\tau)^{2\alpha-2} d\tau \right)^{\frac{1}{2}} \\
&\quad + \frac{\alpha^2}{\Gamma^2(\alpha)} \left(\int_{t_0}^{t} \left| f(\tau) g^{'}(\tau) \right|^2 d\tau \right) \left(\int_{t_0}^{t} (t-\tau)^{2\alpha-2} d\tau \right), \\
&\leq (1-\alpha)^2 \left\| f \right\|_\infty^2 \left\| g^{'} \right\|_\infty^2 \\
&\quad + 2(1-\alpha) \left\| f \right\|_\infty \left\| g^{'} \right\|_\infty \frac{\alpha}{\Gamma(\alpha)} \left\| g^{'} \right\|_\infty \left\| f \right\|_{L^2[a,b]} \\
&\quad + \frac{\alpha^2}{\Gamma^2(\alpha)} \left\| g^{'} \right\|_{L^2[a,b]}^2 \left\| f \right\|_{L^2[a,b]}^2 .
\end{aligned} \tag{5.63}
$$

Therefore

$$
\begin{aligned}
\left| {}_{t_0}^{M} J_g^{\alpha} f(t) \right|^2 \quad &\leq \quad (1-\alpha)^2 \left\| f \right\|_\infty^2 \left\| g^{'} \right\|_\infty^2 + 2\frac{\alpha(1-\alpha)}{\Gamma(\alpha)} \left\| f \right\|_\infty \left\| g^{'} \right\|_\infty \left\| g^{'} \right\|_\infty \left\| f \right\|_{L^2[a,b]} \\
&\quad + \frac{\alpha^2}{\Gamma^2(\alpha)} \left\| g^{'} \right\|_{L^2[a,b]}^2 \left\| f \right\|_{L^2[a,b]}^2 .
\end{aligned} \tag{5.64}
$$

References

1. Hardy, G., Littlewood, J.E., Pólya, G.: Inequalities. Cambridge University Press, Cambridge Mathematical Library (1999). ISBN 0-521-05206-8
2. Beckenbach, E.F., Bellman, R.: An Introduction to Inequalities. Random House Inc. (1975). ISBN 0-394-01559-2
3. Drachman, B.C., Cloud, M.J.: Inequalities: With Applications to Engineering. Springer (1998). ISBN 0-387-98404-6
4. Grinshpan, A.Z.: General inequalities, consequences, and applications. Adv. Appl. Math. **34**(1), 71–100 (2005). https://doi.org/10.1016/j.aam.2004.05.00
5. Hardy, G.H., Littlewood, J.E., Pólya, G.: Hölder's Inequality and Its Extensions. 2.7 and 2.8 in Inequalities, 2nd edn., pp. 21–26. Cambridge, England: Cambridge University Press (1988)

6. Aldaz, J.M., Barza, S., Fujii, M., Moslehian, M.S.: Advances in Operator Cauchy-Schwarz inequalities and their reverses. Annals of Functional Analysis **6**(3), 275–295 (2015)
7. Hardy, G.H., Littlewood, J.E., Pólya, G.: "Minkowski's Inequality" and "Minkowski's Inequality for Integrals." §2.11, 5.7, and 6.13 in Inequalities, 2nd edn., pp. 30–32, 123, and 146–150. Cambridge University Press, Cambridge, England (1988)
8. Harnack, A.: Die Grundlagen der Theorie des logarithmischen Potentiales und der eindeutigen Potentialfunktion in der Ebene, Leipzig: V. G. Teubner (1887)

Chapter 6
Existence and Uniqueness of IVP with Global Differentiation on via Picard Iteration

In contrast to traditional differential equations with fractional, fractal–fractional, and other terms, global derivative differential equations offer opportunities for modeling various types of processes. However, because of their nonlinear nature, it is clear that they cannot always be resolved analytically. This is because there aren't many analytical methods available to resolve these equations a sizable class of these derivatives-containing nonlinear equations. As a result, a lot of academics only use numerical methods to arrive at numerical solutions to these issues. One might determine the circumstances under which these equations permit a singular solution, at least in part, thanks to the theory of existence and uniqueness. In fact, there are a number of theories that have been created so far for classical equations [1–6].

The Cauchy–Peano theorem iteration method is one of the older techniques, though. We will accomplish this in this chapter utilizing the well-known Picard iteration. We begin our analysis with the global IVP based on the power law.

6.1 Existence and Uniqueness of IVP with Power Law Decay-Based Global Fractional Derivative via Picard Iteration

In this section, we consider the following IVP:

$$\begin{cases} {}_{t_0}^{C}D_g^{\alpha}y(t) = f(t, y(t)) \text{ if } t > t_0 \\ y(t_0) = y_0 \end{cases}. \tag{6.1}$$

We shall note that the solution of the above does not need to satisfy the initial condition. Since we are dealing with process with nonlocal behaviors, we need to

A. Atangana and İ. Koca, *Fractional Differential and Integral Operators with Respect to a Function*, Industrial and Applied Mathematics,
https://doi.org/10.1007/978-981-97-9951-0_6

set some important conditions for us to reach the desired results. First of all we have that $g'(t) > 0$ and bounded. Thus we will need in addition that $f(t, y(t))$ satisfies the Lipschitz condition and is bounded within $[a, b]$. **First of all let us remember global Lipschitz condition below.**

Definition : The global Lipschitz condition is a mathematical concept primarily used in the analysis of functions, particularly in optimization and numerical analysis. A function is said to satisfy a global Lipschitz condition if there exists a constant L such that the absolute difference between the function's values at any two points is bounded by the distance between those points multiplied by L. Let $f : R^n \to R$ be a function. It satisfies a global Lipschitz condition if there exists a constant L such that for all $x, y \in R^n$ then we get

$$|f(x) - f(y)| \leq L \|x - y\| \tag{6.2}$$

where $\|.\|$ denotes a norm on the vector space R^n.

Now we recall that, in classical case having the following IVP:

$$\begin{cases} y'(t) = f(t, y(t)) \text{ if } t > t_0 \\ y(t_0) = y_0 \quad\quad\quad \text{if } t = 0 \end{cases}, \tag{6.3}$$

the associated Picard iteration is given by

$$y_{n+1}(t) = y(t_0) + \int_{t_0}^{t} f(\tau, y_n(\tau))d\tau. \tag{6.4}$$

Then in the case of our investigation we have that

$$y_{n+1}(t) = \frac{1}{\Gamma(\alpha)} \int_{t_0}^{t} f(\tau, y_n(\tau))(t - \tau)^{\alpha-1} g'(\tau)d\tau. \tag{6.5}$$

We have to show that the above is well-defined, well-posed, uniformly bounded, and uniformly equicontinuous. $\forall t \in [a, b]$, we have

$$\begin{aligned} |y_{n+1}(t)| &\leq \left| \frac{1}{\Gamma(\alpha)} \int_{t_0}^{t} f(\tau, y_n(\tau))(t - \tau)^{\alpha-1} g'(\tau)d\tau \right|, \\ &\leq \frac{1}{\Gamma(\alpha)} \int_{t_0}^{t} |f(\tau, y_n(\tau))| (t - \tau)^{\alpha-1} g'(\tau)d\tau, \\ &\leq \frac{1}{\Gamma(\alpha)} \int_{t_0}^{t} \sup_{l \in [t_0, \tau]} |f(l, y_n(l))| (t - \tau)^{\alpha-1} g'(\tau)d\tau, \end{aligned} \tag{6.6}$$

$$\leq \frac{M}{\Gamma(\alpha)}\int_{t_0}^{t}(t-\tau)^{\alpha-1}\left|g^{'}(\tau)\right|d\tau,$$

$$\leq \frac{M}{\Gamma(\alpha)}\int_{t_0}^{t}(t-\tau)^{\alpha-1}\sup_{l\in[t_0,\tau]}\left|g^{'}(l)\right|d\tau,$$

$$\leq \frac{MM_1}{\Gamma(\alpha)}\frac{(t-t_0)^{\alpha}}{\alpha},$$

$$\leq \frac{MM_1(t-t_0)^{\alpha}}{\Gamma(\alpha+1)}.$$

Since $\forall t\in[a,b]$ then

$$|y_{n+1}(t)| \leq \frac{MM_1(b-t_0)^{\alpha}}{\Gamma(\alpha+1)}. \tag{6.7}$$

But $\forall n>0$,

$$|y_{n+1}(t)| < c, \tag{6.8}$$

where $[y_0-c, y_0+c]$, then, we will need

$$\frac{MM_1(b-t_0)^{\alpha}}{\Gamma(\alpha+1)} < c, \tag{6.9}$$

then

$$b < \left(\frac{\Gamma(\alpha+1)}{MM_1}\right)^{\frac{1}{\alpha}} + t_0. \tag{6.10}$$

Therefore, under the above condition $y_n(t)$ for $n\geq 0$ is uniformly bounded and well-defined. But if $g^{'}(t)$, $f(.,y)$ are in $L^2[a,b]$ then

$$|y_{n+1}(t)| \leq \frac{1}{\Gamma(\alpha)}\left(\int_{t_0}^{t}|f(\tau,y_n(\tau))|^2 d\tau\right)^{\frac{1}{2}}\left(\int_{t_0}^{t}(t-\tau)^{2\alpha-2}\left|g^{'}(\tau)\right|^2 d\tau\right)^{\frac{1}{2}}, \tag{6.11}$$

$$\leq \frac{1}{\Gamma(\alpha)}\|f\|_{L^2[a,b]}\left\|g'\right\|_{\infty}\left(\frac{(t-t_0)^{2\alpha-1}}{2\alpha-1}\right)^{\frac{1}{2}}.$$

We will need $(2\alpha-1)>0$.

If $f(t, y(t))$ is globally Lipschitz, we have that

$$\begin{aligned} |y_n(t)| &\leq \left| \frac{1}{\Gamma(\alpha)} \int_{t_0}^{t} f(\tau, y_{n-1}(\tau))(t-\tau)^{\alpha-1} g^{'}(\tau) d\tau \right|, \\ &\leq \frac{1}{\Gamma(\alpha)} \int_{t_0}^{t} \left| g^{'}(\tau) \right| |f(\tau, y_{n-1}(\tau))| (t-\tau)^{\alpha-1} d\tau, \\ &\leq \frac{1}{\Gamma(\alpha)} \int_{t_0}^{t} \left| g^{'}(\tau) \right| c \left(1 + |y_{n-1}(\tau)|\right) (t-\tau)^{\alpha-1} d\tau, \\ &\leq \frac{c \left\| g' \right\|_{\infty}}{\Gamma(\alpha)} \frac{(t-t_0)^{\alpha}}{\alpha} + \frac{c \left\| g' \right\|_{\infty}}{\Gamma(\alpha)} \int_{t_0}^{t} |y_{n-1}(\tau)| (t-\tau)^{\alpha-1} d\tau. \end{aligned} \tag{6.12}$$

On the other hand if we define

$$y_n(t) = \frac{1}{\Gamma(\alpha)} \int_{t_0}^{t} f(\tau, y_n(\tau))(t-\tau)^{-\alpha} g^{'}(\tau) d\tau. \tag{6.13}$$

We can now repeat the derivation in Eq. (6.11) as follows:

$$\begin{aligned} |y_n(t)| &\leq \frac{1}{\Gamma(\alpha)} \int_{t_0}^{t} |f(\tau, y_n(\tau))| (t-\tau)^{\alpha-1} g^{'}(\tau) d\tau \\ &\leq \frac{1}{\Gamma(\alpha)} \int_{t_0}^{t} c \left(1 + |y_n(\tau)|\right) (t-\tau)^{\alpha-1} g^{'}(\tau) d\tau \\ &\leq \frac{1}{\Gamma(\alpha)} \int_{t_0}^{t} c(t-\tau)^{\alpha-1} g^{'}(\tau) d\tau + \frac{c}{\Gamma(\alpha)} \int_{t_0}^{t} |y_n(\tau)| (t-\tau)^{\alpha-1} g^{'}(\tau) d\tau \\ &\leq \frac{\left\| g' \right\|_{\infty} c (t-t_0)^{\alpha}}{\Gamma(\alpha+1)} + \frac{c \left\| g' \right\|_{\infty}}{\Gamma(\alpha)} \int_{t_0}^{t} |y_n(\tau)| (t-\tau)^{\alpha-1} g^{'}(\tau) d\tau. \end{aligned} \tag{6.14}$$

Using the Gronwall inequality leads to

$$|y_n(t)| \leq \frac{\left\| g' \right\|_{\infty} c (t-t_0)^{\alpha}}{\Gamma(\alpha+1)} \exp\left[\frac{\left\| g' \right\|_{\infty} c (t-t_0)^{\alpha}}{\Gamma(\alpha+1)} \right]. \tag{6.15}$$

Let t_1 and $t_2 \in [a, b]$, $\forall n \geq 0$ we have that if $t_1 > t_2$

$$|y_n(t_1) - y_n(t_2)| = \frac{1}{\Gamma(\alpha)} \left| \begin{array}{l} \int_{t_0}^{t_1} f(\tau, y_{n-1}(\tau))(t_1-\tau)^{\alpha-1} g'(\tau) d\tau \\ - \int_{t_0}^{t_2} f(\tau, y_{n-1}(\tau))(t_2-\tau)^{\alpha-1} g'(\tau) d\tau \end{array} \right|, \tag{6.16}$$

$$= \frac{1}{\Gamma(\alpha)} \left| \begin{array}{l} \int_{t_0}^{t_2} f(\tau, y_{n-1}(\tau))(t_1-\tau)^{\alpha-1} g'(\tau) d\tau \\ - \int_{t_0}^{t_2} f(\tau, y_{n-1}(\tau))(t_2-\tau)^{\alpha-1} g'(\tau) d\tau \\ + \int_{t_2}^{t_1} f(\tau, y_{n-1}(\tau))(t_1-\tau)^{\alpha-1} g'(\tau) d\tau \end{array} \right|,$$

$$\leq \frac{1}{\Gamma(\alpha)} \int_{t_0}^{t_2} |f(\tau, y_{n-1}(\tau))| \left\{ (t_1-\tau)^{\alpha-1} - (t_2-\tau)^{\alpha-1} \right\} \left| g'(\tau) \right| d\tau$$

$$+ \frac{1}{\Gamma(\alpha)} \int_{t_2}^{t_1} |f(\tau, y_{n-1}(\tau))| (t_1-\tau)^{\alpha-1} \left| g'(\tau) \right| d\tau,$$

$$\leq \frac{M_f M_{g'}}{\Gamma(\alpha)} \left\{ \frac{(t_1-t_0)^\alpha}{\alpha} - \frac{(t_2-t_0)^\alpha}{\alpha} - \frac{(t_1-t_2)^\alpha}{\alpha} \right\}$$

$$+ \frac{M_f M_{g'}}{\Gamma(\alpha)} \left\{ \frac{(t_1-t_2)^\alpha}{\alpha} \right\},$$

$$\leq \frac{M_f M_{g'}}{\Gamma(\alpha+1)} \left\{ (t_1-t_0)^\alpha - (t_2-t_0)^\alpha \right\}.$$

We know that $\forall t \in [a, b]$, $(t - t_0)^\alpha$ is Differentiable; therefore, due to the Mean value theorem, we have there exists $l_1 \in [t_2 - t_0, t_1 - t_0]$ such that

$$\alpha(c - t_0)^{\alpha-1}(t_1 - t_2) = (t_1 - t_0)^\alpha - (t_2 - t_0)^\alpha. \tag{6.17}$$

Therefore

$$|y_n(t_1) - y_n(t_2)| \leq \frac{M_f M_{g'}}{\Gamma(\alpha+1)} \left(\alpha(c - t_0)^{\alpha-1}(t_1 - t_2) \right), \tag{6.18}$$

$$\leq K\,|t_1 - t_2| < \varepsilon.$$

Therefore

$$\delta < \frac{\varepsilon\Gamma(\alpha+1)}{\alpha M_f M_{g'}(c-t_0)^{\alpha-1}}. \tag{6.19}$$

Under the condition, the $y_n(t)$ is uniformly equicontinuous. Using the Cauchy–Peano theorem there exists a subsequence of $y_n(t)$ say $(y_{nl})_{n>0}$ such that (y_{nl}) converges uniformly to a function say to $\overline{y}(t)$. In addition having $f(t, y(t))$ Lipschitz then, the solution is unique.

We shall show that $(y_0(t), y_1(t), ..., y_n(t))$ converges on a suitable interval. We do this by writing

$$\begin{aligned} y_n(t) &= y_1(t) + (y_2(t) - y_1(t)) + \ldots + (y_n(t) - y_{n-1}(t)), \\ y_n(t) &= \sum_{j=1}^{n} \left(y_j(t) - y_{j-1}(t)\right). \end{aligned} \tag{6.20}$$

We have that

$$\lim_{n\to\infty} y_n(t) = \sum_{j=1}^{\infty} \left(y_j(t) - y_{j-1}(t)\right). \tag{6.21}$$

We have that

$$\begin{aligned} \left|y_j(t) - y_{j-1}(t)\right| &= \left|\frac{1}{\Gamma(\alpha)} \int_{t_0}^{t} (t-\tau)^{\alpha-1} g'(\tau) \left(f(\tau, y_{j-1}(\tau)) - f(\tau, y_{j-2}(\tau))\right) d\tau\right|, \\ &\leq \frac{1}{\Gamma(\alpha)} \int_{t_0}^{t} \left|g'(\tau)\right| (t-\tau)^{\alpha-1} \left|f(\tau, y_{j-1}(\tau)) - f(\tau, y_{j-2}(\tau))\right| d\tau, \\ &\leq \frac{1}{\Gamma(\alpha)} \int_{t_0}^{t} \sup_{l\in[t_0,\tau]} \left|g'(l)\right| L \left|\left(y_{j-1}(\tau) - y_{j-2}(\tau)\right)\right| (t-\tau)^{\alpha-1} d\tau, \\ &\leq \frac{ML}{\Gamma(\alpha)} \int_{t_0}^{t} \left|\left(y_{j-1}(\tau) - y_{j-2}(\tau)\right)\right| (t-\tau)^{\alpha-1} d\tau. \end{aligned} \tag{6.22}$$

For $j = 2$ we have

$$|y_2(t) - y_1(t)| \leq \frac{ML}{\Gamma(\alpha)} \int_{t_0}^{t} |(y_1(\tau) - y_0(\tau))| (t-\tau)^{\alpha-1} d\tau, \tag{6.23}$$

where

$$|(y_1(t) - y_0(t))| = \left| \frac{1}{\Gamma(\alpha)} \int_{t_0}^{t} g^{'}(\tau) f(y_0(\tau), \tau)(t-\tau)^{\alpha-1} d\tau - y_0 \right|, \quad (6.24)$$
$$\leq \frac{1}{\Gamma(\alpha)} \int_{t_0}^{t} \sup_{l \in [t_0, \tau]} \left| g^{'}(l) \right| M (t-\tau)^{\alpha-1} d\tau + |y_0|,$$
$$\leq \frac{M_g M}{\Gamma(\alpha)} \frac{(t-t_0)^{\alpha}}{\alpha} = \frac{\overline{M_g M}\,(t-t_0)^{\alpha}}{\Gamma(\alpha+1)}.$$

$$|y_2(t) - y_1(t)| \leq \frac{M_g L}{\Gamma(\alpha)} \int_{t_0}^{t} |y_1(\tau) - y_0(\tau)| (t-\tau)^{\alpha-1} d\tau, \quad (6.25)$$
$$\leq \frac{M_g^2 \overline{M} L}{\Gamma(\alpha)\Gamma(\alpha+1)} \frac{(t-t_0)^{\alpha+1}}{\alpha+1},$$
$$\leq \frac{M_g^2 \overline{M} L}{\Gamma(\alpha)\Gamma(\alpha+2)} (t-t_0)^{\alpha+1},$$

and

$$|y_3(t) - y_2(t)| \leq \frac{M_g^3 \overline{M} L^2}{\Gamma(\alpha)\Gamma(\alpha+2)\Gamma(\alpha+3)} (t-t_0)^{\alpha+2}, \quad (6.26)$$

$$\vdots$$

$$\vdots$$

$$\left| y_{j-1}(t) - y_{j-2}(t) \right| \leq \frac{M_g^{j-1} \overline{M} L^{j-2}}{\prod_{l=0}^{j} \Gamma(\alpha+l)} (t-t_0)^{\alpha+j-2}. \quad (6.27)$$

Therefore

$$\left| y_j(t) - y_{j-1}(t) \right| \leq \frac{M_g^{j} \overline{M} L^{j-1}}{\prod_{l=0}^{j} \Gamma(\alpha+l)} (t-t_0)^{\alpha+j-1}. \quad (6.28)$$

Therefore

$$
\begin{aligned}
\sum_{j=1}^{\infty}\left|y_j(t)-y_{j-1}(t)\right| &\leq \sum_{j=1}^{\infty}\frac{M_g^j\overline{M}L^{j-1}}{\prod_{l=0}^{j}\Gamma(\alpha+l)}(t-t_0)^{\alpha+j-1}, \qquad (6.29)\\
&\leq \sum_{j=1}^{\infty}\frac{M_g^j\overline{M}L^{j-1}}{(j-1)!}(t-t_0)^{\alpha}(t-t_0)^{j-1},\\
&\leq \sum_{j=0}^{\infty}\frac{M_g^{j+1}L^j}{j!}(t-t_0)^{\alpha}(t-t_0)^{j},\\
&\leq M_g(t-t_0)\exp\left(M_gL(t-t_0)\right),\\
&\leq M_gb\exp\left(M_gLb\right)
\end{aligned}
$$

which implies the series converges. This completes the proof that the series converges. We can then conclude that

$$\overline{y}(t)=\lim_{n\to\infty}y_n(t). \qquad (6.30)$$

We shall now show that the function $\overline{y}(t)$ satisfies an equation. We know that

$$\overline{y}(t)=\frac{1}{\Gamma(\alpha)}\int_{t_0}^{t}f(\tau,\overline{y}(\tau))(t-\tau)^{\alpha-1}g'(\tau)d\tau, \qquad (6.31)$$

introducing the Picard iteration yields

$$
\begin{aligned}
\overline{y}_{n+1}(t) &= \frac{1}{\Gamma(\alpha)}\int_{t_0}^{t}f(\tau,\overline{y}_n(\tau))(t-\tau)^{\alpha-1}g'(\tau)d\tau, \qquad (6.32)\\
\lim_{n\to\infty}\overline{y}_{n+1}(t) &= \lim_{n\to\infty}\frac{1}{\Gamma(\alpha)}\int_{t_0}^{t}f(\tau,\overline{y}_n(\tau))(t-\tau)^{\alpha-1}g'(\tau)d\tau,\\
y(t) &= \lim_{n\to\infty}\frac{1}{\Gamma(\alpha)}\int_{t_0}^{t}f(\tau,\overline{y}_n(\tau))(t-\tau)^{\alpha-1}g'(\tau)d\tau.
\end{aligned}
$$

To obtain the desired result we have to show that

$$
\begin{aligned}
&\lim_{n\to\infty}\frac{1}{\Gamma(\alpha)}\int_{t_0}^{t}f(\tau,\overline{y}_n(\tau))(t-\tau)^{\alpha-1}g'(\tau)d\tau \qquad (6.33)\\
&=\frac{1}{\Gamma(\alpha)}\int_{t_0}^{t}f(\tau,\overline{y}(\tau))(t-\tau)^{\alpha-1}g'(\tau)d\tau,
\end{aligned}
$$

or

$$\lim_{n\to\infty}\frac{1}{\Gamma(\alpha)}\left|\begin{array}{l}\int_{t_0}^{t} f(\tau,\overline{y}_n(\tau))(t-\tau)^{\alpha-1}g^{'}(\tau)d\tau \\ -\int_{t_0}^{t} f(\tau,\overline{y}(\tau))(t-\tau)^{\alpha-1}g^{'}(\tau)d\tau\end{array}\right| \tag{6.34}$$

$$\leq \lim_{n\to\infty}\frac{1}{\Gamma(\alpha)}\int_{t_0}^{t}(t-\tau)^{\alpha-1}\left|g^{'}(\tau)\right|\left|f(\tau,\overline{y}_n(\tau))-f(\tau,\overline{y}(\tau))\right|d\tau,$$

$$\leq \lim_{n\to\infty}\frac{1}{\Gamma(\alpha)}\int_{t_0}^{t}(t-\tau)^{\alpha-1}\sup_{l\in[t_0,\tau]}\left|g^{'}(l)\right|L\left|\overline{y}_n(\tau)-\overline{y}(\tau)\right|d\tau,$$

$$\leq M_{g'}L\lim_{n\to\infty}\int_{t_0}^{t}(t-\tau)^{\alpha-1}\left|\overline{y}_n(\tau)-\overline{y}(\tau)\right|d\tau.$$

But we have that

$$\overline{y}(\tau)=\sum_{j=1}^{\infty}\left[y_j(\tau)-y_{j-1}(\tau)\right]. \tag{6.35}$$

Therefore

$$|y(\tau)-y_n(\tau)|=\sum_{l=n+1}^{\infty}|y_l(\tau)-y_{l-1}(\tau)|. \tag{6.36}$$

But using what we prove before we have

$$|y_l(\tau)-y_{l-1}(\tau)|\leq\frac{M_g^l L^{l-1}\overline{M}\,(t-t_0)^{\alpha+l-1}}{(l-1)!}. \tag{6.37}$$

Thus

$$\left|\overline{y}(\tau)-\overline{y}_n(\tau)\right|=\sum_{l=n+1}^{\infty}\frac{M_g^l L^{l-1}\overline{M}\,(t-t_0)^{\alpha+l-1}}{(l-1)!}. \tag{6.38}$$

So we have

$$\int_{t_0}^{t} (t-\tau)^{\alpha-1} \left|\overline{y}(\tau) - \overline{y}_n(\tau)\right| d\tau \leq \int_{t_0}^{t} (t-\tau)^{\alpha-1} \sum_{l=n+1}^{\infty} \frac{M_g^l L^{l-1} \overline{M} (\tau - t_0)^{\alpha+l-1}}{(l-1)!} d\tau, \tag{6.39}$$

$$\leq \sum_{l=n+1}^{\infty} \frac{M_g^l L^{l-1} \overline{M} (b - t_0)^{\alpha+l-1}}{(l-1)!} \int_{t_0}^{t} (t-\tau)^{\alpha-1} d\tau,$$

$$\leq \sum_{l=n+1}^{\infty} \frac{M_g^l L^{l-1} \overline{M} (b - t_0)^{\alpha+l-1}}{(l-1)!} \frac{(t-t_0)^{\alpha}}{\alpha},$$

$$\leq \sum_{l=n+1}^{\infty} \frac{M_g^l L^{l-1} \overline{M} (b - t_0)^{\alpha+l}}{(l-1)!\alpha},$$

$$\leq \frac{\overline{M}}{\alpha L} \sum_{l=n+1}^{\infty} \frac{M_g^l L^l b^{\alpha+l}}{(l-1)!},$$

$$\leq \frac{\overline{M} M_g^l L b^{\alpha+1}}{\alpha L} \sum_{l=n+1}^{\infty} \frac{M_g^{l-1} L^{l-1} b^{\alpha+l-1}}{(l-1)!},$$

$$< \frac{\overline{M} M_g^l b^{\alpha+1}}{\alpha} \sum_{l=n+1}^{\infty} \frac{M_g^{l-1} L^{l-1} b^{l-1}}{(l-1)!}.$$

Notice that

$$\sum_{l=n+1}^{\infty} \frac{\left(M_g L b\right)^{l-1}}{(l-1)!} = \sum_{l=1}^{\infty} \frac{\left(M_g L b\right)^{l-1}}{(l-1)!} - \sum_{l=1}^{n} \frac{\left(M_g L b\right)^{l-1}}{(l-1)!}, \tag{6.40}$$

$$= \exp(M_g L b) - \sum_{l=1}^{n} \frac{\left(M_g L b\right)^{l-1}}{(l-1)!}.$$

But

$$\lim_{n\to\infty} \sum_{l=n+1}^{\infty} \frac{\left(M_g L b\right)^{l-1}}{(l-1)!} = \exp(M_g L b) - \exp(M_g L b), \tag{6.41}$$

$$= 0.$$

On the other hand we had that

$$\left| \begin{array}{l} \int_{t_0}^{t} f(\tau, \overline{y}_n(\tau))(t-\tau)^{\alpha-1} g'(\tau) d\tau \\ - \int_{t_0}^{t} f(\tau, \overline{y}(\tau))(t-\tau)^{\alpha-1} g'(\tau) d\tau \end{array} \right| \leq \frac{M M_g b^{\alpha+l}}{\alpha} \sum_{l=n+1}^{\infty} \frac{\left(M_g L b\right)^{l-1}}{(l-1)!}, \tag{6.42}$$

$$\lim_{n\to\infty}\left|\begin{array}{l}\int\limits_{t_0}^{t} f(\tau,\overline{y}_n(\tau))(t-\tau)^{\alpha-1}g'(\tau)d\tau \\ -\int\limits_{t_0}^{t} f(\tau,\overline{y}(\tau))(t-\tau)^{\alpha-1}g'(\tau)d\tau\end{array}\right| \le \frac{MM_g b^{\alpha+l}}{\alpha}\lim_{n\to\infty}\sum_{l=n+1}^{\infty}\frac{(M_g Lb)^{l-1}}{(l-1)!}=0. \tag{6.43}$$

Therefore

$$\int\limits_{t_0}^{t} f(\tau,\overline{y}(\tau))(t-\tau)^{\alpha-1}g'(\tau)d\tau=\lim_{n\to\infty}\int\limits_{t_0}^{t} f(\tau,\overline{y}_n(\tau))(t-\tau)^{\alpha-1}g'(\tau)d\tau. \tag{6.44}$$

We shall show that $y(t)$ is continuous let t_1 and t_2 such that $t_1 > t_2$,

$$\begin{aligned} y(t_1)-y(t_2) &= \frac{1}{\Gamma(\alpha)}\int\limits_{t_0}^{t_1} f(\tau,y(\tau))(t_1-\tau)^{\alpha-1}g'(\tau)d\tau \\ &\quad-\frac{1}{\Gamma(\alpha)}\int\limits_{t_0}^{t_2} f(\tau,y(\tau))(t_2-\tau)^{\alpha-1}g'(\tau)d\tau, \end{aligned} \tag{6.45}$$

$$|y(t_1)-y(t_2)| \le \frac{1}{\Gamma(\alpha)}\left|\begin{array}{l}\int\limits_{t_0}^{t_2} f(\tau,y(\tau))\left\{(t_1-\tau)^{\alpha-1}-(t_2-\tau)^{\alpha-1}\right\}g'(\tau)d\tau \\ \quad+\int\limits_{t_2}^{t_1} f(\tau,y(\tau))(t_1-\tau)^{\alpha-1}g'(\tau)d\tau\end{array}\right|, \tag{6.46}$$

$$\begin{aligned} &\le \begin{array}{l}\frac{1}{\Gamma(\alpha)}\int\limits_{t_0}^{t_2} |f(\tau,y(\tau))|\left\{(t_1-\tau)^{\alpha-1}-(t_2-\tau)^{\alpha-1}\right\}\left|g'(\tau)\right|d\tau \\ \quad+\frac{1}{\Gamma(\alpha)}\int\limits_{t_2}^{t_1} |f(\tau,y(\tau))|(t_1-\tau)^{\alpha-1}\left|g'(\tau)\right|d\tau,\end{array} \\ &\le \frac{M_f M_g}{\Gamma(\alpha)}\int\limits_{t_0}^{t_2}\left\{(t_1-\tau)^{\alpha-1}-(t_2-\tau)^{\alpha-1}\right\}d\tau \\ &\quad+\frac{M_f M_g}{\Gamma(\alpha)}\int\limits_{t_2}^{t_1}(t_1-\tau)^{\alpha-1}d\tau, \end{aligned}$$

$$\leq \frac{M_f M_g}{\alpha\Gamma(\alpha)} \{(t_1 - t_0)^\alpha - (t_2 - t_0)^\alpha - (t_2 - t_2)^\alpha + (t_1 - t_2)^\alpha\},$$

$$\leq \frac{M_f M_g}{\Gamma(\alpha+1)} \{(t_1 - t_0)^\alpha - (t_2 - t_0)^\alpha\},$$

$$|y(t_1) - y(t_2)| \leq \frac{M_f M_g}{\Gamma(\alpha+1)} \{(t_1 - t_0)^\alpha - (t_2 - t_0)^\alpha\}.$$

Since $(t - t_0)^\alpha$ is differentiable $\forall t \in [a, b]$, therefore there exists $\overline{a}_1 \in [t_2 - t_0, t_1 - t_0]$ such that by the

$$(t_1 - t_0)^\alpha - (t_2 - t_0)^\alpha = (t_1 - t_2)\alpha \left(\overline{a}_1 - t_0\right)^{\alpha-1}. \tag{6.47}$$

Thus

$$\begin{aligned} |y(t_1) - y(t_2)| &\leq \frac{M_f M_g}{\Gamma(\alpha+1)}(t_1 - t_2)\alpha \left(\overline{a}_1 - t_0\right)^{\alpha-1}, \\ |y(t_1) - y(t_2)| &< \varepsilon. \end{aligned} \tag{6.48}$$

So we get

$$\delta < \frac{\varepsilon\Gamma(\alpha+1)}{\alpha M_f M_g \left(\overline{a}_1 - t_0\right)^{\alpha-1}} \tag{6.49}$$

which completes the proof.

We now show that $y(t)$ is unique. We assume there exists another function $y_1(t)$ such that

$$y_1(t) = \frac{1}{\Gamma(\alpha)} \int_{t_0}^{t} f(\tau, y_1(\tau))(t_1 - \tau)^{\alpha-1} g'(\tau) d\tau, \tag{6.50}$$

then

$$|y(t) - y_1(t)| = u(t) \leq \frac{1}{\Gamma(\alpha)} \int_{t_0}^{t} |f(\tau, y_1(\tau)) - f(\tau, y(\tau))| (t - \tau)^{\alpha-1} \left|g'(\tau)\right| d\tau, \tag{6.51}$$

$$\leq \frac{LM'_g}{\Gamma(\alpha)} \int_{t_0}^{t} |y_1(\tau) - y(\tau)| (t - \tau)^{\alpha-1} d\tau,$$

$$\leq \frac{LM'_g}{\Gamma(\alpha)} \int_{t_0}^{t} u(\tau)(t - \tau)^{\alpha-1} d\tau,$$

$$\leq \varepsilon + \frac{LM'_g}{\Gamma(\alpha)} \int_{t_0}^{t} u(\tau)(t - \tau)^{\alpha-1} d\tau.$$

By the Gronwall inequality we have

$$\begin{aligned} u(t) &\leq \varepsilon \exp\left[\frac{LM_g'}{\Gamma(\alpha)}\int_{t_0}^{t}(t-\tau)^{\alpha-1}d\tau\right], \\ &\leq \varepsilon \exp\left[\frac{LM_g'}{\Gamma(\alpha+1)}(t-t_0)\right], \end{aligned} \tag{6.52}$$

$$\begin{aligned} u(t) &\leq \lim_{\varepsilon\to 0}\varepsilon \exp\left[\frac{LM_g'}{\Gamma(\alpha+1)}(t-t_0)\right], \\ &\leq 0. \end{aligned} \tag{6.53}$$

Therefore

$$|y(t)-y_1(t)| \leq 0 \to y(t)=y_1(t). \tag{6.54}$$

Therefore y is unique, which completes the proof.

6.2 Existence and Uniqueness of IVP with Caputo–Fabrizio Decay-Based Global Fractional Derivative via Picard Iteration

In this section, we consider the following IVP:

$$\begin{cases} {}_{t_0}^{RCF}D_g^{\alpha}y(t) = f(t, y(t)) \text{ if } t > t_0 \\ y(t_0) = y_0 \end{cases}. \tag{6.55}$$

By the fundamental theorem of Calculus, the above can be transformed to

$$y(t) = y(t_0) + (1-\alpha)f(t, y(t)) + \alpha\int_{t_0}^{t} f(\tau, y(\tau)d\tau. \tag{6.56}$$

We shall show that the above equation has a unique solution using the Picard iteration. To achieve this, some important steps need to be as following.

(a) Construct a sequence of functions $\{y_0(t), y_1(t), ..., y_n(t)\}$ called Picard iteration, which approximates a solution of our equation.

(b) Show that the sequence converges via the Cauchy–Peano theorem.

(c) Show that the sequence of functions converges and define

$$y(t) = \lim_{n\to\infty} y_n(t). \tag{6.57}$$

(d) Show that the function $y(t)$ satisfies the equation.
(e) Show that it is continuous.
(f) Show that is unique.
From the above, the Picard iteration is given by

$$y_n(t) = y(0) + (1-\alpha)g'(t)f(t, y_{n-1}(t)) + \alpha \int_{t_0}^{t} g'(\tau)f(\tau, y_{n-1}(\tau)d\tau, \quad (6.58)$$

noting $\forall n \geq 0$,

$$y_n(t_0) = y_0. \quad (6.59)$$

Here to be specific we denote B the rectangle and its interior $\forall(t, y)$ such that

$$t_0 - a \leq t \leq t_0 + a,\, y_0 - b \leq y \leq y_0 - b. \quad (6.60)$$

By hypothesis we assumed that f was bounded, thus $|f|$ also is bounded and thus has a maximum with B.

$$M = \sup_{(t,y)\in B} |f(t, y(t))|. \quad (6.61)$$

$$\begin{aligned}
|y_n(t)| &= \left| y(t_0) + (1-\alpha)g'(t)f(t, y_{n-1}(t)) + \alpha \int_{t_0}^{t} g'(\tau)f(\tau, y_{n-1}(\tau)d\tau \right|, \quad (6.62)\\
&\leq |y(t_0)| + (1-\alpha)\left|g'(t)\right|\left|f(t, y_{n-1}(t))\right| + \alpha \int_{t_0}^{t} \left|g'(\tau)\right|\left|f(\tau, y_{n-1}(\tau)\right| d\tau,\\
&\leq |y(t_0)| + (1-\alpha) \sup_{t\in[t_0-a,t_0+a]} \left|g'(t)\right| \sup_{(t,y_{n-1})\in B} \left|f(t, y_{n-1}(t))\right|\\
&\quad \alpha \int_{t_0}^{t} \sup_{l\in[t_0,\tau]} \left|g'(l)\right| \sup_{(l,y_{n-1})\in B} \left|f(l, y_{n-1}(l)\right| d\tau,\\
&\leq |y(t_0)| + M_f M_{g'}(1-\alpha) + \alpha M_f M_{g'}(t-t_0).
\end{aligned}$$

$\forall t \in [t_0 - a, t_0 + a]$,

$$|y_n(t) - y(t_0)| \leq M_f M_{g'}(1-\alpha) + \alpha M_f M_g(t-t_0). \quad (6.63)$$

This shows that $\forall n \geq 0$,

$$\begin{aligned}
|y_{n+1}(t) - y(t_0)| &\leq M_f M_{g'}(1-\alpha) + \alpha M_f M_g(t-t_0) \quad (6.64)\\
&\leq M_f M_{g'}\left(1-\alpha+\alpha(t-t_0)\right).
\end{aligned}$$

Our next attempt is to prove that the Picard iteration converges in this case too.

The aim is to show that $y_n(t)$ since bounded on suitable interval converges that is to say $\{y_0(t), y_1(t), ..., y_n(t)\}$ converges on that intervals. Noting that

$$y_n(t) = y_0(t) + (y_1(t) - y_0(t)) + (y_2(t) - y_1(t)) + \cdots + (y_n(t) - y_{n-1}(t)), \tag{6.65}$$

$$y_n(t) = y_0(t) + \sum_{j=1}^{n} \left(y_j(t) - y_{j-1}(t)\right). \tag{6.66}$$

Clearly, we should have that

$$\lim_{n\to\infty} y_n(t) = y_0(t) + \sum_{j=1}^{\infty} \left(y_j(t) - y_{j-1}(t)\right). \tag{6.67}$$

Indeed having

$$\sum_{j=1}^{\infty} \left(y_j(t) - y_{j-1}(t)\right), \tag{6.68}$$

converging will imply that

$$(y_0(t), y_1(t), ...). \tag{6.69}$$

Also if $(y_n(t))$ converges then $y_n(t)$ converges an absolutely convergent sequence. Note that

$$\lim_{n\to\infty} |y_n(t)| \leq |y_0(t)| + \sum_{j=1}^{\infty} \left|y_j(t) - y_{j-1}(t)\right|. \tag{6.70}$$

We shall then evaluate

$$\left|y_j(t) - y_{j-1}(t)\right| = \left| \begin{array}{c} (1-\alpha)\left[f(t, y_{j-1}(t) - f(t, y_{j-2}(t)\right] g'(t) \\ +\alpha \int_{t_0}^{t} f(\tau, y_{j-1}(\tau) g'(\tau) d\tau \\ -\alpha \int_{t_0}^{t} f(\tau, y_{j-2}(\tau) g'(\tau) d\tau \end{array} \right|. \tag{6.71}$$

Here we have that $f(t, y(t))$ is Lipschitz on y and also differentiable. These two are important for this demonstration.

$$\left|y_j(t) - y_{j-1}(t)\right| \leq \left| \begin{array}{c} (1-\alpha)\left|f(t, y_{j-1}(t) - f(t, y_{j-2}(t)\right| \left|g'(t)\right| \\ +\alpha \int_{t_0}^{t} \left|f(\tau, y_{j-1}(\tau) - f(\tau, y_{j-2}(\tau)\right| \left|g'(\tau)\right| d\tau \end{array} \right|. \tag{6.72}$$

We use the Lipschitz condition of f to obtain

$$\begin{aligned}
\left|y_j(t) - y_{j-1}(t)\right| &\leq \Bigg|(1-\alpha)L\left|y_{j-1}(t) - y_{j-2}(t)\right|\left|g'(t)\right| \\
&\quad + \alpha L \int_{t_0}^{t} \left|y_{j-1}(\tau) - y_{j-2}(\tau)\right|\left|g'(\tau)\right| d\tau\Bigg|, \qquad (6.73) \\
\left|y_j(t) - y_{j-1}(t)\right| &\leq \Bigg|(1-\alpha)LM_{g'}\left|y_{j-1}(t) - y_{j-2}(t)\right| \\
&\quad + \alpha LM_{g'} \int_{t_0}^{t} \left|y_{j-1}(\tau) - y_{j-2}(\tau)\right| d\tau\Bigg|.
\end{aligned}$$

For $j = 1$, we have

$$\begin{aligned}
|y_1 - y_0| &\leq (1-\alpha)\left|g'(t)\right||f(t, y_0(t))| + \alpha \int_{t_0}^{t} |f(\tau, y_0(\tau))|\left|g'(\tau)\right| d\tau \qquad (6.74) \\
&\leq (1-\alpha)M_{g'}M_f + \alpha M_{g'}M_f(t - t_0) \\
&\leq M_{g'}M_f 1 - \alpha + \alpha(t - t_0))
\end{aligned}$$

$$\begin{aligned}
|y_2 - y_1| &\leq (1-\alpha)\left|g'(t)\right||f(t, y_1(t)) - f(t, y_0(t))| \qquad (6.75) \\
&\quad + \alpha \int_{t_0}^{t} |f(t, y_1(\tau)) - f(\tau, y_0(\tau))|\left|g'(\tau)\right| d\tau, \\
&\leq (1-\alpha)M_{g'}L|y_1 - y_0| + \alpha M_{g'}L \int_{t_0}^{t} |y_1 - y_0| d\tau, \\
&\leq (1-\alpha)M_{g'}L(M_{g'}M_f(1 - \alpha + \alpha(t - t_0))) \\
&\quad + \alpha M_{g'}L \int_{t_0}^{t} \left[M_{g'}M_f(1 - \alpha + \alpha(t - t_0))\right] d\tau, \\
&\leq (1-\alpha)M_{g'}L(M_{g'}M_f(1 - \alpha + \alpha(t - t_0))) \\
&\quad + \alpha M_{g'}L\left((1-\alpha)(t - t_0) + \frac{\alpha}{2}(t - t_0)^2\right),
\end{aligned}$$

$$\begin{aligned}
&\leq (1-\alpha)M_{g'}^2 LM_f + (1-\alpha)\, M_{g'} L\alpha(t-t_0) \\
&\quad + \alpha\,(1-\alpha)\, M_{g'} L(t-t_0) + \frac{\alpha^2}{2} M_{g'} L(t-t_0)^2, \\
&\leq (1-\alpha)M_{g'}^2 LM_f + 2\,(1-\alpha)\, M_{g'} L\alpha(t-t_0) + \frac{M_{g'} L}{2} \left(\alpha(t-t_0)\right)^2.
\end{aligned}$$

$$\begin{aligned}
|y_3 - y_2| &\leq (1-\alpha)\left|g'(t)\right| |f(t, y_2(t)) - f(t, y_1(t))| \\
&\quad + \alpha \int_{t_0}^{t} |f(t, y_2(\tau)) - f(\tau, y_1(\tau))| \left|g'(\tau)\right| d\tau, \\
&\leq (1-\alpha) M_{g'} L\, |y_2 - y_1| + \alpha M_{g'} L \int_{t_0}^{t} |y_2 - y_1|\, d\tau, \\
&\leq (1-\alpha) M_{g'} L \left\{ \begin{array}{c} (1-\alpha)M_{g'}^2 LM_f + 2\,(1-\alpha)\, M_{g'} L\alpha(t-t_0) \\ + M_{g'} \frac{L}{2} \left(\alpha(t-t_0)\right)^2 \end{array} \right\} \\
&\quad + \alpha M_{g'} L \left\{ \begin{array}{c} (1-\alpha)M_{g'}^2 LM_f(t-t_0) + 2\,(1-\alpha)\, M_{g'} L\alpha \frac{(t-t_0)^2}{2} \\ + M_{g'} \frac{L}{6} \left(\alpha^2 (t-t_0)^3\right) \end{array} \right\} \\
&\leq (1-\alpha)^2 M_{g'}^2 L^2 M_f + 2\,(1-\alpha)^2\, M_{g'}^2 L^2 \alpha(t-t_0) \\
&\quad + \frac{(1-\alpha)}{2} M_{g'}^2 L^2 \left(\alpha^2 (t-t_0)^2\right) + \alpha(1-\alpha) M_f M_{g'}^3 L^2 (t-t_0) \\
&\quad + 2\alpha^2 (1-\alpha) M_{g'}^2 L^2 \frac{(t-t_0)^2}{2} + \alpha^3 M_{g'}^2 \frac{L^2}{6} (t-t_0)^3, \\
&\leq a_0 + a_1(t-t_0) + a_2 \frac{(t-t_0)^2}{2} + a_3 \frac{(t-t_0)^3}{6}.
\end{aligned} \tag{6.76}$$

Therefore in general

$$\left|y_j(t) - y_{j-1}(t)\right| \leq \bar{a}_0 + \bar{a}_1(t-t_0) + \bar{a}_2 \frac{(t-t_0)^2}{2} + \bar{a}_3 \frac{(t-t_0)^3}{3!} + \ldots + \bar{a}_j \frac{(t-t_0)^j}{j!}. \tag{6.77}$$

If we take

$$a = \max_{0 \leq l \leq j} \left|\bar{a}_j\right|, \tag{6.78}$$

then indeed a is a constant independent of t and t_0 then

$$\begin{aligned}
\left|y_j(t) - y_{j-1}(t)\right| &\leq a \left(1 + (t-t_0) + \frac{(t-t_0)^2}{2} + \ldots + \frac{(t-t_0)^j}{j!}\right) \\
&\leq a \sum_{l=0}^{n} \frac{(t-t_0)^l}{l!}, \text{ for } \forall_t \in [a, b].
\end{aligned} \tag{6.79}$$

$$\sum_{l=1}^{\infty}\left|y_j(t)-y_{j-1}(t)\right| \le a\sum_{l=1}^{n}\frac{(t-t_0)^l}{l!}, \text{ for } \forall_t \in [a,b], \tag{6.80}$$

$$\le a\left[\sum_{l=0}^{n}\frac{(t-t_0)^l}{l!}-1\right],$$

$$\le a\left[\exp(t-t_0)-1\right].$$

This completes the proof. Therefore,

$$y(t) = \lim y_n(t). \tag{6.81}$$

We shall show that

$$y(t) = \lim_{n\to\infty} y_n(t) \tag{6.82}$$

satisfies the equation. We shall recall that

$$y_{n+1}(t) = y(t_0) + (1-\alpha)g'(t)f(t,y_n(t)) + \alpha\int_{t_0}^{t} g'(\tau)f(\tau,y_n(\tau)d\tau, \tag{6.83}$$

$$\lim_{n\to\infty} y_{n+1}(t) = (1-\alpha)g'(t)\lim_{n\to\infty} f(t,y_n(t)) + \lim_{n\to\infty}\alpha\int_{t_0}^{t} g'(\tau)f(\tau,y_n(\tau)d\tau, \tag{6.84}$$

to conclude that the sequence satisfies the equation, we need to show that

$$\lim_{n\to\infty}\left((1-\alpha)g'(t)f(t,y_n(t)) + \alpha\int_{t_0}^{t} g'(\tau)f(\tau,y_n(\tau)d\tau\right) \tag{6.85}$$

$$= (1-\alpha)g'(t)f(t,y(t)) + \alpha\int_{t_0}^{t} g'(\tau)f(\tau,y(\tau)d\tau.$$

Therefore, we shall show that

$$\lim_{n\to\infty}\left|\begin{array}{c}(1-\alpha)g'(t)\left(f(t,y_n(t))-f(t,y(t))\right)\\ +\alpha\int_{t_0}^{t} g'(\tau)\left(f(\tau,y_n(\tau)-f(\tau,y(\tau))\right)d\tau\end{array}\right| = 0, \text{ for } \forall_t \in [a,b]. \tag{6.86}$$

However, using the properties of norm, we have

$$\lim_{n\to\infty}\left|\begin{array}{l}(1-\alpha)g'(t)\left(f(t,y_n(t))-f(t,y(t))\right)\\ +\alpha\int\limits_{t_0}^{t}g'(\tau)\left(f(\tau,y_n(\tau)-f(\tau,y(\tau))\right)d\tau\end{array}\right| \tag{6.87}$$

$$\leq \lim_{n\to\infty}(1-\alpha)M_{g'}L\left|y_n(t)-y(t)\right|+\alpha M_{g'}L\lim_{n\to\infty}\int\limits_{t_0}^{t}\left|y_n(\tau)-y(\tau)\right|d\tau.$$

But we have to remember that

$$y_n(\tau)=y_0(\tau)+\sum_{j=1}^{n}\left|y_j(\tau)-y_{j-1}(\tau)\right| \tag{6.88}$$

while

$$y(\tau)=y_0(\tau)+\sum_{j=1}^{\infty}\left|y_j(\tau)-y_{j-1}(\tau)\right|. \tag{6.89}$$

Therefore

$$y_n(\tau)-y(\tau)=\sum_{j=n+1}^{\infty}\left|y_j(\tau)-y_{j-1}(\tau)\right|, \tag{6.90}$$

$$\begin{aligned}\left|y_n(t)-y(t)\right| &= \left|\sum_{l=n+1}^{\infty}\left[y_j(t)-y_{j-1}(t)\right]\right|, \\ &\leq a\sum_{l=n+1}^{\infty}\frac{(t-t_0)^j}{j!}.\end{aligned} \tag{6.91}$$

On the other hand we have that

$$\alpha\int\limits_{t_0}^{t}\sum_{l=n+1}^{\infty}\frac{(\tau-t_0)^j}{j!}d\tau=\alpha a\sum_{l=n+1}^{\infty}\frac{(t-t_0)^{j+1}}{(j+1)!}. \tag{6.92}$$

Therefore

$$\begin{aligned}\alpha\int_{t_0}^{t}|y_n(\tau)-y(\tau)|\,d\tau &\leq \alpha a\sum_{l=n+1}^{\infty}\frac{(t-t_0)^{j+1}}{(j+1)!}, \\ &\leq \alpha a\sum_{l=1}^{\infty}\frac{(t-t_0)^{j+1}}{(j+1)!}-\alpha a\sum_{l=1}^{n}\frac{(t-t_0)^{j+1}}{(j+1)!}, \\ &\leq \alpha a\left(\sum_{l=1}^{\infty}\frac{(t-t_0)^{j+1}}{(j+1)!}-\sum_{l=1}^{n}\frac{(t-t_0)^{j+1}}{(j+1)!}\right).\end{aligned} \tag{6.93}$$

Then replacing yields

$$\begin{aligned}&\lim_{n\to\infty}\left|\begin{array}{l}(1-\alpha)g'(t)\left(f(t,y_n(t))-f(t,y(t))\right)\\ +\alpha\displaystyle\int_{t_0}^{t}g'(\tau)\left(f(\tau,y_n(\tau)-f(\tau,y(\tau))\right)d\tau\end{array}\right| \\ &\leq (1-\alpha)M_{g'}L+\lim_{n\to\infty}a\left[\sum_{l=1}^{\infty}\frac{(t-t_0)^{j}}{j!}-\sum_{l=1}^{n}\frac{(t-t_0)^{j}}{j!}\right] \\ &\quad+\lim_{n\to\infty}\alpha aL\left[\sum_{l=1}^{\infty}\frac{(t-t_0)^{j+1}}{(j+1)!}-\sum_{l=1}^{n}\frac{(t-t_0)^{j+1}}{(j+1)!}\right],\end{aligned} \tag{6.94}$$

$$\lim_{n\to\infty}\left|\begin{array}{l}(1-\alpha)g'(t)\left(f(t,y_n(t))-f(t,y(t))\right)\\ +\alpha\displaystyle\int_{t_0}^{t}g'(\tau)\left(f(\tau,y_n(\tau)-f(\tau,y(\tau))\right)d\tau\end{array}\right|\leq 0. \tag{6.95}$$

$\forall_n \geq 0$ and $\forall_t \in [a,b]$ therefore

$$\begin{aligned}&(1-\alpha)g'(t)f(t,y(t))+\alpha\int_{t_0}^{t}g'(\tau)f(\tau,y_{n-1}(\tau)d\tau \\ &=\lim_{n\to\infty}\left((1-\alpha)g'(t)f(t,y_n(t))\right)+\alpha\lim_{n\to\infty}\left(\int_{t_0}^{t}g'(\tau)f(\tau,y_n(\tau)d\tau\right)\end{aligned} \tag{6.96}$$

which completes the proof.

We shall show that the function is also continuous. We shall adopt two ways. Let t_1, t_2 such that $t_1 > t_2$. We evaluate

$$\lim_{n\to\infty} |y_n(t_1) - y_n(t_2)| = |y(t_1) - y(t_2)|, \tag{6.97}$$

$$\begin{aligned} |y(t_1) - y(t_2)| &= \left| \sum_{j=1}^{\infty} [y_j(t_1) - y_{j-1}(t_1)] - \sum_{j=1}^{\infty} [y_j(t_2) - y_{j-1}(t_2)] \right|, \\ &= \alpha a \sum_{j=1}^{\infty} \frac{(t_1 - t_0)^j}{j!} - \alpha a \sum_{j=1}^{\infty} \frac{(t_2 - t_0)^j}{j!}, \\ &\leq \alpha a \sum_{j=1}^{\infty} \frac{(t_1 - t_0)^j - (t_2 - t_0)^j}{j!}. \end{aligned} \tag{6.98}$$

But the function $(t - t_0)^j$ is differentiable on t_1 and t_2; therefore, by the mean value theorem, there exists $l \in [t_2, t]$ such that

$$(l - t_0)^{j-1} j(t_1 - t_2) = (t_1 - t_0)^j - (t_2 - t_0)^j, \tag{6.99}$$

therefore

$$\begin{aligned} |y(t_1) - y(t_2)| &\leq \alpha a \sum_{j=1}^{\infty} \frac{(l - t_0)^{j-1} j(t_1 - t_2)}{j!}, \\ &\leq \alpha a \sum_{j=1}^{\infty} \frac{(l - t_0)^{j-1}(t_1 - t_2)}{(j-1)!}, \\ &\leq \alpha a \sum_{j=0}^{\infty} \frac{(l - t_0)^{j}(t_1 - t_2)}{j!}, \\ &\leq \alpha a \exp(l - t_0)(t_1 - t_2). \end{aligned} \tag{6.100}$$

But

$$|y(t_1) - y(t_2)| < \varepsilon. \tag{6.101}$$

We need to have

$$\delta < \frac{\varepsilon}{\alpha a \exp(l - t_0)}. \tag{6.102}$$

Therefore

$$|y(t_1) - y(t_2)| < \varepsilon \tag{6.103}$$

then we have

$$\delta < \frac{\varepsilon}{\alpha a \exp(l - t_0)} \tag{6.104}$$

which completes the proof.

Alternatively, we show that $\forall \varepsilon > 0$ there exists a $\delta > 0$ such that

$$|h| < \delta \to |y(t+h) - y(t)| < \varepsilon. \tag{6.105}$$

$$\begin{aligned} y(t+h) - y(t) &= (1-\alpha)\left[f(t+h, y(t+h)) - f(t, y(t))\right] \\ &+ \alpha \left[\int_{t_0}^{t+h} f(\tau, y(\tau)d\tau - \int_{t_0}^{t} f(\tau, y(\tau)d\tau \right], \end{aligned} \tag{6.106}$$

$$\begin{aligned} |y(t+h) - y(t)| &\leq (1-\alpha)\left|f(t+h, y(t+h)) - f(t, y(t))\right| \\ &+ \alpha \left| \int_{t_0}^{t} f(\tau, y(\tau)d\tau + \int_{t}^{t+h} f(\tau, y(\tau)d\tau - \int_{t_0}^{t} f(\tau, y(\tau)d\tau \right|, \\ &\leq (1-\alpha)\left|f(t+h, y(t+h)) - f(t, y(t))\right| + \alpha \int_{t_0}^{t+h} \left|f(\tau, y(\tau)\right| d\tau, \\ &\leq (1-\alpha)\left|f(t+h, y(t+h)) - f(t, y(t))\right| + \alpha M_f h, \\ &\leq (1-\alpha)Ih + \alpha M_f h. \end{aligned} \tag{6.107}$$

Since $f(t, y(t))$ is differentiable with respect to t_0, t, and $y(t)$, we can find $\zeta \in [t, t+h]$ such that the above is obtained

$$|y(t+h) - y(t)| \leq \left((1-\alpha)I + \alpha M_f\right) h < \varepsilon \tag{6.108}$$

and we have

$$h < \frac{1}{(1-\alpha)I + \alpha M_f} \tag{6.109}$$

which completes the proof.

Now we shall show that $y(t)$ is unique. Let $\zeta(t)$ be another solution satisfying the equation then let

$$u(t) = |y(t) - \zeta(t)|. \tag{6.110}$$

We aim to show that $u(t)$ is zero $\forall_t \in [a, b]$

But

$$\begin{aligned} |y(t) - \zeta(t)| &\leq (1-\alpha)\left|f(t, y(t)) - f(t, \zeta(t))\right| \\ &+ \alpha \int_{t_0}^{t} \left|f(\tau, y(\tau) - f(\tau, \zeta(\tau)\right| d\tau. \end{aligned} \tag{6.111}$$

Using either the Lipschitz condition of f or its differentiability yields

$$|y(t)-\zeta(t)| \;=\; u(t) \leq (1-\alpha)L\,|y(t)-\zeta(t)| + \alpha L \int_{t_0}^{t} |y(\tau)-\zeta(\tau)|\,d\tau, \tag{6.112}$$

$$u(t) \;\leq\; (1-\alpha)Lu(t) + \alpha L \int_{t_0}^{t} u(\tau)d\tau,$$

$$(1-(\alpha-1)L)u(t) \;\leq\; \alpha L \int_{t_0}^{t} u(\tau)d\tau, \tag{6.113}$$

$$u(t) \;\leq\; \frac{\alpha L}{1+\alpha L - L} \int_{t_0}^{t} u(\tau)d\tau.$$

We need $1+\alpha L - L > 0$. Under the above condition, we have that

$$u(t) \leq \varepsilon + \frac{\alpha L}{1+\alpha L - L} \int_{t_0}^{t} u(\tau)d\tau \;\; \forall \varepsilon > 0. \tag{6.114}$$

By the Gronwall inequality, we get

$$u(t) \leq \varepsilon \exp\left[\frac{\alpha L}{1+\alpha L - L}(t-t_0)\right] \tag{6.115}$$

taking $\varepsilon = 0$ yields

$$u(t) = 0 \rightarrow y(t) = \zeta(t), \forall_t \in [a,b] \tag{6.116}$$

which completes the proof.

6.3 Existence and Uniqueness of IVP with Mittag–Leffler Decay-Based Global Fractional Derivative via Picard Iteration

In this section, we consider the following IVP:

$$\begin{cases} {}^{ABR}_{t_0}D^{\alpha}_{g} y(t) = f(t, y(t)) \text{ if } t > t_0 \\ y(t_0) = y_0 \end{cases}. \tag{6.117}$$

By the fundamental theorem of Calculus, the above can be transformed to

$$y(t) = y(t_0) + (1-\alpha)g'(t)f(t,y(t)) + \frac{\alpha}{\Gamma(\alpha)}\int_{t_0}^{t} f(\tau, y(\tau)(t-\tau)^{\alpha-1}g'(\tau)d\tau, \tag{6.118}$$

$$y(t_0) = y_0.$$

We follow the same routine as presented before, but this time we have a complex solution. We build the Picard iteration

$$y_n(t) = y(t_0) + (1-\alpha)g'(t)f(t,y_{n-1}(t)) + \frac{\alpha}{\Gamma(\alpha)}\int_{t_0}^{t} f(\tau, y_{n-1}(\tau)(t-\tau)^{\alpha-1}g'(\tau)d\tau, \tag{6.119}$$

$\forall(t,n) \in [a,b] \times N$

$$|y_n(t) - y(t_0)| = \left|(1-\alpha)g'(t)f(t,y_{n-1}(t)) + \frac{\alpha}{\Gamma(\alpha)}\int_{t_0}^{t} f(\tau, y_{n-1}(\tau)(t-\tau)^{\alpha-1}g'(\tau)d\tau\right|, \tag{6.120}$$

$$\leq M_{g'}(1-\alpha)\left|f(t,y_{n-1}(t)\right| + \frac{\alpha}{\Gamma(\alpha)}M_{g'}\int_{t_0}^{t}\left|f(\tau, y_{n-1}(\tau)\right|(t-\tau)^{\alpha-1}d\tau,$$

$$\leq (1-\alpha)M_{g'}M_f + \frac{\alpha M_{g'}M_f}{\Gamma(\alpha+1)}(t-t_0)^{\alpha},$$

$$\leq a_0 + \frac{a_1(t-t_0)^{\alpha}}{\Gamma(\alpha)}.$$

On the other hand

$$|y_n(t)| \leq \overline{a}_0 + \frac{a_1(t-t_0)^{\alpha}}{\Gamma(\alpha)} \tag{6.121}$$

where

$$\overline{a}_0 = a_0 + |y_0|. \tag{6.122}$$

Conditions for well-posedness and boundedness of the sequence can be drawn. We shall show that $\{y_0(t), y_1(t), ..., y_n(t)\}$ converges on suitable intervals. Noting that

$$y_n(t) = y_0(t) + (y_1(t) - y_0(t)) + (y_2(t) - y_1(t)) + ... + (y_n(t) - y_{n-1}(t)), \tag{6.123}$$

$$y_n(t) = y_0(t) + \sum_{j=1}^{n}\left(y_j(t) - y_{j-1}(t)\right).$$

In this case there is boundedness condition on $f(t, y(t))$

$$\lim_{n\to\infty} y_n(t) = y_0(t) + \sum_{j=1}^{\infty} \left(y_j(t) - y_{j-1}(t)\right), \tag{6.124}$$

$$\lim_{n\to\infty} |y_n(t)| = |y_0(t)| + \sum_{j=1}^{\infty} \left|y_j(t) - y_{j-1}(t)\right|.$$

We evaluate

$$\begin{aligned} \left|y_j(t) - y_{j-1}(t)\right| &\leq (1-\alpha)M_{g'}\left[f(t, y_{j-1}(t)) - f(t, y_{j-2}(t))\right] \\ &+ \frac{\alpha M_{g'}}{\Gamma(\alpha)}\int_{t_0}^{t}\left[f(\tau, y_{j-1}(\tau)) - f(\tau, y_{j-2}(\tau))\right](t-\tau)^{\alpha-1}g'(\tau)d\tau, \\ &\leq (1-\alpha)M_{g'}L\left|y_{j-1}(t) - y_{j-2}(t)\right| + \frac{\alpha M_{g'}L}{\Gamma(\alpha)}\int_{t_0}^{t}\left|y_{j-1}(\tau) - y_{j-2}(\tau)\right|(t-\tau)^{\alpha-1}d\tau. \end{aligned} \tag{6.125}$$

Noting that

$$\begin{aligned} |y_1 - y_0| &\leq (1-\alpha)M_{g'}\left|f(t, y_0(t))\right| + \frac{\alpha M_{g'}}{\Gamma(\alpha)}\int_{t_0}^{t}\left|f(\tau, y_0(\tau))\right|(t-\tau)^{\alpha-1}d\tau, \\ &\leq (1-\alpha)M_{g'}M_f + \frac{\alpha M_{g'}}{\Gamma(\alpha+1)}M_f(t-t_0)^{\alpha}, \\ &\leq a_0 + a_1\frac{(t-t_0)^{\alpha}}{\Gamma(\alpha)}. \end{aligned} \tag{6.126}$$

$$\begin{aligned} |y_2 - y_1| &\leq (1-\alpha)M_{g'}\left|f(t, y_1(t)) - f(t, y_0(t))\right| \\ &+ \frac{\alpha M_{g'}}{\Gamma(\alpha)}\int_{t_0}^{t}\left|f(t, y_1(\tau)) - f(\tau, y_0(\tau))\right|(t-\tau)^{\alpha-1}d\tau, \\ &\leq (1-\alpha)M_{g'}L\left|y_1 - y_0\right| + \frac{\alpha M_{g'}}{\Gamma(\alpha)}L\int_{t_0}^{t}\left|y_1 - y_0\right|(t-\tau)^{\alpha-1}d\tau, \\ &\leq a_0 + (1-\alpha)M_{g'}L\left(a_0 + a_1\frac{(t-t_0)^{\alpha}}{\Gamma(\alpha)}\right) \\ &+ \frac{\alpha M_{g'}}{\Gamma(\alpha)}L\int_{t_0}^{t}\left(a_0 + a_1\frac{(\tau-t_0)^{\alpha}}{\Gamma(\alpha)}\right)(t-\tau)^{\alpha-1}d\tau, \end{aligned} \tag{6.127}$$

$$\begin{aligned}
&\leq a_0 + (1-\alpha)M_{g'}La_0 + (1-\alpha)M_{g'}La_1\frac{(t-t_0)^{\alpha}}{\Gamma(\alpha)} \\
&+ \frac{\alpha M_{g'}La_0}{\Gamma(\alpha)}(t-t_0)^{\alpha} + \frac{a_1\alpha M_{g'}}{\Gamma(\alpha)}L\int_{t_0}^{t}(\tau-t_0)(t-\tau)^{\alpha-1}d\tau, \\
&\leq a_0 + (1-\alpha)M_{g'}La_0 + (1-\alpha)M_{g'}La_1\frac{(t-t_0)^{\alpha}}{\Gamma(\alpha)} \\
&+ \frac{\alpha M_{g'}La_0}{\Gamma(\alpha)}(t-t_0)^{\alpha} + \frac{a_1\alpha M_{g'}}{\Gamma(\alpha)}L\frac{(t-t_0)^{\alpha+1}}{\alpha}, \\
&\leq \widetilde{a}_0 + \widetilde{a}_1\frac{(t-t_0)^{\alpha}}{\Gamma(\alpha)} + a_2\frac{(t-t_0)^{\alpha+1}}{\Gamma(\alpha+1)}.
\end{aligned}$$

Therefore

$$\begin{aligned}
|y_2(t)-y_1(t)| &\leq \widetilde{a}_0 + \widetilde{a}_1\frac{(t-t_0)^{\alpha}}{\Gamma(\alpha)} + a_2\frac{(t-t_0)^{\alpha+1}}{\Gamma(\alpha+1)}, \\
\left|y_j(\tau)-y_{j-1}(\tau)\right| &\leq \widetilde{a}_0 + \widetilde{a}_1\frac{(t-t_0)^{\alpha}}{\Gamma(\alpha)} + a_2\frac{(t-t_0)^{\alpha+1}}{\Gamma(\alpha+1)} \\
&+ ... + \widetilde{a}_j\frac{(t-t_0)^{\alpha+j}}{\Gamma(\alpha+j)},
\end{aligned}$$

which is

$$\left|y_j(t)-y_{j-1}(t)\right| \leq \sum_{n=0}^{j}\widetilde{a}_{n+1}\frac{(t-t_0)^{n+\alpha}}{\Gamma(n+\alpha)} + \widetilde{a}_0. \tag{6.128}$$

Let

$$\overline{a} = \max_{0\leq n\leq j}|\widetilde{a}_{n+1}|, \tag{6.129}$$

then

$$\left|y_j(t)-y_{j-1}(t)\right| \leq \overline{a}\sum_{n=0}^{j}\frac{(t-t_0)^{n+\alpha}}{\Gamma(n+\alpha)} + \overline{a}. \tag{6.130}$$

$$\sum_{j=0}^{\infty}\left|y_j(\tau)-y_{j-1}(\tau)\right| \leq \overline{a}\sum_{n=0}^{\infty}\frac{(t-t_0)^{n+\alpha}}{\Gamma(n+\alpha)} + \overline{a}. \tag{6.131}$$

Noting that $t \in [a, b]$ therefore

$$\sum_{j=0}^{\infty}\left|y_j(\tau)-y_{j-1}(\tau)\right| \leq \overline{a}\sum_{n=0}^{\infty}\frac{b^{n+\alpha}}{\Gamma(n+\alpha)} + \overline{a} \tag{6.132}$$

which converges toward a Mittag–Leffler function or a stretched exponential function. This completes the proof. We shall next show that

$$\lim_{n\to\infty} y_n(t) = y(t). \tag{6.133}$$

We recall that

$$y_{n+1}(t) = (1-\alpha)f(t, y_n(t)) + y(t_0) + \frac{\alpha}{\Gamma(\alpha)}\int_{t_0}^{t} f(\tau, y_n(\tau)(t-\tau)^{\alpha-1}d\tau, \tag{6.134}$$

$$\lim_{n\to\infty} y_{n+1}(t) = (1-\alpha)\lim_{n\to\infty} f(t, y_n(t)) + y(t_0) + \lim_{n\to\infty}\frac{\alpha}{\Gamma(\alpha)}\int_{t_0}^{t} f(\tau, y_n(\tau))(t-\tau)^{\alpha-1}d\tau. \tag{6.135}$$

On the left-hand side, we have

$$y(t) = \lim_{n\to\infty} y_{n+1}(t). \tag{6.136}$$

But to reach the goal, we need to show that

$$\lim_{n\to\infty}\left((1-\alpha)f(t, y_n(t)) + \frac{\alpha}{\Gamma(\alpha)}\int_{t_0}^{t} f(\tau, y_n(\tau)(t-\tau)^{\alpha-1}d\tau\right) \tag{6.137}$$

$$= (1-\alpha)f(t, y(t)) + \frac{\alpha}{\Gamma(\alpha)}\int_{t_0}^{t} f(\tau, y(\tau))(t-\tau)^{\alpha-1}d\tau.$$

This is equivalent to show that

$$\lim_{n\to\infty}\left|\begin{array}{c}(1-\alpha)\left(f(t, y_n(t)) - f(t, y(t))\right) \\ +\frac{\alpha}{\Gamma(\alpha)}\int_{t_0}^{t}\left(f(\tau, y_n(\tau)) - f(\tau, y(\tau))\right)(t-\tau)^{\alpha-1}d\tau\end{array}\right| = 0. \tag{6.138}$$

But using inequalities, we have that

$$\lim_{n\to\infty}\left|\begin{array}{c}(1-\alpha)\left(f(t,y_n(t))-f(t,y(t))\right)\\+\frac{\alpha}{\Gamma(\alpha)}\int\limits_{t_0}^{t}\left(f(\tau,y_n(\tau))-f(\tau,y(\tau))\right)(t-\tau)^{\alpha-1}d\tau\end{array}\right|, \tag{6.139}$$

$$\begin{aligned}&\leq \lim_{n\to\infty}\left((1-\alpha)\left(f(t,y_n(t))-f(t,y(t))\right)\right)\\&+\lim_{n\to\infty}\frac{\alpha}{\Gamma(\alpha)}\int\limits_{t_0}^{t}\left(f(\tau,y_n(\tau))-f(\tau,y(\tau))\right)(t-\tau)^{\alpha-1}d\tau.\end{aligned}$$

Using the Lipschitz condition of $f(t,y)$ with respect to y we have

$$|f(t,y_n(t))-f(t,y(t))|\leq L\,|y_n(t)-y(t)|\,. \tag{6.140}$$

However

$$\begin{aligned}|y(t)-y_n(t)| &= \sum_{j=n+1}^{\infty}\left|y_j(t)-y_{j-1}(t)\right|\leq \overline{a}\sum_{j=n+1}^{\infty}\frac{(t-t_0)^{j+\alpha}}{\Gamma(j+\alpha)}+\overline{a}, \quad (6.141)\\&\leq \overline{a}\sum_{j=n+1}^{\infty}\frac{(b-t_0)^{j+\alpha}}{\Gamma(j+\alpha)}+\overline{a}\end{aligned}$$

on the other hand

$$\begin{aligned}\int\limits_{t_0}^{t}|y(\tau)-y_n(\tau)|\,(t-\tau)^{\alpha-1}d\tau &\leq \overline{a}\int\limits_{t_0}^{t}\sum_{j=n+1}^{\infty}\frac{(t-t_0)^{j+\alpha}}{\Gamma(j+\alpha)}(t-\tau)^{\alpha-1}d\tau+\overline{a}\int\limits_{t_0}^{t}(t-\tau)^{\alpha-1}d\tau, \quad (6.142)\\&\leq \overline{a}\sum_{j=n+1}^{\infty}\int\limits_{t_0}^{t}\frac{(t-t_0)^{j+1}}{\Gamma(j+\alpha)}(t-\tau)^{\alpha-1}d\tau+\overline{a}\frac{(t-t_0)^{\alpha}}{\alpha},\\&\leq \overline{a}\sum_{j=n+1}^{\infty}\frac{1}{\Gamma(j+\alpha)}\int\limits_{t_0}^{t}(t-t_0)^{j+1}(t-\tau)^{\alpha-1}d\tau+\frac{\overline{a}}{\alpha}(t-t_0)^{\alpha},\\&\leq \overline{a}\sum_{j=n+1}^{\infty}\frac{(b-t_0)^{j+1}}{\Gamma(j+\alpha)}\frac{(t-t_0)^{\alpha}}{\alpha}+\frac{\overline{a}}{\alpha}(t-t_0)^{\alpha},\\&\leq \frac{\overline{a}}{\alpha}(t-t_0)^{\alpha}\left(\sum_{j=n+1}^{\infty}\frac{(b-t_0)^{j+1}}{\Gamma(j+\alpha)}+1\right).\end{aligned}$$

Nevertheless

$$\overline{a}\sum_{j=n+1}^{\infty}\frac{(t-t_0)^{j+\alpha}}{\Gamma(j+\alpha)} = \overline{a}\sum_{j=0}^{\infty}\frac{(t-t_0)^{j+\alpha}}{\Gamma(j+\alpha)}-\overline{a}\sum_{j=0}^{n}\frac{(t-t_0)^{j+\alpha}}{\Gamma(j+\alpha)}, \tag{6.143}$$

$$\lim_{n\to\infty}\left(\overline{a}\sum_{j=0}^{\infty}\frac{(t-t_0)^{j+\alpha}}{\Gamma(j+\alpha)}-\overline{a}\sum_{j=0}^{n}\frac{(t-t_0)^{j+\alpha}}{\Gamma(j+\alpha)}\right) = 0.$$

Also

$$\frac{\overline{a}}{\alpha}(t-t_0)^{\alpha}\left(\left(\sum_{j=0}^{\infty}\frac{(b-t_0)^{j+1}}{\Gamma(j+\alpha)}+1\right)-\left(\sum_{j=0}^{n}\frac{(b-t_0)^{j+\alpha}}{\Gamma(j+\alpha)}+1\right)\right), \tag{6.144}$$

$$\begin{aligned}&\lim_{n\to\infty}\left(\frac{\overline{a}}{\alpha}(t-t_0)^{\alpha}\left(\left(\sum_{j=0}^{\infty}\frac{(b-t_0)^{j+1}}{\Gamma(j+\alpha)}+1\right)-\left(\sum_{j=0}^{\infty}\frac{(b-t_0)^{j+\alpha}}{\Gamma(j+\alpha)}+1\right)\right)\right),\\ &= 0.\end{aligned}$$

Therefore

$$\lim_{n\to\infty}\left|\begin{array}{c}(1-\alpha)g'(t)\left(f(t,y_n(t))-f(t,y(t))\right)\\ +\frac{\alpha}{\Gamma(\alpha)}\int_{t_0}^{t}\left(f(\tau,y_n(\tau))-f(\tau,y(\tau))\right)g'(\tau)(t-\tau)^{\alpha-1}d\tau\end{array}\right| = 0. \tag{6.145}$$

We can conclude that

$$\begin{aligned}&\lim_{n\to\infty}(1-\alpha)g'(t)f(t,y_n(t))+\frac{\alpha}{\Gamma(\alpha)}\int_{t_0}^{t}f(\tau,y_n(\tau))g'(\tau)(t-\tau)^{\alpha-1}d\tau \qquad (6.146)\\ &= (1-\alpha)g'(t)f(t,y(t))+\frac{\alpha}{\Gamma(\alpha)}\int_{t_0}^{t}f(\tau,y(\tau))g'(\tau)(t-\tau)^{\alpha-1}d\tau,\\ &= y(t),\end{aligned}$$

which completes the proof. We shall now show that $y(t)$ is continuous. Let then consider t_1 and t_2 that $t_1 > t_2$. We aim to show that there exist $\delta > 0$ such that if

$$|t_1-t_2| < \delta, \tag{6.147}$$

then

$$|y(t_1) - y(t_2)| < \varepsilon. \tag{6.148}$$

We know that

$$\begin{aligned} y(t_1) &= y(t_0) + (1-\alpha)g'(t_1)f(t_1, y(t_1)) + \frac{\alpha}{\Gamma(\alpha)}\int_{t_0}^{t_1} f(\tau, y(\tau))g'(\tau)(t_1-\tau)^{\alpha-1}d\tau, \\ &= \lim_{n\to\infty}\left(y(t_0) + (1-\alpha)g'(t_1)f(t_1, y_n(t_1)) + \frac{\alpha}{\Gamma(\alpha)}\int_{t_0}^{t_1} f(\tau, y_n(\tau))g'(\tau)(t_1-\tau)^{\alpha-1}d\tau\right), \\ y(t_2) &= y(t_0) + (1-\alpha)g'(t_2)f(t_2, y(t_2)) + \frac{\alpha}{\Gamma(\alpha)}\int_{t_0}^{t_2} f(\tau, y(\tau))g'(\tau)(t_2-\tau)^{\alpha-1}d\tau, \\ &= \lim_{n\to\infty}\left(y(t_0) + (1-\alpha)g'(t_2)f(t_2, y_n(t_2)) + \frac{\alpha}{\Gamma(\alpha)}\int_{t_0}^{t_2} f(\tau, y_n(\tau))g'(\tau)(t_2-\tau)^{\alpha-1}d\tau\right). \end{aligned} \tag{6.149}$$

$$\begin{aligned} |y(t_1) - y(t_2)| &= \lim_{n\to\infty} |y_{n+1}(t_1) - y_{n+1}(t_2)|, \\ &\le \lim_{n\to\infty} ((1-\alpha)\,|f(t_1, y_n(t_1)) - |\,f(t_2, y_n(t_2)))\,|g'(t_1)| \\ &\quad + \frac{\alpha}{\Gamma(\alpha)}\int_{t_0}^{t_2} \left[(t_1-\tau)^{\alpha-1} - (t_2-\tau)^{\alpha-1}\right] |f(\tau, y_n(\tau))|\, g'(\tau)d\tau \\ &\quad + \frac{\alpha}{\Gamma(\alpha)}\int_{t_2}^{t_1} |f(\tau, y_n(\tau))|\, g'(\tau)(t_1-\tau)^{\alpha-1}d\tau, \\ &\le \lim_{n\to\infty} (1-\alpha)L\,|y_n(t_1) - y_n(t_2)|\, \left\|g'\right\|_\infty \\ &\quad + \frac{\alpha}{\Gamma(\alpha)}\left\|g'\right\|_\infty \|f\|_\infty \int_{t_0}^{t_2} \left[(t_1-\tau)^{\alpha-1} - (t_2-\tau)^{\alpha-1}\right] d\tau \\ &\quad + \frac{\alpha}{\Gamma(\alpha)}\left\|g'\right\|_\infty \|f\|_\infty \int_{t_2}^{t_1} (t_1-\tau)^{\alpha-1}d\tau, \end{aligned} \tag{6.150}$$

$$\begin{aligned} &\leq \lim_{n\to\infty}(1-\alpha)L\left|y_n(t_1)-y_n(t_2)\right|\left\|g'\right\|_\infty \\ &\quad+\frac{\alpha}{\Gamma(\alpha)}\left\|g'\right\|_\infty\|f\|_\infty\left[\frac{(t_1-t_0)^\alpha}{\alpha}-\frac{(t_2-t_0)^\alpha}{\alpha}-\frac{(t_1-t_2)^\alpha}{\alpha}+\frac{(t_1-t_2)^\alpha}{\alpha}\right], \\ &\leq \lim_{n\to\infty}(1-\alpha)L\left|y_n(t_1)-y_n(t_2)\right|\left\|g'\right\|_\infty \\ &\quad+\frac{\alpha}{\Gamma(\alpha)}\left\|g'\right\|_\infty\|f\|_\infty\left[\frac{(t_1-t_0)^\alpha}{\alpha}-\frac{(t_2-t_0)^\alpha}{\alpha}\right]. \end{aligned} \tag{6.151}$$

So we have

$$\begin{aligned} |y(t_1)-y(t_2)|\left[1-(1-\alpha)L\left\|g'\right\|_\infty\right] &\leq \frac{\left\|g'\right\|_\infty\|f\|_\infty}{\Gamma(\alpha)}\left[\frac{(t_1-t_0)^\alpha}{\alpha}-\frac{(t_2-t_0)^\alpha}{\alpha}\right], \\ |y(t_1)-y(t_2)| &\leq \frac{\left\|g'\right\|_\infty\|f\|_\infty}{\left[1-(1-\alpha)L\left\|g'\right\|_\infty\right]}\left[\frac{(t_1-t_0)^\alpha}{\alpha}-\frac{(t_2-t_0)^\alpha}{\alpha}\right]. \end{aligned} \tag{6.152}$$

Condition is that $1-(1-\alpha)L\left\|g'\right\|_\infty > 0$, under that condition, we have

$$|y(t_1)-y(t_2)| \leq \frac{M_{g'}M_f}{\left[1-(1-\alpha)L\|g'\|_\infty\right]}\left[\frac{(t_1-t_0)^\alpha}{\alpha}-\frac{(t_2-t_0)^\alpha}{\alpha}\right]. \tag{6.153}$$

But $(t-t_0)^\alpha$ is differentiable $\forall t\in[t_2-t_0,t_1-t_0]$. Therefore, we can define $l\in[t_2-t_0,t_1-t_0]$ such that

$$(t_1-t_0)\alpha(l-t_0)^{\alpha-1}=(t_1-t_0)^\alpha-(t_2-t_0)^\alpha. \tag{6.154}$$

Then

$$|y(t_1)-y(t_2)| \leq \frac{M_{g'}M_f(t_1-t_2)\alpha(l-t_0)^{\alpha-1}}{\left[1-(1-\alpha)L\|g'\|_\infty\right]} < \varepsilon, \tag{6.155}$$

We need

$$\delta < \frac{\varepsilon\left[1-(1-\alpha)L\left\|g'\right\|_\infty\right]}{\alpha M_{g'}M_f(t_1-t_2)(l-t_0)^{\alpha-1}} \tag{6.156}$$

which completes the proof.

Alternatively, we can show that

$$|h|<\delta \rightarrow |y(t+h)-y(t)|<\varepsilon. \tag{6.157}$$

$$|y(t+h)-y(t)| = |(1-\alpha)f(t+h,y(t+h))-f(t,y(t))| \left\|g'\right\|_{\infty} + \left| \begin{array}{l} \frac{\alpha}{\Gamma(\alpha)} \int\limits_{t_0}^{t+h} f(\tau,y(\tau))g'(\tau)(t+h-\tau)^{\alpha-1}d\tau \\ -\frac{\alpha}{\Gamma(\alpha)} \int\limits_{t_0}^{t} f(\tau,y(\tau))g'(\tau)(t-\tau)^{\alpha-1}d\tau \end{array} \right| \tag{6.158}$$

$$\leq (1-\alpha)\,|f(t+h,y(t+h))-f(t,y(t))|$$

$$+\frac{\alpha}{\Gamma(\alpha)} \int\limits_{t_0}^{t} |f(\tau,y(\tau))|\left|g'(\tau)\right| \left\{(t+h-\tau)^{\alpha-1}-(t-\tau)^{\alpha-1}\right\} d\tau \tag{6.159}$$

$$+\frac{\alpha}{\Gamma(\alpha)} \int\limits_{t}^{t+h} |f(\tau,y(\tau))|\left|g'(\tau)\right| (t+h-\tau)^{\alpha-1} d\tau$$

$$\leq (1-\alpha)L\left\|g'\right\|_{\infty} |y(t+h)-y(t)|$$

$$+\frac{\alpha\left\|g'\right\|_{\infty}\|f\|_{\infty}}{\Gamma(\alpha)}\left\{\frac{(t+h-t_0)^{\alpha}}{\alpha}-\frac{(t-t_0)^{\alpha}}{\alpha}-\frac{(t+h-t)^{\alpha}}{\alpha}+\frac{(t+h-t)^{\alpha}}{\alpha}\right\}, \tag{6.160}$$

$$\leq (1-\alpha)LM_{g'}\,|y(t+h)-y(t)|+\frac{\alpha M_{g'}M_f}{\Gamma(\alpha)}\left\{(t+h-t_0)^{\alpha}-(t-t_0)^{\alpha}\right\},$$

$$\leq (1-\alpha)LM_{g'}\,|y(t+h)-y(t)|+\frac{\alpha^2 M_{g'}M_f}{\Gamma(\alpha+1)}(l-t_0)^{\alpha-1}h.$$

Therefore

$$|y(t+h)-y(t)| \leq \frac{\alpha^2 M_{g'}M_f}{\Gamma(\alpha+1)}\frac{(l-t_0)^{\alpha-1}h}{\left[1-(1-\alpha)LM_{g'}\right]} < \varepsilon. \tag{6.161}$$

We need

$$\delta < \frac{\varepsilon\Gamma(\alpha)\left[1-(1-\alpha)LM_{g'}\right]}{\alpha M_{g'}M_f(l-t_0)^{\alpha-1}} \tag{6.162}$$

which completes the proof. We can conclude that $y(t)$ is continuous $t \in [a,b]$. We shall now prove $y(t)$ is unique, we choose then $y_1(t)$ to be a solution to an equation we shall prove that $\forall t \in [a,b]$,

$$|y_1(t)-y(t)| = 0. \tag{6.163}$$

$$\begin{aligned}\|y_1(t) - y(t)\| &= (1-\alpha)\left|(f(t, y_1(t)) - f(t, y(t)))\, g'(t)\right| \\ &+ \tfrac{\alpha}{\Gamma(\alpha)} \int_{t_0}^{t} f(\tau, y_1(\tau))(t-\tau)^{\alpha-1} g'(\tau) d\tau \\ &- \tfrac{\alpha}{\Gamma(\alpha)} \int_{t_0}^{t} f(\tau, y(\tau))(t-\tau)^{\alpha-1} g'(\tau) d\tau,\end{aligned} \tag{6.164}$$

$$\begin{aligned} &\leq (1-\alpha)\left\|g'\right\|_\infty |f(t, y_1(t)) - f(t, y(t))| \quad (6.165)\\ &+ \frac{\alpha}{\Gamma(\alpha)} \int_{t_0}^{t} |f(\tau, y_1(\tau)) - f(\tau, y(\tau))|\,(t-\tau)^{\alpha-1} g'(\tau) d\tau, \\ &\leq (1-\alpha)L\left\|g'\right\|_\infty |y_1(t) - y(t)| + \frac{\alpha L M_{g'}}{\Gamma(\alpha)} \int_{t_0}^{t} |y_1 - y|\,(t-\tau)^{\alpha-1} d\tau \\ &\leq (1-\alpha)L M_{g'} |y_1(t) - y(t)| + \frac{\alpha L M_{g'}}{\Gamma(\alpha)} \int_{t_0}^{t} |y_1 - y|\,(t-\tau)^{\alpha-1} d\tau,\end{aligned}$$

$$\begin{aligned}\|y_1(t) - y(t)\| &\leq \frac{\alpha L M_{g'}}{\Gamma(\alpha)\left[1 - (1-\alpha)L M_{g'}\right]} \int_{t_0}^{t} |y_1 - y|\,(t-\tau)^{\alpha-1} d\tau, \quad (6.166)\\ \|y_1(t) - y(t)\| &\leq \varepsilon + \frac{\alpha L M_{g'}}{\Gamma(\alpha)\left[1 - (1-\alpha)L M_{g'}\right]} \int_{t_0}^{t} |y_1 - y|\,(t-\tau)^{\alpha-1} d\tau.\end{aligned}$$

By the Gronwall inequality, we get

$$\|y_1(t) - y(t)\| \leq \varepsilon \exp\left[\frac{\alpha L M_{g'}}{\Gamma(\alpha)\left[1 - (1-\alpha)L M_{g'}\right]} \frac{(t-t_0)^\alpha}{\alpha}\right], \tag{6.167}$$

with $\varepsilon = 0$, we have that

$$\|y_1(t) - y(t)\| = 0, \forall t \in [a, b] \tag{6.168}$$

therefore

$$y_1(t) = y(t), \forall t \in [a, b] \tag{6.169}$$

which proves the uniqueness.

References

1. Lindelöf, E.: Sur l'application de la méthode des approximations successives aux équations différentielles ordinaires du premier ordre. Comptes rendus hebdomadaires des séances de l'Acad (1894)
2. Peano, G.: Sull'integrabilità delle equazioni differenziali del primo ordine. Atti Accad. Sci. Torino. **21**, 437–445 (1886)
3. Peano, G.: Demonstration de l'intégrabilité des é (1890)
4. Kirk, William A., Sims, Brailey: Handbook of Metric Fixed Point Theory. Kluwer Academic, London (2001)978-0-7923-7073-4
5. Gronwall, Thomas H.: Note on the derivatives with respect to a parameter of the solutions of a system of differential equations. Ann. Math. **20**(2), 292–296 (1919)
6. Rosa, A.: An episodic history of the staircased iteration diagram. Antiquitates Mathematicae **15**, 3–90 (2001). https://doi.org/10.14708/am.v15i1.7056

Chapter 7
Existence and Uniqueness via Carathéodory Approach

In this chapter, we devote our attention to using the Caratheodory principle to derive conditions under which these global fractional differential equations admit unique solution. In general, Caratheodory's existence theorem implies that a given nonlinear differential equation admits a solution under some mild conditions. The literature says that these conditions are indeed the extension of well-known Peano's existence theorem. We recall that Peano's theorem requests that function $f(t, y(t))$ of the right-hand side of the differential equation to be continuous and this way generalized by Caratheodory to accommodate a large number of functions even those which are discontinuous pointwise [3]. In mathematical terms, consider the following ordinary differential equation where the differential operator is the classical derivative

$$\begin{aligned} y'(t) &= f(t, y(t)), \\ y(t_0) &= y_0. \end{aligned} \tag{7.1}$$

Here the function $f(t, y(t))$ is defined in a rectangle domain $D : \{(t, y) : |t - t_0| \leq a,\ |y - y_0| \leq b\}$ with the following conditions.

(1) $f(t, y(t))$ is continuous for every fixed t in y.

(2) $f(t, y(t))$ is measurable in t for every fixed y.

(3) $\exists m(t)$ function which is Lebesque integrable defined on $[t_0 - a, t_0 + a] \to [0, \infty]$ such that $\forall (t, y(t)) \in D$

$$|f(t, y(t))| \leq m(t). \tag{7.2}$$

The above conditions guarantee the existence of a solution in extended sense in a neighborhood of the initial condition. On the other hand to obtain the uniqueness, it is required that the mapping f satisfied the Caratheodory principle on D and additionally there should exist a Lebesgue integrable function $M : [t_0 - a, t_0 + a] \to [0, \infty]$

A. Atangana and İ. Koca, *Fractional Differential and Integral Operators with Respect to a Function*, Industrial and Applied Mathematics,
https://doi.org/10.1007/978-981-97-9951-0_7

such that when we take $(t, y_1), (t, y_2) \in D$ the following inequality should be satisfied:

$$|f(t, y_1) - f(t, y_2)| \leq M(t) |y_1 - y_2|. \tag{7.3}$$

Then, one can find a unique solution $y(t_0) = y(t, t_0, y_0)$ to the initial value problem

$$\begin{aligned} y'(t) &= f(t, y(t)), \\ y(t_0) &= y_0. \end{aligned} \tag{7.4}$$

7.1 Existence and Uniqueness for IVP with the Classical Global Derivative via Carathéodory

In this section, we present the extension of this theorem when the classical derivative is replaced by D_g derivative. Therefore, consider the following nonlinear equation:

$$\begin{cases} D_g y(t) = f(t, y(t)), \text{ if } t \geq t_0 \\ y(0) = y_0, \text{ if } t = t_0 \end{cases} \tag{7.5}$$

- $g^{'}(t)$ is positive and continuous,
- $g^{'}(t)$ is bounded,
- $f(t, y(t))$ is a Caratheodory mapping.

Theorem 7.1 *Let $I_1 = [0, 1]$, $D = I_1 \times R, \forall (t, y) \in D$, the function f satisfies the Caratheodory principle. In addition if there exists an absolutely continuous function $M \in L'[I, R]$ such that $\forall (t, y) \in D$*

$$|f(t, y(t))| \leq M(t) (1 + |y|), \tag{7.6}$$

then there exists an absolutely continuous function $v(t)$ such that

$$v(t) = v(0) + \int_0^t f(\tau, v(\tau)) g^{'}(\tau) d\tau. \tag{7.7}$$

The above theorem shall be the topic of our discussion. Let $I_b = [0, b]$ with $b > 0$ and let $v_n \to v$ in $C[I_b, R]$. The uniform convergence norm is defined on $C[I_b, R]$. Let

$$d_n = \sup_{t \in I_b} |v_n(t) - v(t)| \tag{7.8}$$

and

$$t^n = t - d_n \; \forall t \in I_b. \tag{7.9}$$

If $f(t, y)$ satisfies the theorem then immediately we have $f(t, v_n(t)) \to f(t, v(t))$ on I_b pointwise. The aim is to show that

$$f\left(t^n, v_n(t^n)\right) \to f(t, v(t)) \tag{7.10}$$

pointwise on I_b. If $f(t, y(t))$ satisfies the condition of the theorem, the above is satisfied pointwise but an additional requirement see [1] is that all v_n are Lipschitz continuous with the some Lipschitz constant in fact

$$K = \max_{1 \leq j \leq n} |K_n| \tag{7.11}$$

where K_n is the Lipschitz constant of v_n. The assertion is fully correct if in addition, for all sequences chosen from a certain convex compact space of $C[I_b, R]$. For this to be fullfit, we shall need additional definitions; see [2, 3].

Definition 7.1 Let $b > 0$, $I_b = [0, b]$ and $q^1, \ldots. q^m, l^1, \ldots, l^m \in L_g^1[I_b, R]$ where

$$L_g' = \left\{ f \,\middle|\, fg' \in L'[I_b, R] \right\}. \tag{7.12}$$

The set

$$G_g(q, l) : \left\{ h : h \in L_g'[I, R]; q^j(t) \leq h^j(t) \leq l^j(t), \forall t \in I_b, 1 \leq j \leq m \right\}. \tag{7.13}$$

Indeed $G_g(q, l)$ is convex set since whenever one consider h_1 and $\overline{h}$ in $G_g(q, l)$.
$\forall t \in [0, 1]$,

$$\overline{h}_1(t) = h(t)(1 - t) + \overline{h}(t) \in G_g(q, l). \tag{7.14}$$

Definition 7.2 Let $G_g(q, l)$ be defined as above. Let $f : I_b \times R \to R$ be such that

$$f\left(t, \int_0^t h(\tau) g'(\tau) d\tau\right) \in G_g(q, l), \tag{7.15}$$

$\forall h \in G_g(q, l)$ then f is said to be G_g integrable on I_b.

We shall need extra definitions before we can start the derivation of the proof.

Definition 7.3 Let f be G_g integrable on I_b and let $h \in G_g$. Let the sequence

$$v_n(t) = \int_0^t h_n(\tau) g'(\tau) d\tau, \tag{7.16}$$

$h_n \in G_g$ be such that $v_n(t) \to v(t)$ uniformly with

$$v(t) = \int_0^t b(\tau) g^{'}(\tau) d\tau \tag{7.17}$$

on I_b. We choose t^n as defined before $t^n = t - d_n$. If for every sequence of function $f(t^n, v_n(t)) \rightarrow f(t, v(t))$ then f is said to be G_g regular on I.

Definition 7.4 If $G_g = G_g(q, l)$ and $M \in L_g^{'}[I, R]$ is such that $|q(t)| \leq M(t)$ and $|l(t)| \leq M(t)$ $\forall t \in I_b$ then G_g is said to be dominated by M.

Theorem 7.2 *Let $b > 0$ and let $I_b = [0, b]$. Then if f be G_g-regular on I_b. Then one can find at least one absolute continuous v satisfying on equation.*

Proof We define the following set:

$$\Lambda = \left\{ v : v(t) = y_0 + \int_0^t h(\tau) g^{'}(\tau) d\tau \right\}. \tag{7.18}$$

We let additionally

$$\Gamma v(t) = y_0 + \int_0^t f(\tau, v(\tau)) g^{'}(\tau) d\tau, \; v \in \Lambda. \tag{7.19}$$

Since $f \in G_g\text{-}integrable$, then $\Gamma v \in \Lambda$, indeed Λ is also a convex set. To continue an adequate choice of a lower semi-continuous $M \in L_g^{'}[I_b, R]$ that will dominate G_g. However, the dominance of M on G_g implies that every function in the Λ which is the closure of Λ must be in Λ itself. This leads to the conclusion that Λ is compact. Let the next assignment is to demonstrate that Γ is continuous then with the help of Schauder'sfixed-point theorem the demonstration will be completed. If f could be Lipschitzian, then one could directly have $\forall v, \overline{v} \in G_g$, we could have

$$\begin{aligned} |\Gamma v - \Gamma \overline{v}| &\leq \int_0^t \left| g^{'} \right| |f(\tau, v) - f(\tau, \overline{v})| \, d\tau, \qquad (7.20) \\ |\Gamma v - \Gamma \overline{v}| &\leq \|\Gamma v - \Gamma \overline{v}\|_\infty \leq \left\| g^{'} \right\|_\infty L \, \|v - \overline{v}\|_\infty t. \end{aligned}$$

Such that $\forall \varepsilon > 0$ with

$$\left\| g^{'} \right\|_\infty L \, \|v - \overline{v}\|_\infty T < \varepsilon, \tag{7.21}$$

we could have

$$\|v - \overline{v}\|_\infty < \delta < \frac{\varepsilon}{\left\| g^{'} \right\|_\infty LT}. \tag{7.22}$$

Since this is not the case here, we will let $v_n \to v$, $v_n \in \Lambda$ we will then show that $\Gamma v_n \to \Gamma v$.

That will be the same to show that for any subsequence of Γv_n say for example Γv_{nl} converges to Γv. We consider t^n to be $t^n = t - d_n$ as defined before. We consider the subsequence to be the sequence itself and this will reduce the bound of the notation. To continue, we recall the Fatou Lemma which establishes inequality associated to the Lebesgue integral of the limit inferior of a sequence of function to the limit of these functions, the statement of the theorem is given.

Theorem 7.3 *Let a measure space (R, F, u) and a set $x \in F$, let (u_n) be a sequence of (F, B_{R^+})-measurable nonnegative function $u_n : x \to [0, \infty]$. Define the function on $u : x \to [0, \infty]$ by setting*

$$u(t) = \lim_{n\to\infty} f_n(t), \forall t \in x. \tag{7.23}$$

Then $v \in (F, B_{R^+})$-measurable and also

$$\int_x f d\mu \leq \lim_{n\to\infty} \inf \int_x u_n d\mu, \tag{7.24}$$

where the integrable may be infinite. From the above theorem together with the dominance of M on G_g, we have that

$$\int_0^t \underline{\lim} g'(\tau)\left(M\left(\tau^n\right) - f^l\left(\tau^n, v_n\left(\tau^n\right)\right)\right) d\tau, \tag{7.25}$$

$$\leq \underline{\lim} \int_0^t g'(\tau)\left(M\left(\tau^n\right) - f^l\left(\tau^n, v_n\left(\tau^n\right)\right)\right) d\tau 1 \leq l \leq n.$$

We now let

$$z_n(\tau) = f^l\left(\tau^n, v_n\left(\tau^n\right)\right) \tag{7.26}$$

it follows that $z_n \to z = f(\tau, v(\tau))$ and that $M(\tau) \leq \underline{\lim} M\left(\tau^n\right)$. We therefore have

$$\underline{\lim} \int_0^t g'(\tau) M\left(\tau^n\right) d\tau \leq \lim \int_0^t g'(\tau) M\left(\tau^n\right) d\tau = \int_0^t g'(\tau) M(\tau) d\tau. \tag{7.27}$$

We show via Fatou Lemma that

$$\overline{\lim} \int_0^t z_n(\tau) d\tau \leq \int_0^t z(\tau) d\tau. \tag{7.28}$$

Finally

$$\lim \int_0^t z_n(\tau)d\tau = \int_0^t z(\tau)d\tau. \tag{7.29}$$

Then $\Gamma v_n \to \Gamma v$ *and is uniform, so* Γ *is continuous. The application of the Schauder fixed-point theorem helps complete the proof.*

Theorem 7.4 *Let* $f : I_b \times R \to R$ *be such that*

$$|f(t, y)| \le K(t)\,(1 + |y|) \tag{7.30}$$

$\forall (t, y) \in D = I_b \times R$ *for some* $K \in L_g'(I_b, R)$ *and that* $y \to f(t, y)$ *is continuous for every fixed* $t \in I_b$ *and* $G_g(p, q)$ *such that* f *is* G_g*-regular on* I.

Proof Let $b > 0$ and $I_b = [0, b]$. Define $K_1 = \sup_{t\in[0,b]} |K(t)|$, indeed if $v \in C(I_b \times R)$ then $f(t, v(t))$ is measurable on I_b. We defined

$$q^l(t) = p^l(t) = 2K_1(t)\,\|g'\|_\infty \exp\left(2m\,\|g'\|_\infty \int_0^t K_1(\tau)d\tau\right), 1 \le l \le b. \tag{7.31}$$

We have used p, q in $G_g(p, q)$. If $h \in G_g$, we have

$$\begin{aligned} \left| f(t, \int_0^t g'(\tau)h(\tau)\,d\tau) \right| &\le K(t)\left(1 + \left|\int_0^t g'(\tau)h(\tau)\,d\tau\right|\right) \\ &\le K(t)\left(1 + 2m\,\|g'\|_\infty \left|\int_0^t K_1(\tau)\exp\left(2\,\|g'\|_\infty \int_0^t K_1(\lambda)d\lambda\right)d\tau\right|\right) \\ &\le K(t)\left(1 + \exp\left(2m\,\|g'\|_\infty \int_0^t K_1(\tau)d\tau\right) - 1\right) \\ &\le 2K_1(t)\exp\left(2m\,\|g'\|_\infty \int_0^t K_1(\tau)d\tau\right). \end{aligned} \tag{7.32}$$

This shows that $f \in G_g - integrable$ on I_b. We let

$$c = \int_{I_b} |K(\tau)|\,d\tau. \tag{7.33}$$

We now investigate $f(t, y)$ restricted to $I_b \times \{y; |y| \le L\}$. Then $\exists B \subset I_b$ a compact subset such that the restrict on f to $B \times \{y; |y| \le L\}$ is continuous and such that the measure of the points of I_b not in B is very minimal than any previously associated positive value

$$v_n(t) = \int_0^t g'(\tau)h_n(\tau)\,d\tau \to v(t) = \int_0^t g'(\tau)h(\tau)\,d\tau \tag{7.34}$$

uniformly on I_b, $h_n \in G_g$, $h \in G_g$. Then we have from the remark presented above that

$$f\left(t^n, v_n\left(t^n\right)\right) \to f\left(t, v(t)\right) \text{ on } I_b. \tag{7.35}$$

7.2 Existence and Uniqueness for IVP with the Caputo Global Derivative via Carathéodory

In this section, we should present the local and global existence and uniqueness of the solution of a general nonlinear initial bounded value problem with the Caputo global derivative.

$$\begin{cases} {}^{RL}_{t_0}D^{\alpha}_{g}y(t) = f(t, y(t)), \text{ if } t > 0 \\ y(0) = y_0 \text{ if } t = t_0. \end{cases} \tag{7.36}$$

We shall start with the local existence. To achieve this, we start with the following theorem.

Theorem 7.5 *Let denote*

$$\begin{aligned} I_b &= [0, b], E = \{y \in R \mid |y - y_0| \le c\}, \\ D &= \{(t, y) \in R \times R \,|t \in I_b, y \in E\} \end{aligned}$$

(1) $f(t, y)$ is Lebesgue measurable with respect to $t \in I_b$.
(2) $f(t, y)$ is continuous with respect to $y \in E$.
(3) $g'(t)$ is continuous and is $L^2[I_b]$.
(4) There exists a real-valued function $\Omega(t) \in L^2[I_b]$ such that

$$|f(t, y)| \le \Omega(t) \tag{7.37}$$

for almost every $t \in I_b$ and all $y \in E$. Then $\forall \alpha > \frac{1}{2}$, there exists a solution of the IVP mentioned above on the interval $[0, t_0 + h]$ for $\forall h > 0$. The proof for Caputo case meaning when $g'(t) = 1$ can be found in [2]. We shall show the extension of this proof in this case.

We have to show that $g'(\tau)(t-\tau)^{\alpha-1} f(\tau, y(\tau))$ is Lebesgue integrable with respect to $\tau \in [t_0, t_0 + h]$, $t \le t_0 + h \le b$, $\forall t \in I_b$. This will be achieved under the condition that $y(t)$ is of course Lebesgue measurable on $[t_0, t_0 + h]$. The first way is to choose $y(t)$ will be a constant, $y(t) = \lambda\ (t_0 \le t \le t_0 + h)$. This will imply that $f(t, \lambda)$ is Lebesgue measurable be the case of the first hypothesis of the theorem.

The second choice will be to consider a step function, which will be to say the function $y(t)$ is constant within the piece of intervals. Therefore, we can subdivide the interval $[t_0, t_0 + h]$ in small pieces and each one of them $y(t)$ is constant $t_0 = t_1 < t_2 < ... < t_n = t_0 + h$, in $\left[t_j, t_{j+1}\right]$, $y(t) = \lambda_j$. Indeed $\left[t_j, t_{j+1}\right]$ could be any interval. Indeed as in the case of constant, we have by the first hypothesis that $f(t, y(t)) = \lambda_j\ \forall t \in \left[t_j, t_{j+1}\right]$, which implies that $f(t, y(t))$ is Lebesgue integrable on $[t_0, t_0 + h]$.

In general if $y(t)$ is Lebesgue measurable on $[t_0, t_0 + h]$ one can find a sequence of step functions say for example $(y_i(t))$ $(i = 1, 2, ...)$ such that $y_i(t) \to y(t)$ almost everywhere when $i \to \infty$. Therefore with the second hypothesis of the theorem

$$f(t, y_i(t)) \to f(t, y(t)) \tag{7.38}$$

almost everywhere whenever $i \to \infty$. Therefore, we have that

$$\lim_{i\to\infty} f(t, y_i(t)) = f(t, y(t)) \tag{7.39}$$

is also Lebesgue measurable on $[t_0, t_0 + h]$. Since $g^i(t)$ is also Lebesgue measurable on $[t_0, t_0 + h]$, we shall have that

$$\lim_{i\to\infty} g^i(t) f(t, y_i(t)) = g(t) f(t, y(t)) \tag{7.40}$$

therefore as limit $g(t) f(t, y(t))$ is Lebesgue measurable on $[t_0, t_0 + h]$. It is also important to recall that

$$\left|g'(\tau)(t-\tau)^{\alpha-1} f(\tau, y(\tau))\right| \le g'(\tau)\Omega(t)(t-\tau)^{\alpha-1}. \tag{7.41}$$

If we let $\overline{\Omega}(t) = g'(\tau)\Omega(t)$, we will have that $\overline{\Omega}(t)$ is a real-valued function that is $L^2[I_b]$ such that

$$\left|g'(\tau)(t-\tau)^{\alpha-1} f(\tau, y(\tau))\right| \le \overline{\Omega}(t)(t-\tau)^{\alpha-1} \tag{7.42}$$

almost everywhere whenever $\tau \le t$ with $\tau, t \in I_b$. Indeed

$$\begin{aligned} \left|\int_{t_0}^{t_0+h} (t-\tau)^{\alpha-1} d\tau\right|^2 &\le \int_{t_0}^{t_0+h} (t-\tau)^{2\alpha-2} d\tau, \\ &\le \frac{(t-t_0)^{2\alpha-1}}{2\alpha-1} - \frac{(t-t_0-h)^{2\alpha-1}}{2\alpha-1} \end{aligned} \tag{7.43}$$

with $\alpha > \frac{1}{2}$ we can see that

$$(t-\tau)^{\alpha-1} \in L^2[I_b]. \tag{7.44}$$

Noting that

$$\begin{aligned}\left|\int_{t_0}^{t} g'(\tau)(t-\tau)^{\alpha-1} f(\tau,y(\tau))d\tau\right| &\leq \int_{t_0}^{t} \left|g'(\tau)f(\tau,y(\tau))\right|(t-\tau)^{\alpha-1}d\tau, \quad (7.45)\\ &\leq \int_{t_0}^{t} \left|g'(\tau)f(\tau,y(\tau))\right|^2 d\tau \int_{t_0}^{t} \frac{(t-\tau)^{2\alpha-2}}{\Gamma^2(\alpha)}d\tau,\\ &\leq \int_{t_0}^{t} \left|\overline{\Omega}(\tau)\right|^2 d\tau \frac{(t-t_0)^{2\alpha-1}}{(2\alpha-1)\Gamma^2(\alpha)},\\ &\leq \left\|\overline{\Omega}\right\|_{L^2[t_0,t]} \frac{(t-t_0)^{2\alpha-1}}{(2\alpha-1)\Gamma^2(\alpha)}.\end{aligned}$$

If $p > 2$, then we require that $\alpha > \frac{p-1}{p}$ such that $\overline{\Omega} \in L^p[t_0, t]$ and $\frac{(t-t_0)^{p\alpha-p+1}}{p\alpha-p+1}$.

$$\left|\int_{t_0}^{t} g'(\tau)(t-\tau)^{\alpha-1} f(\tau,y(\tau))d\tau\right| \leq \left\|\overline{\Omega}\right\|_{L^p[t_0,t]} \frac{(t-t_0)^{p\alpha-p+1}}{p\alpha-p+1}. \quad (7.46)$$

We should note that the equality was obtained via Cauchy–Schwarz and Hölder inequalities. The first step of the proof is concluded.

In the following step, we shall construct a sequence, in particular the original Tonelli or Picard will be constructed, we will then demonstrate their boundness equicontinuity. Noting that $\Omega^2(t)$ is completely continuous, then $\forall M > 0$, we can find $h' > 0$ that satisfies

$$\int_{t_0}^{t_0+h} \Omega^2(\tau)d\tau \leq M. \quad (7.47)$$

That will be

$$\begin{aligned}\int_{t_0}^{t_0+h} \Omega^2(\tau)\left(g'(\tau)\right)^2 d\tau &\leq \int_{t_0}^{t_0+h} \Omega^2(\tau)d\tau \int_{t_0}^{t_0+h} \left(g'(\tau)\right)^2 d\tau \quad (7.48)\\ &\leq MM_{g'} = \overline{M}.\end{aligned}$$

Whenever

$$h < \min\left\{h', \left(\frac{c^2(2\alpha-1)\Gamma^2(\alpha)}{\overline{M}}\right)^{\frac{1}{2\alpha-1}}\right\}. \quad (7.49)$$

As soon as the proper h is chosen, a sequence $(y_n(t))_{n=1}^{\infty}$ can be defined using either Picard or Tonelli [4, 5]. With Tonelli, we have

$$\begin{cases} y_0, & t_0 \leq t \leq t_0 + \frac{h}{n} \\ \frac{1}{\Gamma(\alpha)} \int\limits_{t_0}^{t_0 - \frac{h}{n}} g'(\tau)\,(t-\tau)^{\alpha-1} f(\tau, y_n(\tau))d\tau, & t_0 - \frac{h}{n} \leq t \leq t_0 + \frac{h}{n}. \end{cases} \tag{7.50}$$

For Picard, we have that

$$\begin{cases} y_n(t) = \frac{1}{\Gamma(\alpha)} \int\limits_{t_0}^{t} g'(\tau)\,(t-\tau)^{\alpha-1} f(\tau, y_{n-1}(\tau))d\tau \\ y(t_0) = y_0 \end{cases} \tag{7.51}$$

with Picard let $t_1 > t_2$, we shall have

$$\begin{aligned} |y_n(t_1) - y_n(t_2)| &= \frac{1}{\Gamma(\alpha)} \left| \begin{array}{l} \int\limits_{t_0}^{t_1} g'(\tau)\,(t_1-\tau)^{\alpha-1} f(\tau, y_{n-1}(\tau))d\tau - \\ \int\limits_{t_0}^{t_2} g'(\tau)\,(t_2-\tau)^{\alpha-1} f(\tau, y_{n-1}(\tau))d\tau \end{array} \right| \\ &= \frac{1}{\Gamma(\alpha)} \left| \begin{array}{l} \int\limits_{t_0}^{t_2} g'(\tau)\,(t_1-\tau)^{\alpha-1} f(\tau, y_{n-1}(\tau))d\tau \\ + \int\limits_{t_2}^{t_1} g'(\tau)\,(t_1-\tau)^{\alpha-1} f(\tau, y_{n-1}(\tau))d\tau \\ - \int\limits_{t_0}^{t_2} g'(\tau)\,(t_2-\tau)^{\alpha-1} f(\tau, y_{n-1}(\tau))d\tau \end{array} \right| \\ &\leq \frac{1}{\Gamma(\alpha)} \int\limits_{t_0}^{t_2} g'(\tau)\left|f(\tau, y_{n-1}(\tau))\right| \left\{(t_1-\tau)^{\alpha-1} - (t_2-\tau)^{\alpha-1}\right\} d\tau \\ &\quad + \frac{1}{\Gamma(\alpha)} \int\limits_{t_2}^{t_1} g'(\tau)\,(t_1-\tau)^{\alpha-1} \left|f(\tau, y_{n-1}(\tau))\right| d\tau \\ &\leq \frac{\overline{M}}{\Gamma(\alpha)} \left(\int\limits_{t_0}^{t_2} \left\{(t_1-\tau)^{\alpha-1} - (t_2-\tau)^{\alpha-1}\right\}^2 d\tau + \int\limits_{t_2}^{t_1} (t_1-\tau)^{2\alpha-2} d\tau \right) \\ &\leq \frac{2\overline{M}}{\Gamma(\alpha)} \left(\frac{(t_1-t_0)^{2\alpha-1}}{2\alpha-1} - \frac{(t_1-t_2)^{2\alpha-1}}{2\alpha-1} - \frac{(t_2-t_0)^{2\alpha-1}}{2\alpha-1} + \frac{(t_1-t_2)^{2\alpha-1}}{2\alpha-1} \right) \\ &\leq \frac{2\overline{M}}{\Gamma(\alpha)\,(2\alpha-1)} \left\{(t_1-t_0)^{2\alpha-1} - (t_2-t_0)^{2\alpha-1}\right\}. \end{aligned} \tag{7.52}$$

Note that $(t-t_0)^{2\alpha-1}$ is differentiable within $[t_2-t_0, t_1-t_0]$; therefore, by the Mean value theorem, $\exists l \in [t_2-t_0, t_1-t_0]$ such that

$$\begin{aligned} |y_n(t_1)-y_n(t_2)| &\leq \frac{\overline{M}\,(2\alpha-1)\,(l-t_0)^{2\alpha-2}}{\Gamma(\alpha)\,(2\alpha-1)}\,(t_1-t_2)\,, \\ &\leq \frac{\overline{M}\,(l-t_0)^{2\alpha-2}}{\Gamma(\alpha)}\,(t_1-t_2)\,, \end{aligned} \tag{7.53}$$

$$|y_n(t_1)-y_n(t_2)| < \varepsilon, \tag{7.54}$$

$\forall\varepsilon > 0$, we will have

$$\delta < \frac{\varepsilon\Gamma(\alpha)\,(l-t_0)^{2-2\alpha}}{\overline{M}} \tag{7.55}$$

which shows the equicontinuity of the sequence $y_n(t)$ on $[t_0, t_0+h]$.

In the case of Tonelli, we shall have from the definition, we have that when $t \in \left[t_0+\frac{h}{n}, t_0+\frac{h}{2n}\right]$, we will have $t_0 \leq t-\frac{h}{n} \leq t_0+\frac{h}{n}$. This with the design of the sequence leads us to the fact that $f(\tau, y_n(\tau)) = f(\tau, y_0)$ whenever $t_0 \leq \tau \leq t-\frac{h}{n}$. Therefore also we shall have that

$$g'(\tau)f(\tau, y_0) = g'(\tau)f(\tau, y_n(\tau)) \tag{7.56}$$

within this interval. This show that $g'(\tau)f(\tau, y_n(\tau))$ is Lebesgue measurable, therefore also $g'(\tau)f(\tau, y_n(\tau))\,(t-\tau)^{\alpha-1}$ is Lebesgue measurable and integrable on $\left[t_0, t-\frac{h}{n}\right]$. We can now discuss this within the following interval $\left[t_0+\frac{h}{n}, t_0+\frac{h}{2n}\right]$. Here two different cases should be presented. The first case will be to consider $t_1 \leq t_0+t_0+\frac{h}{n} \leq t_2 \leq t_0+\frac{h}{2n}$. Our aim here is to show the equicontinuity of the defined sequence.

$$\begin{aligned} |y_n(t_2)-y_n(t_1)| &\leq \int_{t_0}^{t_2-\frac{h}{n}} \frac{(t_2-\tau)^{\alpha-1}}{\Gamma(\alpha)}\,|g'(\tau)|\,|f(\tau, y_n(\tau))|\,d\tau, \\ &\leq \int_{t_0}^{t_2-\frac{h}{n}} \frac{(t_2-\tau)^{\alpha-1}}{\Gamma(\alpha)}\,\sup_{l\in[t_0,\tau]}|g'(l)|\,M(\tau)d\tau, \\ &\leq \int_{t_0}^{t_2-\frac{h}{n}} \frac{(t_2-\tau)^{\alpha-1}}{\Gamma(\alpha)}\,d\tau M_{g'}M \\ &\leq \frac{\overline{M}}{\Gamma(\alpha+1)}\left\{(t_2-t_0)^{\alpha}-\left(t_2-t_2+\frac{h}{n}\right)^{\alpha}\right\} \end{aligned} \tag{7.57}$$

$$\leq \frac{\overline{M}}{\Gamma(\alpha+1)}\left\{(t_2-t_0)^\alpha-\left(\frac{h}{n}\right)^\alpha\right\}$$
$$\leq \frac{\overline{M}}{\Gamma(\alpha+1)}\left\{\left(t_2-\left(t_0+\frac{h}{n}\right)+\frac{h}{n}\right)^\alpha-\left(\frac{h}{n}\right)^\alpha\right\}$$
$$\leq \frac{\overline{M}}{\Gamma(\alpha+1)}\left(t_2-\left(t_0+\frac{h}{n}\right)\right)^\alpha$$
$$\leq \frac{\overline{M}}{\Gamma(\alpha+1)}\left(t_2-\left(t_0+\frac{h}{n}\right)\right)^\alpha<\varepsilon$$

$\forall\varepsilon>0$, we will have

$$\delta<\left(\frac{\varepsilon\Gamma(\alpha+1)}{\overline{M}}\right)^{\frac{1}{\alpha}}. \tag{7.58}$$

However, if we consider the fact that $\overline{M}$ is $L^p\,[I_b]$, then we shall have

$$|y_n(t_2)-y_n(t_1)| \leq \frac{1}{\Gamma(\alpha)}\int_{t_0}^{t_2-\frac{h}{n}}(t_2-\tau)^{\alpha-1}M(\tau)\left|g'(\tau)\right|d\tau \tag{7.59}$$
$$\leq \frac{1}{\Gamma(\alpha)}\int_{t_0}^{t_2-\frac{h}{n}}(t_2-\tau)^{\alpha-1}M(\tau)M_{g'}(\tau)d\tau$$
$$\leq \frac{1}{\Gamma(\alpha)}\int_{t_0}^{t_2-\frac{h}{n}}(t_2-\tau)^{\alpha-1}\overline{M}(\tau)d\tau$$
$$\leq \frac{\sqrt{\overline{M}}}{\Gamma(\alpha)\sqrt{2\alpha-1}}\left[(t_2-t_0)^{2\alpha-1}-\left(\frac{h}{n}\right)^{2\alpha-1}\right]^{\frac{1}{2}}$$
$$\leq \frac{\sqrt{\overline{M}}}{\Gamma(\alpha)\sqrt{2\alpha-1}}\left[\left(t_2-\left(t_0+\frac{h}{n}\right)+\frac{h}{n}\right)^{2\alpha-1}-\left(\frac{h}{n}\right)^{2\alpha-1}\right]^{\frac{1}{2}}$$
$$\leq \frac{\sqrt{\overline{M}}}{\Gamma(\alpha)\sqrt{2\alpha-1}}\left[t_2-\left(t_0+\frac{h}{n}\right)\right]^{\alpha-\frac{1}{2}}.$$

Therefore $\forall\varepsilon>0$ such that

$$|y_n(t_2)-y_n(t_1)|<\varepsilon \tag{7.60}$$

we shall have

$$\delta<\left(\frac{\varepsilon^2\Gamma^2(\alpha)\,(2\alpha-1)}{\overline{M}}\right)^{\frac{1}{2\alpha-1}}. \tag{7.61}$$

But when $\left(t_0 + \frac{h}{n}\right) \leq t_1 \leq t_2 \leq \left(t_0 + \frac{2h}{n}\right)$, we have

$$
\begin{aligned}
J_1 &= \int_{t_0}^{t_1-\frac{h}{n}} \left[(t_1-\tau)^{\alpha-1} - (t_2-\tau)^{\alpha-1}\right]^2 d\tau \\
&= \int_{t_0}^{t_1-\frac{h}{n}} \left[(t_1-\tau)^{2\alpha-2} - 2(t_1-\tau)^{\alpha-1}(t_2-\tau)^{\alpha} + (t_2-\epsilon)^{2\alpha-2}\right] d\tau \\
&= \int_{t_0}^{t_1-\frac{h}{n}} \left[(t_2-\tau)^{\alpha-1} - (t_1-\tau)^{\alpha-1}\right]^2 d\tau
\end{aligned}
\tag{7.62}
$$

$t_2 > t_1 \to t_2 - \tau > t_1 - \tau$, $\forall \epsilon \in \left[t_0, t_1 - \frac{h}{n}\right]$; therefore, we have using the fact that $\forall_{a,b} \in R^+$ such that $b > a$, we have

$$
b^2 - a^2 > (b-a)^2, \tag{7.63}
$$

therefore

$$
\begin{aligned}
J_1 &\leq \int_{t_0}^{t_1-\frac{h}{n}} \left[(t_2-\tau)^{2\alpha-2} - (t_1-\tau)^{2\alpha-2}\right] d\tau \\
&\leq \frac{(t_2-t_0)^{2\alpha-1}}{2\alpha-1} - \frac{\left(t_2-t_1+\frac{h}{n}\right)^{2\alpha-1}}{2\alpha-1} + \frac{(t_1-t_0)^{2\alpha-1}}{2\alpha-1} - \frac{\left(t_1-t_1+\frac{h}{n}\right)^{2\alpha-1}}{2\alpha-1} \\
&\leq \frac{(t_2-t_0)^{2\alpha-1}}{2\alpha-1} - \frac{\left(t_2-t_1+\frac{h}{n}\right)^{2\alpha-1}}{2\alpha-1} - \frac{\left(\frac{h}{n}\right)^{2\alpha-1}}{2\alpha-1} + \frac{(t_1-t_0)^{2\alpha-1}}{2\alpha-1} \\
&\leq \frac{1}{2\alpha-1}(\Lambda_1 + \Lambda_2)
\end{aligned}
\tag{7.64}
$$

where

$$
\begin{aligned}
\Lambda_1 &= (t_2-t_0)^{2\alpha-1} - (t_1-t_0)^{2\alpha-1}, \\
\Lambda_2 &= \left(t_2 - t_1 + \frac{h}{n}\right)^{2\alpha-1} - \left(\frac{h}{n}\right)^{2\alpha-1}.
\end{aligned}
\tag{7.65}
$$

Therefore, we shall have $\forall \varepsilon > 0$, $\exists \delta > 0$ such that $(t_2 - t_1) < \delta_1$.

$$
\Lambda_1 < \frac{(2\alpha-1)\left[\varepsilon\Gamma(\alpha)\right]^2}{8\overline{M}}. \tag{7.66}
$$

Also we can find δ_2 such that $(t_2 - t_1) < \delta_2$

$$\Lambda_2 \leq (t_2 - t_1)^{2\alpha - 1} < \frac{(2\alpha - 1)\,[\varepsilon \Gamma\,(\alpha)]^2}{8M}. \tag{7.67}$$

It is therefore necessary to choose $(t_2 - t_1)$, $\overline{\delta} = \min\{\delta_1, \delta_2, \delta\}$ whereas

$$\begin{aligned} J_2 &= \frac{1}{\Gamma(\alpha)} \int_{t_0}^{t_1 - \frac{h}{n}} \left[(t_2 - \tau)^{\alpha-1} - (t_1 - \tau)^{\alpha-1}\right] \left|g'(\tau)\right| \left|f(\tau, y_n(\tau))\right| d\tau \qquad (7.68) \\ &\leq \frac{1}{\Gamma(\alpha)} \int_{t_0}^{t_1 - \frac{h}{n}} \left[(t_2 - \tau)^{\alpha-1} - (t_1 - \tau)^{\alpha-1}\right] M(\tau) M_{g'}(\tau) d\tau \\ &\leq \frac{1}{\Gamma(\alpha)} \int_{t_0}^{t_1 - \frac{h}{n}} \left[(t_2 - \tau)^{\alpha-1} - (t_1 - \tau)^{\alpha-1}\right] \overline{M}(\tau) d\tau \\ &\leq \frac{\sqrt{\overline{M}}}{\Gamma(\alpha)} J_1^{\frac{1}{2}} \\ &\leq \sqrt{\frac{\overline{M}}{2\alpha - 1}} \frac{(\Lambda_1 + \Lambda_2)^{\frac{1}{2}}}{\Gamma(\alpha)} \\ &\leq \frac{\varepsilon}{2}. \end{aligned}$$

On the other hand however

$$\begin{aligned} J_3 &= \frac{1}{\Gamma(\alpha)} \int_{t_1 - \frac{h}{n}}^{t_2 - \frac{h}{n}} (t_2 - \tau)^{\alpha-1} \left|g'(\tau)\right| \left|f(\tau, y_n(\tau))\right| d\tau \qquad (7.69) \\ &\leq \frac{1}{\Gamma(\alpha)} \int_{t_1 - \frac{h}{n}}^{t_2 - \frac{h}{n}} (t_2 - \tau)^{\alpha-1} M(\tau) M_{g'}(\tau) d\tau \\ &\leq \frac{1}{\Gamma(\alpha)} \int_{t_1 - \frac{h}{n}}^{t_2 - \frac{h}{n}} (t_2 - \tau)^{\alpha-1} \overline{M}(\tau) d\tau \\ &\leq \sqrt{\frac{\overline{M}}{2\alpha - 1}} \frac{(\Lambda_2)^{\frac{1}{2}}}{\Gamma(\alpha)}. \end{aligned}$$

This also implies that $\forall\varepsilon > 0$ such

$$\begin{aligned} J_3 &\le \frac{\varepsilon}{2\sqrt{2}}, \\ |y_n(t_1) - y_n(t_2)| &\le J_2 + J_3 < \varepsilon, \\ (t_2 - t_1) &< \overline{\delta}. \end{aligned} \tag{7.70}$$

Again we obtain the same result leading us to conclude that the sequence defined via Tonelli approach $y_n(t)$ is equicontinuous with respect to $t \in [t_0, t_0 + \frac{2h}{n}]$, $\forall n \ge 1$. We shall now show that the sequence is $y_n(t)$ is well-defined. To do this, we shall evaluate

$$|y_n(t) - y_0| = 0 \tag{7.71}$$

when $t \in [t_0, t_0 + \frac{h}{n}]$.

But when $t \in \left[t_0 + \frac{h}{l}, t_0 + \frac{2h}{n}\right]$,

$$\begin{aligned} |y_n(t)| &\le \frac{1}{\Gamma(\alpha)} \int_{t_0}^{t-\frac{h}{l}} (t-\tau)^{\alpha-1} M(\tau) \left|g'(\tau)\right| d\tau \\ &\le \frac{1}{\Gamma(\alpha)} \int_{t_0}^{t-\frac{h}{l}} (t-\tau)^{\alpha-1} M(\tau) M_{g'}(\tau) d\tau \\ &\le \frac{1}{\Gamma(\alpha)} \int_{t_0}^{t-\frac{h}{l}} (t-\tau)^{\alpha-1} \overline{M}(\tau) d\tau \\ &\le \frac{1}{\Gamma(\alpha)} \sqrt{\frac{\overline{M}}{2\alpha-1}} \left[(t-t_0)^{2\alpha-1} - \left(\frac{h}{l}\right)^{2\alpha-1}\right]^{\frac{1}{2}} \\ &\le \frac{1}{\Gamma(\alpha)} \sqrt{\frac{\overline{M}}{2\alpha-1}} h^{\alpha-\frac{1}{2}} \\ &\le c \end{aligned} \tag{7.72}$$

which shows that $\forall t \in I_b$, $(t, y_n(t)) \in D$. The procedure can be repeated until we cover the whole interval, and we can conclude that $(y_n(t))_n$ is uniformly equicontinuous and bounded. Using the Arzela–Ascoli Lemma, there exists a subsequence $(y_{nl}(t))$ of $(y_n(t))$ that converges uniformly to $y(t)$. We shall show that $y(t)$ is the solution to the equation. From the second hypothesis of the theorem, we can find $\forall \eta > 0$, $\exists \overline{N}$, such that whenever $l > \overline{N}$,

$$|f(t, y_{nl}(t)) - f(t, y(t))| < \frac{\Gamma(\alpha+1)\eta}{2h^{\alpha}}. \tag{7.73}$$

Therefore

$$
\begin{aligned}
g'(t)\left|f\left(t, y_{nl}(t)\right)-f\left(t, y(t)\right)\right| &< \frac{g'(\tau)\Gamma\left(\alpha+1\right)\eta}{2h^{\alpha}} \\
&< \frac{M_{g'}\Gamma\left(\alpha+1\right)\eta}{2h^{\alpha}}.
\end{aligned}
\tag{7.74}
$$

It is then shown that $y(t)$ satisfies the equation. We then have that

$$
\begin{aligned}
&= \frac{1}{\Gamma(\alpha)}\left|\begin{array}{l}\displaystyle\int_{t_0}^{t-\frac{h}{l}} (t-\tau)^{\alpha-1} g'(\tau) f\left(\tau, y_{nl}(\tau)\right) d\tau \\ \displaystyle-\int_{t_0}^{t} (t-\tau)^{\alpha-1} g'(\tau) f\left(\tau, y(\tau)\right) d\tau\end{array}\right| \qquad (7.75) \\
&= \frac{1}{\Gamma(\alpha)}\left|\begin{array}{l}\displaystyle\int_{t_0}^{t} (t-\tau)^{\alpha-1} g'(\tau)\left[f\left(\tau, y_{nl}(\tau)\right)-f\left(\tau, y(\tau)\right)\right] d\tau \\ \displaystyle\qquad -\int_{t-\frac{h}{l}}^{t_0} (t-\tau)^{\alpha-1} g'(\tau) f\left(\tau, y_{nl}(\tau)\right) d\tau\end{array}\right| \\
&\leq \frac{1}{\Gamma(\alpha)}\int_{t_0}^{t} (t-\tau)^{\alpha-1} M_{g'}(\tau)\left|f\left(\tau, y_{nl}(\tau)\right)-f\left(\tau, y(\tau)\right)\right| d\tau \\
&\quad + \frac{1}{\Gamma(\alpha)}\int_{t-\frac{h}{l}}^{t_0} (t-\tau)^{\alpha-1} M_{g'}(\tau)\left|f\left(\tau, y_{nl}(\tau)\right)\right| d\tau \\
&\leq \frac{\Gamma(\alpha+1)\eta h^{\alpha}}{2h^{\alpha}\Gamma(\alpha+1)} < \frac{\eta}{2}+\frac{\eta}{2}.
\end{aligned}
$$

Then we have also that $\overline{N}_1 = \left(\frac{2c}{\eta}\right)^{\frac{2}{2\alpha-1}}$ such that $l > \overline{N}_1$. We know $N_3 = \max\left\{\overline{N}, \overline{N}_1\right\}$. We arrived at

$$
y(t) = \frac{1}{\Gamma(\alpha)}\int_{t_0}^{t} g'(\tau)\left(t-\tau\right)^{\alpha-1} f(\tau, y(\tau)) d\tau \tag{7.76}
$$

which completes the proof and the existence of a unique local solution is achieved.

Theorem 7.6 *Denote by* $I_b = [0, b]$, $E = \{y \in R \,||y - y_0| \le c\}$, $D = \{(t, y) \in R \times R \,|t \in I_b, y \in E\}$.

(1) $f(t, y)$ is Lebesgue measurable with respect to t on I_b.
(2) $f(t, y)$ is continuous with respect $y \in E$.
(3) There exists a real-valued function $\Omega(t) \in L^2(I_b)$ such that

$$|f(t, y)| \le \Omega(t) \tag{7.77}$$

for almost every $t \in I_b$ and all $y \in E$.

(4) There exists real-valued function $\Omega(t) \in L^4(I_b)$ such that

$$|f(t, y) - f(t, \overline{y})| \le \Omega(t)\,|y - \overline{y}| \tag{7.78}$$

for almost every $t \in I_b$ and all $y, \overline{y} \in D$. There exists a unique solution of the initial value problem on $[t_0, t_0 + h]$ with some positive number h if $g'(t) \in L^4(I_b)$.

Proof The property of $\eta^4(t)$ allows a positive number β such that $\exists h^* > 0$, $\forall t \in [t_0, t_0 + h]$

$$\begin{aligned} \int_{t_0}^{t} |\eta(\tau)|^4 \left|g'(\tau)\right|^4 d\tau &\le \int_{t_0}^{t_0+h^*} \eta^4 \left(g'(\tau)\right)^4 d\tau \\ &\le \int_{t_0}^{t_0+h^*} \eta^4 d\tau \int_{t_0}^{t_0+h^*} \left(g'(\tau)\right)^4 d\tau \\ &\le \beta^2 \left\|g'\right\|^4_{L^4(I_b)}. \end{aligned} \tag{7.79}$$

We now denote the space of continuous functions on Γ by $C(I_b)$.

$$\Lambda_h = \{y \in C[t_0, t_0 + h], \,||y(t)| < c, \forall t \in [t_0, t_0 + h]\}. \tag{7.80}$$

$h > 0$, $\min\{h', h^*\}$ where h' is the one that was obtained from the theorem above.

$$\begin{aligned} \Gamma(y) &= \mu, \\ \mu(t) &= \frac{1}{\Gamma(\alpha)} \int_{t_0}^{t} g'(\tau)(t - \tau)^{\alpha-1} f(\tau, y(\tau)) d\tau. \end{aligned} \tag{7.81}$$

Let $t_0 \le t_1 \le t_2 \le t_0 + h$ then

$$\le \sqrt{\frac{\overline{M}}{2\alpha - 1}} \left\{ \frac{\left[(l - t_0)^{2\alpha-2}(t_2 - t_1) + (t_2 - t_1)^{\alpha}\right]^{\frac{1}{2}} + (t_2 - t_1)^{\frac{\alpha}{2}}}{\Gamma(\alpha)} \right\} \tag{7.82}$$

$$\le 2\left(\max\left\{(l - t_0)^{2\alpha-2}, 1\right\}\right)^{\frac{1}{2}} \left\{ \begin{array}{c} \left((t_2 - t_1)^{\frac{1}{2}} + (t_2 - t_1)^{\frac{\alpha}{2}}\right) \\ + (t_2 - t_1)^{\frac{\alpha}{2}} \end{array} \right\} \frac{\sqrt{\overline{M}}}{\sqrt{2\alpha - 1}\Gamma(\alpha)}.$$

We know $\underset{\alpha \in (\frac{1}{2}, 1]}{max} \left\{(t_2 - t_1)^{\frac{1}{2}}, (t_2 - t_1)^{\frac{\alpha}{2}}\right\}$.

Then

$$\mu(t_2) - \mu(t_1) \le \sqrt{\frac{\overline{M}}{2\alpha - 1}} 4 \left(\underset{\alpha \in (\frac{1}{2}, 1)}{max} \left\{(l - t_0)^{2\alpha-2}, 1\right\} \right) \times \underset{\alpha \in (\frac{1}{2}, 1]}{max} \left\{(t_2 - t_1)^{\frac{1}{2}}, (t_2 - t_1)^{\frac{\alpha}{2}}\right\} \tag{7.83}$$

If then

$$\underset{\alpha \in (\frac{1}{2}, 1]}{max} \left\{(t_2 - t_1)^{\frac{1}{2}}, (t_2 - t_1)^{\frac{\alpha}{2}}\right\} = (t_2 - t_1)^{\frac{\alpha}{2}}, \tag{7.84}$$

δ can be obtained the same if $(t_2 - t_1)^{\frac{1}{2}}$ and this shows $\mu(t)$ is continuous on I_b.

$$\begin{aligned} |\mu(t_2) - \mu(t_1)| &\le \frac{1}{\Gamma(\alpha)} \int_{t_0}^{t_1} \left\{(t_2 - \tau)^{\alpha-1} - (t_1 - \tau)^{\alpha-1}\right\} \left|g'(\tau)\right| |f(\tau, \mu(\tau))| \, d\tau \\ &+ \frac{1}{\Gamma(\alpha)} \int_{t_1}^{t_2} (t_2 - \tau)^{\alpha-1} \left|g'(\tau)\right| |f(\tau, \mu(\tau))| \, d\tau \\ &\le \sqrt{\frac{\overline{M}}{2\alpha - 1}} \frac{(\Lambda_1 + \Lambda_2^*)^{\frac{1}{2}} + \Lambda_2^{\frac{1}{2}}}{\Gamma(\alpha)}, \end{aligned} \tag{7.85}$$

where

$$\begin{aligned} \Lambda_1 &= (t_2 - t_0)^{2\alpha-1} - (t_1 - t_0)^{2\alpha-1}, \\ \Lambda_2^* &= (t_2 - t_1)^{\alpha}. \end{aligned} \tag{7.86}$$

The function $(t-t_0)^{2\alpha-1}$ is differentiable between $[t_1-t_0, t_2-t_0]$ by the mean value theorem and there exist $l \in [t_1-t_0, t_2-t_0]$ such that $(2\alpha-1)(l-t_0)^{2\alpha-2}(t_2-t_1)$. Therefore

$$|\mu(t_2)-\mu(t_1)| \leq \sqrt{\frac{\overline{M}}{2\alpha-1}}\frac{\left[(2\alpha-1)(l-t_0)^{2\alpha-2}(t_2-t_1)+(t_2-t_1)^{\alpha}\right]^{\frac{1}{2}}}{\Gamma(\alpha)} \tag{7.87}$$

Since $l \geq \alpha > \frac{1}{2}$, $(2\alpha-1) \leq 1$.

$$\begin{aligned} |\mu(t)| &\leq \frac{1}{\Gamma(\alpha)}\int_{t_0}^{t}(t-\tau)^{2\alpha-2} M M_{g'} d\tau, \\ &\leq \frac{1}{\Gamma(\alpha)}\sqrt{\frac{\overline{M}}{2\alpha-1}} h^{\alpha-\frac{1}{2}} \leq c. \end{aligned} \tag{7.88}$$

This implies that

$$\Gamma(y) = \mu \in \Lambda_h. \tag{7.89}$$

Let us consider then μ_1 and $\mu_2 \in \Lambda_h$. We pick $\lambda > 0$, then

$$\begin{aligned} &\exp[-\lambda(t-t_0)]\,|\Gamma(\mu_1)-\Gamma(\mu_2)| \\ &\leq \frac{1}{\Gamma(\alpha)}\int_{t_0}^{t}(t-\tau)^{\alpha-1}\exp[-\lambda(t-t_0)]\,g'(\tau)\,|f(\tau, y_1(\tau))-f(\tau, y_2(\tau))|\,d\tau \\ &\leq \frac{1}{\Gamma(\alpha)}\int_{t_0}^{t}(t-\tau)^{\alpha-1} e^{-\lambda(t-\tau)}\eta(\tau)\, e^{-\lambda(\tau-t_0)}\,|\mu_1(\tau)-\mu_2(\tau)|\,d\tau \\ &\leq \|\mu_1-\mu_2\|_{\lambda}\left\|(t-\tau)^{\alpha-1}\right\|_{L^2[t_0,t]}\left\|e^{-\lambda(t-\tau)}\right\|_{L^4[t_0,t]}\|\eta(\tau)\|_{L^4[t_0,t]}\left\|g'\right\|_{L^4[t_0,t]} \\ &\leq \|\mu_1-\mu_2\|_{\lambda}\frac{\left(1-e^{-4\lambda h}\right)^{\frac{1}{4}}}{\Gamma(\alpha)\lambda^{\frac{1}{4}}}\sqrt{\frac{\beta}{4\alpha-2}} M_{g'} h^{\alpha}. \end{aligned} \tag{7.90}$$

Noting that

$$\|\mu\|_{\lambda} = \sup_{t\in[t_0,t_0+h]}\left\{e^{-\lambda(t-t_0)}\|\mu\|\right\}, \forall \mu \in \Lambda_h. \tag{7.91}$$

Therefore, we have that

$$|\Gamma(\mu_1)-\Gamma(\mu_2)| \leq \left(1-e^{-4\lambda h}\right)^{\frac{1}{4}} M_{g'}\|\mu_1-\mu_2\|_{\lambda} \tag{7.92}$$

where

$$\lambda > \frac{h^{4\alpha-2}}{\Gamma^4(\alpha)} \left(\frac{\beta}{4\alpha - 2} \right)^2 M_{g'}. \tag{7.93}$$

Λ_h is a Banach space induced by the norm $\|.\|_\lambda$. The contractive mapping principle leads to the existence of unique solution $y^*(t)$.

$$y^*(t) = \frac{1}{\Gamma(\alpha)} \int_{t_0}^{t} g'(\tau) (t-\tau)^{\alpha-1} f(\tau, y^*(\tau)) d\tau \tag{7.94}$$

of which completes the proof.

The global existence shall also be discussed into the following section.

Theorem 7.7 *Let assume that $f(t, y)$ is Carathéodory in the global space and*

$$|f(t, y)| \leq a + b\,|y|, \tag{7.95}$$

$\forall t \in R$ and $y \in R$, $a, b > 0$. Then there exists function $y(t) \in R$ as a solutionof our equation.

Proof Let $f(t, y)$ satisfies the hypothesis then

$$\overline{D} = \{(t, y) \in R \times R\, \| t - t_0| \leq b,\ |y - y_0| \leq a\} \tag{7.96}$$

$\forall t_0 \in R$ and $y_0 \in R$, then the solution is at least a solution $y(t)$ of the equation on $[t_0, t_0 + h]$.

$$\begin{aligned} |y(t)| &\leq \frac{1}{\Gamma(\alpha)} \int_{t_0}^{t} g'(\tau) (t-\tau)^{\alpha-1} (a + by(\tau))\, d\tau \\ &\leq \frac{1}{\Gamma(\alpha)} \int_{t_0}^{t} g'(\tau) (t-\tau)^{\alpha-1} a d\tau \\ &\quad + \frac{1}{\Gamma(\alpha)} \int_{t_0}^{t} g'(\tau) (t-\tau)^{\alpha-1} b\, |y(\tau)|\, d\tau \\ &\leq \frac{M_{g'}}{\Gamma(\alpha+1)} a\, (t-t_0)^\alpha + \frac{bM_{g'}}{\Gamma(\alpha)} \int_{t_0}^{t} (t-\tau)^{\alpha-1} |y(\tau)|\, d\tau. \end{aligned} \tag{7.97}$$

We have that $\Phi(t) = |y(t)|$.

$$\Phi(t) \leq v(t) + \int_{t_0}^{t} \overline{z}(\tau) \Phi(\tau) d\tau. \tag{7.98}$$

$v(t)$ needs all the requirements of the Gronwall; therefore,

$$\begin{aligned}\Phi(t) &\leq v(t)\exp\left[\int_{t_0}^{t}\overline{z}(\tau)d\tau\right] \\ &\leq v(t)\exp\left[\frac{M_{g'}b(t-t_0)^{\alpha}}{\Gamma(\alpha+1)}\right] \\ &\leq \frac{M_g a(t-t_0)^{\alpha}}{\Gamma(\alpha+1)}\exp\left[\frac{M_{g'}b(t-t_0)^{\alpha}}{\Gamma(\alpha+1)}\right] \\ &< \infty\end{aligned} \tag{7.99}$$

which completes the proof.

7.3 Existence and Uniqueness for IVP with the Caputo–Fabrizio Global Derivative via Carathéodory

In this section the IVP with the Caputo–Fabrizio global derivative is considered.

$$\begin{cases} {}_{t_0}^{CF}D_g^{\alpha}y(t) = f(t, y(t)) & \text{if } t > 0 \\ y(t_0) = y_0 & \text{if } t = 0 \end{cases}. \tag{7.100}$$

By the fundamental theorem of Calculus, the above can be transformed to

$$\begin{aligned} y(t) &= y(t_0) + (1-\alpha)g'(t)f(t, y(t)) + \alpha\int_{t_0}^{t} g'(\tau)f(\tau, y(\tau))d\tau \text{ if } t > 0, \\ y(t_0) &= y_0. \end{aligned} \tag{7.101}$$

We shall present the existence and uniqueness of the above equation using the Carathéodory function within $[t_0, t_0 + h]$.

Theorem 7.8 *Let $f(t, y(t))$ be Lipschitz within $[t_0, T]$, then the following inequality is obtained* $\forall y, \overline{y} \in E$, $\forall t \in [t_0, T]$

$$|y(t) - \overline{y}(t)| = 0 \tag{7.102}$$

if $g(t) = t$ and $1 - (1-\alpha)L > 0$.

Proof Let $t \in [t_0, T]$ and $y, \overline{y} \in E$ then

$$
\begin{aligned}
|y(t) - \overline{y}(t)| &= \left| \begin{array}{c} (1-\alpha) f(t, y(t)) - (1-\alpha) f(t, \overline{y}(t)) \\ +\alpha \int\limits_{t_0}^{t} f(\tau, y(\tau)) d\tau - \alpha \int\limits_{t_0}^{t} f(\tau, \overline{y}(\tau)) d\tau \end{array} \right| \qquad (7.103) \\
&\leq (1-\alpha) |f(t, y(t)) - f(t, \overline{y}(t))| \\
&\quad + \alpha \int\limits_{t_0}^{t} |f(\tau, y(\tau)) - f(\tau, \overline{y}(\tau))| d\tau \\
&\leq (1-\alpha) L |y - \overline{y}| + \alpha L \int\limits_{t_0}^{t} |y(\tau) - \overline{y}(\tau)| d\tau \\
&\leq (1-\alpha) L \Phi(t) + \alpha L \int\limits_{t_0}^{t} \Phi(\tau) d\tau.
\end{aligned}
$$

If we let

$$\Phi(t) = |y(t) - \overline{y}(t)|. \qquad (7.104)$$

Since $1 - (1-\alpha) L > 0$, then we have

$$\Phi(t) \leq \frac{\alpha}{(1 - (1-\alpha) L)} \int\limits_{t_0}^{t} \Phi(\tau) d\tau. \qquad (7.105)$$

With the help of the Gronwall inequality, we get

$$\Phi(t) \leq \varepsilon \exp\left[\frac{\alpha}{(1 - (1-\alpha) L)} (t - t_0)\right] \qquad (7.106)$$

where indeed $\varepsilon = 0$ therefore $\forall t \in [t_0, T], \forall y, \overline{y} \in E$ if $f(t, y(t))$ is Lipschitz with respect to y then

$$\Phi(t) = |y(t) - \overline{y}(t)| = 0. \qquad (7.107)$$

If $g(t) \neq t$, such that $g'(t)$ is continuous and positive $\forall t \in [t_0, T]$ then we shall have

$$
\begin{aligned}
|y(t) - \overline{y}(t)| &= \left| \begin{array}{c} (1-\alpha) g'(t) f(t, y(t)) - (1-\alpha) g'(t) f(t, \overline{y}(t)) \\ +\alpha \int\limits_{t_0}^{t} g'(\tau) f(\tau, y(\tau)) d\tau - \alpha \int\limits_{t_0}^{t} g'(\tau) f(\tau, \overline{y}(\tau)) d\tau \end{array} \right| \qquad (7.108) \\
&\leq (1-\alpha) g'(t) |f(t, y(t)) - f(t, \overline{y}(t))| + \alpha \int\limits_{t_0}^{t} g'(\tau) |f(\tau, y(\tau)) - f(\tau, \overline{y}(\tau))| d\tau
\end{aligned}
$$

$$\leq (1-\alpha) L \left|g'(t)\right| |y-\overline{y}| + \alpha L \int_{t_0}^{t} g'(\tau) |y(\tau) - \overline{y}(\tau)| \, d\tau$$

$$\leq (1-\alpha) \left|g'(t)\right| L\Phi(t) + \alpha L \int_{t_0}^{t} g'(\tau)\Phi(\tau) \, d\tau,$$

taking advance of the boundness of the derivative $g'(t)$, we get

$$\begin{aligned} \Phi(t) &\leq \frac{\alpha L}{\left(1-(1-\alpha) M_{g'} L\right)} \int_{t_0}^{t} g'(\tau)\Phi(\tau) \, d\tau \\ &\leq \varepsilon \exp\left[\frac{\alpha L}{\left(1-(1-\alpha) M_{g'} L\right)} (g(t) - g(t_0))\right] \\ &\leq 0. \end{aligned}$$

Since $\varepsilon = 0$.

However in this case the function satisfies the Carathéodory principle, we shall obtain alternative results using different approaches. We start with the existence. We choose $f : [t_0, T] \times R \to R$, continuous. To establish the Carathéodory–Tonelli, we define

$$y_n(t) = \begin{cases} y_0, & \text{if } t_0 \leq t \leq t_0 + \frac{h}{n} \\ y_0 + (1-\alpha) g'(t) f(t, y_n(t)) \\ +\alpha \displaystyle\int_{t_0}^{t-\frac{h}{n}} g'(\tau) f(\tau, y_n(\tau)) d\tau & \text{if } t_0 + \frac{h}{n} \leq t \leq T \end{cases}. \tag{7.109}$$

We can choose to have h smaller or take it to be 1, such that if $t_0 = 0$ we get

$$y_n(t) = \begin{cases} y_0, & \text{if } 0 \leq t \leq \frac{1}{n} \\ y_0 + (1-\alpha) g'(t) f(t, y_n(t)) \\ +\alpha \displaystyle\int_{0}^{t-\frac{1}{n}} g'(\tau) f(\tau, y_n(\tau)) d\tau & \text{if } \frac{1}{n} \leq t \leq 1 \end{cases}. \tag{7.110}$$

Here it is clear that the computation of $y_n(t)$ for $t \in \left[\frac{j}{n}, \frac{j+1}{n}\right]$ $(1 \leq j \leq n-1)$ requires the establishment of $y_n(t)$ with $\left[\frac{j-1}{n}, \frac{j}{n}\right]$. The procedure was called the method of backward stop. The defined sequence is equicontinuous equibounded at least in an $[0, l]$ with $0 \leq l \leq 1$; therefore, the Ascoli–Arzela compactness condition will obtain a subsequence (y_{ni}) of (y_n) that converges uniformly to the solution of the equation, with an additional condition that $g'(t)$ is bounded. We shall show a detailed proof of these statements. The results hold if $t \in \left[0, \frac{1}{n}\right]$, our concern is which $t \in \left[\frac{1}{n} \leq t \leq 1\right]$. Let then $t \in \left[\frac{1}{n}, T\right]$,

$$
\begin{aligned}
y_n(t) &\leq |y_n(t)| = \left| \begin{array}{l} y_0 + (1-\alpha)\, g'(t) f(t, y_n(t)) \\ +\alpha \int\limits_{t_0}^{t} g'(\tau) f(\tau, y_n(\tau)) d\tau \end{array} \right| \qquad (7.111) \\
&\leq |y(0)| + (1-\alpha)\, g'(t)\, |f(t, y_n(t))| + \alpha \int\limits_{t_0}^{t-\frac{1}{n}} g'(\tau) f(\tau, y_n(\tau)) d\tau \\
&\leq |y(0)| + (1-\alpha)\, M_{g'} (1 + |y_n(t)|)\, M(t) + \alpha \int\limits_{t_0}^{t-\frac{1}{n}} M_{g'} (1 + |y_n(\tau)|)\, M(\tau) d\tau \\
&\leq |y(0)| + M_{g'} (1-\alpha)\, M(t) + M_{g'} (1-\alpha)\, M(t)\, |y_n(t)| + \alpha M_{g'} \int\limits_{0}^{t-\frac{1}{n}} M(\tau) d\tau \\
&\quad + \alpha M_{g'} \int\limits_{0}^{t-\frac{1}{n}} |y_n(\tau)|\, M(\tau) d\tau \\
y_n(t) &\leq \frac{y(0)}{1 - M_{g'} (1-\alpha)\, M(t)\, (1-\alpha)} + \frac{(1-\alpha)\, M_{g'} M(t)}{1 - M_{g'} (1-\alpha)\, M(t)\, (1-\alpha)} \\
&\quad + \frac{\alpha M_{g'}}{1 - M_{g'} (1-\alpha)\, M(t)\, (1-\alpha)} \int\limits_{0}^{t-\frac{1}{n}} |y_n(\tau)|\, M(\tau) d\tau.
\end{aligned}
$$

Let

$$\overline{K} = \min_{t \in [0,1]} \left\{1 - M_{g'} (1-\alpha)\, M(t)\, (1-\alpha)\right\}$$

then we have

$$
\begin{aligned}
|y_n(t)| &\leq \frac{y(0)}{\overline{K}} + \frac{(1-\alpha)\, M_{g'} M(t)}{\overline{K}} + \frac{\alpha M_{g'}}{\overline{K}} \int\limits_{0}^{t-\frac{1}{n}} |y_n(\tau)|\, M(\tau) d\tau, \qquad (7.112) \\
&\leq v(t) + \int\limits_{0}^{t-\frac{1}{n}} \frac{\alpha M_{g'}}{\overline{K}} |y_n(\tau)|\, M(\tau) d\tau, \\
&\leq v(t) \exp\left[\int\limits_{0}^{t-\frac{1}{n}} \frac{\alpha M_{g'}}{\overline{K}} M(\tau) d\tau\right] \\
&\leq v(t) \exp\left[\frac{\alpha M_{g'}}{\overline{K}} \left(\overline{M}(t - \frac{1}{n}) - \overline{M}(0)\right)\right].
\end{aligned}
$$

If $\overline{M}(t)$ is the integral of $M(t)$. Therefore $\forall n \geq 0$

$$|y_n(t)| \leq v(t) \exp\left[\frac{\alpha M_{g'}}{\overline{K}}\overline{M}\left(t - \frac{1}{n}\right)\right]. \tag{7.113}$$

This shows that $y_n(t)$ is equibounded and indeed $|y_n(t)| < \beta$. We now show that $(y_n(t))$ is equicontinuous. By taking $t_1 > t_2$

$$\begin{aligned}
|y_n(t_1) - y_n(t_2)| &\leq (1-\alpha)\,|f(t_1, y_n(t_1)) - f(t_2, y_n(t_2))| \\
&\quad + \alpha \int_{t_2-\frac{1}{n}}^{t_1-\frac{1}{n}} g'(\tau)\,|f(\tau, y_n(\tau))|\,d\tau \\
&\leq (1-\alpha)\,|f(t_1, y_n(t_1)) - f(t_2, y_n(t_2))| \\
&\quad + \alpha \int_{t_2-\frac{1}{n}}^{t_1-\frac{1}{n}} M_{g'} M(\tau)\,(1 + |y_n(\tau)|)\,d\tau \\
&\leq (1-\alpha)\,|f(t_1, y_n(t_1)) - f(t_2, y_n(t_2))| \\
&\quad + \alpha M_{g'} \int_{t_2-\frac{1}{n}}^{t_1-\frac{1}{n}} M(\tau) d\tau + \alpha M_{g'} \int_{t_2-\frac{1}{n}}^{t_1-\frac{1}{n}} M(\tau)\,|y_n(\tau)|\,d\tau \\
&\leq (1-\alpha)\,|f(t_1, y_n(t_1)) - f(t_2, y_n(t_2))| \\
&\quad + \alpha M_{g'} M\,(t_1 - t_2) + \alpha M_{g'} M \int_{t_2-\frac{1}{n}}^{t_1-\frac{1}{n}} |y_n(\tau)|\,d\tau.
\end{aligned} \tag{7.114}$$

Since $M(t)$ is $L'[0, T]$ but also absolutely continuous. We have proven that $y_n(t)$ is bounded uniformly therefore $\forall n \geq 0$, $\exists M$ such that whenever $n > M$, $|y_n(t_1)| < M_y$ $\forall t \in [0, T]$ therefore

$$\begin{aligned}
|y_n(t_1) - y_n(t_2)| &\leq (1-\alpha)\,|f(t_1, y_n(t_1)) - f(t_2, y_n(t_2))| \\
&\quad + \alpha M_{g'} M\,(t_1 - t_2) + \alpha M_{g'} M M_y\,(t_1 - t_2) \\
&\leq (1-\alpha)\,|f(t_1, y_n(t_1)) - f(t_2, y_n(t_2))| + A\,(t_1 - t_2)
\end{aligned} \tag{7.115}$$

where

$$A = \alpha M_{g'} M + \alpha M_{g'} M M_y. \tag{7.116}$$

$$\begin{aligned} |y_n(t_1) - y_n(t_2)| &\leq (1-\alpha)\left\{\begin{array}{l} m(t_1) + m(t_1)|y_n(t_1)| \\ +m(t_2) + m(t_2)|y_n(t_2)| \end{array}\right\} + A(t_1 - t_2) \\ &\leq (1-\alpha)\left\{\left(1+M_y\right)m(t_1) + \left(1+M_y\right)m(t_2)\right\} + A(t_1 - t_2) \\ &\leq (1-\alpha)\left(1+M_y\right)\{m(t_1) + m(t_2)\} + A(t_1 - t_2) \\ &\leq B + A(t_1 - t_2). \end{aligned} \tag{7.117}$$

Since $m(t_1)$ and $m(t_2)$ are positive fixed constant, take $max\{B, A\} = \overline{A}$

$$|y_n(t_1) - y_n(t_2)| \leq \overline{A}\{1 + (t_1 - t_2)\}. \tag{7.118}$$

Due to order in R^+, $\exists\overline{\lambda} \in R^+ \backslash \{0\}$ such that

$$1 + (t_1 - t_2) \leq \lambda(t_1 - t_2). \tag{7.119}$$

Theorem 7.9 *The sequence Carathéodory–Tonelli,* $(y_n)_{n\in N}$ *given above converges to a global solution* y *of problem if* $g'(t)$ *is constant in* $[0, 1]$.

(1) $f(t, y_0) = 0$ for every $t \in [0, 1]$ and $g'(t)$ is constant.

(2) $f(t, y)$ is increasing with respect to y and $f(t, y) \geq 0$ $\forall(t, y) \in [0, 1] \times R$ which implies that $(y_n)_{n\in N}$ converges to the lower integral of the problem uniformly.

(3) $f(t, y)$ is increasing with respect to y and $f(t, y) \leq 0$ $\forall(t, y) \in [0, 1] \times R$ which implies that $(y_n)_{n\in N}$ converges to the upper integral of the problem uniformly.

(4) $(y_n)_{n\in N}$ converges in a point to the value α, such that (t_0, α) is not a Peano point for ${}_{t_0}^{CF}D_g^{\alpha}y(t) = f(t, y(t))$ therefore

$$|y_n(t_1) - y_n(t_2)| \leq \overline{A}\lambda(t_1 - t_2). \tag{7.120}$$

Thus $\forall\varepsilon > 0$ such that $\exists M > 0$, $\forall n \geq 1$

$$|y_n(t_1) - y_n(t_2)| < \varepsilon \to \delta < \frac{\varepsilon}{\overline{A}\lambda}. \tag{7.121}$$

This shows that $(y_n(t))$ is equicontinuous, which then completes the first part which is the existence of the solution under the Carathéodory principle. We present next alternative conditions for the convergence of the sequence within $[0, 1]$. We define $f : [0, 1] \times R \to R$ is continuous, the Carathéodory–Tonelli sequence is defined as

$$y_n(t) = \begin{cases} y_0, & \text{if } 0 \leq t \leq \frac{1}{n} \\ y_0 + (1-\alpha)g'(t)f(t, y_n(t)) \\ +\alpha \displaystyle\int_0^{t-\frac{1}{n}} g'(\tau)f(\tau, y_n(\tau))d\tau & \text{if } \frac{1}{n} \leq t \leq 1 \end{cases}. \tag{7.122}$$

Proof From the hypothesis

(1) We have that every y_n is constant solution of the problem since indeed for a fixed

$$\alpha \in [0,1], \int_0^{t-\frac{1}{n}} f(\tau, y_n(\tau))d\tau \tag{7.123}$$

is constant therefore since $g'(t)$ is constant too, we have that

$$y_n(t) = (1-\alpha)\, g'(t) f(t, y_n(t)) + \alpha \int_0^{t-\frac{1}{n}} g'(\tau) f(\tau, y_n(\tau))d\tau \tag{7.124}$$

is constant too.

(2) $\forall n \in N$ and $t \in \left[0, \frac{1}{n}\right]$ we have $y_n(t) \leq y_{n+1}(t)$. f is nonnegative and increases with respect to y therefore within $\left[\frac{1}{n}, \frac{2}{n}\right]$ we shall have

$$\begin{aligned} y_{n+1}(t) - y_n(t) &= (1-\alpha)\left[f(t, y_{n+1}(t)) - f(t, y_n(t))\right] g'(t) \\ &\quad + \alpha \int_0^{t-\frac{1}{n+1}} g'(\tau) f(\tau, y_{n+1}(\tau))d\tau - \alpha \int_0^{t-\frac{1}{n}} g'(\tau) f(\tau, y_n(\tau))d\tau \\ &\geq (1-\alpha)\left\{f(t, y_{n+1}(t)) - f(t, y_n(t))\right\} g'(t) \\ &\quad + \alpha \int_0^{t-\frac{1}{n}} g'(\tau)\left[f(\tau, y_{n+1}(\tau)) - f(\tau, y_n(\tau))\right] d\tau \\ &\geq 0. \end{aligned} \tag{7.125}$$

This can be done with the whole interval $[0, 1]$ therefore the sequence $(y_n)_{n\in N}$ is increasing. Let now y be any solution to the problem. We have $\forall n \in N$ and $t \in \left[0, \frac{1}{n}\right]$

$$\begin{aligned} y(t) &= y_0 + (1-\alpha)\, g'(t) f(t, y(t)) + \alpha \int_{t_0}^{t} g'(\tau) f(\tau, y(\tau))d\tau \\ &\geq y_0 = y_n(t). \end{aligned} \tag{7.126}$$

But $t \in \left[\frac{1}{n}, \frac{2}{n}\right]$, then we shall have

$$\begin{aligned} y(t) - y_n(t) &\geq (1-\alpha)\, g'(t)\, (f\,(t, y(t)) - f\,(t, y_n(t))) \\ &\quad + \alpha \int_{t_0}^{t-\frac{1}{n}} g'(\tau)\, (f(\tau, y(\tau)) - f(\tau, y_n(\tau))\, d\tau \\ &\geq 0. \end{aligned} \tag{7.127}$$

Therefore by repetition we shall have that $\forall t \in [0, 1]$

$$y_n(t) \leq y(t). \tag{7.128}$$

This shows that $(y_n)_{n\in N}$ converges toward the lower integral of the problem. Indeed the 3 condition leads to $y_n(t)$ converges toward the upper interval of the problem. Lastly since (t_0, α) is not a Peano point then every convergent subsequence of the sequence $(y_n)_{n\in N}$ must then converge to the unique solution.

$$y(t) = y_0 + (1-\alpha)\, g'(t) f(t, y(t)) + \alpha \int_{t_0}^{t} g'(\tau) f(\tau, y(\tau)) d\tau. \tag{7.129}$$

The compactness of $(y_n)_{n\in N}$ implies that (y_n) converges. It is then concluded that the Carathéodory–Tonelli sequence for IVP with the Caputo–Fabrizio global derivative converges to a global solution y of the problem if the following requirements hold.

(1) $f : I_b \times R \to R$ increases with respect to y.
(2) $f(t, y) \geq 0$ in an open neigbourhood I of $(0, y_0)$.
(3) The points of ∂I are not Peano-points for the equation ${}_{t_0}^{CF}D_g^{\alpha} y(t)$, when $g'(t)$ meets the requirement presented above.

References

1. Atangana, A.: Extension of rate of change concept: from local to nonlocal operators with applications. Results Phys. **19**, 103515 (2020). https://doi.org/10.1016/j.rinp.2020.103515
2. Atangana, A., Khan, M.A.: Analysis of fractional global differential equations with power law. AIMS Math. **8**(10), 24699–24725 (2023)
3. Persson, J.: A generalization of Caratheodory's existence theorem for ordinary differential equations. J. Math. Anal. Appl. **49**, 496–503 (1975). https://doi.org/10.1016/0022-247X(75)90192-4
4. Robin, W.A.: Solving differential equations using modified Picard iteration. Int. J. Math. Educ. Sci. Technol. **41**(5) (2010)
5. Tonelli, L.: Sulle equazioni funzionali deltipo di Volterra. Bull. Calcutta Math. Soc. **20**, 31–48 (1928)

Chapter 8
Existence and Uniqueness Analysis of Nonlocal Global Differential Equations with Expectation Approach

We used an iterative technique in the previous chapter to present the circumstances in which global fractional differential equations and the associated global integral equations admit unique solutions. The Picard technique and the Peano–Cauchy existence theorem were specifically applied. The circumstances under which these equations admit unique solutions are established in this chapter using several notions. In particular, we'll assume in this chapter that the nonlinear linear function on the right side of these common Cauchy problems satisfies the requirement for linear growth. Furthermore, it will be assumed that the function is bounded and meets the Lipschitz requirement. These two methods have been successfully applied over the past few decades and have been expanded to include nonlocal differential equations [1–4].

8.1 Cauchy Problems with Classical Global Derivative

The global derivative gives birth to new class of ordinary and partial differential equations, thus in that chapter, we present a detailed analysis of existence and uniqueness and numerical analysis of this new class of ordinary differential equations. We consider the following general Cauchy problem:

$$\begin{cases} D_g y(t) = f(t, y(t)), \ t > t_0 \\ y(t_0) = y_0. \end{cases} \tag{8.1}$$

In this chapter, we assume that $g(t)$ is differentiable and $g^{'}(t) \neq 0$. $f(t, y(t))$ is a nonlinear function twice differentiable.

A. Atangana and İ. Koca, *Fractional Differential and Integral Operators with Respect to a Function*, Industrial and Applied Mathematics,
https://doi.org/10.1007/978-981-97-9951-0_8

Applying the global integral or the Stieltjes–Riemann integral on equation yields

$$y(t) = y(t_0) + \int_{t_0}^{t} f(\tau, y(\tau))g'(\tau)d\tau. \tag{8.2}$$

We defined the following norm:

$$\|\varphi\|_\infty = \sup_{t\in D_\varphi} |\varphi(t)|\,. \tag{8.3}$$

We start our investigation with the Picard–Lindelof theorem.

We let

$$\Pi_{a,b} = \overline{I_a(t_0)} \times \overline{B_b(y_0)} \tag{8.4}$$

where

$$\begin{aligned} \overline{I_a(t_0)} &= [t_0 - a, t_0 + a], \\ \overline{B_b(y_0)} &= [y_0 - b, y_0 + b]. \end{aligned} \tag{8.5}$$

$\Pi_{a,b}$ is a compact cylinder and product of two compact subsets of $\mathbb{R}$. We assume that the nonlinear function $f(t, y(t))$ is bounded. This implies that $\Phi = \sup_{\Pi_{a,b}} |f|$. To continue, we assume that f satisfies the Lipschitz condition for the second variable.

Then, we can find a constants K such that

$$|f(t, y_1) - f(t, y_2)| < K\,|y_1 - y_2|\,. \tag{8.6}$$

The Banach fixed-point theorem will be applied using the metric on $\Pi_{a,b}$ which is induced by the uniform norm defined earlier

$$\Lambda : \Pi_{a,b} \to \Pi_{a,b} \tag{8.7}$$

is the map we defined by

$$\Lambda y_1(t) = y_0 + \int_{t_0}^{t} f(\tau, y(\tau))g'(\tau)d\tau. \tag{8.8}$$

Our aim is to show that the above operator maps a complete nonempty metric space y into itself; in addition, it is a contraction. In this case the compact $\overline{B_b(y_0)}$ is closed in the space of continuous with under y_0.

We aim to show that

$$\|y - y_0\|_\infty \leq b \tag{8.9}$$

then

$$\begin{aligned}
\|\Lambda y - y_0\| &= \left| \int_{t_0}^{t} f(\tau, y(\tau)) g'(\tau) d\tau \right|, \\
&\leq \int_{t_0}^{t} |f(\tau, y(\tau))| \left| g'(\tau) \right| d\tau, \\
&< \int_{t_0}^{t} \sup_{l \in (t_0, \tau]} |f(l, y(l))| \, g'(\tau) d\tau, \\
&< \sup_{t \in \overline{I_b(t_0)}} |f(t, y(t))| \int_{t_0}^{t} g'(\tau) d\tau, \\
&< L \left(g(t) - g(t_0) \right) < b.
\end{aligned} \tag{8.10}$$

Such that a restriction can be imposed on the function $g(t)$ as

$$|\Lambda y(t) - y_0| < L \sup_{t \in D_g} |g(t) - g(t_0)| < b. \tag{8.11}$$

Thus

$$L < \frac{b}{\sup_{t \in D_g} |g(t) - g(0)|}. \tag{8.12}$$

Our last proof is to show that the Λ is a contraction, thus we take y_1 and y_2

$$\begin{aligned}
|\Lambda y_1 - \Lambda y_2| &= \left| \int_{t_0}^{t} \left(f(\tau, y_1(\tau)) - f(\tau, y_2(\tau)) \right) g'(\tau) d\tau \right|, \\
&< \int_{t_0}^{t} |f(\tau, y_1(\tau)) - f(\tau, y_2(\tau))| \left| g'(\tau) \right| d\tau,
\end{aligned} \tag{8.13}$$

using the Lipschitz condition of f yields

$$|\Lambda y_1 - \Lambda y_2| < \int_{t_0}^{t} K \, |y_1 - y_2| \, g'(\tau) d\tau. \tag{8.14}$$

Since $g^{'}(\tau)$ is positive

$$\sup_{t\in(y_1-y_2)} |\Lambda y_1 - \Lambda y_2| < K \sup_{t\in|y_1-y_2|} |y_1 - y_2| (g(t) - g(t_0)), \tag{8.15}$$

$$\begin{aligned} \|\Lambda y_1 - \Lambda y_2\|_\infty &< K \|y_1 - y_2\|_\infty (g(t) - g(t_0)) \\ &< K \|y_1 - y_2\|_\infty \sup_{t\in D_g} |g(t) - g(t_0)| \\ &< K \|y_1 - y_2\|_\infty \|g - g_0\|_\infty . \end{aligned} \tag{8.16}$$

We shall yield contraction if

$$K \|g - g_0\|_\infty < 1 \tag{8.17}$$

meaning

$$K < \frac{1}{\|g - g_0\|_\infty}. \tag{8.18}$$

We have now demonstrated that the considered operator is a contraction on the Banach spaces that is induced by the uniform norm. Nevertheless to achieve the condition of the fixed-point theorem, we need to have that

$$\|g - g_0\|_\infty < \left\{\frac{1}{K}, \frac{b}{L}\right\}. \tag{8.19}$$

We will also verify that for some $l \in N$,

$$\left\|\Lambda^l y_1(t) - \Lambda^l y_2(t)\right\|_\infty \leq \frac{K^l (g(t) - g(t_0))^l}{l!} \|y_1 - y_2\|_\infty \;\; \forall t \in [t_0 - a, t_0 + a]. \tag{8.20}$$

We present the proof using the induction principle.

$$\left|\Lambda^l y_1(t) - \Lambda^l y_2(t)\right| = \left|\int_{t_0}^{t} \left(f(\tau, \Lambda^{l-1} y_1(\tau)) - f(\tau, \Lambda^{l-1} y_2(\tau))\right) g^{'}(\tau) d\tau\right|. \tag{8.21}$$

When $l = 1$,

$$\left\|\Lambda^l y_1(t) - \Lambda^l y_2(t)\right\|_\infty < K \|y_1 - y_2\|_\infty (g(t) - g(t_0)). \tag{8.22}$$

We assume the formula is true for $l-1$, thus at l, we have

$$\left|\Lambda^l y_1(t)-\Lambda^l y_2(t)\right| = \left|\int_{t_0}^{t}\left(f(\tau,\Lambda^{l-1}y_1(\tau))-f(\tau,\Lambda^{l-1}y_2(\tau))\right)g^{'}(\tau)d\tau\right|, \tag{8.23}$$

$$\leq \int_{t_0}^{t}\left|f(\tau,\Lambda^{l-1}y_1(\tau))-f(\tau,\Lambda^{l-1}y_2(\tau))\right|\left|g^{'}(\tau)\right|d\tau,$$

$$\leq K\int_{t_0}^{t}\left\|\Lambda^{l-1}y_1-\Lambda^{l-1}y_2\right\|_\infty\left|g^{'}(\tau)\right|d\tau,$$

$$\leq K\int_{t_0}^{t}\frac{K^{l-1}\left|g(t)-g(t_0)\right|^{l-1}}{(l-1)!}\left|g^{'}(\tau)\right|\|y_1-y_2\|_\infty d\tau,$$

$$\leq K^{l}\int_{t_0}^{t}\frac{(g(\tau)-g(t_0))^{l-1}}{(l-1)!}(g(\tau)-g(t_0))^{'}\|y_1-y_2\|_\infty d\tau,$$

$$\leq \frac{K^{l}}{(l-1)!}\frac{(g(t)-g(t_0))^{l}}{l}\|y_1-y_2\|_\infty,$$

$$\leq \frac{K^{l}}{l!}(g(t)-g(t_0))^{l}\|y_1-y_2\|_\infty.$$

Therefore

$$\left\|\Lambda^l y_1-\Lambda^l y_2\right\|_\infty \leq \frac{K^{l}}{l!}(g(t)-g(t_0))^{l}\|y_1-y_2\|_\infty. \tag{8.24}$$

Having imposed the condition on the function $g(t)$, for a large value of l, we have that

$$\frac{K^{l}}{l!}(g(t)-g(t_0))^{l}<1. \tag{8.25}$$

Secondly, if the nonlinear function $f(t,y(t))$ satisfies the following conditions:

(1) $|f(t,y(t))|^2<\left(1+|y|^2\right)K$,

(2) $|f(t,y_1(t))-f(t,y_2(t))|^2<\overline{K}\,|y_1-y_2|^2$.

The following inequalities can be obtained:

$$|\Lambda y(t)|^2 = \left| y_0 + \int_{t_0}^{t} f(\tau, y(\tau)) g^{'}(\tau) d\tau \right|^2, \tag{8.26}$$

$$\leq 2|y_0|^2 + 2\|g'\|_\infty (g(t) - g(t_0)) \int_{t_0}^{t} |f(\tau, y(\tau))|^2 d\tau$$

$$\leq 2|y_0|^2 + 4\|g'\|_\infty K (g(t) - g(t_0)) \int_{t_0}^{t} \left(1 + |y|^2\right) d\tau,$$

$$\leq 2|y_0|^2 + 4\|g'\|_\infty K (g(t) - g(t_0)) \int_{t_0}^{t} \left(1 + \sup_{l \in (t_0, \tau]} |y(l)|^2\right) d\tau,$$

$$\leq 2|y_0|^2 + 4\|g'\|_\infty K \left(1 + \|y\|_\infty^2\right)(t - t_0)(g(t) - g(t_0)).$$

Since $g(t)$ is differentiable, using the Mean Value theorem, we get $c \in [t_0, t]$ such that

$$g^{'}(c) = \frac{g(t) - g(t_0)}{(t - t_0)}. \tag{8.27}$$

Thus

$$|\Lambda y(t)|^2 \leq 2|y_0|^2 + 4g^{'}(c)\|g'\|_\infty K \left(1 + \|y\|_\infty^2\right)(t - t_0)^2 \tag{8.28}$$

or

$$|\Lambda y(t)|^2 \leq 2|y_0|^2 + 4g^{'}(c)\|g'\|_\infty (T - t_0)^2 K. \tag{8.29}$$

Second

$$|\Lambda y_1 - \Lambda y_2|^2 = \left| \int_{t_0}^{t} \left(f(\tau, y_1(\tau)) - f(\tau, y_2(\tau))\right) g^{'}(\tau) d\tau \right|^2, \tag{8.30}$$

$$\leq 2\|g'\|_\infty (g(t) - g(t_0)) \int_{t_0}^{t} |f(\tau, y_1(\tau)) - f(\tau, y_2(\tau))|^2 d\tau$$

$$\leq 2\|g'\|_\infty (g(t) - g(t_0)) \overline{K} \int_{t_0}^{t} |y_1(\tau) - y_2(\tau)|^2 d\tau,$$

$$\leq 2\|g'\|_\infty (g(t) - g(t_0)) \overline{K} \int_{t_0}^{t} \sup_{l \in (t_0, \tau]} |y_1(l) - y_2(l)|^2 d\tau,$$

$$\leq 2\|g'\|_\infty (g(t) - g(t_0)) \overline{K} \|y_1 - y_2\|_\infty^2 (t - t_0).$$

Therefore

$$
\begin{aligned}
|\Lambda y_1 - \Lambda y_2|^2 &\leq 2 \left\| g' \right\|_\infty (g(t) - g(t_0)) \overline{K} \left\| y_1 - y_2 \right\|_\infty^2 (t - t_0) \\
&\leq 2 \left\| g' \right\|_\infty (g(T) - g(t_0)) \overline{K} \left\| y_1 - y_2 \right\|_\infty^2 (T - t_0) \\
&\leq K_1 \left\| y_1 - y_2 \right\|_\infty^2 .
\end{aligned}
\tag{8.31}
$$

With the above conditions, the considered Cauchy problem admits at most one solution.

8.2 Cauchy Problems for Global Derivative with Power Kernel

In this chapter, we consider the Cauchy problems for global derivative with power kernel. To show this let us consider the following general Cauchy problem:

$$
\begin{cases}
{}_0^C D_g^\alpha y(t) = f(t, y(t)), \ t > t_0 \\
y(t_0) = y_0.
\end{cases}
\tag{8.32}
$$

In this chapter, we assume that $g(t)$ is differentiable and $g'(t) \neq 0$. $f(t, y(t))$ is a nonlinear function and twice differentiable then

$$
\begin{cases}
{}_0^C D_t^\alpha y(t) = g'(t) f(t, y(t)), \ t > t_0 \\
y(t_0) = y_0.
\end{cases}
\tag{8.33}
$$

Applying the global Riemann integral on equation yields

$$
y(t) = \frac{1}{\Gamma(\alpha)} \int_{t_0}^{t} f(\tau, y(\tau)) g'(\tau)(t - \tau)^{\alpha - 1} d\tau.
\tag{8.34}
$$

Also we defined the following norm:

$$
\|y\|_\infty = \sup_{t \in D_y} |y(t)| .
\tag{8.35}
$$

Let us give now sufficient conditions for which the Cauchy problems for global derivative with power kernel has a unique equation if the nonlinear function $f(t, y(t))$ satisfies the following conditions: [5]

(1) $|f(t, y(t))|^2 < \left(1 + |y|^2\right) K$ (Linear Growth Condition),

(2) $|f(t, y_1(t)) - f(t, y_2(t))|^2 < \overline{K} |y_1 - y_2|^2$ (Lipschitz Condition).

If $g(t)$ is differentiable and $g'(t) \neq 0$, $f(t, y(t))$ is a nonlinear function and twice differentiable, then the Cauchy problem has a unique solution in $L^2([t_0, T], R)$.

Assume that the linear growth condition is satisfied, then, $\forall_l \geq 1$, we can use the stopping periodic

$$\lambda_l = \inf\{T,\ \inf\{t \in [t_0, T] : |y(t)| > l\}\}. \tag{8.36}$$

So we have that $\lim\limits_{l\to\infty} \lambda_l = T$. Then we can define the sequence

$$y_l(t) = y(\inf(t, \lambda_l)),\ \ \forall_t \in [t_0, T]. \tag{8.37}$$

This proceeds and $y_l(t)$ meets the requirements that

$$y_l(t) = \frac{1}{\Gamma(\alpha)} \int_{t_0}^{t} I_{[t_0,\lambda_l]}(\tau) g^{'}(\tau) f(\tau, y_l(\tau))(t-\tau)^{\alpha-1} d\tau, \tag{8.38}$$

$$|y_l(t)|^2 \leq \left| \int_{t_0}^{t} I_{[t_0,\lambda_l]}(\tau) g^{'}(\tau) f(\tau, y_l(\tau)) \frac{(t-\tau)^{\alpha-1}}{\Gamma(\alpha)} d\tau \right|^2 . \tag{8.39}$$

By the Cauchy–Schwarz inequality, we can write

$$|y_l(t)|^2 \leq \int_{t_0}^{t} g^{2'}(\tau) \frac{(t-\tau)^{2\alpha-2}}{\Gamma^2(\alpha)} d\tau \int_{t_0}^{t} |f(\tau, y_l(\tau))|^2 d\tau. \tag{8.40}$$

Using the linear growth condition of $f(t, y(t))$, we have

$$|y_l(t)|^2 \leq \frac{\left\|g'\right\|_\infty^2}{\Gamma^2(\alpha)} \frac{(t-t_0)^{2\alpha-1}}{(2\alpha-1)} \int_{t_0}^{t} K\left(1 + |y_l(\tau)|^2\right) d\tau$$

and

$$\sup_{t_0 \leq k \leq t} |y_l(k)|^2 \leq \frac{\left\|g'\right\|_\infty^2}{\Gamma^2(\alpha)} \frac{(t-t_0)^{2\alpha-1}}{(2\alpha-1)} K \int_{t_0}^{t} \left(1 + \sup_{t_0 \leq r \leq \tau} |y_l(r)|^2\right) d\tau. \tag{8.41}$$

Applying the expectation formula on both sides, we have

$$E\left(\sup_{t_0 \leq k \leq t} |y_l(k)|^2\right) \leq \frac{\left\|g'\right\|_\infty^2}{\Gamma^2(\alpha)} \frac{(t-t_0)^{2\alpha-1}}{(2\alpha-1)} K \int_{t_0}^{t} \left(1 + E\left(\sup_{t_0 \leq r \leq \tau} |y_l(r)|^2\right)\right) d\tau. \tag{8.42}$$

If we add 1 on both sides, then we have

$$1+E\left(\sup_{t_0\leq k\leq t}|y_l(k)|^2\right)\leq 1+\frac{\|g'\|_\infty^2}{\Gamma^2(\alpha)}\frac{(t-t_0)^{2\alpha-1}}{(2\alpha-1)}K\int_{t_0}^{t}\left(1+E\left(\sup_{t_0\leq r\leq \tau}|y_l(r)|^2\right)\right)d\tau. \tag{8.43}$$

Using the Gronwall inequality, we will get

$$E\left(\sup_{t_0\leq t\leq T}|y_l(t)|^2\right)\leq \exp\left[KI_t^\alpha g'(t)(T-t_0)\right]. \tag{8.44}$$

So we can obtain the uniqueness.

Let y_1 and y_2 be the solution of the Cauchy problem. Then $y_1(.),y_2(.)\in L^2([t_0,T],R)$. So we have

$$|y_1(t)-y_2(t)|^2\leq\left|\int_{t_0}^{t}g'(\tau)(f(\tau,y_1(\tau))-f(\tau,y_2(\tau)))\frac{(t-\tau)^{\alpha-1}}{\Gamma(\alpha)}d\tau\right|^2, \tag{8.45}$$

then applying the Hölder inequality and adding 1 on both sides, we will get

$$\sup_{t_0\leq t\leq T}|y_1(t)-y_2(t)|^2\leq \overline{K}I_t^\alpha g'(t)\int_{t_0}^{t}\sup_{t_0\leq r\leq\tau}|y_1(r)-y_2(r)|^2d\tau, \tag{8.46}$$

$$1+\sup_{t_0\leq t\leq T}|y_1(t)-y_2(t)|^2\leq 1+\overline{K}I_t^\alpha g'(t)\int_{t_0}^{t}\sup_{t_0\leq r\leq\tau}|y_1(r)-y_2(r)|^2d\tau,$$

using the expectation formula again

$$E\left(\sup_{t_0\leq t\leq T}|y_1(t)-y_2(t)|^2\right)\leq \overline{K}I_t^\alpha g'(t)\int_{t_0}^{t}E\left(\sup_{t_0\leq r\leq\tau}|y_1(r)-y_2(r)|^2\right)d\tau, \tag{8.47}$$

where

$$I_t^\alpha g'(t)=\frac{\|g'\|_\infty^2}{\Gamma^2(\alpha)}\frac{(T-t_0)^{2\alpha-1}}{(2\alpha-1)} \tag{8.48}$$

since $2\alpha>1$.

With the Gronwall inequality finally we get

$$E\left(\sup_{t_0\leq t\leq T}|y_1(t)-y_2(t)|^2\right)=0,\forall t\in[t_0,t]. \tag{8.49}$$

So we have

$$y_1(t) = y_2(t), \forall t \in [t_0, t]. \tag{8.50}$$

8.2.1 Existence of Solution

Now we define the Picard iteration for obtaining the existence of solution. $\forall_l \geq 1$, we must prove that $y_l(t) \in L^2([t_0, T], R)$. Considering the recursive formula on l, we get following equality for solution of Cauchy problem.

$$y_l(t) = \frac{1}{\Gamma(\alpha)} \int_{t_0}^{t} f(\tau, y_{n-1}(\tau)) g'(\tau)(t-\tau)^{\alpha-1} d\tau. \tag{8.51}$$

For $l = 1$ we get

$$y_1(t) = \frac{1}{\Gamma(\alpha)} \int_{t_0}^{t} f(\tau, y_0(\tau)) g'(\tau)(t-\tau)^{\alpha-1} d\tau \tag{8.52}$$

where y_0 is initial condition and belongs to $L^2([t_0, T], R)$. So we write

$$|y_1(t)|^2 = \left| \frac{1}{\Gamma(\alpha)} \int_{t_0}^{t} f(\tau, y_0(\tau)) g'(\tau)(t-\tau)^{\alpha-1} d\tau \right|^2, \tag{8.53}$$

$$|y_1(t)|^2 \leq K I_t^\alpha g'(t) \int_{t_0}^{t} \left(1 + |y_0|^2\right) d\tau.$$

Finally, we get

$$E\left(\sup_{t_0 \leq t \leq T} |y_1(t)|^2\right) \leq K I_t^\alpha g'(t)(T - t_0)\left(1 + |y_0|^2\right). \tag{8.54}$$

So $y_1(t) \in L^2([t_0, T], R)$. We continue with $\forall_l \geq 1$, $y_l(t) \in L^2([t_0, T], R)$, then we must show that $y_{l+1}(t) \in L^2([t_0, T], R)$.

$$y_{l+1}(t) = \frac{1}{\Gamma(\alpha)} \int_{t_0}^{t} f(\tau, y_l(\tau)) g'(\tau)(t-\tau)^{\alpha-1} d\tau. \tag{8.55}$$

Then we have

$$|y_{l+1}(t)|^2 = \left| \frac{1}{\Gamma(\alpha)} \int_{t_0}^{t} f(\tau, y_l(\tau)) g'(\tau)(t-\tau)^{\alpha-1} d\tau \right|^2, \tag{8.56}$$

$$|y_{l+1}(t)|^2 \leq K I_t^{\alpha} g'(t) \int_{t_0}^{t} \left(1 + |y_l(\tau)|^2\right) d\tau$$

and

$$E\left(\sup_{t_0 \leq t \leq T} |y_{l+1}(t)|^2\right) \leq K I_t^{\alpha} g'(t) \int_{t_0}^{t} \left(1 + E\left(\sup_{t_0 \leq r \leq \tau} |y_l(r)|^2\right)\right) d\tau. \tag{8.57}$$

By hypothesis, $y_l(t) \in L^2([t_0, T], R)$ and $E\left(\sup_{t_0 \leq t \leq T} |y_l(t)|^2 \leq M\right)$.

$$E\left(\sup_{t_0 \leq t \leq T} |y_{l+1}(t)|^2\right) \leq K I_t^{\alpha} g'(t)(1 + M)(T - t_0). \tag{8.58}$$

Since

$$I_t^{\alpha} g'(t) = \frac{\|g'\|_{\infty}^2 (T - t_0)^{2\alpha - 1}}{\Gamma^2(\alpha)(2\alpha - 1)} < \infty \tag{8.59}$$

then

$$E\left(\sup_{t_0 \leq t \leq T} |y_{l+1}(t)|^2\right) < \infty. \tag{8.60}$$

Thus $y_{l+1}(t) \in L^2([t_0, T], R)$. With the idea of inductive formula, $\forall_l \geq 0$, $y_l(t) \in L^2([t_0, T], R)$.

$$|y_1(t) - y_0(t)|^2 = |y_1(t) - y_0|^2 \leq 2|y_0|^2 + 2\left| \int_{t_0}^{t} g'(\tau) f(\tau, y_0(\tau)) \frac{(t-\tau)^{\alpha-1}}{\Gamma(\alpha)} d\tau \right|^2, \tag{8.61}$$

$$|y_1(t) - y_0|^2 \leq 2|y_0|^2 + 2K \|g'\|_{\infty} (g(T) - g(t_0)) \left(1 + |y_0|^2\right) \frac{T^{\alpha}}{\Gamma(\alpha + 1)}.$$

Then we get

$$E\left(\sup_{t_0\leq t\leq T}|y_1(t)-y_0|^2\right)\leq 2E|y_0|^2+2K\left\|g'\right\|_\infty(g(T)-g(t_0))\left(1+E|y_0|^2\right)\frac{T^\alpha}{\Gamma(\alpha+1)}. \tag{8.62}$$

So

$$E\left(\sup_{t_0\leq t\leq T}|y_1(t)-y_0|^2\right)\leq\gamma. \tag{8.63}$$

Now $\forall_l\geq 1$,

$$|y_{l+1}(t)-y_l(t)|^2\leq\overline{K}I_t^\alpha g'(t)\int_{t_0}^{t}|y_l(\tau)-y_{l-1}(\tau)|^2\,d\tau, \tag{8.64}$$

$$E\left(\sup_{t_0\leq k\leq t}|y_{l+1}(k)-y_l(k)|^2\right)\leq\overline{K}I_t^\alpha g'(t)\int_{t_0}^{t}E\,|y_l(\tau)-y_{l-1}(\tau)|^2\,d\tau,$$

$$E\left(|y_{l+1}(t)-y_l(t)|^2\right)\leq\overline{K}I_t^\alpha g'(t)\frac{(t-t_0)^l}{n!}.$$

So we have

$$E\left(|y_{l+1}(t)-y_l(t)|^2\right)\leq\gamma\frac{(\beta(t-t_0))^l}{l!}. \tag{8.65}$$

For $l=0$, the inequality holds.

Now we continue our proofs at t_{l+1}, then we will have the following:

$$E\left(|y_{l+2}(t)-y_{l+1}(t)|^2\right)\leq\overline{K}I_t^\alpha g'(t)\int_{t_0}^{t}E\,|y_{l+1}(\tau)-y_l(\tau)|^2\,d\tau, \tag{8.66}$$

$$<\beta\int_{t_0}^{t}E\,|y_{l+1}(\tau)-y_l(\tau)|^2\,d\tau.$$

$$E\left(|y_{l+2}(t)-y_{l+1}(t)|^2\right)\leq\beta\int_{t_0}^{t}\gamma\frac{(\beta(t-t_0))^l}{l!}, \tag{8.67}$$

$$\leq\beta\frac{(\gamma(t-t_0))^{l+1}}{(l+1)!}.$$

So we can easily say that the inequality obtained for $l + 1$.

$$\sum_{l=0}^{\infty} \beta \frac{(\gamma (t - t_0))^{l+1}}{(l + 1)!} = \beta \exp[\gamma (t - t_0)] < \infty, \forall t \in [0, T]. \tag{8.68}$$

The Borel–Contelli lemma helps to find a positive integer number $l_0 = l_0(\varepsilon), \forall \varepsilon \in \Pi$ that

$$\sup_{t_0 \leq t \leq T} |y_{l+1}(t) - y_l(t)|^2 \leq \frac{1}{2^l}, \; l \geq l_0. \tag{8.69}$$

It continues that the sum

$$y_0(t) + \sum_{k=0}^{l-1} \left[y_{k+1}(t) - y_k(t) \right] = y_l(t) \tag{8.70}$$

converges uniformly in $[0, T]$. Now, if we take

$$\lim_{n \to \infty} y_l(t) = y(t). \tag{8.71}$$

Also the sequence $\{y_l(t)\}_{l \geq 1}$ is a Cauchy sequence in $L^2[0, T]$. It follows that $y_l(t) \to y(t)$. With the growth property, we say that $y(t) \in L^2[0, T]$. We now must show that $y(.)$ satisfies the following equation:

$$\left| \int_0^t [f(\tau, y_l(\tau)) - f(\tau, y(\tau)] (t - \tau)^{\alpha - 1} d\tau \right|^2 \leq \int_0^t |y_l(\tau) - y(\tau)|^2 d\tau. \tag{8.72}$$

Taking when $n \to \infty$ we have that

$$y(t) = \frac{1}{\Gamma(\alpha)} \int_0^t f(\tau, y(\tau)) g^{'}(\tau)(t - \tau)^{\alpha - 1} d\tau, \; \forall t \in [0, T]. \tag{8.73}$$

8.3 Cauchy Problems for Global Derivative with Exponential Kernel

In this chapter, we consider the Cauchy problems for global derivative with exponential kernel. To show this let us consider the following general Cauchy problem:

$$\begin{cases} {}_0^{CF} D_g^{\alpha} y(t) = f(t, y(t)), \\ y(0) = y_0. \end{cases} \tag{8.74}$$

Again, we assume that $g(t)$ is differentiable and $g^{'}(t) \neq 0$. $f(t, y(t))$ is a nonlinear function and twice differentiable then

$$\begin{cases} {}_{0}^{CF}D_{t}^{\alpha}y(t) = g^{'}(t)f(t, y(t)), \\ y(0) = y_0. \end{cases} \tag{8.75}$$

Now applying the Caputo–Fabrizio integral, this implies

$$y(t) = y_0 + \frac{1-\alpha}{M(\alpha)}g^{'}(t)f(t, y(t)) + \frac{\alpha}{M(\alpha)}\int_{t_0}^{t} g^{'}(\tau)f(\tau, y(\tau))d\tau. \tag{8.76}$$

Now we give the proof of existence and uniqueness of the solution of the Cauchy problem for global derivative with exponential kernel. While doing proof we use the linear growth condition and Lipschitz condition for $f(t, y(t))$.

We again defined a stopping time as presented in the section given before [5].

$\forall_l \geq 1$, we defined the following sequence:

$$y_l(t) = y_0 + \frac{1-\alpha}{M(\alpha)}g^{'}(t)f(t, y_l(t)) + \frac{\alpha}{M(\alpha)}\int_{t_0}^{t} g^{'}(\tau)f(\tau, y_l(\tau))d\tau. \tag{8.77}$$

Then

$$|y_l(t)|^2 \leq 3|y_0|^2 + 3\left|\frac{1-\alpha}{M(\alpha)}g^{'}(t)f(t, y_l(t))\right|^2 + 3\left|\int_{t_0}^{t}\frac{\alpha g^{'}(\tau)}{M(\alpha)}f(\tau, y_l(\tau))d\tau\right|^2, \tag{8.78}$$

then using the linear growth hypothesis gives

$$\begin{aligned} |y_l(t)|^2 \leq & 3|y_0|^2 + 3K\left|g^{'}(t)\right|^2\left(1+|y_l(t)|^2\right)\left(\frac{1-\alpha}{M(\alpha)}\right)^2 \\ & + 3K\left\|g'\right\|_{\infty}(g(t)-g(t_0))\frac{\alpha^2}{M^2(\alpha)}\int_{t_0}^{t}\left(1+|y_l(\tau)|^2\right)d\tau, \end{aligned} \tag{8.79}$$

$$\begin{aligned} \max_{t_0 \leq k \leq t}|y_l(k)|^2 \leq & 3|y_0|^2 + 3K\left|g^{'}(t)\right|^2\left(1+\max_{t_0 \leq k \leq t}|y_l(k)|^2\right)\left(\frac{1-\alpha}{M(\alpha)}\right)^2 \\ & + 3K\left\|g'\right\|_{\infty}(g(t)-g(t_0))\frac{\alpha^2}{M^2(\alpha)}\int_{t_0}^{t}\left(1+\max_{t_0 \leq r \leq \tau}|y_l(r)|^2\right)d\tau, \end{aligned} \tag{8.80}$$

by applying the expectation, we get

$$E\left(\max_{t_0\le k\le t}|y_l(k)|^2\right)\le 3|y_0|^2+3K\left|g^{'}(t)\right|^2\left(1+E\left(\max_{t_0\le k\le t}|y_l(k)|^2\right)\right)\left(\frac{1-\alpha}{M(\alpha)}\right)^2 \\ +3K\left\|g'\right\|_\infty(g(t)-g(t_0))\frac{\alpha^2}{M^2(\alpha)}\int_{t_0}^{t}\left(1+E\left(\max_{t_0\le r\le\tau}|y_l(r)|^2\right)\right)d\tau, \tag{8.81}$$

if we move equations into a more acceptable position, then we will get

$$E\left(\max_{t_0\le k\le t}|y_l(k)|^2\right)\le 3|y_0|^2+3K\left\|g^{'}\right\|_\infty^2\left(1+E\left(\max_{t_0\le k\le t}|y_l(k)|^2\right)\right)\left(\frac{1-\alpha}{M(\alpha)}\right)^2 \\ +3K\left\|g'\right\|_\infty(g(t)-g(t_0))\frac{\alpha^2}{M^2(\alpha)}\int_{t_0}^{t}\left(1+E\left(\max_{t_0\le r\le\tau}|y_l(r)|^2\right)\right)d\tau. \tag{8.82}$$

By addition of 1 on both sides, we have

$$1+E\left(\max_{t_0\le k\le t}|y_l(k)|^2\right)\le 1+3|y_0|^2+3K\left\|g^{'}\right\|_\infty^2\left(1+E\left(\max_{t_0\le k\le t}|y_l(k)|^2\right)\right)\left(\frac{1-\alpha}{M(\alpha)}\right)^2 \\ +3K\left\|g'\right\|_\infty(g(t)-g(t_0))\frac{\alpha^2}{M^2(\alpha)}\int_{t_0}^{t}\left(1+E\left(\max_{t_0\le r\le\tau}|y_l(r)|^2\right)\right)d\tau. \tag{8.83}$$

So finally we get

$$E\left(\max_{t_0\le k\le t}|y_l(k)|^2\right)\le 1+3|y_0|^2+3K\left\|g^{'}\right\|_\infty^2\left(1+E\left(\max_{t_0\le k\le t}|y_l(k)|^2\right)\right)\left(\frac{1-\alpha}{M(\alpha)}\right)^2 \\ +\exp\left[3K\left\|g'\right\|_\infty(g(T)-g(t_0))\frac{\alpha^2}{M^2(\alpha)}(T-t_0)\right]. \tag{8.84}$$

Let $y_1(.)$ and $y_2(.)$ be the solution of the Cauchy problem. Then $y_1(.),y_2(.)\in L^2([t_0,T],R)$. So we have

$$|y_1(t)-y_2(t)|^2\le 2\left|\frac{1-\alpha}{M(\alpha)}g^{'}(t)\left(f(t,y_1(t))-f(t,y_2(t))\right)\right|^2 \\ +2\left|\int_{t_0}^{t}\frac{\alpha g^{'}(\tau)}{M(\alpha)}\left(f(\tau,y_1(\tau))-f(\tau,y_2(\tau))\right)d\tau\right|^2, \tag{8.85}$$

using the Lipschitz condition for function $f(t, y(t))$ and the Hölder inequality, then we have

$$
\begin{aligned}
|y_1(t) - y_2(t)|^2 \leq & 2\overline{K}\left(\frac{1-\alpha}{M(\alpha)}\right)^2 \left|g^{'}(t)\right|^2 |y_1(t) - y_2(t)|^2 \\
& + 2\overline{K}\left(\frac{\alpha}{M(\alpha)}\right)^2 (g(T) - g(t_0)) \left\|g'\right\|_\infty \int_{t_0}^{t} |y_1(\tau) - y_2(\tau)|^2 \, d\tau,
\end{aligned} \tag{8.86}
$$

thus

$$
\begin{aligned}
E\left(\sup_{t_0 \leq k \leq t} |y_1(k) - y_2(k)|^2\right) \leq & 2\overline{K}\left(\frac{1-\alpha}{M(\alpha)}\right)^2 \left|g^{'}(t)\right|^2 E\left(\sup_{t_0 \leq k \leq t} |y_1(k) - y_2(k)|^2\right) \\
& + 2\overline{K}\left(\frac{\alpha}{M(\alpha)}\right)^2 (g(T) - g(t_0)) \left\|g'\right\|_\infty \int_{t_0}^{t} E\left(\sup_{t_0 \leq r \leq \tau} |(y_1(r) - y_2(r))|^2\right) d\tau, \\
\leq & 2\overline{K}\left(\frac{1-\alpha}{M(\alpha)}\right)^2 \left\|g^{'}\right\|_\infty^2 E\left(\sup_{t_0 \leq k \leq t} |y_1(k) - y_2(k)|^2\right) \\
& + 2\overline{K}\left(\frac{\alpha}{M(\alpha)}\right)^2 (g(T) - g(t_0)) \left\|g'\right\|_\infty \int_{t_0}^{t} E\left(\sup_{t_0 \leq r \leq \tau} |(y_1(r) - y_2(r))|^2\right) d\tau,
\end{aligned} \tag{8.87}
$$

$$
\begin{aligned}
E\left(\sup_{t_0 \leq t \leq T} |y_1(t) - y_2(t)|^2\right) \leq & 2\overline{K}\left(\frac{1-\alpha}{M(\alpha)}\right)^2 \left\|g^{'}\right\|_\infty^2 E\left(\sup_{t_0 \leq t \leq T} |y_1(t) - y_2(t)|^2\right) \\
& + 2\overline{K}\left(\frac{\alpha}{M(\alpha)}\right)^2 (g(T) - g(t_0)) \left\|g'\right\|_\infty \int_{t_0}^{t} E\left(\sup_{t_0 \leq r \leq \tau} |(y_1(r) - y_2(r))|^2\right) d\tau,
\end{aligned} \tag{8.88}
$$

$$
\begin{aligned}
& \left(1 - 2\overline{K}\left(\frac{1-\alpha}{M(\alpha)}\right)^2 \left\|g^{'}\right\|_\infty^2\right) E\left(\sup_{t_0 \leq t \leq T} |y_1(t) - y_2(t)|^2\right) \\
& \leq 2\overline{K}\left(\frac{\alpha}{M(\alpha)}\right)^2 (g(T) - g(t_0)) \left\|g'\right\|_\infty \int_{t_0}^{t} E\left(\sup_{t_0 \leq r \leq \tau} |(y_1(r) - y_2(r))|^2\right) d\tau,
\end{aligned} \tag{8.89}
$$

$$
E\left(\sup_{t_0 \leq t \leq T} |y_1(t) - y_2(t)|^2\right) \leq \frac{2\overline{K}\left(\frac{\alpha}{M(\alpha)}\right)^2 (g(T) - g(t_0)) \left\|g'\right\|_\infty \int_{t_0}^{t} E\left(\sup_{t_0 \leq r \leq \tau} |(y_1(r) - y_2(r))|^2\right) d\tau}{\left(1 - 2\overline{K}\left(\frac{1-\alpha}{M(\alpha)}\right)^2 \left\|g'\right\|_\infty^2\right)} \tag{8.90}
$$

$$
E\left(\sup_{t_0 \leq t \leq T} |y_1(t) - y_2(t)|^2\right) \leq \varpi \int_{t_0}^{t} E\left(\sup_{t_0 \leq r \leq \tau} |(y_1(r) - y_2(r))|^2\right) d\tau, \tag{8.91}
$$

where

$$\varpi = \frac{2\overline{K}\left(\frac{\alpha}{M(\alpha)}\right)^2 (g(T) - g(t_0)) \left\|g'\right\|_\infty}{\left(1 - 2\overline{K}\left(\frac{1-\alpha}{M(\alpha)}\right)^2 \left\|g'\right\|_\infty^2\right)} \tag{8.92}$$

with following condition:

$$2\overline{K}\left(\frac{1-\alpha}{M(\alpha)}\right)^2 \left\|g^{'}\right\|_\infty^2 \neq 1. \tag{8.93}$$

Applying the Gronwall inequality

$$E\left(\sup_{t_0 \le t \le T} |y_1(t) - y_2(t)|^2\right) = 0, \forall t \in [t_0, T]. \tag{8.94}$$

So we have

$$y_1(.) = y_2(.)\ \forall t \in [t_0, T]. \tag{8.95}$$

8.3.1 Existence of Solution

In this subsection, we give a proof of existence of the solution via using the Picard recursive approach.

$$y_l(t) = \frac{1-\alpha}{M(\alpha)} g^{'}(t) f(t, y_{l-1}(t)) + \frac{\alpha}{M(\alpha)} \int_{t_0}^{t} g^{'}(\tau) f(\tau, y_{l-1}(\tau)) d\tau. \tag{8.96}$$

Now we have to show that $\forall_l \ge 0$, $y_l(t) \in L^2([t_0, T], R)$. $y_0(.) = y_0$ is the initial condition by definition $y_l(t) \in L^2([t_0, T], R)$.

For $l = 1$,

$$y_1(t) = \frac{1-\alpha}{M(\alpha)} g^{'}(t) f(t, y_0(t)) + \frac{\alpha}{M(\alpha)} \int_{t_0}^{t} g^{'}(\tau) f(\tau, y_0(\tau)) d\tau. \tag{8.97}$$

Thus, we get

$$\begin{aligned} E\left(\sup_{t_0 \le t \le T} |y_1(t)|^2\right) \le & 2K\left(\frac{1-\alpha}{M(\alpha)}\right)^2 \sup_{t_0 \le t \le T} \left|g^{'}(t)\right|^2 \left(1 + E|y_0|^2\right) \\ & + 2K\left(\frac{\alpha}{M(\alpha)}\right)^2 \left(1 + E|y_0|^2\right)(T - t_0)\left\|g'\right\|_\infty (g(T) - g(t_0)), \end{aligned} \tag{8.98}$$

$$E\left(\sup_{t_0\le t\le T}|y_1(t)|^2\right)\le\left(2K\left(1+E|y_0|^2\right)\right)\left(\begin{array}{c}\left(\frac{1-\alpha}{M(\alpha)}\right)^2\sup_{t_0\le t\le T}\left|g'(t)\right|^2\\+\left(\frac{\alpha}{M(\alpha)}\right)^2\left\|g'\right\|_\infty(g(T)-g(t_0))\end{array}\right). \tag{8.99}$$

We know that $y_0\in L^2([t_0,T],R)$, thus $E|y_0|^2<\infty$.

In that case

$$E\left(\sup_{t_0\le t\le T}|y_1(t)|^2\right)<\infty. \tag{8.100}$$

We assume that $\forall_l\ge 1,\ y_l(t)\in L^2([t_0,T],R)$. Now we must show that $y_{l+1}(t)\in L^2([t_0,T],R)$.

$$y_{l+1}(t)=\frac{1-\alpha}{M(\alpha)}g'(t)f(t,y_l(t))+\frac{\alpha}{M(\alpha)}\int_{t_0}^{t}g'(\tau)f(\tau,y_l(\tau))d\tau. \tag{8.101}$$

So we get

$$\begin{aligned}|y_{l+1}(t)|^2\le&2K\left(\frac{1-\alpha}{M(\alpha)}\right)^2\sup_{t_0\le t\le T}\left|g'(t)\right|^2\left(1+|y_l(t)|^2\right)\\&+2K\left(\frac{\alpha}{M(\alpha)}\right)^2\left\|g'\right\|_\infty(g(T)-g(t_0))\int_{t_0}^{t}\left(1+|y_l(\tau)|^2\right)d\tau,\end{aligned} \tag{8.102}$$

$$\begin{aligned}E\left(\sup_{t_0\le t\le T}|y_{l+1}(t)|^2\right)\le&2K\left(\frac{1-\alpha}{M(\alpha)}\right)^2\left\|g'\right\|_\infty^2\left(1+E\left(\sup_{t_0\le t\le T}|y_l(t)|^2\right)\right)\\&+2K\left(\frac{\alpha}{M(\alpha)}\right)^2\left\|g'\right\|_\infty(g(T)-g(t_0))\int_{t_0}^{t}\left(1+E\left(\sup_{t_0\le r\le\tau}|y_l(r)|^2\right)\right)d\tau.\end{aligned} \tag{8.103}$$

By inductive hypothesis, $y_l(t)\in L^2([t_0,T],R)$ therefore

$$E\left(\sup_{t_0\le t\le T}|y_l(t)|^2\right)\le M. \tag{8.104}$$

So we obtain

$$E\left(\sup_{t_0\le t\le T}|y_{l+1}(t)|^2\right)\le 2K(1+M)\left\{\begin{array}{c}\left(\frac{1-\alpha}{M(\alpha)}\right)^2\left\|g'\right\|_\infty^2\\ +\left(\frac{\alpha}{M(\alpha)}\right)^2\left\|g'\right\|_\infty(g(T)-g(t_0))(T-t_0)\end{array}\right\} \tag{8.105}$$

$$<\infty.$$

Therefore $y_{l+1}(t)\in L^2([t_0,T],R)$.

By inductive rules, we can say that $\forall_l\ge 0,\ y_l(t)\in L^2([t_0,T],R)$.

We now analyze

$$E\left(|y_1(t)-y_0|^2\right)\le 3K\left(\frac{1-\alpha}{M(\alpha)}\right)^2\left\|g'\right\|_\infty^2\left(1+E\left(|y_0|^2\right)\right)+3E\left(|y_0|^2\right) \tag{8.106}$$
$$+3(T-t_0)K\left(1+E\left(|y_0|^2\right)\right)\left(\frac{\alpha}{M(\alpha)}\right)^2\left\|g'\right\|_\infty(g(T)-g(t_0)),$$

and

$$E\left(|y_1(t)-y_0|^2\right)\le\gamma_1, \tag{8.107}$$

where

$$\gamma_1= 3K\left(\frac{1-\alpha}{M(\alpha)}\right)^2\left\|g'\right\|_\infty^2\left(1+E\left(|y_0|^2\right)\right)+3E\left(|y_0|^2\right) \tag{8.108}$$
$$+3(T-t_0)K\left(1+E\left(|y_0|^2\right)\right)\left(\frac{\alpha}{M(\alpha)}\right)^2\left\|g'\right\|_\infty(g(T)-g(t_0)).$$

Now $\forall_l\ge 1$,

$$E\left(\sup_{t_0\le t\le T}|y_1(t)|^2\right)\le\left(2K\left(1+E|y_0|^2\right)\right)\left\{\begin{array}{c}\left(\frac{1-\alpha}{M(\alpha)}\right)^2\sup_{t_0\le t\le T}\left|g'(t)\right|^2\\ +\left(\frac{\alpha}{M(\alpha)}\right)^2(T-t_0)\left\|g'\right\|_\infty(g(T)-g(t_0))\end{array}\right\}. \tag{8.109}$$

$$\left|y_{l+1}(t)-y_l(t)\right|^2\le 2\overline{K}\left(\frac{1-\alpha}{M(\alpha)}\right)^2\left|g'(t)\right|^2\left|y_l(t)-y_{l-1}(t)\right|^2 \tag{8.110}$$
$$+2\overline{K}\left(\frac{\alpha}{M(\alpha)}\right)^2\left\|g'\right\|_\infty(g(T)-g(t_0))\int_{t_0}^{t}\left|y_l(\tau)-y_{l-1}(\tau)\right|^2d\tau.$$

Considering induction for $\forall_l\ge 0$, we get

$$E\left(\sup_{t_0\le t\le T}|y_{l+1}(t)-y_l(t)|^2\right)\le\gamma_1\frac{(\beta_1(t-t_0))^l}{l!},\ \ t_0\le t\le T. \tag{8.111}$$

We accept that the inequality time for $\forall l\ge 1$. We must show its proof at t_{l+1}.

At t_{l+1}, we have

$$E\left(\sup_{t_0\le t\le T}|y_{l+2}(t)-y_{l+1}(t)|^2\right)\le\gamma_1\frac{(\beta_1(t-t_0))^l}{(l+1)!}(t-t_0)\le\gamma_1\frac{(\beta_1(t-t_0))^{l+1}}{(l+1)!},\ t_0\le t\le T. \tag{8.112}$$

So at $l+1$, the inequality is valid via inductive principle.

We can conclude that the Borel–Contelli lemma helps to find a positive integer number $l_0=l_0(\varepsilon)$, $\forall\varepsilon\in\Pi$ that

$$\sup_{t_0\le t\le T}|y_{l+1}(t)-y_l(t)|^2\le\frac{1}{2^l},\ l\ge l_0. \tag{8.113}$$

It continues that the sum

$$y_0(t)+\sum_{k=0}^{l-1}\left[y_{k+1}(t)-y_k(t)\right]=y_l(t) \tag{8.114}$$

converges uniformly in $[0,T]$. Now, if we take

$$\lim_{n\to\infty}y_l(t)=y(t). \tag{8.115}$$

Thus, we have

$$E\,|y_{l+1}(t)-y(t)|^2\le\beta_1\,|y_l(t)-y(t)|^2\,. \tag{8.116}$$

Taking as $l\to\infty$, the right side of equality goes to zero, so we obtain

$$y(t)=\frac{1-\alpha}{M(\alpha)}g'(t)f(t,y(t))+\frac{\alpha}{M(\alpha)}\int_{t_0}^{t}g'(\tau)f(\tau,y(\tau))d\tau. \tag{8.117}$$

This completes the proof.

8.4 Cauchy Problems for Global Derivative with Mittag–Leffler Kernel

In this chapter, we consider the Cauchy problems for global derivative with Mittag–Leffler kernel. To show this let us consider the following general Cauchy problem:

$$\begin{cases}{}_0^{ABC}D_g^{\alpha}y(t)=f(t,y(t)),\\ y(0)=y_0.\end{cases} \tag{8.118}$$

Again, we assume that $g(t)$ is differentiable and $g^{'}(t) \neq 0$. $f(t, y(t))$ is a nonlinear function and twice differentiable then

$$\begin{cases} {}_{0}^{ABC}D_{t}^{\alpha} y(t) = g^{'}(t) f(t, y(t)). \\ y(0) = y_0. \end{cases} \tag{8.119}$$

Now applying the Atangana–Baleanu integral, this implies

$$y(t) = \frac{1-\alpha}{AB(\alpha)} g^{'}(t) f(t, y(t)) + \frac{\alpha}{\Gamma(\alpha) AB(\alpha)} \int_{t_0}^{t} g^{'}(\tau) f(\tau, y(\tau))(t-\tau)^{\alpha-1} d\tau. \tag{8.120}$$

Now we give the proof of existence and uniqueness of the solution of the Cauchy problem for global derivative with Mittag–Leffler kernel. While doing proof we use the linear growth condition and Lipschitz condition for $f(t, y(t))$.

We again defined a stopping time as presented in section given by before [5]. We define

$$I_{t}^{\alpha} g^{'}(t) = \frac{\left\|g'\right\|_{\infty}^{2} (T-t_0)^{2\alpha-1}}{\Gamma^{2}(\alpha)\,(2\alpha-1)} \tag{8.121}$$

$\forall_l \geq 1$, we defined the following sequence:

$$y_l(t) = \frac{1-\alpha}{AB(\alpha)} g^{'}(t) f(t, y_l(t)) + \frac{\alpha}{\Gamma(\alpha) AB(\alpha)} \int_{t_0}^{t} g^{'}(\tau) f(\tau, y_l(\tau))(t-\tau)^{\alpha-1} d\tau \tag{8.122}$$

and

$$|y_l(t)|^2 \leq 2 \left| \frac{1-\alpha}{AB(\alpha)} g^{'}(t) f(t, y_l(t)) \right|^2 + 2 \left| \frac{\alpha}{\Gamma(\alpha) AB(\alpha)} \int_{t_0}^{t} g^{'}(\tau) f(\tau, y_l(\tau))(t-\tau)^{\alpha-1} d\tau \right|^2. \tag{8.123}$$

Using the linear growth hypothesis and Hölder inequality, we have

$$\begin{aligned} |y_l(t)|^2 \leq & 2K \left| g^{'}(t) \right|^2 \left(1 + |y_l(t)|^2\right) \left(\frac{1-\alpha}{AB(\alpha)}\right)^2 \\ & + 2K \frac{\alpha}{AB(\alpha)\Gamma(\alpha)} I_{t}^{\alpha} g^{'}(t) \int_{t_0}^{t} \left(1 + |y_l(\tau)|^2\right) d\tau, \end{aligned} \tag{8.124}$$

$$\max_{t_0 \le k \le t} |y_l(k)|^2 \le 2K \left|g^{'}(t)\right|^2 \left(1 + \max_{t_0 \le k \le t} |y_l(k)|^2\right) \left(\frac{1-\alpha}{AB(\alpha)}\right)^2 + 2K \frac{\alpha}{AB(\alpha)\Gamma(\alpha)} I_t^{\alpha} g^{'}(t) \int_{t_0}^{t} \left(1 + \max_{t_0 \le r \le \tau} |y_l(r)|^2\right) d\tau. \tag{8.125}$$

By applying the expectation, we get

$$E\left(\max_{t_0 \le k \le t} |y_l(k)|^2\right) \le 2K \left|g^{'}(t)\right|^2 \left(1 + E\left(\max_{t_0 \le k \le t} |y_l(k)|^2\right)\right) \left(\frac{1-\alpha}{AB(\alpha)}\right)^2 + 2K \frac{\alpha}{AB(\alpha)\Gamma(\alpha)} I_t^{\alpha} g^{'}(t) \int_{t_0}^{t} \left(1 + E\left(\max_{t_0 \le r \le \tau} |y_l(r)|^2\right)\right) d\tau. \tag{8.126}$$

If we move equations into a more acceptable position, then we will get

$$E\left(\max_{t_0 \le k \le t} |y_l(k)|^2\right) \le 2K \left\|g^{'}\right\|_{\infty}^2 \left(1 + E\left(\max_{t_0 \le k \le t} |y_l(k)|^2\right)\right) \left(\frac{1-\alpha}{AB(\alpha)}\right)^2 + 2K \frac{\alpha}{AB(\alpha)\Gamma(\alpha)} I_t^{\alpha} g^{'}(t) \int_{t_0}^{t} \left(1 + E\left(\max_{t_0 \le r \le \tau} |y_l(r)|^2\right)\right) d\tau. \tag{8.127}$$

By addition of 1 on both sides, we have

$$1 + E\left(\max_{t_0 \le k \le t} |y_l(k)|^2\right) \le 2K \left\|g^{'}\right\|_{\infty}^2 \left(1 + E\left(\max_{t_0 \le k \le t} |y_l(k)|^2\right)\right) \left(\frac{1-\alpha}{AB(\alpha)}\right)^2 + 2K \frac{\alpha}{AB(\alpha)\Gamma(\alpha)} I_t^{\alpha} g^{'}(t) \int_{t_0}^{t} \left(1 + E\left(\max_{t_0 \le r \le \tau} |y_l(r)|^2\right)\right) d\tau. \tag{8.128}$$

So finally we get

$$E\left(\max_{t_0 \le k \le t} |y_l(k)|^2\right) \le 1 + 2K \left\|g^{'}\right\|_{\infty}^2 \left(1 + E\left(\max_{t_0 \le k \le t} |y_l(k)|^2\right)\right) \left(\frac{1-\alpha}{AB(\alpha)}\right)^2 + \exp\left[2K \frac{\alpha}{AB(\alpha)\Gamma(\alpha)} I_t^{\alpha} g^{'}(t) (T - t_0)\right]. \tag{8.129}$$

Let $y_1(.)$ and $y_2(.)$ be the solution of the Cauchy problem. Then $y_1(.), y_2(.) \in L^2([t_0, T], R)$. So we have

$$|y_1(t) - y_2(t)|^2 \leq 2\left|\frac{1-\alpha}{AB(\alpha)} g^{'}(t)\left(f(t, y_1(t)) - f(t, y_2(t))\right)\right|^2 + 2\left|\int_{t_0}^{t} \frac{\alpha g^{'}(\tau)}{AB(\alpha)\Gamma(\alpha)}\left(f(\tau, y_1(\tau)) - f(\tau, y_2(\tau))\right) d\tau\right|^2, \tag{8.130}$$

using the Lipschitz conditon for function $f(t, y(t))$ and the Hölder inequality, then we have

$$|y_1(t) - y_2(t)|^2 \leq 2\overline{K}\left(\frac{1-\alpha}{AB(\alpha)}\right)^2 \left|g^{'}(t)\right|^2 |y_1(t) - y_2(t)|^2 + 2\overline{K}\left(\frac{\alpha}{AB(\alpha)\Gamma(\alpha)}\right)^2 I_t^{\alpha} g^{'}(t) \int_{t_0}^{t} |y_1(\tau) - y_2(\tau)|^2 d\tau, \tag{8.131}$$

thus

$$\begin{aligned} E\left(\sup_{t_0 \leq k \leq t} |y_1(k) - y_2(k)|^2\right) \leq & 2\overline{K}\left(\frac{1-\alpha}{AB(\alpha)}\right)^2 \left|g^{'}(t)\right|^2 E\left(\sup_{t_0 \leq k \leq t} |y_1(k) - y_2(k)|^2\right) \\ & + 2\overline{K}\left(\frac{\alpha}{AB(\alpha)\Gamma(\alpha)}\right)^2 I_t^{\alpha} g^{'}(t) \int_{t_0}^{t} E\left(\sup_{t_0 \leq r \leq \tau} |(y_1(r) - y_2(r))|^2\right) d\tau, \\ \leq & 2\overline{K}\left(\frac{1-\alpha}{AB(\alpha)}\right)^2 \left\|g^{'}\right\|_{\infty}^2 E\left(\sup_{t_0 \leq k \leq t} |y_1(k) - y_2(k)|^2\right) \\ & + 2\overline{K}\left(\frac{\alpha}{AB(\alpha)\Gamma(\alpha)}\right)^2 I_t^{\alpha} g^{'}(t) \int_{t_0}^{t} E\left(\sup_{t_0 \leq r \leq \tau} |(y_1(r) - y_2(r))|^2\right) d\tau, \end{aligned} \tag{8.132}$$

$$\begin{aligned} & \left(1 - 2\overline{K}\left(\frac{1-\alpha}{AB(\alpha)}\right)^2 \left\|g^{'}\right\|_{\infty}^2\right) E\left(\sup_{t_0 \leq t \leq T} |y_1(t) - y_2(t)|^2\right) \\ & \leq 2\overline{K}\left(\frac{\alpha}{AB(\alpha)\Gamma(\alpha)}\right)^2 I_t^{\alpha} g^{'}(t) \int_{t_0}^{t} E\left(\sup_{t_0 \leq r \leq \tau} |(y_1(r) - y_2(r))|^2\right) d\tau, \end{aligned} \tag{8.133}$$

$$E\left(\sup_{t_0 \leq t \leq T} |y_1(t) - y_2(t)|^2\right) \leq \frac{2\overline{K}\left(\frac{\alpha}{AB(\alpha)\Gamma(\alpha)}\right)^2 I_t^{\alpha} g^{'}(t) \int_{t_0}^{t} E\left(\sup_{t_0 \leq r \leq \tau} |(y_1(r) - y_2(r))|^2\right) d\tau}{\left(1 - 2\overline{K}\left(\frac{1-\alpha}{AB(\alpha)}\right)^2 \left\|g^{'}\right\|_{\infty}^2\right)} \tag{8.134}$$

$$E\left(\sup_{t_0\leq t\leq T}|y_1(t)-y_2(t)|^2\right)\leq\varpi_1\int_{t_0}^{t}E\left(\sup_{t_0\leq r\leq \tau}|(y_1(r)-y_2(r))|^2\right)d\tau, \quad (8.135)$$

where

$$\varpi_1=\frac{2\overline{K}\left(\frac{\alpha}{AB(\alpha)\Gamma(\alpha)}\right)^2 I_t^{\alpha}g'(t)}{\left(1-2\overline{K}\left(\frac{1-\alpha}{AB(\alpha)}\right)^2\|g'\|_{\infty}^2\right)} \quad (8.136)$$

with the following condition:

$$2\overline{K}\left(\frac{1-\alpha}{AB(\alpha)}\right)^2\left\|g'\right\|_{\infty}^2\neq 1. \quad (8.137)$$

Applying the Gronwall inequality

$$E\left(\sup_{t_0\leq t\leq T}|y_1(t)-y_2(t)|^2\right)=0,\ \forall t\in[t_0,T]. \quad (8.138)$$

So we have

$$y_1(.)=y_2(.)\ \forall t\in[t_0,T]. \quad (8.139)$$

Our proof is completed for existence of solution.

8.4.1 Existence of Solution

Same as earlier chapters, we put Picard recursive approach.

$$y_l(t)=\frac{1-\alpha}{AB(\alpha)}g'(t)f(t,y_{l-1}(t))+\frac{\alpha}{AB(\alpha)\Gamma(\alpha)}\int_{t_0}^{t}g'(\tau)f(\tau,y_{l-1}(\tau))(t-\tau)^{\alpha-1}d\tau, \quad (8.140)$$

Now we have to show that $\forall_l\geq 0$, $y_l(t)\in L^2([t_0,T],R)$. $y_0(.)=y_0$ is the initial condition by definition $y_l(t)\in L^2([t_0,T],R)$.

For $l=1$,

$$y_1(t)=\frac{1-\alpha}{AB(\alpha)}g'(t)f(t,y_0(t))+\frac{\alpha}{AB(\alpha)\Gamma(\alpha)}\int_{t_0}^{t}g'(\tau)f(\tau,y_0(\tau))(t-\tau)^{\alpha-1}d\tau. \quad (8.141)$$

Thus we get

$$E\left(\sup_{t_0\le t\le T}|y_1(t)|^2\right) \le 2K\left(\frac{1-\alpha}{AB(\alpha)}\right)^2 \sup_{t_0\le t\le T}\left|g^{'}(t)\right|^2\left(1+E|y_0|^2\right) + 2K\left(\frac{\alpha}{AB(\alpha)\Gamma(\alpha)}\right)^2 I_t^{\alpha}g^{'}(t)\int_{t_0}^{t}\left(1+E|y_0|^2\right)d\tau, \tag{8.142}$$

$$E\left(\sup_{t_0\le t\le T}|y_1(t)|^2\right) \le 2K\left(\frac{1-\alpha}{AB(\alpha)}\right)^2 \left\|g^{'}\right\|_{\infty}^2\left(1+E|y_0|^2\right) + 2K\left(\frac{\alpha}{AB(\alpha)\Gamma(\alpha)}\right)^2 I_t^{\alpha}g^{'}(t)\left(1+E|y_0|^2\right)(T-t_0). \tag{8.143}$$

We know that $y_0\in L^2([t_0,T],R)$, Thus $E|y_0|^2<\infty$.
In that case

$$E\left(\sup_{t_0\le t\le T}|y_1(t)|^2\right) < \infty. \tag{8.144}$$

We assume that $\forall_l\ge 1,\ y_l(t)\in L^2([t_0,T],R)$. Now we must show that $y_{l+1}(t)\in L^2([t_0,T],R)$.

$$y_{l+1}(t) = \frac{1-\alpha}{AB(\alpha)}g^{'}(t)f(t,y_l(t)) + \frac{\alpha}{AB(\alpha)\Gamma(\alpha)}\int_{t_0}^{t}g^{'}(\tau)f(\tau,y_l(\tau))(t-\tau)^{\alpha-1}d\tau. \tag{8.145}$$

So we get

$$|y_{l+1}(t)|^2 \le 2K\left(\frac{1-\alpha}{AB(\alpha)}\right)^2 \sup_{t_0\le t\le T}\left|g^{'}(t)\right|^2\left(1+|y_l(t)|^2\right) + 2K\left(\frac{\alpha}{AB(\alpha)\Gamma(\alpha)}\right)^2 I_t^{\alpha}g^{'}(t)\int_{t_0}^{t}\left(1+|y_l(\tau)|^2\right)d\tau, \tag{8.146}$$

$$E\left(\sup_{t_0\le t\le T}|y_{l+1}(t)|^2\right) \le 2K\left(\frac{1-\alpha}{AB(\alpha)}\right)^2 \left\|g^{'}\right\|_{\infty}^2\left(1+E\left(\sup_{t_0\le t\le T}|y_l(t)|^2\right)\right) + 2K\left(\frac{\alpha}{AB(\alpha)\Gamma(\alpha)}\right)^2 I_t^{\alpha}g^{'}(t)\int_{t_0}^{t}\left(1+E\left(\sup_{t_0\le r\le \tau}|y_l(r)|^2\right)\right)d\tau. \tag{8.147}$$

By inductive hypothesis, $y_l(t) \in L^2([t_0, T], R)$ therefore $E\left(\sup_{t_0 \le t \le T} |y_l(t)|^2\right) \le M$.

So we obtain

$$E\left(\sup_{t_0 \le t \le T} |y_{l+1}(t)|^2\right) \le 2K(1+M)\left\{\begin{array}{c}\left(\frac{1-\alpha}{AB(\alpha)}\right)^2 \left\|g'\right\|_\infty^2 \\ +\left(\frac{\alpha}{AB(\alpha)\Gamma(\alpha)}\right)^2 I_t^\alpha g'(t)(T-t_0)\end{array}\right\} \tag{8.148}$$

$$< \infty.$$

Therefore

$$y_{l+1}(t) \in L^2([t_0, T], R).$$

By inductive rules, we can say that $\forall_l \ge 0,\ y_l(t) \in L^2([t_0, T], R)$.
We now analyze

$$\begin{aligned} E\left(|y_1(t) - y_0|^2\right) \le & 2K\left(\frac{1-\alpha}{AB(\alpha)}\right)^2 \left\|g'\right\|_\infty^2 \left(1 + E\left(|y_0|^2\right)\right) \\ & + 2(T-t_0)K\left(1 + E\left(|y_0|^2\right)\right)\left(\frac{\alpha}{AB(\alpha)\Gamma(\alpha)}\right)^2 I_t^\alpha g'(t), \end{aligned} \tag{8.149}$$

and

$$E\left(|y_1(t) - y_0|^2\right) \le \gamma_2, \tag{8.150}$$

where

$$\begin{aligned} \gamma_1 = & 2K\left(\frac{1-\alpha}{AB(\alpha)}\right)^2 \left\|g'\right\|_\infty^2 \left(1 + E\left(|y_0|^2\right)\right) \\ & + 2(T-t_0)K\left(1 + E\left(|y_0|^2\right)\right)\left(\frac{\alpha}{AB(\alpha)\Gamma(\alpha)}\right)^2 I_t^\alpha g'(t). \end{aligned} \tag{8.151}$$

Now $\forall_l \ge 1$,

$$E\left(\sup_{t_0 \le t \le T} |y_1(t)|^2\right) \le \left(2K\left(1 + E|y_0|^2\right)\right)\left\{\begin{array}{c}\left(\frac{1-\alpha}{AB(\alpha)}\right)^2 \sup_{t_0 \le t \le T} \left|g'(t)\right|^2 \\ +\left(\frac{\alpha}{AB(\alpha)\Gamma(\alpha)}\right)^2 (T-t_0) I_t^\alpha g'(t)\end{array}\right\}. \tag{8.152}$$

$$|y_{l+1}(t) - y_l(t)|^2 \leq 2\overline{K}\left(\frac{1-\alpha}{AB(\alpha)}\right)^2 \left|g^{'}(t)\right|^2 |y_l(t) - y_{l-1}(t)|^2 + 2\overline{K}\left(\frac{\alpha}{AB(\alpha)\Gamma(\alpha)}\right)^2 I_t^{\alpha} g^{'}(t) \int_{t_0}^{t} |y_l(\tau) - y_{l-1}(\tau)|^2 d\tau. \tag{8.153}$$

Considering induction for $\forall_l \geq 0$, we get

$$E\left(\sup_{t_0 \leq t \leq T} |y_{l+1}(t) - y_l(t)|^2\right) \leq \gamma_2 \frac{(\beta_2 (t-t_0))^l}{l!}, \quad t_0 \leq t \leq T. \tag{8.154}$$

We accept that the inequality time for $\forall l \geq 1$. We must show its proof at t_{l+1}.

At t_{l+1}, we have

$$E\left(\sup_{t_0 \leq t \leq T} |y_{l+2}(t) - y_{l+1}(t)|^2\right) \leq \gamma_2 \frac{(\beta_2 (t-t_0))^l}{(l+1)!}(t-t_0) \leq \gamma_2 \frac{(\beta_2 (t-t_0))^{l+1}}{(l+1)!}, \quad t_0 \leq t \leq T. \tag{8.155}$$

So at $l+1$, the inequality is valid via inductive principle.

We can conclude that the Borel–Contelli lemma helps to find a positive integer number

$$l_0 = l_0(\varepsilon), \tag{8.156}$$

$\forall \varepsilon \in \Pi$ that

$$\sup_{t_0 \leq t \leq T} |y_{l+1}(t) - y_l(t)|^2 \leq \frac{1}{2^l}, \quad l \geq l_0. \tag{8.157}$$

It continues that the sum

$$y_0(t) + \sum_{k=0}^{l-1} \left[y_{k+1}(t) - y_k(t)\right] = y_l(t) \tag{8.158}$$

converges uniformly in $[0, T]$. Now, if we take

$$\lim_{n \to \infty} y_l(t) = y(t). \tag{8.159}$$

Thus we have

$$E\,|y_{l+1}(t) - y_l(t)|^2 \leq \beta_2\,|y_l(t) - y(t)|^2. \tag{8.160}$$

Taking as $l \to \infty$, the right side of equality goes to zero, so we obtain

$$y(t) = \frac{1-\alpha}{AB(\alpha)} g^{'}(t) f(t, y(t)) + \frac{\alpha}{AB(\alpha)\Gamma(\alpha)} \int_{t_0}^{t} g^{'}(\tau) f(\tau, y(\tau))(t-\tau)^{\alpha-1} d\tau. \tag{8.161}$$

This completes the proof.

References

1. Xu, D., Yang, Z., Huang, Y.: Existence-uniqueness and continuation theorems for stochastic functional differential equations. J. Differ. Equ. **245**(6), 1681–1703 (2008)
2. Caraballo, T., Chueshov, I.D., Marín-Rubio, P., Real, J.: Existence and asymptotic behaviour for stochastic heat equations with multiplicative noise in materials with memory. Discret. Contin. Dyn. Syst. **18**(2–3), 253–270 (2007)
3. Wei, F., Wang, K.: The existence and uniqueness of the solution for stochastic functional differential equations with infinite delay. J. Math. Anal. Appl. **331**(1), 516–531 (2007)
4. Jiang, F., Shen, Y.: A note on the existence and uniqueness of mild solutions to neutral stochastic partial functional differential equations with non-Lipschitz coefficients. Comput. Math. Appl. **61**(6), 1590–1594 (2011)
5. Atangana, A.: Extension of rate of change concept: from local to nonlocal operators with applications. Results Phys. **19**, 103515 (2020)

Chapter 9
Chaplygin's Method for Global Differential Equations

In this chapter, we are interested to establish the existence and uniqueness solution of the Cauchy problem with Riemann–Stieltjes derivative using the well-known Chaplygin's method [1–4]. We shall also present a discussion underpinning the initial condition dependency as well as parameters.

9.1 Chaplygin's Method and Initial Condition Dependence: Classical Case

We shall consider the following Cauchy problem:

$$\begin{cases} {}_{t_0}D_g y(t) = f(t, y(t)) \text{ if } t \in [t_0, T] \\ y(t_0) = y_0 \end{cases}. \tag{9.1}$$

We have that $g'(t)$ is positive continuous and bounded. The above can be reformulated as

$$\begin{aligned} y'(t) &= g'(t) f(t, y(t)), \\ y(t_0) &= y_0. \end{aligned} \tag{9.2}$$

Theorem 9.1 *Let $f \in C[R, R_0]$; we have*

$$R_0 = \{t \,||t - t_0| < a \; ; \; y \,||y - y_0| < b\}. \tag{9.3}$$

We assume that f is bounded on R_0 and $\alpha = \min\{a, b\}$ where $\overline{M} = M_f . M_{g'}$.

A. Atangana and İ. Koca, *Fractional Differential and Integral Operators with Respect to a Function*, Industrial and Applied Mathematics,
https://doi.org/10.1007/978-981-97-9951-0_9

We assume that f_y, f_{yy} *exist and* $f_{yy} > 0$ *in* R_0. *We let* $u_0 = u_0(t)$, $v_0 = v_0(t)$ *be differentiable for* $t_0 \leq t \leq t_0 + \alpha$ *such that* $(t, u_0(t))$, $(t, v_0(t)) \in R_0$ *and*

$$D_g u_0(t) < f(t, u(t)), \; u_0(t_0) = y_0, \qquad (9.4)$$
$$D_g v_0(t) > f(t, v(t)), \; v_0(t_0) = y_0.$$

Then, we can find a Chaplygin sequence $\{u_n(t), v_n(t)\}$ *such that*

$$u_n(t) < u_{n+1}(t) < y(t) < v_{n+1}(t) < v_n(t), \qquad (9.5)$$
$$\forall t \in [t_0, t_0 + \alpha].$$

$$u_n(t_0) = y_0 = v_n(t_0) \qquad (9.6)$$

where $y(t)$ *is unique solution of*

$$\begin{cases} {}_{t_0}D_g y(t) = f(t, y(t)) \text{ if } t \in [t_0, T] \\ y(t_0) = y_0 \end{cases} \qquad (9.7)$$

which exists on $[t_0, t_0 + \alpha]$. *In addition* $u_n(t)$ *and* $v_n(t)$ *tend uniformly to* $y(t)$ *on* $[t_0, t_0 + \alpha]$ *as* $n \to \infty$. *If, in addition for a suitable constant* β,

$$0 \leq v_0(t) - u_0(t) \leq \beta \qquad (9.8)$$

then

$$|u_n(t) - v_n(t)| \leq \frac{2\beta}{2^{2n}}, \quad t \in [t_0, t_0 + \alpha]. \qquad (9.9)$$

Proof By hypothesis $\exists$ M_f and $M_{g'}$ positive constant such that $\forall t \in [t_0, t_0 + \alpha]$,

$$|f(t, y)| < M_f \qquad (9.10)$$

and

$$\|g'\|_\infty = M_{g'} < \infty. \qquad (9.11)$$

By assumptions, we have

$$u_0(t) < y(t) < v_0(t), \; t \in [t_0, t_0 + \alpha]. \qquad (9.12)$$

The following functions are defined:

$$f_1(t, y; u_0, v_0) = f(t, u_0(t)) + f_y(t, u_0(t))(y(t) - u_0(t)), \qquad (9.13)$$
$$f_2(t, y; u_0, v_0) = f(t, u_0(t)) + \frac{f(t, u_0(t)) - f(t, v_0(t))}{u_0(t) - v_0(t)}(y(t) - u_0(t)).$$

We first note that when $t = t_{0,}$ we have

$$f_1(t_0, y; u_0, v_0) = f(t_0, u_0(t_0)) = f(t_0, y_0), \qquad (9.14)$$
$$f_2(t_0, y; u_0, v_0) = f(t_0, v_0(t_0)) = f(t_0, y_0),$$

therefore

$$f_1(t_0, y; u_0, v_0) = f_2(t_0, y; u_0, v_0) = f(t_0, y_0). \qquad (9.15)$$

Let $u_1(t)$ and $v_1(t)$ be the solutions of the linear differential equations

$$D_g u_1(t) = f_1(t, u_1(t); u_0, v_0), u_1(t_0) = y_0, \qquad (9.16)$$
$$D_g v_1(t) = f_2(t, v_1(t); u_0, v_0), v_1(t_0) = y_0.$$

Indeed there exist on $[t_0, t_0 + \alpha]$. By the definition of f and the inequality of the theorem, we have that

$$u_0(t) < u_0(t_0) + \int_{t_0}^{t} f(\tau, u_0(\tau))g'(\tau)d\tau, \;\; u_0(t_0) = y_0, \qquad (9.17)$$

$$= u_0(t_0) + \int_{t_0}^{t} f_1(\tau, u_0(\tau); u_0, v_0)\, g'(\tau)d\tau.$$

Therefore, we yield the following remark:

$$u_0(t) < u_1(t), \;\; t \in [t_0, t_0 + \alpha]. \qquad (9.18)$$

We use a similar routine to arrive at

$$v_1(t) < v_0(t), \;\; t \in [t_0, t_0 + \alpha]. \qquad (9.19)$$

The next step will consist of showing that $u_1(t)$ and $v_1(t)$ satisfy the differential inequalities in the theorem. We have from the hypothesis that $f_y(t, y)$ is strictly increasing; using the definition of f_1 and the defined linear differential equation, with the mean value theorem, we have that

$$u_1(t) = u_1(t_0) + \int_{t_0}^{t} f_1(\tau, u_1(\tau); u_0, v_0)\, g'(\tau)d\tau, \qquad (9.20)$$

$$\leq u_1(t_0) + \int_{t_0}^{t} f_1(\tau, u_1(\tau))\, g'(\tau)d\tau, \;\; t \in [t_0, t_0 + \alpha].$$

We have on the other hand that

$$\begin{aligned} u_0^+(t) &< u_0(t_0) + \int_{t_0}^{t} f\left(\tau, u_0(\tau)\right) g'(\tau) d\tau \\ &= u_0(t_0) + \int_{t_0}^{t} f_2\left(\tau, u_0(\tau); u_0, v_0\right) g'(\tau) d\tau \\ &= v_1(t). \end{aligned} \tag{9.21}$$

We have that

$$u_0(t) < v_1(t) \ \forall t \in [t_0, t_0 + \alpha]. \tag{9.22}$$

We have that

$$\begin{aligned} \frac{f_1\left(t, y; u_0, v_0\right) - f\left(t, u_0(t)\right)}{y - u_0} &= f_y\left(t, u_0(t)\right), \\ \frac{f_1\left(t, v_0; u_0, v_0\right) - f\left(t, v_0(t)\right)}{v_0 - u_0} &= f_y\left(t, v_0(t)\right). \end{aligned} \tag{9.23}$$

Then

$$f_y\left(t, u_0(t)\right) < \frac{f\left(t, u_0(t)\right) - f\left(t, v_0(t)\right)}{u_0(t) - v_0(t)} \tag{9.24}$$

and

$$\begin{aligned} f\left(t, v_1(t)\right) &= f\left(t, u_0(t)\right) + f_y\left(t, u_0(t)\right)\left[v_1(t) - u_0(t)\right] \\ &+ \frac{1}{2} f_{yy}\left(t, \eta\right)\left[v_1(t) - u_0(t)\right]^2; \ u_0(t) < \eta < v_1(t). \end{aligned} \tag{9.25}$$

Applying the mean value theorem yields

$$f_{yy}\left(t, \eta\right) > 0 \tag{9.26}$$

leading to

$$\begin{aligned} v_1(t) &= v_1(t_0) + \int_{t_0}^{t} f_2\left(\tau, v_1(\tau); u_0, v_0\right) g'(\tau) d\tau, \\ &> v_1(t_0) + \int_{t_0}^{t} f\left(\tau, v_1(\tau)\right) g'(\tau) d\tau, \ t \in [t_0, t_0 + \alpha]. \end{aligned} \tag{9.27}$$

The functions $u_1(t)$, $y(t)$ and $v_1(t)$ properties lead to

$$u_1(t) < y(t) < v_1(t), \ t \in [t_0, t_0 + \alpha]. \tag{9.28}$$

With what was presented before, we have

$$u_0(t) < u_1(t) < y(t) < v_1(t) < v_0(t) \ t \in [t_0, t_0 + \alpha]. \tag{9.29}$$

We can now define on mapping Λ to couple (u_0, v_0) to get (u_1, v_1) as

$$\begin{aligned} (u_1, v_1) &= \Lambda\left[(u_0, v_0)\right], \\ (u_2, v_2) &= \Lambda\left[(u_1, v_1)\right]. \end{aligned} \tag{9.30}$$

By repetition of the application of such mapping, we obtain

$$(u_{n+1}, v_{n+1}) = \Lambda\left[(u_n, v_n)\right] \tag{9.31}$$

of functions that satisfy the following relation:

$$\begin{aligned} u_n(t) &< u_n(t_0) + \int_{t_0}^{t} f\left(\tau, u_n(\tau)\right) g'(\tau) d\tau, \ u_n(t_0) = y_0 \\ v_n(t) &> v_n(t_0) + \int_{t_0}^{t} f\left(\tau, v_n(\tau)\right) g'(\tau) d\tau, \ v_n(t_0) = y_0, \\ u_n(t) &< u_{n+1}(t) < y(t) < v_{n+1}(t) < v_n(t) \ t \in [t_0, t_0 + \alpha], \\ u_{n+1}(t) &= u_{n+1}(t_0) + \int_{t_0}^{t} f_1\left(\tau, u_{n+1}(\tau); u_n(\tau), v_n(\tau)\right) g'(\tau) d\tau, \\ v_{n+1}(t) &= v_{n+1}(t_0) + \int_{t_0}^{t} f_2\left(\tau, v_{n+1}(\tau); u_n(\tau), v_n(\tau)\right) g'(\tau) d\tau. \end{aligned} \tag{9.32}$$

Clearly $\{u_n\}$ and $\{v_n\}$ are monotonic and uniformly bounded on $[t_0, t_0 + \alpha]$. They are equicontinuous since $\forall n$ fixed u_n, v_n are solutions of a linear equation. They converge toward $y(t)$ whenever $n \to \infty$. Let

$$\Omega = \sup_{\substack{u_0(t) \le y \le v_0(t) \\ t_0 \le t \le t_0 + \alpha}} \left| f_y(t, y) \right| \tag{9.33}$$

and

$$\overline{\Omega} = \sup_{\substack{u_0(t) \le y \le v_0(t) \\ t_0 \le t \le t_0+\alpha}} \left| f_{yy}(t,y) \right| . \tag{9.34}$$

We now assume that $\forall t \in [t_0, t_0+\alpha]$,

$$0 \le v_0(t) - u_0(t) \le \frac{1}{2\overline{\Omega}\alpha \exp\left(\alpha\Omega M_g\right)} = \lambda. \tag{9.35}$$

Therefore for $n = 0$, we have

$$v_0(t) - u_0(t) \le \left(2\overline{\Omega}\alpha \exp\left(\alpha\Omega\right)\right)^{-1} = \lambda. \tag{9.36}$$

We assume that the assertion is true for n fixed

$$|u_n(t) - v_n(t)| \le \frac{2\lambda}{2^{2n}}. \tag{9.37}$$

From the definition, we have that

$$\begin{aligned} v'_{n+1}(t) - u'_{n+1}(t) &= g'(t)\left[\frac{f(t,u_n(t)) - f(t,v_n(t))}{u_n(t) - v_n(t)}\right](v_{n+1}(t) - v_n(t)) \\ &\quad - f_x(t,u_n(t))(u_{n+1}(t) - u_n(t)) \\ &= g'(t) f_y(t,\eta)(v_{n+1}(t) - v_n(t)) + (u_{n+1}(t) - u_n(t)) \\ &\quad \left[f_y(t,\eta) - f_y(t,u_n(t))\right] \end{aligned} \tag{9.38}$$

where

$$u_n(t) < \eta < v_n(t). \tag{9.39}$$

But we have that

$$f_y(t,\eta) - f_y(t,u_n(t)) = f_{yy}(t,\eta)(\eta - u_n(t)) \tag{9.40}$$

where

$$u_n(t) < \eta. \tag{9.41}$$

We then have that

$$\left|v'_{n+1}(t) - u'_{n+1}(t)\right| < \overline{\Omega}\left|v_{n+1}(t) - u_{n+1}(t)\right| + \Omega\left|\eta - u_n(t)\right|\left|u_{n+1}(t) - u_n(t)\right| . \tag{9.42}$$

In addition, we have

$$|\eta - u_n(t)| \le |u_n(t) - v_n(t)| \tag{9.43}$$

and

$$|u_{n+1}(t) - v_n(t)| \leq |u_n(t) - v_n(t)| \,. \tag{9.44}$$

Putting all together, we obtain

$$D^+ |v_{n+1}(t) - u_{n+1}(t)| \leq M_{g'}\overline{\Omega} |v_{n+1}(t) - u_{n+1}(t)| + \Omega \frac{2^2\lambda^2}{2^{2n+1}} \tag{9.45}$$

and integrating yields

$$|v_{n+1}(t) - u_{n+1}(t)| < M_{g'}\overline{\Omega} \frac{2^2\lambda^2}{2^{2n+1}} \int_{t_0}^{t} \exp\left(\Omega\left(t - \tau\right)\right) d\tau. \tag{9.46}$$

We know that

$$\int_{t_0}^{t} \exp\left(\Omega\left(t - \tau\right)\right) d\tau \leq \alpha e^{\Omega\alpha} \tag{9.47}$$

therefore, we get

$$|v_{n+1}(t) - u_{n+1}(t)| < \frac{2\lambda}{2^{2n+1}}. \tag{9.48}$$

Therefore by induction, we can conclude that the relation is true for $\forall n$. Therefore

$$\begin{aligned} |y(t) - u_n(t)| &\leq \tfrac{2\lambda}{2^{2n}}, \\ |y(t) - v_n(t)| &\leq \tfrac{2\lambda}{2^{2n}}, \end{aligned} \tag{9.49}$$

which completes the proof. The next step is to demonstrate the lower and upper Chaplygin's sequence (u_n) and (v_n) below. We shall again consider the following equation:

$$D_g y(t) = f(t, y(t)) \; t \in [t_0, t_0 + \alpha] \,. \tag{9.50}$$

Theorem 9.2 *Let $f \in C\left[R_0, R\right]$, where R_0 is defined as before. Let*

$$\|f\left(t, y\right)\|_\infty \leq M_f \tag{9.51}$$

and

$$\left\|g'\right\|_\infty \leq M_g \tag{9.52}$$

on R_0. We assume that $f(t, y)$ is quasi-monotone nondecreasing in y $\forall t \in [t_0, t_0 + a]$ and $\frac{\partial f(t,y)}{\partial y}$ exists and is continuous on R_0. We assume $u_0(t)$ to be continuously differentiable on $[t_0, t_0 + \alpha]$, where

$$\alpha = \min\left\{a, \frac{b}{\overline{M}}\right\}, \ \overline{M} = M_{g'}M_f \ ;(t, u_0(t)) \in R_0 \tag{9.53}$$

and

$$u_0'(t) < g'(t)f(t, u_0(t)), \ u_0(t_0) = y_0. \tag{9.54}$$

Furthermore, we let

$$f(t, y) + f_y(t, y)(z - y) < f(t, y) \text{ if } z < y. \tag{9.55}$$

There exists a Chaplygin sequence $\{u_n(t)\}$ *such that* $u_n(t_0) = y_0$:

$$u_n(t) < u_{n+1}(t) < y(t), \ t \in [t_0, t_0 + \alpha] \tag{9.56}$$

where $y(t)$ *is the solution of the equation on* $[t_0, t_0 + \alpha]$ *and*

$$\lim_{n\to\infty} u_n(t) = y(t) \tag{9.57}$$

uniformly on $[t_0, t_0 + \alpha]$.

Proof By hypothesis, we have that $\frac{\partial f(t,y)}{\partial y} \geq 0$. This follows from the quasi-monotonicity property of $f(t, y)$. We further have that

$$u_0(t) < y(t) \ \forall t \in [t_0, t_0 + \alpha]. \tag{9.58}$$

The corresponding linear equation on the case of the lower solution is given as

$$z' = g'(t)\left[f(t, u_0(t)) + \frac{\partial f(t, u_0(t))}{\partial y}(z - u_0(t))\right] = \widetilde{f}(t, z; u_0(t)); \ z(t_0) = y_0. \tag{9.59}$$

By the mean of quasi-monotonicity of f with respect to z, $\widetilde{f}(t, z; u_0(t))$ also is quasi-monotone with respect to z. It follows that

$$u_n(t) < v_1(t) \tag{9.60}$$

where $v_1(t)$ is the solution of the equation. This by assumption implies that

$$\begin{aligned} u_1'(t) &< g'(t)f(t, u_1(t); u_0(t)), \\ D_g u_1(t) &< f(t, u_1(t); u_0(t)) \\ &< f(t, u_1(t)). \end{aligned} \tag{9.61}$$

Therefore

$$u_1(t) < y(t), \ \forall t \in [t_0, t_0 + \alpha]. \tag{9.62}$$

Then indeed we have

$$u_0(t) < u_1(t) < y(t), \ \forall t \in [t_0, t_0 + \alpha]. \tag{9.63}$$

The mapping

$$\Lambda\left[u_0(t)\right] = u_1(t) \tag{9.64}$$

helps establish the method of Chaplygin to have

$$u_n(t) < u_{n+1}(t) < y(t). \tag{9.65}$$

9.2 Dependence on Parameters and Initial Condition

In this section, we show that the solution of a classical differential equation with the Riemann–Stieltjes integral depends on the parameters and initial condition. We shall consider the following initial value problem:

$$\begin{cases} {}_{t_0}D_g y(t) = f(t, y(t)) \text{ if } t \in [t_0, T] \\ y(t_0) = y_0 \end{cases}. \tag{9.66}$$

We assume that $g'(t)$ is continuous and bounded on the prescribed interval. We assume that $y(t, t_0, y_0)$ is the solution of the above equation, where y_0 is the initial condition. We assume that $\overline{y}_0$ is another initial condition; for simplicity we assume that

$$\overline{y}_0 = y_0 + \xi \tag{9.67}$$

with associated solution

$$\begin{aligned} |\overline{y}(t) - y(t)| &\leq |\xi| + \int_{t_0}^{t} g'(\tau) \left| f(\tau, y(\tau)) - f(\tau, \overline{y}(\tau)) \right| d\tau, \\ &\leq |\xi| + \int_{t_0}^{t} g'(\tau) L \left| y - \overline{y} \right| d\tau, \\ \Phi(t) &\leq |\xi| + \int_{t_0}^{t} g'(\tau) L \Phi(\tau) d\tau. \end{aligned} \tag{9.68}$$

By the Gronwall inequality, we have

$$\Phi(t) \leq |\xi| \exp\left[L\left(g(t) - g(t_0)\right)\right] \tag{9.69}$$

where

$$\Phi(t) = |\overline{y}(t) - y(t)| . \tag{9.70}$$

We note that this was achieved under the condition that f is Lipschitz with respect to y for a fixed $t \in (t_0, t_0 + \alpha]$. However if this condition is not met, we have the following.

Lemma 9.1 *Let $f \in C[R_0, R]$ and*

$$H(t, r) = \sup_{\|y - y_0\| \leq y(t, t_0, y_0)} |f(t, y)| . \tag{9.71}$$

We assume that $r^* (t, t_0, 0)$ is the maximal solution of

$$D_g y(t) = H(t, y) \tag{9.72}$$

through $(t_0, 0)$. Let $y(t, t_0, y_0)$ be the solution of the equation under investigation. Then

$$|y(t, t_0, y_0) - y_0| \leq r^* (t, t_0, 0) \;\; t \geq t_0. \tag{9.73}$$

Proof We have that

$$\begin{aligned} |y(t, t_0, y_0) - y_0| &= \left| \int_{t_0}^{t} g'(\tau) f(\tau, y(\tau, t_0, y_0)) \, d\tau \right| \\ &\leq \int_{t_0}^{t} |g'(\tau)| \, |f(\tau, y(\tau, t_0, y_0))| \, d\tau \\ &\leq \int_{t_0}^{t} |g'(\tau)| \sup_{\|y - y_0\| \leq y(t, t_0, y_0)} |f(\tau, y)| \, d\tau \\ &\leq \int_{t_0}^{t} g'(\tau) \, |f(\tau, |y(\tau, t_0, y_0) - y_0|)| \, d\tau. \end{aligned} \tag{9.74}$$

Theorem 9.3 *We assume that $f \in C[R_0, R]$ and that $\forall (t, y), (t, z) \in R_0$*

$$|f(t, y) - f(t, z)| \leq \overline{g}(t, |x - y|), \;\; \overline{g} \in C[R_0^+, R^+]. \tag{9.75}$$

We assume that $v(t) = 0$ is the unique solution. The differential equation

$$D_g v = \overline{g}(t, v), \tag{9.76}$$

such that $v(t_0) = 0$, *then if the solutions* $v(t, t_0, v_0)$ *of the above equation through* $\forall(t_0, v_0)$ *are continuous with respect to the initial conditions* (t_0, v_0) *the solutions* $y(t, t_0, v_0)$ *are unique and continuous with respect to* (t_0, y_0).

Proof Indeed previously, we have presented the uniqueness of the solutions. Our next investigation shall be the continuity. Let $y(t, t_0, y_0)$ and $\overline{y}(t, t_0, \overline{y}_0)$ be the solutions of our equation with initial data (t_0, y_0), $\left(t_0, \overline{y}_0\right)$ respectively. We consider the following function:

$$\Omega(t) = \left|y(t, t_0, y_0) - \overline{y}(t, t_0, \overline{y}_0)\right|. \tag{9.77}$$

The condition of the theorem implies that

$$\Omega(t) \leq \int_0^t g'(\tau)\overline{g}(\tau, \Omega(\tau))d\tau. \tag{9.78}$$

Therefore by the maximal solution principle, we have that

$$\Omega(t) \leq r(t, t_0, \left|y_0 - \overline{y}_0\right|);\ \ t \geq t_0 \tag{9.79}$$

such that

$$v(t_0) = \left|y_0 - \overline{y}_0\right|. \tag{9.80}$$

But the solutions of our equations are continuous with respect to the initial date (t_0, y_0), $\left(t_0, \overline{y}_0\right)$, then we have that

$$\lim_{y_0 \to \overline{y}_0} r(t, t_0, \left|y_0 - \overline{y}_0\right|) = r(t, t_0, 0). \tag{9.81}$$

However within the framework of the theorem $r(t, t_0, 0) = 0$. This therefore leads to the conclusion that

$$\lim_{y_0 \to \overline{y}_0} y(t, t_0, y_0) = \overline{y}(t, t_0, \overline{y}_0); \tag{9.82}$$

with the above result, we can conclude that $y(t, t_0, y_0)$ is continuous with respect to the parameter y_0. We shall now show the same with the parameter t. We now assume that $y(t, t_0, y_0)$ and $\overline{y}(t, t_0, \overline{y}_0)$ with $t_1 > t_0$ are solutions of an equation via (t_0, y_0), (t_1, y_0), respectively, again here

$$\Omega(t) \leq \int_{t_0}^t g'(\tau)\overline{g}(\tau, \Omega(\tau))d\tau. \tag{9.83}$$

We have on the other hand that

$$\Omega(t_1) = |y(t_1, t_0, y_0) - y(t_1, t_1, y_0)| \\ = |y(t_1, t_0, y_0) - y_0|. \tag{9.84}$$

Therefore with the maximal solution principle, we shall have

$$\Omega(t_1) \le r^*(t_1, t_0, 0) \tag{9.85}$$

and

$$m(t) \le \widetilde{r}(t), \ \forall t > t_1 \tag{9.86}$$

where

$$\widetilde{r}(t) = \widetilde{r}(t, t_1, r^*(t_1, t_0, 0)) \tag{9.87}$$

is the maximal solution of

$$D_g v = \overline{g}(t, v) \tag{9.88}$$

through $(t_1, r^*(t_1, t_0, 0))$. Since $r^*(t_1, t_0, 0) = 0$, this leads to.

Therefore by hypothesis $\widetilde{r}(t, t_1, r^*(t_1, t_0, 0))$ is identically zero; this concludes the theorem that $y(t, t_0, y_0)$ is continuous with respect to t.

9.3 Chaplygin's Method and Initial Condition Dependence: Caputo Case

In this section, we shall extend the sequential method to the nonlinear ordinary differential equations with Caputo global derivative. The mathematical equation under consideration here is given by

$$\begin{cases} {}_{t_0}^{C}D_g^{\alpha} y(t) = f(t, y(t)) & \text{if } t \in [t_0, t_0 + a] \\ y(t_0) = y_0 & \text{if } t = t_0 \end{cases} \tag{9.89}$$

which can be converted to

$$\begin{cases} y(t) = \frac{1}{\Gamma(\alpha)} \displaystyle\int_{t_0}^{t} g'(\tau) f(\tau, y(\tau)) (t - \tau)^{\alpha - 1} d\tau & \text{if } t > t_0 \\ y(t_0) = y_0 & \text{if } t = t_0 \end{cases}. \tag{9.90}$$

The properties of $g'(t)$ are the same like before.

Theorem 9.4 *Let* $f \in C[R, R_0]$, *where*

$$R_0 = \{(t, y) \,||y - y_0| < b,\ t_0 \le t \le t_0 + a\}. \tag{9.91}$$

We let $M \ge |f(t, y)|$ *on* R_0 *and* $\lambda = \min\{a, \overline{a}\}$. *We assume that* f_y, f_{yy} *exist and* $f_{yy} > 0$ *in* R_0. *Let* $u_0(t)$ *and* $v_0(t)$ *be differentiable for* $t_0 \le t \le t_0 + \lambda$ *such that* $(t, u_0(t))$, $(t, v_0(t)) \in R_0$ *and*

$$\begin{cases} u_0(t) < \frac{1}{\Gamma(\alpha)} \int\limits_{t_0}^{t} g'(\tau) f(\tau, u_0(\tau)) (t - \tau)^{\alpha - 1} \, d\tau & \text{if } t > t_0 \\ u(t_0) = u_0 & \text{if } t = t_0 \end{cases}. \tag{9.92}$$

$$\begin{cases} v_0(t) > \frac{1}{\Gamma(\alpha)} \int\limits_{t_0}^{t} g'(\tau) f(\tau, v_0(\tau)) (t - \tau)^{\alpha - 1} \, d\tau & \text{if } t > t_0 \\ v(t_0) = v_0 & \text{if } t = t_0 \end{cases}. \tag{9.93}$$

Then, there exists a Chaplygin sequence $\{u_n(t), v_n(t)\}$ *such that*

$$\begin{gathered} u_n(t) < u_{n+1}(t) < y(t) < v_{n+1}(t) < v_n(t), \\ \forall t \in [t_0, t_0 + \lambda],\ u_n(t_0) = y_0 = v_n(t_0) \end{gathered} \tag{9.94}$$

where $y(t)$ *is the unique solution of*

$$\begin{cases} y(t) = \frac{1}{\Gamma(\alpha)} \int\limits_{t_0}^{t} g'(\tau) f(\tau, y(\tau)) (t - \tau)^{\alpha - 1} \, d\tau & \text{if } t > t_0 \\ y(t_0) = y_0 & \text{if } t = t_0 \end{cases} \tag{9.95}$$

existing on $[t_0, t_0 + \lambda]$. *Additionally* $u_n(t)$, $v_n(t)$ *tend uniformly to* $y(t)$ *on* $[t_0, t_0 + \lambda]$ *as* $n \to \infty$. *If in additional for a suitable* γ

$$0 \le v_0(t) < u_0(t) \le \gamma \tag{9.96}$$

then

$$|u_n(t) - v_n(t)| \le \frac{2\gamma}{2^{2n}}\ t \in [t_0, t_0 + \lambda]. \tag{9.97}$$

Proof We shall start our proof by finding $\bar{a}$,

$$\begin{aligned} |y(t)| &= \left| \frac{1}{\Gamma(\alpha)} \int_{t_0}^{t} g'(\tau) f(\tau, y(\tau)) (t-\tau)^{\alpha-1} d\tau \right|, \\ &\leq \frac{1}{\Gamma(\alpha)} \int_{t_0}^{t} \left|g'(\tau)\right| |f(\tau, y(\tau))| (t-\tau)^{\alpha-1} d\tau, \\ &\leq \frac{1}{\Gamma(\alpha)} \int_{t_0}^{t} \max_{l\in[t_0,\tau]} \left|g'(l)\right| \max_{l\in[t_0,\tau]} |f(l, y(l))| (t-\tau)^{\alpha-1} d\tau, \\ |y(t)| &\leq \frac{1}{\Gamma(\alpha+1)} M_{g'} M_f \left(t^\alpha - t_0^\alpha\right), \\ &\leq \frac{M_{g'} M_f}{\Gamma(\alpha+1)} a^\alpha < b, \end{aligned} \tag{9.98}$$

$$a < \left(\frac{b\Gamma(\alpha+1)}{M_{g'} M_f}\right)^{\frac{1}{\alpha}}. \tag{9.99}$$

Therefore in this case, we have

$$\bar{a} = \left(\frac{b\Gamma(\alpha+1)}{M_{g'} M_f}\right)^{\frac{1}{\alpha}}. \tag{9.100}$$

We need then

$$\lambda = \min\left\{a, \left(\frac{b\Gamma(\alpha+1)}{M_{g'} M_f}\right)^{\frac{1}{\alpha}}\right\}. \tag{9.101}$$

We can now start the demonstration. We have by the hypothesis that

$$\begin{aligned} u_0(t) &< \frac{1}{\Gamma(\alpha)} \int_{t_0}^{t} g'(\tau) f(\tau, u_0(\tau)) (t-\tau)^{\alpha-1} d\tau, \\ &< \frac{1}{\Gamma(\alpha)} \int_{t_0}^{t} g'(\tau) f(\tau, y(\tau)) (t-\tau)^{\alpha-1} d\tau, \\ &< y(t). \end{aligned} \tag{9.102}$$

Therefore $\forall t \in [t_0, t_0 + \lambda]$

$$u_0(t) < y(t). \tag{9.103}$$

On the other hand

$$
\begin{aligned}
v_0(t) &> \frac{1}{\Gamma(\alpha)} \int_{t_0}^{t} g'(\tau) f(\tau, v_0(\tau)) (t-\tau)^{\alpha-1} \, d\tau, \qquad (9.104)\\
&> \frac{1}{\Gamma(\alpha)} \int_{t_0}^{t} g'(\tau) f(\tau, y(\tau)) (t-\tau)^{\alpha-1} \, d\tau,\\
&> y(t).
\end{aligned}
$$

Therefore $\forall t \in [t_0, t_0 + \lambda]$

$$
y(t) < v_0(t). \qquad (9.105)
$$

Now, we have on the one hand that

$$
v_0(t) > \frac{1}{\Gamma(\alpha)} \int_{t_0}^{t} g'(\tau) f(\tau, v_0(\tau)) (t-\tau)^{\alpha-1} \, d\tau, \qquad (9.106)
$$

$$
u_0(t) < \frac{1}{\Gamma(\alpha)} \int_{t_0}^{t} g'(\tau) f(\tau, u_0(\tau)) (t-\tau)^{\alpha-1} \, d\tau.
$$

Thus

$$
-u_0(t) > -\frac{1}{\Gamma(\alpha)} \int_{t_0}^{t} g'(\tau) f(\tau, u_0(\tau)) (t-\tau)^{\alpha-1} \, d\tau, \qquad (9.107)
$$

then

$$
v_0(t) - u_0(t) > \frac{1}{\Gamma(\alpha)} \int_{t_0}^{t} g'(\tau) (t-\tau)^{\alpha-1} \left(f(\tau, v_0(\tau)) - f(\tau, u_0(\tau)) \right) d\tau, \qquad (9.108)
$$

$$
v_0(t) - u_0(t) > \frac{1}{\Gamma(\alpha)} \int_{t_0}^{t} g'(\tau) (t-\tau)^{\alpha-1} f_y(\tau, \xi) \left(v_0(\tau) - u_0(\tau) \right) d\tau.
$$

The above is obtained since the function $f(t, y)$ is differentiable on R_0, thus using the mean value theorem where

$$
u_0(t) < \xi < v_0(t), \forall t \in [t_0, t_0 + \lambda].
$$

$$v_0(t) - u_0(t) > \frac{1}{\Gamma(\alpha)} \int_{t_0}^{t} g'(\tau)\,(t-\tau)^{\alpha-1} \min_{l\in[t_0,\tau]} \left(f_y\,(l,\xi)\right) d\tau, \tag{9.109}$$

$$> \frac{\overline{M}_{f_y}}{\Gamma(\alpha)} \int_{t_0}^{t} g'(\tau)\,(t-\tau)^{\alpha-1} \left(v_0(\tau) - u_0(\tau)\right) d\tau,$$

$$> \frac{\overline{M}_{f_y}\overline{M}_{g'}}{\Gamma(\alpha)} \int_{t_0}^{t} (t-\tau)^{\alpha-1} \left(v_0(\tau) - u_0(\tau)\right) d\tau,$$

$$\overline{M}_{g'} = \min_{t\in[t_0,t_0+\lambda]} \left|g'(\tau)\right|, \tag{9.110}$$

$$v_0(t) - u_0(t) > \frac{\overline{M}}{\Gamma(\alpha)} \int_{t_0}^{t} (t-\tau)^{\alpha-1} \left(v_0(\tau) - u_0(\tau)\right) d\tau.$$

Therefore $\forall t \in [t_0, t_0 + \lambda]$ indeed $v_0(t) > u_0(t)$. Therefore, we have

$$u_0(t) < y(t) < v_0(t). \tag{9.111}$$

We now consider the following functions:

$$f_1\,(t, y; u_0, v_0) = f\,(t, u_0(t)) + f_y\,(t, u_0(t))\,(y(t) - u_0(t))\,, \tag{9.112}$$

$$f_2\,(t, y; u_0, v_0) = f\,(t, u_0(t)) + \frac{f\,(t, u_0(t)) - f\,(t, v_0(t))}{u_0(t) - v_0(t)}\,(y(t) - u_0(t))\,.$$

That will be used to define u_1 and v_1. But in general, we shall have

$$f_1\,(t, y; u_n(t), v_n(t)) = f\,(t, u_n(t)) + f_y\,(t, u_n(t))\,(y(t) - u_n(t))\,, \tag{9.113}$$

$$f_2\,(t, y; u_n(t), v_n(t)) = f\,(t, u_n(t)) + \frac{f\,(t, u_n(t)) - f\,(t, v_n(t))}{u_n(t) - v_n(t)}\,(y(t) - u_n(t))\,,$$

for v_{n+1} and u_{n+1}. In general when $t = t_0$, we have that

$$f_1(t_0, y, u_n, v_n) = f_2(t_0, y, u_n, v_n). \tag{9.114}$$

It is assumed that $u_1(t)$ and $v_1(t)$ are solution of the linear equation

$$u_1(t) = \frac{1}{\Gamma(\alpha)} \int_{t_0}^{t} g'(\tau) f_1\,(\tau, u_1(\tau); u_0, v_0)\,(t-\tau)^{\alpha-1}\,d\tau, \tag{9.115}$$

$$v_1(t) = \frac{1}{\Gamma(\alpha)} \int_{t_0}^{t} g'(\tau) f_2\,(\tau, v_1(\tau); u_0, v_0)\,(t-\tau)^{\alpha-1}\,d\tau.$$

In general we shall have

$$u_{n+1}(t) = \frac{1}{\Gamma(\alpha)} \int_{t_0}^{t} g'(\tau) f_1(\tau, u_{n+1}(\tau); u_0, v_0) (t-\tau)^{\alpha-1} \, d\tau, \tag{9.116}$$

$$v_{n+1}(t) = \frac{1}{\Gamma(\alpha)} \int_{t_0}^{t} g'(\tau) f_2(\tau, v_{n+1}(\tau); u_0, v_0) (t-\tau)^{\alpha-1} \, d\tau,$$

which exist on $[t_0, t_0 + \lambda]$. However from

$$\begin{aligned} u_0(t) &< \frac{1}{\Gamma(\alpha)} \int_{t_0}^{t} g'(\tau) f(\tau, u_0(\tau)) (t-\tau)^{\alpha-1} \, d\tau, \\ &< \frac{1}{\Gamma(\alpha)} \int_{t_0}^{t} g'(\tau) f_1(\tau, u_0(\tau); u_0, v_0) (t-\tau)^{\alpha-1} \, d\tau, \\ &= u_1(t). \end{aligned} \tag{9.117}$$

Therefore $\forall t \in [t_0, t_0 + \lambda]$ we have $u_0(t) < u_1(t)$. On the other hand, we have

$$\begin{aligned} v_0(t) &> \frac{1}{\Gamma(\alpha)} \int_{t_0}^{t} g'(\tau) f(\tau, v_0(\tau)) (t-\tau)^{\alpha-1} \, d\tau, \\ &> \frac{1}{\Gamma(\alpha)} \int_{t_0}^{t} g'(\tau) f_1(\tau, v_0(\tau); u_0, v_0) (t-\tau)^{\alpha-1} \, d\tau, \\ &= v_1(t). \end{aligned} \tag{9.118}$$

Therefore $\forall t \in [t_0, t_0 + \lambda]$ we have $v_1(t) < v_0(t)$:

$$\begin{aligned} u_0(t) &< \frac{1}{\Gamma(\alpha)} \int_{t_0}^{t} g'(\tau) f(\tau, u_0(\tau)) (t-\tau)^{\alpha-1} \, d\tau, \\ &< \frac{1}{\Gamma(\alpha)} \int_{t_0}^{t} g'(\tau) f_2(\tau, u_0(\tau); u_0, v_0) (t-\tau)^{\alpha-1} \, d\tau, \\ &< v_1(t) \end{aligned} \tag{9.119}$$

$$v_1(t) = \frac{1}{\Gamma(\alpha)} \int_{t_0}^{t} g'(\tau)(t-\tau)^{\alpha-1} \left[\begin{array}{c} f(\tau, u_0(\tau)) \\ + \frac{f(\tau,u_0(\tau)) - f(\tau,v_0(\tau))}{u_0(\tau)-v_0(\tau)} (v_1(\tau) - u_0(\tau)) \end{array} \right] d\tau. \tag{9.120}$$

We have that

$$f_y(t, u_0(t)) < \frac{f(t, u_0(t)) - f(t, v_0(t))}{u_0(t) - v_0(t)}. \tag{9.121}$$

But

$$\begin{aligned} f(t, v_1(t)) =& f(t, v_0(t)) + f_y(t, u_0(t))(v_1(t) - u_0(t)) \\ &+ \frac{1}{2} f_{xy}(t, \xi)(v_1(t) - u_0(t))^2, \ u_0(t) < \xi < v_1(t). \end{aligned} \tag{9.122}$$

We also have that

$$f_{yy}(t, \xi) > 0, \tag{9.123}$$

therefore, we have

$$\begin{aligned} v_1(t) &> \frac{1}{\Gamma(\alpha)} \int_{t_0}^{t} g'(\tau) f_2(\tau, v_1(\tau); u_0, v_0)(t-\tau)^{\alpha-1} d\tau, \\ &> \frac{1}{\Gamma(\alpha)} \int_{t_0}^{t} g'(\tau) f(\tau, v_1(\tau))(t-\tau)^{\alpha-1} d\tau, \\ &> \frac{1}{\Gamma(\alpha)} \int_{t_0}^{t} g'(\tau) f(\tau, y(\tau))(t-\tau)^{\alpha-1} d\tau, \\ &> y(t). \end{aligned} \tag{9.124}$$

We have that $\forall t \in [t_0, t_0 + \lambda]$

$$u_0(t) < u_1(t) < y(t) < v_1(t) < v_0(t). \tag{9.125}$$

We repeat the same process, using the function $f_1(t, u_{n+1}(t); u_n(t), v_n(t))$ and $f_2(t, v_{n+1}(t); u_n(t), v_n(t))$.

Then we have a Chaplygin's sequence as functions with the following properties:

$$u_n(t) < \frac{1}{\Gamma(\alpha)} \int_{t_0}^{t} g'(\tau) f(\tau, u_n(\tau))(t-\tau)^{\alpha-1} d\tau, \tag{9.126}$$

$$v_n(t) > \frac{1}{\Gamma(\alpha)} \int_{t_0}^{t} g'(\tau) f(\tau, v_n(\tau)) (t-\tau)^{\alpha-1} d\tau,$$

$$u_n(t) < u_{n+1}(t) < y(t) < v_{n+1}(t) < v_n(t), \ \forall t \in [t_0, t_0 + \lambda]. \tag{9.127}$$

$$u_{n+1}(t) = \frac{1}{\Gamma(\alpha)} \int_{t_0}^{t} g'(\tau) f_1 \left(\tau, u_{n+1}(\tau); u_n(\tau), v_n(\tau)\right) (t-\tau)^{\alpha-1} d\tau, \tag{9.128}$$

$$v_{n+1}(t) = \frac{1}{\Gamma(\alpha)} \int_{t_0}^{t} g'(\tau) f_2 \left(\tau, v_{n+1}(\tau); u_n(\tau), v_n(\tau)\right) (t-\tau)^{\alpha-1} d\tau.$$

We have that $\{u_n\}$ and $\{v_n\}$ are monotonic, uniformly bounded on $[t_0, t_0 + \lambda]$. They are equicontinuous and converge toward $y(t)$ as $n \to \infty$. For the second part we have for $n = 0$

$$v_0(t) - u_0(t) \leq \gamma. \tag{9.129}$$

We assume that the formula is correct for any fixed n, that is

$$|v_n(t) - u_n(t)| < \frac{2\gamma}{2^{2^n}}; \tag{9.130}$$

when at $n + 1$, we have

$$\begin{aligned} &|v_{n+1}(t) - u_{n+1}(t)| \\ &= \frac{1}{\Gamma(\alpha)} \left| \int_{t_0}^{t} g'(\tau) \begin{pmatrix} \frac{f(\tau, u_n(\tau)) - f(\tau, v_n(\tau))}{u_n(\tau) - v_n(\tau)} \\ \times (v_{n+1}(\tau) - u_n(\tau)) - f_y(\tau, u_n(\tau)) (u_{n+1}(\tau) - u_n(\tau)) \end{pmatrix} (t-\tau)^{\alpha-1} d\tau \right| \\ &= \frac{1}{\Gamma(\alpha)} \left| \begin{array}{c} \int_{t_0}^{t} f_y(t, \xi) (v_{n+1}(\tau) - u_{n+1}(\tau)) \\ + (u_{n+1}(\tau) - u_n(\tau)) \left(f_y(\tau, \xi) - f_y(\tau, u_n(\tau))\right) (t-\tau)^{\alpha-1} d\tau \end{array} \right| \end{aligned} \tag{9.131}$$

due to the mean value theorem.

Since

$$f_y(t, \xi) - f_y(t, u_n(t)) = f_{yy}(t, \bar{\xi}) (\xi - u_n(t)) \tag{9.132}$$

where

$$u_n(t) < \bar{\xi} < \xi \tag{9.133}$$

replacing using the fact that

$$\begin{aligned} |\xi - u_n(t)| &\leq |u_n(t) - v_n(t)|, \\ |u_{n+1}(t) - u_n(t)| &\leq |u_n(t) - v_n(t)|. \end{aligned} \tag{9.134}$$

We have that

$$\left|u_{n+1}(t)-v_{n+1}(t)\right| = \frac{1}{\Gamma(\alpha)}\left|\int_{t_0}^{t} g'(\tau)(t-\tau)^{\alpha-1}\begin{pmatrix}\Omega_1\left|v_{n+1}(\tau)-u_{n+1}(\tau)\right| \\ +\frac{2\gamma}{2^{2^{n+1}}}\Omega_2\end{pmatrix}d\tau\right|, \tag{9.135}$$

$$\begin{aligned}
&\leq \frac{\Omega_1}{\Gamma(\alpha)}\int_{t_0}^{t} g'(\tau)(t-\tau)^{\alpha-1}\left|v_{n+1}(\tau)-u_{n+1}(\tau)\right|d\tau \\
&+\frac{2^2\gamma^2\Omega_2}{2^{2^{n+1}}}\frac{a^{\alpha}M_{g'}}{\Gamma(\alpha+1)}, \\
&\leq \frac{\Omega_1 M_{g'}}{\Gamma(\alpha)}\int_{t_0}^{t}(t-\tau)^{\alpha-1}\left|v_{n+1}(\tau)-u_{n+1}(\tau)\right|d\tau \\
&+\frac{2^2\gamma^2\Omega_2 a^{\alpha}M_{g'}}{2^{2^{n+1}}\Gamma(\alpha+1)}.
\end{aligned}$$

By the mean of the Gronwall inequality, we have that

$$\begin{aligned}
\left|u_{n+1}(t)-v_{n+1}(t)\right| &\leq \frac{2^2\gamma^2\Omega_2 a^{\alpha}M_{g'}}{2^{2^{n+1}}\Gamma(\alpha+1)}\exp\left[\frac{\Omega_1 M_{g'}}{\Gamma(\alpha+1)}\left(t^{\alpha}-t_0^{\alpha}\right)\right], \\
&\leq \frac{2^2\gamma^2\Omega_2 a^{\alpha}M_{g'}}{2^{2^{n+1}}\Gamma(\alpha+1)}\exp\left[\frac{\Omega_1 M_{g'}}{\Gamma(\alpha+1)}a^{\alpha}\right].
\end{aligned} \tag{9.136}$$

To meet the requirement of the theorem, we will need

$$\gamma = \left(\frac{2\Omega_2 a^{\alpha}M_{g'}}{\Gamma(\alpha+1)}\exp\left[\frac{\Omega_1 M_{g'}a^{\alpha}}{\Gamma(\alpha+1)}\right]\right)^{-1} \tag{9.137}$$

such that

$$\left|u_{n+1}(t)-v_{n+1}(t)\right| < \frac{2\gamma}{2^{2^{n+1}}} \tag{9.138}$$

which completes the proof; therefore $\forall n$, we have

$$\begin{aligned}
\left|y(t)-u_n(t)\right| &\leq \tfrac{2\gamma}{2^{2^n}}, \\
\left|y(t)-v_n(t)\right| &\leq \tfrac{2\gamma}{2^{2^n}}.
\end{aligned} \tag{9.139}$$

We shall now show a lead demonstration of the upper Chaplygin's sequence $\{v_n(t)\}$ with the same extra condition.

Theorem 9.5 *Let* $f \in C[R, R_0]$, *where*

$$R_0 = \{(t,y)\,||y-y_0| < b,\ t_0 \leq t \leq t_0+a\}. \tag{9.140}$$

We let $M \geq |f(t, y)|$ on R_0. We assume $f(t, y)$ is quasi-monotone increasing in y for all $t \in [t_0, t_0 + a]$ and that $\frac{\partial f(t,y)}{\partial y}$ exists and is continuous on R_0. We let $v_0(t)$ be a continuously differentiable function on $[t_0, t_0 + \lambda]$, where

$$\lambda = \min\left\{a, \left(\frac{b\Gamma(\alpha+1)}{M_{g'}M_f}\right)^{\frac{1}{\alpha}}\right\}, \quad (t, v_0(t)) \in R_0, \tag{9.141}$$

$$v_0(t) > \frac{1}{\Gamma(\alpha)} \int_{t_0}^{t} g'(\tau) f(\tau, v_0(\tau)) (t-\tau)^{\alpha-1} \, d\tau, \tag{9.142}$$

$$v_0(t_0) = v_0.$$

Additionally, let

$$f(t, y) + f_y(t, y)(z - y) > f(t, z), \quad \text{if } y < z. \tag{9.143}$$

Then there exists a Chaplygin's sequence $\{v_n(t)\}$ such that $v_n(t_0) = y_0$.

$$y(t) < v_{n+1}(t) < v_n(t), \quad t \in [t_0, t_0 + \lambda] \tag{9.144}$$

where $y(t)$ is the solution of

$$z(t) = \frac{1}{\Gamma(\alpha)} \int_{t_0}^{t} g'(\tau) f(\tau, z(\tau)) (t-\tau)^{\alpha-1} \, d\tau, \tag{9.145}$$

existing on $[t_0, t_0 + \lambda]$ and

$$\lim_{n \to \infty} v_n(t) = y(t) \tag{9.146}$$

uniformly on $[t_0, t_0 + \lambda]$.

Proof We have that f is quasi-monotonic; moreover we have

$$\begin{aligned} v_0(t) &> \frac{1}{\Gamma(\alpha)} \int_{t_0}^{t} g'(\tau) f(\tau, v_0(\tau)) (t-\tau)^{\alpha-1} \, d\tau, \\ &> \frac{1}{\Gamma(\alpha)} \int_{t_0}^{t} g'(\tau) f(\tau, y(\tau)) (t-\tau)^{\alpha-1} \, d\tau, \\ &= y(t). \end{aligned} \tag{9.147}$$

Therefore $\forall t \in [t_0, t_0 + \lambda]$

$$v_0(t) > y(t). \tag{9.148}$$

We have the following defined linear differential equation:

$$\begin{aligned}
\overline{y}(t) &= \frac{1}{\Gamma(\alpha)} \int_{t_0}^{t} g'(\tau)(t-\tau)^{\alpha-1} \left(f(\tau, v_0(\tau)) + \frac{\partial f(\tau, u_0(\tau))}{\partial y} (\overline{y} - u_0(\tau)) \right) d\tau, \\
\overline{y}(t_0) &= y_0, \\
\overline{y}(t) &= \frac{1}{\Gamma(\alpha)} \int_{t_0}^{t} g'(\tau)(t-\tau)^{\alpha-1} \overline{f}(\tau, \overline{y}; v_0(\tau)).
\end{aligned} \tag{9.149}$$

$f(t, y; v_0(\tau))$ is also quasi-monotone with respect to $\overline{y}$. Hence we have that

$$v_n(t) < v_0(t) \; \forall t \in [t_0, t_0 + \lambda]. \tag{9.150}$$

The hypothesis implies that

$$\begin{aligned}
v_1(t) &> \frac{1}{\Gamma(\alpha)} \int_{t_0}^{t} g'(\tau) f(\tau, v_1(\tau)) (t-\tau)^{\alpha-1} d\tau, \\
&> \frac{1}{\Gamma(\alpha)} \int_{t_0}^{t} g'(\tau) f(\tau, y(\tau)) (t-\tau)^{\alpha-1} d\tau, \\
&> y(t).
\end{aligned} \tag{9.151}$$

Therefore we have that $\forall t \in [t_0, t_0 + \lambda]$

$$y(t) < v_1(t) < v_0(t). \tag{9.152}$$

We can then set a mapping $\overline{\Lambda}$ verifying

$$v_n(t) = \overline{\Lambda}(v_0(t)), \tag{9.153}$$

and by repeating we obtain

$$\overline{\Lambda}(v_n(t)) = v_{n+1}(t) \tag{9.154}$$

with

$$y(t) < v_{n+1}(t) < v_n(t), \; t_0 \in [t_0, t_0 + \lambda] \tag{9.155}$$

which completes the proof.

9.4 Dependence on Parameters and Initial Condition

We are interested to investigate the problem of continuity and differentiability of solution $y(t, t_0, y_0)$ of the equation. We shall consider the following initial value problem:

$$\begin{aligned} {}_{t_0}^{RL}D_g^{\alpha} y(t) &= f(t, y(t)) \\ y(t_0) &= y_0 \end{aligned} \tag{9.156}$$

with respect to the initial data (t_0, y_0) . We start with the following lemma.

Lemma 9.2 *Let $f \in C[I \times R, R]$ and let*

$$H(t, r) = \sup_{|y-y_0| \leq r} |f(t, y)| . \tag{9.157}$$

We assume that $r^*(t, t_0, 0)$ is the maximal solution of

$${}_{t_0}^{RL}D_g^{\alpha} z(t) = H(t, z) \tag{9.158}$$

via $(t_0, 0)$. Let $y(t, t_0, y_0)$ be any solution of

$$\begin{aligned} {}_{t_0}^{RL}D_g^{\alpha} y(t) &= f(t, y(t)) \\ y(t_0) &= y_0 \end{aligned} , \tag{9.159}$$

then

$$|y(t, t_0, y_0)| \leq r^*(t, t_0, 0). \tag{9.160}$$

Proof We define $n(t) = |y(t, t_0, y_0)|$, then

$$\begin{aligned} n(t) &\leq \frac{1}{\Gamma(\alpha)} \int_{t_0}^{t} g'(\tau) |f(\tau, y(\tau, t_0, y_0))| (t-\tau)^{\alpha-1} d\tau, \\ &\leq \frac{1}{\Gamma(\alpha)} \int_{t_0}^{t} g'(\tau) \sup_{|y| \leq n(\tau)} |f(\tau, y(\tau))| (t-\tau)^{\alpha-1} d\tau, \\ &\leq \frac{1}{\Gamma(\alpha)} \int_{t_0}^{t} g'(\tau) H(\tau, n(\tau)) (t-\tau)^{\alpha-1} d\tau. \end{aligned} \tag{9.161}$$

By the maximal solution principle, we have that

$$n(t) = |y(t, t_0, y_0)| \leq r^*(t, t_0, 0); \ \ t \geq t_0 \tag{9.162}$$

which proves the lemma.

Theorem 9.6 *Let $f \in C[I \times R, R]$ and for $(t, y), (t, z) \in I \times R$,*

$$|f(t, y) - f(t, z)| \leq h(t, |y - z|) \tag{9.163}$$

where the function $h \in C[I \times R, R]$. Assume that $v(t) = 0$ is the unique solution of the differential equation

$$ {}_{t_0}^{RL}D_g^{\alpha} v(t) = g(t, v(t)) \tag{9.164}$$

such that $v(t_0) = 0$. Then if the solutions $v(t, t_0, v_0)$ of

$$ {}_{t_0}^{RL}D_g^{\alpha} v(t) = g(t, v(t)) \tag{9.165}$$

via every point (t_0, v_0) are continuous with respect to initial conditions (t_0, v_0), the solution of $y(t, t_0, y_0)$, $y(t_0) = y$ is unique and continuous with respect to the initial data (t_0, y_0).

Proof We have shown using Chaplygin's approach that the equation has a unique solution. We shall only proceed our investigation by proving the continuity. To do this, we consider $y(t, t_0, y_0)$ and $\overline{y}(t, t_0, \overline{y}_0)$ to be the solutions via (t_0, y_0) and $\left(t_0, \overline{y}_0\right)$. We let

$$n(t) = \left|y(t, t_0, y_0) - \overline{y}(t, t_0, \overline{y}_0)\right|. \tag{9.166}$$

We have from the condition of the theorem that

$$\begin{aligned} n(t) &\leq \frac{1}{\Gamma(\alpha)} \int_{t_0}^{t} g'(\tau) h(\tau, n(\tau)) (t - \tau)^{\alpha - 1} d\tau, \\ &\leq r(t, t_0, \left|y_0 - \overline{y}_0\right|), t > t_0 \end{aligned} \tag{9.167}$$

where $r(t, t_0, \left|y_0 - \overline{y}_0\right|)$ is the maximal solution of

$$ {}_{t_0}^{RL}D_g^{\alpha} v(t) = h(t, v(t)) \tag{9.168}$$

such that

$$v(t_0) = |y(t_0) - \overline{y}(t_0)|. \tag{9.169}$$

Since $v(t, t_0, v_0)$ is continuous by hypothesis, we shall have

$$\lim_{y \to \overline{y}_0} r(t, t_0, \left|y_0 - \overline{y}_0\right|) = r(t, t_0, 0) \tag{9.170}$$

while by hypothesis

$$r(t, t_0, 0) = 0. \tag{9.171}$$

Therefore under this, we shall have

$$\lim_{y_0 \to \overline{y}_0} y(t, t_0, y_0) = \overline{y}(t, t_0, \overline{y}_0). \tag{9.172}$$

This shows that $y(t, t_0, y_0)$ is continuous with respect to y_0. We shall now show that $y(t, t_0, y_0)$ is continuous with respect to t_0. We then choose $y(t, t_0, y_0)$, $y(t, t_1, y_0)$ with $t_1 > t_0$, being solutions via (t_0, y_0) and (t_1, y_0) respectively, then

$$n(t) \leq \frac{1}{\Gamma(\alpha)} \int_{t_0}^{t} g'(\tau) h(\tau, n(\tau)) (t-\tau)^{\alpha-1} d\tau, \tag{9.173}$$

where

$$n(t) = |y(t, t_0, y_0) - \overline{y}(t, t_1, y_0)|. \tag{9.174}$$

Here

$$n(t_1) = |y(t_1, t_0, y_0)| \leq r^*(t_1, t_0, 0). \tag{9.175}$$

Therefore

$$n(t) \leq \widetilde{r}(t), \; t > t_1 \tag{9.176}$$

where

$$\widetilde{r}(t) = \widetilde{r}\left(t, t_1, r^*(t_1, t_0, 0)\right) \tag{9.177}$$

is the maximal solution via $(t_1, r^*(t_1, t_0, 0))$. But we have that $r^*(t_0, t_0, 0) = 0$. Therefore

$$\lim_{t_1 \to t_0} \widetilde{r}\left(t, t_1, r^*(t_1, t_0, 0)\right) = \widetilde{r}(t, t_0, 0) \tag{9.178}$$

and by theorem hypothesis, we have $\widetilde{r}(t, t_0, 0) = 0$. Thus we can conclude that $y(t, t_0, y_0)$ is continuous with respect to t_0.

Theorem 9.7 *Let $f \in C[B, R]$ where B is a defined open (t, y, η) set in R and $\eta = \eta_0$; let in addition $y_0(t) = y(t, t_0, y_0, \eta_0)$ be the solution of*

$$\begin{aligned} {}^{RL}_{t_0}D^{\alpha}_{g} y(t) &= f(t, y(t)) \\ y(t_0) &= y_0 \end{aligned} \tag{9.179}$$

existing for $t \geq t_0$. It is supposed that

$$\lim_{\eta \to \eta_0} f(t, y, \eta) = f(t, y, \eta_0) \tag{9.180}$$

uniformly in (t, y) and for (t, y_1, η), $(t, y_2, \eta) \in B$,

$$|f(t, y_1, \eta) - f(t, y_2, \eta)| < h(t, |y_1 - y_2|) \tag{9.181}$$

where $h \in C\left[I \times R_{+}, R_{+}\right]$. We assume that $v(t)$ is the unique solution of

$$ {}_{t_0}^{RL}D_g^{\alpha} v(t) = h(t, v(t)) \tag{9.182} $$

with $v(t_0) = 0$. $\forall \varepsilon > 0$, $\exists \delta(\varepsilon) > 0$ such that, $\forall \eta$, $|\eta - \eta_0| < \delta(\varepsilon)$, the differential equation

$$ {}_{t_0}^{RL}D_g^{\alpha} y(t) = f(t, y, \eta), \;\; y(t_0) = y_0 \tag{9.183} $$

has a unique solution

$$ y(t) = y(t, t_0, y_0, \eta) \tag{9.184} $$

with the following condition:

$$ |y(t) - y_0(t)| \leq \varepsilon, \;\; t > t_0. \tag{9.185} $$

Proof We have from the theorem that $v(t) = 0$ is the unique solution of

$$ v(t) = \frac{1}{\Gamma(\alpha)} \int_{t_0}^{t} g'(\tau) h(\tau, v(\tau)) (t - \tau)^{\alpha - 1} d\tau. \tag{9.186} $$

Therefore, knowing that $[t_0, t_0 + a] \subset I$ and $\forall \varepsilon > 0$, $\exists \mu = \mu(\varepsilon) > 0$ such that the maximal solution $r(t, t_0, 0, \eta)$ of

$$ \begin{aligned} v(t) = & \frac{1}{\Gamma(\alpha)} \int_{t_0}^{t} g'(\tau)(t-\tau)^{\alpha-1} (h(\tau, v(\tau)) + \mu) d\tau, \\ = & \frac{1}{\Gamma(\alpha)} \int_{t_0}^{t} g'(\tau)(t-\tau)^{\alpha-1} h(\tau, v(\tau)) d\tau \\ & + \frac{\mu}{\Gamma(\alpha)} \int_{t_0}^{t} g'(\tau)(t-\tau)^{\alpha-1} d\tau \\ \leq & \frac{1}{\Gamma(\alpha)} \int_{t_0}^{t} g'(\tau)(t-\tau)^{\alpha-1} h(\tau, v(\tau)) d\tau + \frac{\mu M_{g'}\left(t^{\alpha} - t_0^{\alpha}\right)}{\Gamma(\alpha+1)} \end{aligned} \tag{9.187} $$

exists when $t \in [t_0, t_0 + a]$ and we have

$$ r(t, t_0, 0, \eta) < \varepsilon, \;\; \forall t \in [t_0, t_0 + a]. \tag{9.188} $$

Indeed, because of

$$ \lim_{\eta \to \eta_0} f(t, y, \eta) = f(t, y, \eta_0) \tag{9.189} $$

uniformly given the fact that $\mu > 0$, we can find $\delta = \delta(\mu) > 0$ such that

$$|f(t, y, \eta) - f(t, y, \eta_0)| < \mu \tag{9.190}$$

when only

$$|\eta - \eta_0| < \mu. \tag{9.191}$$

However, let $\varepsilon > 0$; let us define

$$\eta(t) = |y(t) - \overline{y}(t)| \tag{9.192}$$

where $y(t)$ and $\overline{y}(t)$ are solutions of

$$\begin{aligned} y(t) &= \frac{1}{\Gamma(\alpha)} \int_{t_0}^{t} g'(\tau) f(\tau, y, \eta))(t-\tau)^{\alpha-1} d\tau, \\ y(t_0) &= y_0, \\ \overline{y}(t) &= \frac{1}{\Gamma(\alpha)} \int_{t_0}^{t} g'(\tau) f(\tau, y, \eta_0))(t-\tau)^{\alpha-1} d\tau, \\ \overline{y}(t_0) &= y_0, \end{aligned} \tag{9.193}$$

$$\begin{aligned} \eta(t) \leq & \frac{1}{\Gamma(\alpha)} \int_{t_0}^{t} g'(\tau)(t-\tau)^{\alpha-1} h(\tau, \eta(\tau)) d\tau \\ & + \frac{1}{\Gamma(\alpha)} \int_{t_0}^{t} g'(\tau)(t-\tau)^{\alpha-1} |f(\tau, \overline{y}(\tau), \eta)) - f(\tau, \overline{y}(\tau), \eta_0))| d\tau \end{aligned} \tag{9.194}$$

with the above in hand if we have

$$|\eta - \eta_0| < \delta. \tag{9.195}$$

Then from previous results we have

$$\eta(t) \leq \frac{1}{\Gamma(\alpha)} \int_{t_0}^{t} g'(\tau)(t-\tau)^{\alpha-1} h(\tau, \eta(\tau)) d\tau + \frac{\mu M_{g'}\left(t^{\alpha} - t_0^{\alpha}\right)}{\Gamma(\alpha+1)}. \tag{9.196}$$

Therefore

$$\eta(t) \leq r(t, t_0, 0, \eta), t \geq t_0 \tag{9.197}$$

and then

$$|y(t) - \overline{y}(t)| < \varepsilon, t \geq t_0. \tag{9.198}$$

If we have

$$|\eta - \eta_0| < \delta, \tag{9.199}$$

indeed we can see that δ is a function of ε which completes the proof.

9.5 Chaplygin's Method and Initial Condition Dependence: Caputo–Fabrizio Case

In this section, we shall present the existence and uniqueness of a global differential ordinary equation with the Caputo–Fabrizio global derivative using Chaplygin's method. The assumption of the theorem will be as before with an extra condition that

$$\beta = \min\left\{a, \frac{b + M_{g'} M_f(\alpha - 1)}{\alpha M_{g'} M_f}\right\}. \tag{9.200}$$

The equation under consideration is

$$\begin{cases} {}_{t_0}^{CF}D_g^{\alpha} y(t) = f(t, y(t)) & \text{if } t > t_0 \\ y(t_0) = y_0 & \text{if } t = t_0 \end{cases}. \tag{9.201}$$

The above equation is converted to

$$\begin{cases} y(t) = y(0) + (1-\alpha)g'(t)f(t, y(t)) + \alpha \int\limits_{t_0}^{t} g'(\tau)f(\tau, y(\tau))d\tau & \text{if } t > t_0 \\ y(t_0) = y_0 & \text{if } t = t_0 \end{cases}. \tag{9.202}$$

Again here, we have that the functions $u_0(t)$, $v_0(t)$, and $y(t)$ satisfy the following inequality:

$$u_0(t) < y(t) < v_0(t) \ \forall t \in (t_0, t_0 + \lambda]. \tag{9.203}$$

We consider again the following functions:

$$f_1(t, y; u_0, v_0) = f(t, u_0(t)) + f_y(t, u_0(t))(y(t) - u_0(t)), \tag{9.204}$$

$$f_2(t, y; u_0, v_0) = f(t, u_0(t)) + \frac{f(t, u_0(t)) - f(t, v_0(t))}{u_0(t) - v_0(t)}(y(t) - u_0(t)).$$

We let $u_1(t)$, $v_1(t)$ to be the solution of the linear global differential equation with the Caputo–Fabrizio derivative

$$
\begin{aligned}
{}_{t_0}^{CF}D_g^{\alpha}u_1(t) &= f_1\left(t, u_1(t); u_0, v_0\right), \ u_1(t_0) = y\left(t_0\right), \\
{}_{t_0}^{CF}D_g^{\alpha}v_1(t) &= f_2\left(t, v_1(t); u_0, v_0\right), \ u_1(t_0) = y\left(t_0\right),
\end{aligned} \tag{9.205}
$$

differential equations that exist on $(t_0, t_0 + \lambda]$. Note that from the theorem, we have

$$
{}_{t_0}^{CF}D_g^{\alpha}u_0(t) < f\left(t, u_0(t)\right), \ u_0(t_0) = y_0. \tag{9.206}
$$

By the definition of $f_1\left(t, u_0(t); u_0, v_0\right)$

$$
{}_{t_0}^{CF}D_g^{\alpha}u_0(t) < f_1\left(t, u_0(t); u_0, v_0\right), \tag{9.207}
$$

since

$$
f\left(t, u_0(t)\right) = f_1\left(t, u_0(t); u_0, v_0\right). \tag{9.208}
$$

Applying the corresponding integral yields

$$
\begin{aligned}
u_0(t) < u_0(t_0) &+ (1-\alpha)g'(t)f_1\left(t, u_0(t); u_0, v_0\right) \\
&+ \alpha\int_{t_0}^{t} g'(\tau)f_1\left(\tau, u_0(\tau); u_0, v_0\right)d\tau.
\end{aligned} \tag{9.209}
$$

Note that

$$
u_0(t_0) = u_1(t_0) = y_0, \tag{9.210}
$$

thus

$$
\begin{aligned}
u_0(t) <\ & u_1(t_0) + (1-\alpha)g'(t)f_1\left(t, u_0(t)\right) + \alpha\int_{t_0}^{t} g'(\tau)f_1\left(\tau, u_0(\tau)\right)d\tau \\
<\ & u_1(t_0) + (1-\alpha)g'(t)\left[f\left(t, u_0(t)\right) + f_y\left(t, u_0(t)\right)\left(y(t) - u_0(t)\right)\right] \\
& +\alpha\int_{t_0}^{t} g'(\tau)\left[f\left(\tau, u_0(\tau)\right) + f_y\left(\tau, u_0(\tau)\right)\left(y(\tau) - u_0(\tau)\right)\right]d\tau, \\
=\ & u_1(t).
\end{aligned} \tag{9.211}
$$

Therefore $\forall t \in [t_0, t_0 + \lambda]$,

$$
u_0(t) < u_1(t). \tag{9.212}
$$

Now we have that

$$
u_0(t) < u_1(t) < y(t) < v_0(t) \tag{9.213}
$$

and on the other hand we have that

$$
{}_{t_0}^{CF}D_g^{\alpha}v_0(t) > f\left(t, v_0(t)\right), \ v_0(t_0) = y_0. \tag{9.214}
$$

By applying the integral on both sides we obtain

$$v_0(t) > v_0(t_0) + (1-\alpha)g'(t) f_1(t, v_0(t)) + \alpha \int_{t_0}^{t} g'(\tau) f_1(\tau, v_0(\tau))\, d\tau. \quad (9.215)$$

Note that

$$v_0(t_0) = y(t_0) = v_1(t_0), \quad (9.216)$$

thus

$$v_0(t) > v_1(t_0) + (1-\alpha)g'(t) f(t, v_0(t)) + \alpha \int_{t_0}^{t} g'(\tau) f(\tau, v_0(\tau))\, d\tau, \quad (9.217)$$

$$v_0(t) > v_1(t_0) + (1-\alpha)g'(t) f_2(t, u_0(t); u_0, v_0) + \alpha \int_{t_0}^{t} g'(\tau) f_2(\tau, u_0(\tau); u_0, v_0)\, d\tau,$$

$$\begin{aligned} v_0(t) > {} & v_1(t_0) + (1-\alpha)g'(t)\left[f(t, u_0(t)) + \frac{f(t, u_0(t)) - f(t, v_0(t))}{u_0(t) - v_0(t)} (y(t) - u_0(t))\right] \\ & + \alpha \int_{t_0}^{t} g'(\tau) f(\tau, u_0(\tau)) + \frac{f(\tau, u_0(\tau)) - f(\tau, v_0(\tau))}{u_0(\tau) - v_0(\tau)} (y(\tau) - u_0(\tau))\, d\tau, \\ = {} & v_1(t). \end{aligned}$$

We have that $\forall t \in [t_0, t_0 + \lambda]$

$$v_0(t) > v_1(t) \quad (9.218)$$

since

$$\begin{aligned} u_0(t) < {} & u_0(t_0) + (1-\alpha)g'(t) f(t, u_0(t)) + \alpha \int_{t_0}^{t} g'(\tau) f(\tau, u_0(\tau))\, d\tau, \quad (9.219) \\ = {} & u_0(t_0) + (1-\alpha)g'(t) f_2(t, u_0(t); u_0, v_0) \\ & + \alpha \int_{t_0}^{t} g'(\tau) f_2(\tau, u_0(\tau); u_0, v_0)\, d\tau, \\ = {} & v_1(t). \end{aligned}$$

$\forall t \in [t_0, t_0 + \lambda]$ more importantly

$$f_y(t, u_0(t)) < \frac{f(t, u_0(t)) - f(t, v_0(t))}{u_0(t) - v_0(t)} \quad (9.220)$$

and that

$$f(t, v_1(t)) = f(t, u_0(t)) + f_y(t, u_0(t))[v_1(t) - u_0(t)] + \frac{1}{2} f_{yy}(t, \eta)[v_1(t) - u_0(t)] \tag{9.221}$$

where

$$u_0(t) < \eta < v_1(t). \tag{9.222}$$

We apply twice the mean value theorem and use the fact that $f_{yy}(t, \eta) > 0$; we then obtain that

$$v_1(t) > v_1(t_0) + (1-\alpha) g'(t) f(t, v_1(t)) + \alpha \int_{t_0}^{t} g'(\tau) f(\tau, v_1(\tau)) \, d\tau. \tag{9.223}$$

We repeat the process by defining the following:

$$\begin{aligned} g_1(t, y; u_1, v_1) &= f(t, u_1(t)) + f_y(t, u_1(t))(y(t) - u_1(t)), \\ g_2(t, y; u_1, v_1) &= f(t, u_1(t)) + \frac{f(t, u_1(t)) - f(t, v_1(t))}{u_1(t) - v_1(t)}(y(t) - v_1(t)). \end{aligned} \tag{9.224}$$

We also observe that

$$g_1(t_0, y; u_1, v_1) = g_2(t_0, y; u_1, v_1). \tag{9.225}$$

We now assume that u_2 and v_2 are solutions of the following linear global differential equations:

$$\begin{aligned} {}_{t_0}^{CF}D_g^{\alpha} u_2(t) &= g_1(t, u_2; u_1, v_1), u_2(t_0) = y(t_0), \\ {}_{t_0}^{CF}D_g^{\alpha} v_2(t) &= g_1(t, v_2; u_1, v_1), v_2(t_0) = y(t_0), \end{aligned} \tag{9.226}$$

and on the other hand, we have that

$$\begin{aligned} {}_{t_0}^{CF}D_g^{\alpha} u_1(t) &< f(t, u_2(t)), \\ {}_{t_0}^{CF}D_g^{\alpha} v_1(t) &> f(t, v_2(t)). \end{aligned} \tag{9.227}$$

From here, we get that

$$\begin{aligned} & v_2(t) > v_1(t) > v_0(t) > y(t), \\ & u_0(t) < u_1(t) < u_2(t) < y(t), \\ & u_0(t) < u_1(t) < u_2(t) < y(t) \\ & \quad < v_0(t) < v_1(t) < v_2(t), \forall t \in [t_0, t_0 + \lambda]. \end{aligned} \tag{9.228}$$

We can then build a mapping Λ such that

$$
\begin{aligned}
(u_1, v_1) &= \Lambda\left[(u_0, v_0)\right], \\
(u_2, v_2) &= \Lambda\left[(u_1, v_1)\right].
\end{aligned}
\tag{9.229}
$$

By repetition of the application of such mapping, we obtain

$$
(u_{n+1}, v_{n+1}) = \Lambda\left[(u_n, v_n)\right] \tag{9.230}
$$

of the functions that meet the following requirements:

$$
\begin{aligned}
&u_n(t) < u_n(t_0) + (1-\alpha)g'(t)f(t, u_n(t)) + \alpha\int_{t_0}^{t} g'(\tau)f(\tau, u_n(\tau))\,d\tau, \;\; u_n(t_0) = y_0, \\
&v_n(t) > v_n(t_0) + (1-\alpha)g'(t)f(t, v_n(t)) + \alpha\int_{t_0}^{t} g'(\tau)f(\tau, v_n(\tau))\,d\tau, \;\; v_n(t_0) = y_0, \\
&u_n(t) < u_{n+1}(t) < y(t) < v_{n+1}(t) < v_n(t), \forall t \in [t_0, t_0 + \lambda].
\end{aligned}
\tag{9.231}
$$

$$
\begin{aligned}
u_{n+1}(t) &= u_{n+1}(t_0) + (1-\alpha)g'(t)f_1\left(t, u_{n+1}(t); u_n(t), v_n(t)\right) \\
&\quad + \alpha\int_{t_0}^{t} g'(\tau)f_1\left(\tau, u_{n+1}(\tau); u_n(\tau), v_n(\tau)\right)d\tau, \\
v_{n+1}(t) &= v_{n+1}(t_0) + (1-\alpha)g'(t)f_2\left(t, v_{n+1}(t); u_n(t), v_n(t)\right) \\
&\quad + \alpha\int_{t_0}^{t} g'(\tau)f_2\left(\tau, v_{n+1}(\tau); u_n(\tau), v_n(\tau)\right)d\tau.
\end{aligned}
\tag{9.232}
$$

Indeed on $[t_0, t_0 + \lambda]$ the sequences $\{u_n\}; \{v_n\}$ are monotonic and uniformly bounded. Since they are solution of linear equations, the equicontinuity is obtained for any fixed n. Also $u_n(t)$ and $v_n(t)$ converge to $y(t)$ whenever $n \to \infty$. To continue, we set

$$
\Omega = \sup_{\substack{u_0(t) \le y \le v_0(t) \\ t_0 \le t \le t_0 + \alpha}} \left| f_y(t, y) \right| \tag{9.233}
$$

and

$$
\overline{\Omega} = \sup_{\substack{u_0(t) \le y \le v_0(t) \\ t_0 \le t \le t_0 + \alpha}} \left| f_{yy}(t, y) \right| . \tag{9.234}
$$

As we did before

$$0 \leq v_0(t) - u_0(t) \leq \overline{\beta}. \tag{9.235}$$

We assume the formula when n is fixed

$$|u_n(t) - v_n(t)| < \frac{2\overline{\beta}}{2^{2n}}. \tag{9.236}$$

From the formula of $u_{n+1}(t)$ and $v_{n+1}(t)$ and the mean value theorem, we shall have

$$\begin{aligned} v_{n+1}(t) - u_{n+1}(t) &= v_{n+1}(t_0) - u_{n+1}(t_0) \\ &+ (1-\alpha)g'(t)\left[\begin{array}{c} \frac{f(t,u_n(t))-f(t,v_n(t))}{u_n(t)-v_n(t)}\,(v_{n+1}(t) - u_n(t)) \\ -f_y\,(t, u_n(t)) \end{array}\right]\left[u_{n+1}(t) - u_n(t)\right] \\ &+ \alpha \int_{t_0}^{t} g'(\tau)\left[\begin{array}{c} \frac{f(\tau,u_n(\tau))-f(\tau,v_n(\tau))}{u_n(\tau)-v_n(\tau)}\,(v_{n+1}(\tau) - u_n(\tau)) \\ -f_y\,(\tau, u_n(\tau)) \end{array}\right]\left[u_{n+1}(\tau) - u_n(\tau)\right] d\tau \end{aligned} \tag{9.237}$$

$$\begin{aligned} &= (1-\alpha)g'(t)\left[\begin{array}{c} \left[f_y\,(t,\eta)\left[v_{n+1}(t) - u_{n+1}(t)\right]\right] \\ +\left[u_{n+1}(t) - u_n(t)\right]\left[f_y\,(t,\eta) - f_y\,(t, u_n(t))\right] \end{array}\right] \\ &+\alpha \int_{t_0}^{t} g'(\tau)\left[\begin{array}{c} \left[f_y\,(\tau,\eta)\left[v_{n+1}(\tau) - u_{n+1}(\tau)\right]\right] \\ +\left[u_{n+1}(\tau) - u_n(\tau)\right]\left[f_y\,(\tau,\eta) - f_y\,(\tau, u_n(t))\right] \end{array}\right]. \end{aligned} \tag{9.238}$$

Note that

$$u_n(t) < \eta < v_n(t). \tag{9.239}$$

This is obtained since

$$f_y(t,\eta) - f_y(t, u_n(t)) = f_{yy}(t,\overline{\eta})\,(\overline{\eta} - u_n(t)) \tag{9.240}$$

where

$$u_n(t) < \overline{\eta} < \eta, \tag{9.241}$$

and we obtain

$$\begin{aligned} |v_{n+1}(t) - u_{n+1}(t)| \leq\ &(1-\alpha)M_{g'}\left[\begin{array}{c} \Omega\,|v_{n+1}(t) - u_{n+1}(t)| \\ +\overline{\Omega}\,|\eta - u_n(t)|\,|v_{n+1}(t) - u_{n+1}(t)| \end{array}\right] \\ &+\alpha M_{g'}\int_{t_0}^{t}\left[\begin{array}{c} \Omega\,|v_{n+1}(t) - u_{n+1}(t)| \\ +\overline{\Omega}\,|\eta - u_n(t)|\,|v_{n+1}(t) - u_{n+1}(t)| \end{array}\right] d\tau. \end{aligned} \tag{9.242}$$

Indeed, we have that

$$|\eta - u_0(t)| \leq |v_n(t) - u_n(t)| \tag{9.243}$$

and that

$$|u_{n+1}(t) - u_n(t)| \leq |v_n(t) - u_n(t)| \tag{9.244}$$

and replacing this yields

$$\begin{aligned} |v_{n+1}(t) - u_{n+1}(t)| \leq\ & (1-\alpha)M_{g'}\left[\Omega\,|v_{n+1}(t) - u_{n+1}(t)| + \overline{\Omega}\frac{2^2\overline{\beta}^2}{2^{2n+1}}\right] \\ & +\alpha M_{g'}\int_{t_0}^{t}\left[\Omega\,|v_{n+1}(\tau) - u_{n+1}(\tau)| + \overline{\Omega}\frac{2^2\overline{\beta}^2}{2^{2n+1}}\right]d\tau, \\ =\ & (1-\alpha)M_{g'}\Omega\,|v_{n+1}(t) - u_{n+1}(t)| + \overline{\Omega}\frac{(1-\alpha)M_{g'}2^2\overline{\beta}^2}{2^{2n+1}} \\ & +\overline{\Omega}\frac{\alpha M_{g'}2^2\overline{\beta}^2\lambda}{2^{2n+1}} + \alpha\Omega M_{g'}\int_{t_0}^{t}|v_{n+1}(\tau) - u_{n+1}(\tau)|\,d\tau, \end{aligned} \tag{9.245}$$

under the condition that

$$1 + (\alpha - 1)M_{g'}\Omega > 0, \tag{9.246}$$

then, we have

$$\begin{aligned} |v_{n+1}(t) - u_{n+1}(t)| \leq\ & \overline{\Omega}\frac{2^2\overline{\beta}^2}{2^{2n+1}}M_{g'}\frac{(1-\alpha+\alpha\lambda)}{1+(\alpha-1)\,M_{g'}\Omega} \\ & +\frac{\alpha M_{g'}\Omega}{1+(\alpha-1)\,M_{g'}\Omega}\int_{t_0}^{t}|v_{n+1}(\tau) - u_{n+1}(\tau)|\,d\tau \\ \leq\ & \overline{\Omega}\frac{2^2\overline{\beta}^2}{2^{2n+1}}M_{g'}\frac{(1-\alpha+\alpha\lambda)}{1+(\alpha-1)\,M_{g'}\Omega}\int_{t_0}^{t}\exp\left[\frac{\alpha M_{g'}\Omega}{1+(\alpha-1)\,M_{g'}\Omega}(t-\tau)\right]d\tau. \end{aligned} \tag{9.247}$$

We note that

$$\int_{t_0}^{t}\exp\left[\frac{\alpha M_{g'}\Omega}{1+(\alpha-1)\,M_{g'}\Omega}(t-\tau)\right]d\tau < \lambda\exp\left[\frac{\alpha M_{g'}\Omega\lambda}{1+(\alpha-1)\,M_{g'}\Omega}\right]. \tag{9.248}$$

We now get that

$$|v_{n+1}(t) - u_{n+1}(t)| \leq \frac{2\overline{\beta}}{2^{2n+1}}. \tag{9.249}$$

We can now conclude that

$$|y(t) - u_n(t)| \leq \frac{2\overline{\beta}}{2^{2n}}, \tag{9.250}$$
$$|y(t) - v_n(t)| \leq \frac{2\overline{\beta}}{2^{2n}}.$$

This completes the proof. We now consider the following equation:

$${}_{t_0}^{CF}D_g y(t) = f(t, y(t)) \tag{9.251}$$

and demonstrate the lower Chaplygin's sequence $\{u_n(t)\}$.

Theorem 9.8 *We assume that $f \in C[R_0, R]$, where R_0 is defined as before. Let*

$$\|f(t, y)\|_\infty \leq M_f \tag{9.252}$$

and

$$\|g'\|_\infty \leq M_g \tag{9.253}$$

on R_0. We assume that $f(t, y)$ is quasi-monotone nondecreasing in y $\forall t \in [t_0, t_0 + \lambda]$ and $\frac{\partial f(t,y)}{\partial y}$ exists and is continuous on R_0. We assume $u_0(t)$ to be continuously differentiable on $[t_0, t_0 + \alpha]$, where

$$\min\left\{a, \frac{b + M_{g'}(\alpha - 1) M_f}{\alpha M_{g'} M_f}\right\}; \ (t, u_0(t)) \in R_0 \tag{9.254}$$

and

$${}_{t_0}^{CF}D_g^\alpha u_0(t) < f(t, u_0(t)), \ u_0(t_0) = y_0. \tag{9.255}$$

Furthermore, we consider

$$f(t, y) + f_y(t, y)(z - y) < f(t, z) \ \text{if } y < z. \tag{9.256}$$

Then there is $\{u_n(t)\}$ sequence of functions called Chaplygin sequence with

$$u_n(t_0) = y_0 \tag{9.257}$$

and

$$u_n(t) < u_{n+1}(t) < y(t), \forall t \in [t_0, t_0 + \lambda] \tag{9.258}$$

where $y(t)$ is the solution of

$$\begin{cases} {}_{t_0}^{CF}D_g y(t) = f(t, y(t)) & \text{if } t > 0 \\ y(t_0) = y_0 & \text{if } t = 0 \end{cases} \tag{9.259}$$

and

$$\lim_{n\to\infty} u_n(t) = y(t) \tag{9.260}$$

uniformly on $[t_0, t_0 + \lambda]$.

Proof We have clearly that $\forall i \neq l$, $\frac{\partial f_i}{\partial y_l} \geq 0$, which is the direct consequence of the quasi-monotonicity of $f(t, y(t))$. More important we have shown that

$$u_0(t) < y(t), \forall t \in [t_0, t_0 + \lambda]. \tag{9.261}$$

We now have that

$$z(t) = z(t_0) + (1-\alpha)g'(t)\left[f(t, u_0(t)) + \frac{\partial f(t, u_0(t))}{\partial y}[z - u_0(t)]\right] \tag{9.262}$$
$$+\alpha \int_{t_0}^{t} g'(\tau)\left[f(\tau, u_0(\tau)) + \frac{\partial f(\tau, u_0(\tau))}{\partial y}[z - u_0(t)]\right] d\tau.$$

For simplicity, we have

$$z(t) = z(t_0) + (1-\alpha)g'(t)h(t, z; u_0(t)) + \alpha \int_{t_0}^{t} g'(\tau)h(\tau, z; u_0(\tau))\, d\tau \tag{9.263}$$

with

$$z(t_0) = y_0. \tag{9.264}$$

Having that $\frac{\partial f_i}{\partial y_l} \geq 0$, we conclude that $h(t, z; u_0(t))$ is quasi-monotone with respect to z. Therefore with the maximal solution principle we have

$$u_0(t) < u_1(t), \forall t \in [t_0, t_0 + \lambda] \tag{9.265}$$

where $u_1(t)$ is the solution of

$$z(t) = z(t_0) + (1-\alpha)g'(t)h(t, z; u_0(t)) + \alpha \int_{t_0}^{t} g'(\tau)h(\tau, z; u_0(\tau))\, d\tau, \tag{9.266}$$

$$f(t, y) + f_y(t, y)(z - y) < f(t, z) \text{ if } y < z \tag{9.267}$$

implies that

$$u_1(t) = u_1(t_0) + (1-\alpha)g'(t)f(t, u_1(t)) + \alpha \int_{t_0}^{t} g'(\tau)f(\tau, u_1(\tau))d\tau; \tag{9.268}$$

hence we obtain

$$u_1(t) < y(t), \forall t \in [t_0, t_0 + \lambda]; \tag{9.269}$$

putting all together, we have

$$u_0(t) < u_1(t) < y(t), \forall t \in [t_0, t_0 + \lambda]. \tag{9.270}$$

We can now construct the mapping

$$\Lambda\left(u_0(t)\right) = u_1(t) \tag{9.271}$$

which finally helps to establish the low Chaplygin sequence.

9.6 Dependence of Initial Conditions

The primary objective of this section is to thoroughly demonstrate and provide a clear understanding of how the following nonlinear differential equation operates, including its key properties, underlying principles, and potential implications within the given context

$$\begin{cases} {}_{t_0}^{CF}D_g^{\alpha}y(t) = f(t, y(t)) \text{ if } t > t_0 \\ y(t_0) = y_0 \qquad\qquad\quad \text{if } t = t_0 \end{cases} \tag{9.272}$$

has a solution $y(t, t_0, y_0)$ with initial data (t_0, y_0). We shall consider then the problem of continuity and differentiability. Our investigation shall start with the following lemma.

Lemma 9.3 *We assume that $f \in C\left[R_0, R\right]$ and $H(t, q) = \max_{|y-y_0|\leq q} |f(t, y)|$. We consider $r^*(t, t_0, 0)$ to be the maximal solution of the following equation:*

$${}_{t_0}^{CF}D_g^{\alpha}z(t) = H(t, z) \tag{9.273}$$

via the initial data $(t_0, 0)$. Let us assume that $y(t, t_0, y_0)$ is any solution of

$$\begin{cases} {}_{t_0}^{CF}D_g^{\alpha}y(t) = f(t, y(t)) \text{ if } t > t_0 \\ y(t_0) = y_0 \qquad\qquad\quad \text{if } t = t_0 \end{cases} \tag{9.274}$$

then

$$|y(t, t_0, y_0) - y_0| \leq r^*(t, t_0, 0), \;\; t \geq t_0. \tag{9.275}$$

Proof Let us defined

$$\Omega(t) = |y(t, t_0, y_0) - y_0|. \tag{9.276}$$

Then

$$\begin{aligned}\Omega(t) &= |y(t,t_0,y_0)-y_0| = \left|(1-\alpha)g'(t)f(t,y(t))+\alpha\int_{t_0}^{t} g'(\tau)f(\tau,y(\tau))d\tau\right|, \\ &\leq (1-\alpha)\left|g'(t)\right||f(t,y(t))|+\alpha\int_{t_0}^{t}\left|g'(\tau)\right||f(\tau,y(\tau))|\,d\tau, \\ &\leq (1-\alpha)\left|g'(t)\right||f(t,y(t,t_0,y_0))|+\alpha\int_{t_0}^{t}\left|g'(\tau)\right||f(\tau,y(\tau,t_0,y_0))|\,d\tau, \\ &\leq (1-\alpha)\left|g'(t)\right|\max_{|y-y_0|\leq\Omega(t)}|f(t,y)|+\alpha\int_{t_0}^{t}\left|g'(\tau)\right|\max_{|y-y_0|\leq\Omega(\tau)}|f(\tau,y)|\,d\tau, \\ &= (1-\alpha)g'(t)H(t,\Omega(t))+\alpha\int_{t_0}^{t} g'(\tau)H(\tau,\Omega(\tau))\,d\tau.\end{aligned} \tag{9.277}$$

The above implies from the maxima solution principle that

$$\Omega(t) = |y(t,t_0,y_0)-y_0| \leq r^*(t,t_0,0),\ \ t\geq t_0 \tag{9.278}$$

which concludes the proof. Our investigation will follow with this theorem.

Theorem 9.9 *We let $f\in C[R_0,R]$ and for $(t,y),(t,\overline{y})\in R_0$*

$$|f(t,y)-f(t,\overline{y})| \leq p(t,|y-\overline{y}|) \tag{9.279}$$

where $p\in C[R_0,R]$. It is assumed that $u(t)=0$ is the unique solution of the below differential equation

$${}_{t_0}^{CF}D_g^{\alpha}u(t) = p(t,u) \tag{9.280}$$

with $u(t_0)=0$. Then if the solution $u(t,t_0,u_0)$ of

$${}_{t_0}^{CF}D_g^{\alpha}u(t) = p(t,u) \tag{9.281}$$

via every point (t_0,u_0) is continuous with respect to the initial data (t_0,u_0), the solution $y(t,t_0,y_0)$ of

$$\begin{cases} {}_{t_0}^{CF}D_g^{\alpha}y(t) = f(t,y(t)) & if\, t>t_0 \\ y(t_0)=y_0 & if\, t=t_0\end{cases} \tag{9.282}$$

is unique and continuous with respect to the initial data (t_0,y_0).

Proof The problem of existence and uniqueness has been achieved. We shall then devote out attention to the discussion underpinning the continuity of the solution with respect to y_0 and t_0 respectively. To achieve this we assume $y(t, t_0, y_0)$ and $\overline{y}(t, t_0, \overline{y}_0)$ to be the solution of

$$\begin{cases} {}_{t_0}^{CF}D_g^{\alpha} y(t) = f(t, y(t)) & \text{if } t > t_0 \\ y(t_0) = y_0 & \text{if } t = t_0 \end{cases} \tag{9.283}$$

via (t_0, y_0), $\left(t_0, \overline{y}_0\right)$ respectively. Our aim should be that

$$\lim_{\overline{y}_0 \to y_0} \overline{y}(t, t_0, \overline{y}_0) = y(t, t_0, y_0) \tag{9.284}$$

for the continuity with respect to y_0. We define again

$$\Omega(t) = \left|y(t, t_0, y_0) - \overline{y}(t, t_0, \overline{y}_0)\right|. \tag{9.285}$$

The hypothesis of the theorem leads us to

$$\begin{aligned} \Omega(t) &\leq (1-\alpha)\left|g'(t)\right| \left|f(t, y) - f(t, \overline{y})\right| + \alpha \int_{t_0}^{t} g'(\tau) \left|f(\tau, y) - f(\tau, \overline{y})\right| d\tau \\ &\leq (1-\alpha)\left|g'(t)\right| p(t, |y - y_0|) + \alpha \int_{t_0}^{t} g'(\tau) \left|p(\tau, |y - y_0|\right| d\tau. \end{aligned} \tag{9.286}$$

Following the previously obtained result, we shall have that

$$\Omega(t) \leq r(t, t_0, \left|y - \overline{y}_0\right|), t \geq t_0 \tag{9.287}$$

where $r(t, t_0, \left|y - \overline{y}_0\right|)$ is the maximal solution of

$${}_{t_0}^{CF}D_g^{\alpha} u(t) = p(t, u(t)) \tag{9.288}$$

with

$$\left|y_0 - \overline{y}_0\right| = u_0. \tag{9.289}$$

The continuity with respect to the initial data of $u(t, t_0, y_0)$ leads us to

$$\lim_{y_0 \to \overline{y}_0} r(t, t_0, \left|y - \overline{y}_0\right|) = r(t, t_0, 0). \tag{9.290}$$

But according to the theorem hypothesis

$$r(t, t_0, 0) = 0. \tag{9.291}$$

Therefore

$$\begin{aligned}\lim_{y_0\to\overline{y}_0}\Omega(t) &\leq \lim_{y_0\to\overline{y}_0} r(t,t_0,|y-\overline{y}_0|)=0 \\ \lim_{y_0\to\overline{y}_0}\Omega(t)=0 &\to \lim_{y_0\to\overline{y}_0} y(t,t_0,y_0)=\overline{y}(t,t_0,\overline{y}_0)\end{aligned} \tag{9.292}$$

which shows that the solution is continuous with respect to y_0. We now show that it is also with respect to t_0. If then $y(t, t_0, y_0)$ and $\overline{y}(t, t_1, y_0)$, where $t_1 > t_0$ are solutions via (t_0, y_0), (t_1, y_0) respectively again, we define

$$\Omega(t)=|y(t,t_0,y_0)-\overline{y}(t,t_1,y_0)|\,. \tag{9.293}$$

However

$$\begin{aligned}|y(t,t_0,y_0)-\overline{y}(t,t_1,y_0)| = (1-\alpha)\,|g'(t)|\,|f(t,t_0,y_0)-f(t,t_1,y_0)| \\ +\alpha\int_{t_0}^{t} g'(\tau)\,|f(\tau,t_0,y_0)-f(\tau,t_1,y_0)|\,d\tau.\end{aligned} \tag{9.294}$$

We have on the other hand that

$$\Omega(t_1)=|y(t_1,t_0,y_0)-y_0|\,. \tag{9.295}$$

The result of the lemma helps us to have

$$\Omega(t_1)\leq r^*(t_1,t_0,0) \tag{9.296}$$

which leads to

$$\Omega(t)\leq\overline{r}(t)\,,\,t>t_1 \tag{9.297}$$

with

$$\overline{r}(t)=\overline{r}\left(t,t_1,r^*(t_1,t_0,0)\right) \tag{9.298}$$

the maximal solution of

$${}^{CF}_{t_0}D^{\alpha}_{g}u(t)=p(t,u(t)) \tag{9.299}$$

via $(t_1, r^*(t_1, t_0, 0))$. But we had this $r^*(t_0, t_0, 0) = 0$, thus

$$\lim_{t_1\to t_0}\overline{r}\left(t,t_1,r^*(t_1,t_0,0)\right)=\overline{r}(t,t_0,0)=0 \tag{9.300}$$

which concludes the proof.

Theorem 9.10 *Let $f \in C[A, R]$ here; A is an open (t, y, μ) set with $A \subset R$, for $\mu = \mu_0$; we let*

$$y_0(t)=y\left(t,t_0,y_0,\mu_0\right) \tag{9.301}$$

to be the solution of

$$ {}_{t_0}^{CF}D_g^{\alpha} y(t) = f(t, y, \mu_0), \;\; y(t_0) = y_0 \tag{9.302}$$

uniformly in (t, y) *and for* $(t, y, \mu), (t, \overline{y}, \mu) \in A$

$$|f(t, y, \mu) - f(t, \overline{y}, \mu)| \leq p(t, |y - \overline{y}|) \tag{9.303}$$

where $p \in C\left[R_0^+, R^+\right]$; *we assume that* $u(t) = 0$ *is the unique solution of*

$$\begin{aligned} {}_{t_0}^{CF}D_g^{\alpha} u(t) &= p(t, u(t)), \\ u(t_0) &= 0. \end{aligned} \tag{9.304}$$

Then $\forall \varepsilon > 0$, $\exists \delta(\varepsilon) > 0$ *such that* $\forall \mu$,

$$|\mu - \mu_0| < \delta(\varepsilon), \tag{9.305}$$

the differential equation

$$ {}_{t_0}^{CF}D_g^{\alpha} y(t) = f(t, y(t), \mu), \;\; y(t_0) = y_0 \tag{9.306}$$

has a unique solution

$$y(t) = y(t, t_0, y_0, \mu) \tag{9.307}$$

with

$$|y(t) - y_0(t)| < \varepsilon; \;\; t \geq t_0. \tag{9.308}$$

Proof The issue regarding the uniqueness of the solution was thoroughly discussed and elaborated upon in previous studies, utilizing the approach developed by Chaplygin's method. This method provided a systematic framework to address and analyze the problem, offering valuable insights into its resolution. Also from the assumption

$$u(t) = 0 \tag{9.309}$$

is the only solution with the lemma result, for any compact interval $[t_0, t_0 + \lambda]$ and $\forall \varepsilon > 0$, $\exists \overline{\eta} = \overline{\eta}(\varepsilon)$ such that the maximal solution $r(t, t_0, 0, \overline{\eta})$ of

$$ {}_{t_0}^{CF}D_g^{\alpha} u(t) = p(t, u) + \overline{\eta} \tag{9.310}$$

exists within $t_0 \leq t \leq t_0 + a$ with the property that

$$r(t, t_0, 0, \eta) < \varepsilon, \forall t \in [t_0, t_0 + a]. \tag{9.311}$$

Additionally, from the theorem hypothesis, $\forall \overline{\eta} > 0$ we can find a $\delta = \delta(\overline{\eta}) > 0$ such that

$$|f(t, y, \mu) - f(t, y, \mu_0)| \leq \overline{\eta} \tag{9.312}$$

under the condition that

$$|\mu - \mu_0| < \delta. \tag{9.313}$$

Next, $\forall \varepsilon > 0$ we define again

$$\Omega(t) = |y(t) - y_0(t)| \tag{9.314}$$

where $y(t)$ and $y_0(t)$ are solutions of above-presented equations respectively. Therefore by hypothesis, we have that

$$\begin{aligned} \Omega(t) \leq\ & (1-\alpha)g'(t)\{p(t, \Omega(t)) + |f(t, x_0(t), \mu) - f(t, x_0(t), \mu_0)|\} \\ & +\alpha \int_{t_0}^{t} g'(\tau)\{p(\tau, \Omega(\tau)) + f(\tau, x_0(\tau), \mu) - f(\tau, x_0(\tau), \mu_0)\}\, d\tau. \end{aligned} \tag{9.315}$$

From the above if the following inequality

$$|\mu - \mu_0| < \delta \tag{9.316}$$

is set then

$$\begin{aligned} m(t) \leq\ & (1-\alpha)g'(t)\{p(t, \Omega(t)) + \overline{\eta}\} \\ & +\alpha \int_{t_0}^{t} g'(\tau)\{p(\tau, \Omega(\tau)) + \overline{\eta}\}\, d\tau. \end{aligned} \tag{9.317}$$

By the maximal solution principle, we get

$$m(t) \leq r(t, t_0, 0, \overline{\eta}),\ \ t \geq t_0 \tag{9.318}$$

then

$$|y(t) - y_0(t)| < \varepsilon;\ \ t \geq t_0 \tag{9.319}$$

under the condition that

$$|\mu - \mu_0| < \delta. \tag{9.320}$$

Indeed δ depends on ε, because $\overline{\eta}$ depends. This completes the proof.

9.7 Chaplygin's Method and Initial Condition Dependence: Mittag–Leffler Case

Here we derive Chaplygin's sequence for the nonlinear global equation with the derivative being defined as the fractional global derivative with the Mittag-Leffler kernel. Thus we want to derive something that explores the mathematical framework refinement, which in this case is classical equations combined with fractional calculus and Mittag-Leffler function to tackle complexity dynamical systems. Goal of this paper is to introduce the new sequence in details and demonstrate its prominence in solving a variety of nonlinear problems in the field of fractional calculus. The conditions of the theorem are set to be the same; however, we should derive the parameter $\lambda = \min\{a, \delta\}$. To find δ, we recall that the general Cauchy problem under investigation here is given by

$$\begin{cases} {}_{t_0}^{ABC}D_g^{\alpha}y(t) = f(t, y(t)) \text{ if } t \in [t_0, t_0 + a] \\ y(t_0) = y_0 \end{cases}. \tag{9.321}$$

We shall now apply the corresponding integral to have

$$y(t) = (1-\alpha)g'(t)f(t, y(t)) + \frac{\alpha}{\Gamma(\alpha)} \int_{t_0}^{t} g'(\tau)(t-\tau)^{\alpha-1} f(\tau, y(\tau))d\tau. \tag{9.322}$$

We want to have $\forall t \in [t_0, t_0 + a]$,

$$|y(t)| < b, \tag{9.323}$$

therefore

$$\begin{aligned} |y(t)| \leq\ & \left|(1-\alpha)g'(t)f(t, y(t))\right| + \frac{\alpha}{\Gamma(\alpha)} \int_{t_0}^{t} (t-\tau)^{\alpha-1} \left|g'(\tau)f(\tau, y(\tau))\right| d\tau, \\ \leq\ & \sup_{t\in[t_0,t_0+a]} (1-\alpha)\left|g'(t)\right| \sup_{t\in[t_0,t_0+a]} |f(t, y(t))| + \frac{\alpha}{\Gamma(\alpha)} \int_{t_0}^{t} (t-\tau)^{\alpha-1} \left|g'(\tau)\right| |f(\tau, y(\tau))| d\tau, \\ \leq\ & \left\|g'\right\|_{\infty} M_f (1-\alpha) + \frac{\alpha}{\Gamma(\alpha+1)} \left\|g'\right\|_{\infty} M_f (t-t_0)^{\alpha}, \\ \leq\ & M_{g'} M_f \left\{1 - \alpha + \frac{\alpha a^{\alpha}}{\Gamma(\alpha)}\right\} < b, \\ a <\ & \left\{\left(\frac{b}{M_{g'} M_f} + \alpha - 1\right) \frac{\Gamma(\alpha)}{\alpha}\right\}^{\frac{1}{\alpha}}. \end{aligned} \tag{9.324}$$

Therefore we will need to have

$$\lambda = \min\left\{a, \left\{\left(\frac{b\Gamma(\alpha)}{M_{g'} M_f \alpha} + \frac{(\alpha-1)\Gamma(\alpha)}{\alpha}\right)\right\}^{\frac{1}{\alpha}}\right\}. \tag{9.325}$$

We shall now proceed with the demonstration. From the hypothesis we have that

$$ {}_{t_0}^{ABC}D_g^{\alpha} u_0(t) < f(t, u_0(t)), \tag{9.326} $$

$$ \begin{aligned} u_0(t) &< (1-\alpha)g'(t)f(t,u_0(t)) + \frac{\alpha}{\Gamma(\alpha)}\int_{t_0}^{t} g'(\tau)(t-\tau)^{\alpha-1} f(\tau, u_0(\tau))d\tau, \\ &< (1-\alpha)g'(t)f(t,u_0(t)) + \frac{\alpha}{\Gamma(\alpha)}\int_{t_0}^{t} g'(\tau)(t-\tau)^{\alpha-1} f(\tau, u_0(\tau))d\tau, \\ &< (1-\alpha)g'(t)f(t,y(t)) + \frac{\alpha}{\Gamma(\alpha)}\int_{t_0}^{t} g'(\tau)(t-\tau)^{\alpha-1} f(\tau, y(\tau))d\tau, \\ &< y(t). \end{aligned} $$

Therefore $\forall t \in [t_0, t_0+\lambda]$, we have

$$ u_0(t) < y(t). \tag{9.327} $$

On the other hand by hypothesis we have that

$$ {}_{t_0}^{ABC}D_g^{\alpha} v_0(t) > f(t, v_0(t)), \tag{9.328} $$

$$ \begin{aligned} v_0(t) &> (1-\alpha)g'(t)f(t,v_0(t)) + \frac{\alpha}{\Gamma(\alpha)}\int_{t_0}^{t} g'(\tau)(t-\tau)^{\alpha-1} f(\tau, v_0(\tau))d\tau, \\ v_0(t) &> (1-\alpha)g'(t)f(t,y(t)) + \frac{\alpha}{\Gamma(\alpha)}\int_{t_0}^{t} g'(\tau)(t-\tau)^{\alpha-1} f(\tau, y(\tau))d\tau, \\ &> v_1(t). \end{aligned} $$

Thus $\forall t \in [t_0, t_0+\lambda]$, we have that

$$ y(t) < v_0(t). \tag{9.329} $$

Now on the one hand, we have

$$ v_0(t) > (1-\alpha)g'(t)f(t,v_0(t)) + \frac{\alpha}{\Gamma(\alpha)}\int_{t_0}^{t} g'(\tau)(t-\tau)^{\alpha-1} f(\tau, v_0(\tau))d\tau, \tag{9.330} $$

$$ -u_0(t) > -(1-\alpha)g'(t)f(t,u_0(t)) - \frac{\alpha}{\Gamma(\alpha)}\int_{t_0}^{t} g'(\tau)(t-\tau)^{\alpha-1} f(\tau, u_0(\tau))d\tau. $$

Thus

$$v_0(t) - u_0(t) > (1-\alpha)g'(t)\left(f(t, v_0(t)) - f(t, u_0(t))\right) + \frac{\alpha}{\Gamma(\alpha)}\int_{t_0}^{t} g'(\tau)(t-\tau)^{\alpha-1}\left(f(\tau, v_0(\tau)) - f(\tau, u_0(\tau))\right)d\tau. \tag{9.331}$$

We know by hypothesis that the function $f(t, y(t))$ is differentiable with respect to y; therefore by the mean value theorem, we can find

$$u_0(t) < \eta < v_0(t), \forall t \in [t_0, t_0+\lambda]. \tag{9.332}$$

$$f_y(t, \eta)\left(v_0(t) - u_0(t)\right) = f(t, v_0(t)) - f(t, u_0(t)). \tag{9.333}$$

Thus

$$\begin{aligned} v_0(t) - u_0(t) &> (1-\alpha)g'(t)f_y(t,\eta)\left(v_0(t) - u_0(t)\right) \\ &\quad + \frac{\alpha}{\Gamma(\alpha)}\int_{t_0}^{t} g'(\tau)(t-\tau)^{\alpha-1} f_y(\tau,\eta)\left(v_0(\tau) - u_0(\tau)\right)d\tau, \\ &> (1-\alpha)\inf_{t\in[t_0,t_0+\lambda]}\left|g'(t)\right| \inf_{t\in[t_0,t_0+\lambda]}\left|f_y(t,\eta)\left(v_0(t) - u_0(t)\right)\right| \\ &\quad + \frac{\alpha}{\Gamma(\alpha)}\int_{t_0}^{t} \inf_{l\in[t_0,\tau]}\left|g'(\tau)\right|(t-\tau)^{\alpha-1} \inf_{l\in[t_0,\tau]}\left|f_y(\tau,\eta)\left(v_0(\tau) - u_0(\tau)\right)\right|d\tau, \\ &> (1-\alpha)\overline{M}_{g'}\overline{M}_f\left(v_0(t) - u_0(t)\right) + \frac{\alpha\overline{M}_{g'}\overline{M}_f}{\Gamma(\alpha)}\int_{t_0}^{t}(t-\tau)^{\alpha-1}\left(v_0(\tau) - u_0(\tau)\right)d\tau. \end{aligned} \tag{9.334}$$

Assuming that

$$1 + (\alpha-1)\overline{M}_{g'}\overline{M}_f > 0 \tag{9.335}$$

then we have

$$v_0(t) - u_0(t) > \frac{\alpha\overline{M}_{g'}\overline{M}_f}{\Gamma(\alpha)\left\{1 + (\alpha-1)\overline{M}_{g'}\overline{M}_f\right\}} + \int_{t_0}^{t} g'(\tau)(t-\tau)^{\alpha-1}\left(v_0(\tau) - u_0(\tau)\right)d\tau. \tag{9.336}$$

By the Gronwall inequality, we have

$$v_0(t) - u_0(t) \geq \varepsilon \exp\left[\frac{\overline{M}_{g'}\overline{M}_f a^{\alpha}}{\Gamma(\alpha)\left\{1 + (\alpha-1)\overline{M}_{g'}\overline{M}_f\right\}}\right], \forall t \in (t_0, t_0+\lambda]. \tag{9.337}$$

When $t = t_0$, we have that

$$v_0(t_0) = u_0(t_0) = y(t_0). \tag{9.338}$$

We next show that $\forall t \in (t_0, t_0 + \lambda]$,

$$u_0(t) < u_1(t) < y(t) < v_1(t) < v_0(t). \tag{9.339}$$

By hypothesis, we have that

$$\begin{aligned} u_0(t) &< (1-\alpha)g'(t)f(t, u_0(t)) + \frac{\alpha}{\Gamma(\alpha)} \int_{t_0}^{t} g'(\tau)(t-\tau)^{\alpha-1} f(\tau, v_0(\tau))d\tau, \\ &< (1-\alpha)g'(t)g_1(t, u_0(t); u_0, v_0) + \frac{\alpha}{\Gamma(\alpha)} \int_{t_0}^{t} g'(\tau)(t-\tau)^{\alpha-1} g_1(\tau, u_0(\tau); u_0, v_0)d\tau, \\ &< v_1(t). \end{aligned} \tag{9.340}$$

Therefore $\forall t \in (t_0, t_0 + \lambda]$, we have

$$u_0(t) < v_1(t). \tag{9.341}$$

$$v_1(t) = (1-\alpha)g'(t)f_2(t, v_0(t); u_0, v_0) + \frac{\alpha}{\Gamma(\alpha)} \int_{t_0}^{t} g'(\tau)(t-\tau)^{\alpha-1} f_2(\tau, v_0(\tau); u_0, v_0)d\tau \tag{9.342}$$

on the other hand

$$v_0(t) > (1-\alpha)g'(t)f_2(t, v_0(t)) + \frac{\alpha}{\Gamma(\alpha)} \int_{t_0}^{t} g'(\tau)(t-\tau)^{\alpha-1} f_2(\tau, v_0(\tau))d\tau. \tag{9.343}$$

Note that

$$\begin{aligned} g_2(t, v_0(t); u_0, v_0) &= f(t, u_0(t)) + \frac{f(t, u_0(t)) - f(t, v_0(t))}{u_0(t) - v_0(t)} (v_0(t) - u_0(t)) \\ &= f(t, v_0(t)) \\ v_0(t) &> (1-\alpha)g'(t)g_2(t, v_0(t); u_0, v_0) \\ &\quad + \frac{\alpha}{\Gamma(\alpha)} \int_{t_0}^{t} g'(\tau)(t-\tau)^{\alpha-1} g_2(\tau, v_0(\tau); u_0, v_0)d\tau, \\ &> v_1(t). \end{aligned} \tag{9.344}$$

Therefore $\forall t \in (t_0, t_0 + \lambda]$, we have

$$v_0(t) > v_1(t). \tag{9.345}$$

$$u_1(t) = (1-\alpha)g'(t)g_1(t, u_0(t); u_0, v_0) + \frac{\alpha}{\Gamma(\alpha)} \int_{t_0}^{t} g'(\tau)(t-\tau)^{\alpha-1} g_1(\tau, u_0(\tau); u_0, v_0) d\tau, \tag{9.346}$$

and we have that

$$g_1(t, u_1(t); u_0, v_0) = f(t, u_0(t)) + f_{u_1}(t, u_0(t))(u_1(t) - u_0(t)). \tag{9.347}$$

$$\begin{aligned} u_1(t) &= (1-\alpha)g'(t)\left\{f(t, u_0(t)) + f_{u_1}(t, u_0(t))(u_1(t) - u_0(t))\right\} \\ &+ \frac{\alpha}{\Gamma(\alpha)} \int_{t_0}^{t} g'(\tau)(t-\tau)^{\alpha-1} \left\{f(\tau, u_0(\tau)) + f_{u_1}(\tau, u_0(\tau))(u_1(\tau) - u_0(\tau))\right\} d\tau. \end{aligned} \tag{9.348}$$

We have that

$$f_y(t, u_0(t)) < \frac{f(t, u_0(t)) - f(t, v_0(t))}{u_0(t) - v_0(t)}. \tag{9.349}$$

Therefore, we have

$$\begin{aligned} u_1(t) &< (1-\alpha)g'(t)\left\{f(t, u_0(t)) + \frac{f(t, u_0(t)) - f(t, v_0(t))}{u_0(t) - v_0(t)}(u_0(t) - u_0(t))\right\} \\ &+ \frac{\alpha}{\Gamma(\alpha)} \int_{t_0}^{t} g'(\tau)(t-\tau)^{\alpha-1} \left\{f(\tau, u_0(\tau)) + \frac{f(\tau, u_0(\tau)) - f(\tau, v_0(\tau))}{u_0(\tau) - v_0(\tau)}(u_0(\tau) - u_0(\tau))\right\} d\tau, \\ &< (1-\alpha)g'(t)f(t, v_0(t)) + \frac{\alpha}{\Gamma(\alpha)} \int_{t_0}^{t} g'(\tau)(t-\tau)^{\alpha-1} f(\tau, v_0(\tau)) d\tau \\ &< y(t). \end{aligned} \tag{9.350}$$

Therefore $\forall t \in (t_0, t_0 + \lambda]$, we have

$$u_1(t) < y(t). \tag{9.351}$$

For

$$\begin{aligned} v_1(t) &= (1-\alpha)g'(t)\left\{f(t, u_0(t)) + \frac{f(t, u_0(t)) - f(t, v_0(t))}{u_0(t) - v_0(t)}(v_1(t) - u_0(t))\right\} \\ &+ \frac{\alpha}{\Gamma(\alpha)} \int_{t_0}^{t} g'(\tau)(t-\tau)^{\alpha-1} \left\{f(\tau, u_0(\tau)) + \frac{f(\tau, u_0(\tau)) - f(\tau, v_0(\tau))}{u_0(\tau) - v_0(\tau)}(v_1(\tau) - u_0(\tau))\right\} d\tau. \end{aligned} \tag{9.352}$$

On the other hand we have that

$$f(t, v_1(t)) = f(t, u_0(t)) + f_y(t, u_0(t))(v_1(t) - u_0(t)) + \frac{1}{2} f_{yy}(t, \xi)(v_1(t) - u_0(t))^2 \tag{9.353}$$

where $\forall t \in (t_0, t_0 + \lambda]$.

$$u_0(t) < \xi < v_1(t). \tag{9.354}$$

We have from the hypothesis that $f_{yy}(t, \xi) > 0$; therefore using the mean value theorem, we set

$$\begin{aligned} v_1(t) &> (1-\alpha)g'(t)f(t, v_1(t)) + \frac{\alpha}{\Gamma(\alpha)} \int_{t_0}^{t} g'(\tau)(t-\tau)^{\alpha-1} f(\tau, v_1(\tau))\, d\tau \\ &> y(t). \end{aligned} \tag{9.355}$$

Therefore $\forall t \in (t_0, t_0 + \lambda]$

$$v_1(t) > y(t). \tag{9.356}$$

Therefore

$$u_0(t) < u_1(t) < y(t) < v_1(t) < v_0(t), \ \forall t \in (t_0, t_0 + \lambda] \tag{9.357}$$

using the following linear functions

$$\begin{aligned} v_{n+1}(t) &= (1-\alpha)g'(t)g_2(t, v_{n+1}(t); u_{n-1}, v_{n-1}) \\ &\quad + \frac{\alpha}{\Gamma(\alpha)} \int_{t_0}^{t} g'(\tau)(t-\tau)^{\alpha-1} g_2(\tau, v_{n+1}(\tau); u_{n-1}, v_{n-1}) d\tau, \\ v_{n+1}(t_0) &= y_0 \end{aligned} \tag{9.358}$$

$$\begin{aligned} u_{n+1}(t) &= (1-\alpha)g'(t)g_2(t, u_{n+1}(t); u_{n-1}, v_{n-1}) \\ &\quad + \frac{\alpha}{\Gamma(\alpha)} \int_{t_0}^{t} g'(\tau)(t-\tau)^{\alpha-1} g_2(\tau, u_{n+1}(\tau); u_{n-1}, v_{n-1}) d\tau, \\ u_{n+1}(t_0) &= y_0. \end{aligned} \tag{9.359}$$

We can build a mapping Λ such that

$$\begin{aligned} (u_1, v_1) &= \Lambda\left[(u_0, v_0)\right], \\ (u_2, v_2) &= \Lambda\left[(u_1, v_1)\right]. \end{aligned} \tag{9.360}$$

By repetition of the application of such mapping, we obtain

$$(u_{n+1}, v_{n+1}) = \Lambda\left[(u_n, v_n)\right] \tag{9.361}$$

where the above functions have the following properties:

$$u_n(t) < (1-\alpha)g'(t)f(t, u_n(t)) + \frac{\alpha}{\Gamma(\alpha)}\int_{t_0}^{t} g'(\tau)(t-\tau)^{\alpha-1} f(\tau, u_n(\tau))\, d\tau, \quad (9.362)$$

$$u_n(t_0) = y_0,$$

$$v_n(t) > (1-\alpha)g'(t)f(t, v_n(t)) + \frac{\alpha}{\Gamma(\alpha)}\int_{t_0}^{t} g'(\tau)(t-\tau)^{\alpha-1} f(\tau, v_n(\tau))\, d\tau,$$

$$v_n(t_0) = y_0,$$

$$u_n(t) < u_{n+1}(t) < y(t) < v_{n+1}(t) < v_n(t), \ \forall t \in (t_0, t_0 + \lambda] \quad (9.363)$$

using the following linear functions:

$$\begin{aligned} u_{n+1}(t) &= (1-\alpha)g'(t)g_1(t, u_{n+1}(t); u_n, v_n) \\ &\quad + \frac{\alpha}{\Gamma(\alpha)}\int_{t_0}^{t} g'(\tau)(t-\tau)^{\alpha-1} g_1(\tau, u_{n+1}(\tau); u_n, v_n) d\tau, \\ u_{n+1}(t_0) &= y_0. \end{aligned} \quad (9.364)$$

The obtained sequential functions are both monotonic, uniformly bounded on $[t_0, t_0 + \lambda]$, and equicontinuous. For the second part, we have that from the hypothesis

$$0 \le v_0(t) - u_0(t) \le \gamma. \quad (9.365)$$

γ will be obtained at the end. Therefore for $n = 0$ we have that

$$v_0(t) - u_0(t) \le \gamma. \quad (9.366)$$

For a fixed n we assume that

$$|v_n(t) - u_n(t)| \le \frac{2\gamma}{2^{2^n}}. \quad (9.367)$$

Then by definition

$$\begin{aligned} v_{n+1}(t) - u_{n+1}(t) &= (1-\alpha)g'(t)\begin{bmatrix} f_y(t,\xi)\left[v_{n+1}(t) - u_{n+1}(t)\right] \\ + \left[u_{n+1}(t) - u_n(t)\right]\left[f_y(t,\xi) - f_y(t, u_n(t))\right] \end{bmatrix} \\ &\quad + \frac{\alpha}{\Gamma(\alpha)}\int_{t_0}^{t} g'(\tau)(t-\tau)^{\alpha-1}\begin{bmatrix} f_y(\tau,\xi)\left[v_{n+1}(\tau) - u_{n+1}(\tau)\right] \\ + \left[u_{n+1}(\tau) - u_n(\tau)\right]\left[f_y(\tau,\xi) - f_y(\tau, u_n(\tau))\right] \end{bmatrix} d\tau. \end{aligned} \quad (9.368)$$

We have that

$$u_n(t) < \xi < v_n(t), \tag{9.369}$$
$$\left[f_y(t,\xi) - f_y(t,u_n(t))\right] = f_{yy}\left(t,\overline{\xi}\right)[\xi - u_n(t)]$$

with

$$u_n(t) < \overline{\xi} < \xi. \tag{9.370}$$

Therefore

$$\begin{aligned}
|v_{n+1}(t) - u_{n+1}(t)| = \ & (1-\alpha)g'(t)\left\{\Omega_1 |u_{n+1}(t) - v_{n+1}(t)| + \Omega_2 \frac{2^2\gamma^2}{2^{2^{n+1}}}\right\} \\
& + \frac{\alpha}{\Gamma(\alpha)}\int_{t_0}^{t} g'(\tau)(t-\tau)^{\alpha-1}\left\{\Omega_1 |u_{n+1}(\tau) - v_{n+1}(\tau)| + \Omega_2 \frac{2^2\gamma^2}{2^{2^{n+1}}}\right\} d\tau, \\
\leq \ & M_{g'}(1-\alpha)\Omega_2 \frac{2^2\gamma^2}{2^{2^{n+1}}} + \frac{M_{g'}}{\Gamma(\alpha)} \frac{2^2\gamma^2 a^\alpha \Omega_2}{2^{2^{n+1}}} + \Omega_1(1-\alpha)g'(t)|u_{n+1}(t) - v_{n+1}(t)| \\
& + \frac{M_{g'}}{\Gamma(\alpha)}\Omega_1 \int_{t_0}^{t} g'(\tau)(t-\tau)^{\alpha-1}|u_{n+1}(\tau) - v_{n+1}(\tau)|\, d\tau, \\
\leq \ & M_{g'}\Omega_2 \frac{2^2\gamma^2}{2^{2^{n+1}}} \frac{(1-\alpha+\alpha a^\alpha)}{\left(1+(1-\alpha)M_{g'}\Omega_1\right)} \\
& + \int_{t_0}^{t} \frac{g'(\tau)M_{g'}}{\Gamma(\alpha)}(t-\tau)^{\alpha-1} \frac{1}{\left(1+(1-\alpha)M_{g'}\Omega_1\right)} |u_{n+1}(\tau) - v_{n+1}(\tau)|\, d\tau.
\end{aligned} \tag{9.371}$$

By the Gronwall inequality, we have

$$|v_{n+1}(t) - u_{n+1}(t)| \leq \frac{M_{g'}\Omega_2 2^2\gamma^2(1-\alpha+\alpha a^\alpha)}{2^{2^{n+1}}\left(1+(1-\alpha)M_{g'}\Omega_1\right)} \exp\left[\frac{M_{g'}a^\alpha}{\Gamma(\alpha)\left(1+(1-\alpha)M_{g'}\Omega_1\right)}\right]; \tag{9.372}$$

we then choose

$$\gamma = \frac{M_{g'}\Omega_2 2(1-\alpha+\alpha a^\alpha)}{\left(1+(1-\alpha)M_{g'}\Omega_1\right)} \exp\left[\frac{M_{g'}a^\alpha}{\Gamma(\alpha)\left(1+(1-\alpha)M_{g'}\Omega_1\right)}\right] \tag{9.373}$$

such that

$$|v_{n+1}(t) - u_{n+1}(t)| \leq \frac{2\gamma}{2^{2^{n+1}}}, \forall t \in (t_0, t_0+\lambda]. \tag{9.374}$$

Therefore, we have

$$\begin{aligned}
|y(t) - u_n(t)| &\leq \frac{2\gamma}{2^{2^n}}, \\
|y(t) - v_n(t)| &\leq \frac{2\gamma}{2^{2^n}}.
\end{aligned} \tag{9.375}$$

This completes the proof.

We shall now consider again the general Cauchy problem with the global differential operator with the Mittag–Leffler kernel:

$$\begin{cases} {}^{ABC}_{t_0}D^{\alpha}_{g}y(t) = f(t, y(t)) \text{ if } t \in [t_0, t_0 + a] \\ y(t_0) = y_0 \end{cases}. \tag{9.376}$$

The aim is to derive the lower Chaplygin's sequence $\{u_n\}$ under several addition conditions as presented in the theorem below.

Theorem 9.11 *Let the function $f \in C[R, R_0]$, where the set R_0 is defined as before*

$$R_0 = \{(t, y) \,||y - y_0| < b,\ t_0 \le t \le t_0 + a\}. \tag{9.377}$$

We let $M \ge |f(t, y)|$ on R_0. We suppose that $f(t, y(t))$ is quasi-nondecreasing in y $\forall t \in [t_0, t_0 + a]$ and that $\frac{\partial f(t,y)}{\partial y}$ exists and continuous on R_0. We let $u_0(t)$ be continuously differentiable on $[t_0, t_0 + \lambda]$, where $\lambda = \min\{a, \delta\}$, δ is the same as above, $(t, u_0(t)) \in R_0$ and

$$u_0(t) < (1-\alpha)g'(t)f(t, u_0(t)) + \frac{\alpha}{\Gamma(\alpha)}\int_{t_0}^{t} g'(\tau)(t-\tau)^{\alpha-1}f(\tau, u_0(\tau))\,d\tau, \tag{9.378}$$

$$u_0(t_0) = y_0.$$

Additionally, we let

$$f(t, y(t)) + f_y(t, y(t))(y - z) < f(t, z) \text{ if } y < z. \tag{9.379}$$

Then, there exists a Chaplygin's sequence $\{u_n(t)\}$ with $u_n(t_0) = y_0$,

$$u_n(t) < u_{n+1}(t) < y(t),\ t \in [t_0, t_0 + \lambda] \tag{9.380}$$

where $y(t)$ is the solution

$$y(t) < (1-\alpha)g'(t)f(t, y(t)) + \frac{\alpha}{\Gamma(\alpha)}\int_{t_0}^{t} g'(\tau)(t-\tau)^{\alpha-1}f(\tau, y(\tau))\,d\tau, \tag{9.381}$$

existing on $[t_0, t_0 + \lambda]$ and $u_n(t)$ tends to $y(t)$ uniformly on $[t_0, t_0 + \lambda]$.

Proof Firstly we have that $\frac{\partial f(t,y)}{\partial y} \ge 0$ is the direct consequence of quasi-monotonicity of the function $f(t, y(t))$ as defined in the theorem hypothesis. Also we have that

$$u_0(t) < (1-\alpha)g'(t)f(t,u_0(t)) + \frac{\alpha}{\Gamma(\alpha)}\int_{t_0}^{t} g'(\tau)(t-\tau)^{\alpha-1} f(\tau,u_0(\tau))\,d\tau, \tag{9.382}$$

$$< (1-\alpha)g'(t)f(t,y(t)) + \frac{\alpha}{\Gamma(\alpha)}\int_{t_0}^{t} g'(\tau)(t-\tau)^{\alpha-1} f(\tau,y(\tau))\,d\tau,$$

$$= y(t).$$

Therefore $\forall t \in (t_0, t_0+\lambda]$ we have

$$u_0(t) < y(t). \tag{9.383}$$

For simplicity, we shall let

$$\begin{cases} f_1(z,y;u_0(t)) = f(t,u_0(t)) + \frac{\partial f(t,u_0(t))}{\partial y}[z-u_0], \\ z(t_0) = y_0. \end{cases} \tag{9.384}$$

Since f has the quasi-monotonic property with respect to z, it is then direct that f_n has. We have that when $z = u_1(t)$, we have

$$f_1(t,u_1;u_0(t)) = f(t,u_0(t)) + \frac{\partial f(t,u_0(t))}{\partial y}[u_1(t)-u_0(t)], \tag{9.385}$$

therefore by definition, we have that

$$u_1(t) = (1-\alpha)g'(t)f_1(t,u_1(t);u_0(t)) + \frac{\alpha}{\Gamma(\alpha)}\int_{t_0}^{t} g'(\tau)(t-\tau)^{\alpha-1} f_1(\tau,u_1(\tau);u_0(\tau))\,d\tau, \tag{9.386}$$

and replacing $f_1(t,u_1(t);u_0(t))$ by its value yields

$$u_1(t) = (1-\alpha)g'(t)\left\{f(t,u_0(t)) + \frac{\partial f(t,u_0(t))}{\partial y}[u_1(t)-u_0(t)]\right\} \tag{9.387}$$

$$+\frac{\alpha}{\Gamma(\alpha)}\int_{t_0}^{t} g'(\tau)(t-\tau)^{\alpha-1}\left\{f(\tau,u_0(\tau)) + \frac{\partial f(\tau,u_0(\tau))}{\partial y}[u_1(\tau)-u_0(\tau)]\right\}d\tau,$$

and on the other hand, we have that

$$f_1(t,u_0(t);u_0(t)) = f(t,u_0(t)) \tag{9.388}$$

$$u_0(t) < (1-\alpha)g'(t)f(t,u_0(t)) + \frac{\alpha}{\Gamma(\alpha)}\int_{t_0}^{t} g'(\tau)(t-\tau)^{\alpha-1} f(\tau,u_0(\tau))\,d\tau, \tag{9.389}$$

$$< u_1(t).$$

Thus, by the hypothesis, we have

$$u_1(t) < (1-\alpha)g'(t)f(t,u_1(t)) + \frac{\alpha}{\Gamma(\alpha)}\int_{t_0}^{t} g'(\tau)(t-\tau)^{\alpha-1}f(\tau,u_1(\tau))\,d\tau, \tag{9.390}$$
$$< (1-\alpha)g'(t)f(t,y(t)) + \frac{\alpha}{\Gamma(\alpha)}\int_{t_0}^{t} g'(\tau)(t-\tau)^{\alpha-1}f(\tau,y(\tau))\,d\tau,$$
$$= y(t).$$

Therefore $\forall t \in (t_0, t_0+\lambda]$

$$u_0(t) < u_1(t) < y(t), \tag{9.391}$$
$$u(t_0) = u_1(t_0) = y(t_0).$$

As presented in the previous theorem, we can now build mapping Λ such that

$$\Lambda(u_0(t)) = u_1(t), \tag{9.392}$$
$$\Lambda(u_1(t)) = u_2(t),$$
$$\Lambda(u_2(t)) = u_3(t),$$
$$\vdots$$
$$\Lambda(u_n(t)) = u_{n+1}(t),$$

where

$$u_{n+1}(t) < (1-\alpha)g'(t)f(t,u_{n+1}(t)) + \frac{\alpha}{\Gamma(\alpha)}\int_{t_0}^{t} g'(\tau)(t-\tau)^{\alpha-1}f(\tau,u_{n+1}(\tau))\,d\tau, \tag{9.393}$$

$$u_n(t) < u_{n+1}(t) < y(t), \forall t \in (t_0, t_0+\lambda] \tag{9.394}$$

$$u_{n+1}(t) = (1-\alpha)g'(t)f_1(t,u_{n+1}(t);u_n(t)) \tag{9.395}$$
$$+\frac{\alpha}{\Gamma(\alpha)}\int_{t_0}^{t} g'(\tau)(t-\tau)^{\alpha-1}f(\tau,u_{n+1}(\tau);u_n(\tau))\,d\tau,$$
$$u_{n+1}(t_0) = y_0.$$

9.8 Dependence on Parameters and Initial Condition

We aim in this section at showing that the solution of the following Cauchy problem:

$$y(t) = (1-\alpha)g'(t)f(t,y(t)) + \frac{\alpha}{\Gamma(\alpha)}\int_{t_0}^{t} g'(\tau)(t-\tau)^{\alpha-1} f(\tau, y(\tau))\,d\tau, \quad (9.396)$$

$$y(t_0) = y_0$$

depends on the initial conditions and parameters. We shall start with the following lemma.

Lemma 9.4 *Let $f \in C[I \times R, R]$ and let*

$$H(t,r) = \sup_{|y-y_0|\le r} |f(t,y)|. \quad (9.397)$$

We assume that $r^*(t, t_0, 0)$ is the maximal solution of

$$v(t) = (1-\alpha)g'(t)H(t,r) + \frac{\alpha}{\Gamma(\alpha)}\int_{t_0}^{t} g'(\tau)(t-\tau)^{\alpha-1}H(\tau,r)d\tau, \quad (9.398)$$

$$v(t_0) = v_0$$

via $(t_0, 0)$. We consider $y(t, t_0, y_0)$ to be any solution of

$$y(t) = (1-\alpha)g'(t)f(t,y(t)) + \frac{\alpha}{\Gamma(\alpha)}\int_{t_0}^{t} g'(\tau)(t-\tau)^{\alpha-1} f(\tau, y(\tau))\,d\tau, (9.399)$$

$$y(t_0) = y_0.$$

Then

$$|y(t, t_0, y_0) - y_0| \le r^*(t, t_0, 0),\ t \ge t_0. \quad (9.400)$$

Proof We start with defining

$$\Omega(t) = |y(t, t_0, y_0)|. \quad (9.401)$$

Then

$$\Omega(t) \le (1-\alpha)\left|g'(t)\right| f(t, y(t, t_0, y_0)) \quad (9.402)$$

$$+\frac{\alpha}{\Gamma(\alpha)}\int_{t_0}^{t} \left|g'(\tau)\right| (t-\tau)^{\alpha-1} \left|f(\tau, y(\tau, t_0, y_0))\right| d\tau,$$

$$\leq (1-\alpha) \sup_{t\in[t_0,t_0+\lambda]} \left|g'(t)\right| \sup_{|y-y_0|<\Omega(t)} |f(t, y(t))|$$

$$+\frac{\alpha}{\Gamma(\alpha)} \int_{t_0}^{t} \sup_{l\in[t_0,\tau]} \left|g'(l)\right| (t-\tau)^{\alpha-1} \sup_{|y-y_0|<\Omega(\tau)} |f(\tau, y(\tau))| \, d\tau$$

$$< (1-\alpha)H(t,\Omega(t))M_{g'} + \frac{\alpha M_{g'}}{\Gamma(\alpha)} \int_{t_0}^{t} (t-\tau)^{\alpha-1} H(\tau,\Omega(\tau))d\tau.$$

Therefore

$$\Omega(t) \leq r^*(t, t_0, 0), t \geq t_0 \tag{9.403}$$

which completes the proof of the Lemma. We continue with the following theorem.

Theorem 9.12 *Let $f \in C[I \times R, R]$ and for $(t, y), (t, z) \in I \times R$,*

$$|f(t, y) - f(t, z)| \leq h(t, |y - z|) \tag{9.404}$$

where the function $h \in C[I \times R, R]$. We assume that $v(t) = 0$ is the unique solution of the following integral equation:

$$v(t) = (1-\alpha)g'(t)h(t, v(t)) + \frac{\alpha}{\Gamma(\alpha)} \int_{t_0}^{t} g'(\tau)(t-\tau)^{\alpha-1} h(\tau, v(\tau))d\tau, \tag{9.405}$$

$$v(t_0) = v_0.$$

Then having $v(t, t_0, v_0)$ is any solution of the above equation with initial data (t_0, v_0); if $v(t, t_0, v_0)$ is continuous with respect to t_0 and v_0 then the solution is

$$y(t) = (1-\alpha)g'(t)f(t, y(t)) + \frac{\alpha}{\Gamma(\alpha)} \int_{t_0}^{t} g'(\tau)(t-\tau)^{\alpha-1} f(\tau, y(\tau))\, d\tau, \tag{9.406}$$

$$y(t_0) = y_0$$

is unique and continuous with respect to the initial data (t_0, v_0).

Proof The issue of uniqueness has been thoroughly examined therefore the continuity with to y_0 and t_0 with be the topic of our investigation. Let us consider two distinct solutions $y(t, t_0, y_0)$ and $z(t, t_0, z_0)$. Our aim is to show that

$$\lim_{y_0\to z_0} y(t, t_{0,} y_0) = z(t, t_0, z_0), \tag{9.407}$$

$$\lim_{t_0\to t_1} y(t, t_{0,} y_0) = z(t, t_1, y_0).$$

We shall start with the first part. To that end we set

$$\Omega(t) = \left|y(t, t_{0,} y_0) - z(t, t_0, z_0)\right|. \tag{9.408}$$

Then we shall have

$$\begin{aligned}\Omega(t) \leq\ & (1-\alpha)g'(t)\left|f\left(t, y(t, t_{0,} y_0)\right) - f\left(t, z(t, t_0, z_0)\right)\right| \\ & + \frac{\alpha}{\Gamma(\alpha)}\int_{t_0}^{t} g'(\tau)(t-\tau)^{\alpha-1}\left|f\left(\tau, y(\tau, t_{0,} y_0)\right) - f\left(\tau, z(\tau, t_0, z_0)\right)\right| d\tau;\end{aligned} \tag{9.409}$$

using the hypothesis, we have

$$\begin{aligned}\Omega(t) \leq\ & (1-\alpha)g'(t)h\left(t, \left|y(t, t_{0,} y_0) - z(t, t_0, z_0)\right|\right) \\ & + \frac{\alpha}{\Gamma(\alpha)}\int_{t_0}^{t} g'(\tau)(t-\tau)^{\alpha-1}h\left(\tau, \left|y(\tau, t_{0,} y_0) - z(\tau, t_0, z_0)\right|\right) d\tau, \\ \leq\ & (1-\alpha)g'(t)h\left(t, \Omega(t)\right) + \frac{\alpha}{\Gamma(\alpha)}\int_{t_0}^{t} g'(\tau)(t-\tau)^{\alpha-1}h\left(\tau, \Omega(\tau)\right) d\tau, \\ \leq\ & r(t, t_0, |y_0 - z_0|),\ t > t_0\end{aligned} \tag{9.410}$$

where $r(t, t_0, |y_0 - z_0|)$ is the maximal solution of

$$\begin{aligned}& v(t) = (1-\alpha)g'(t)h(t, v(t)) + \frac{\alpha}{\Gamma(\alpha)}\int_{t_0}^{t} g'(\tau)(t-\tau)^{\alpha-1}h(\tau, v(\tau))d\tau, \\ & v(t_0) = v_0.\end{aligned} \tag{9.411}$$

By assumption $v(t, t_{0,} v_0)$ was continuous with respect to the initial data; thus

$$\lim_{y_0 \to z_0} r(t, t_0; |y_0 - z_0|) = r(t, t_{0,} 0) \tag{9.412}$$

while indeed by hypothesis $r(t, t_{0,} 0) = 0$. We can now conclude that

$$\lim_{y_0 \to z_0} y(t, t_{0,} y_0) = z(t, t_0, z_0) \tag{9.413}$$

which indicates the continuity of $y(t, t_{0,} y_0)$ with respect to y_0. We shall show the second part. Now for $t_0 < t_1$, we have that

$$\begin{aligned}\Omega(t) &= |y(t, t_0, y_0) - z(t, t_1, y_0)| \\ \Omega(t_1) &= |y(t_1, t_0, y_0) - y_0|.\end{aligned} \tag{9.414}$$

Thus

$$\Omega(t) \leq (1-\alpha)g'(t)\left|f\left(t, y(t, t_{0,} y_0)\right) - f\left(t, z(t, t_1, y_0)\right)\right| \tag{9.415}$$
$$+\frac{\alpha}{\Gamma(\alpha)}\int_{t_0}^{t} g'(\tau)(t-\tau)^{\alpha-1}\left|f\left(\tau, y(\tau, t_{0,} y_0)\right) - f\left(\tau, z(\tau, t_1, y_0)\right)\right| d\tau;$$

using the hypothesis, we have

$$\Omega(t) \leq (1-\alpha)g'(t)h\left(t, \left|y(t, t_{0,} y_0) - z(t, t_1, y_0)\right|\right) \tag{9.416}$$
$$+\frac{\alpha}{\Gamma(\alpha)}\int_{t_0}^{t} g'(\tau)(t-\tau)^{\alpha-1}h\left(\tau, \left|y(\tau, t_{0,} y_0) - z(\tau, t_1, y_0)\right|\right) d\tau,$$
$$\leq (1-\alpha)g'(t)h\left(t, \Omega(t)\right) + \frac{\alpha}{\Gamma(\alpha)}\int_{t_0}^{t} g'(\tau)(t-\tau)^{\alpha-1}h\left(\tau, \Omega(\tau)\right) d\tau,$$
$$\leq \widetilde{r}(t), \; t > t_1.$$

$$\Omega(t_1) \leq (1-\alpha)g'(t)\left|f\left(t, y(t, t_1, y_0)\right)\right| + \frac{\alpha}{\Gamma(\alpha)}\int_{t_0}^{t} g'(\tau)(t-\tau)^{\alpha-1}\left|f\left(\tau, y(\tau, t_1, y_0)\right)\right| d\tau,$$
$$\leq r^*(t_1, t_{0,}0),$$

with $\widetilde{r}(t, t_1; r^*(t_1, t_{0,}0))$ being the maximal solution of the equation

$$v(t) = (1-\alpha)g'(t)h(t, v(t)) + \frac{\alpha}{\Gamma(\alpha)}\int_{t_0}^{t} g'(\tau)(t-\tau)^{\alpha-1}h(\tau, v(\tau))d\tau, \tag{9.417}$$
$$v(t_0) = v_0$$

via $\left(t_1, r^*(t_1, t_{0,}0)\right)$. However, we recall that $r^*(t_1, t_{0,}0) = 0$. Indeed then

$$\lim_{t_1 \to t_0} \widetilde{r}(t, t_1; r^*(t_1, t_{0,}0)) = \widetilde{r}(t, t_0, 0) = 0 \tag{9.418}$$

according by the hypothesis. Therefore

$$\lim_{t_0 \to t_1} y(t, t_{0,} y_0) = z(t, t_1, y_0)$$

which completes the proof. For the parameters, we investigate via the following theorem.

Theorem 9.13 *Let $f \in C[O, R]$ here; O is an open (t, y, η) set in $R \times R \times R$ and $\eta = \eta_0$; we let*

$$y_0(t) = y(t, t_0, y_0, \eta_0) \tag{9.419}$$

to be the solution of the following equation:

$$y(t) = (1-\alpha)g'(t)f(t, y, \eta_0) + \frac{\alpha}{\Gamma(\alpha)}\int_{t_0}^{t} g'(\tau)(t-\tau)^{\alpha-1}f(\tau, y, \eta_0)d\tau, \tag{9.420}$$

$$y(t_0) = y_0$$

existing for any $t \geq t_0$. We assume that

$$\lim_{\eta\to\eta_0} f(t, y, \eta) = f(t, y, \eta_0). \tag{9.421}$$

Uniformly in (t, y) and for $(t, y_1, \eta), (t, y_2, \eta) \in O$, the following inequality holds:

$$|f(t, y_1, \eta) - f(t, y_2, \eta)| \leq h(t, |y_1 - y_2|); \tag{9.422}$$

h is a function in $C\left[I \times R_+, R_+\right]$. We also assume that $v(t) = 0$ is the unique solution of the equation

$$v(t) = (1-\alpha)g'(t)h(t, v(t)) + \frac{\alpha}{\Gamma(\alpha)}\int_{t_0}^{t} g'(\tau)(t-\tau)^{\alpha-1}h(\tau, v(\tau))d\tau, \tag{9.423}$$

$$v(t_0) = v_0.$$

Then $\forall\varepsilon > 0$, $\exists\delta(\varepsilon) > 0$ such that $\forall\eta$; $|\eta - \eta_0| < \delta(\varepsilon)$ the equation

$$y(t) = (1-\alpha)g'(t)f(t, y, \eta) + \frac{\alpha}{\Gamma(\alpha)}\int_{t_0}^{t} g'(\tau)(t-\tau)^{\alpha-1}f(\tau, y, \eta)d\tau, \tag{9.424}$$

$$y(t_0) = y_0$$

has a unique solution:

$$y(t) = y(t, t_0, y_0, \eta) \tag{9.425}$$

satisfying

$$|y(t) - y_0(t)| < \varepsilon, t \geq t_0. \tag{9.426}$$

Proof The problem of uniqueness has been taken care of f. However by the hypothesis of the theorem, we have $v(t) = 0$. It is the unique solution of the associated equation, thus providing any compatible interval $[t_0, t_0 + a] \subset I$ and $\varepsilon > 0$, we can find a positive number $\mu = \mu(\varepsilon)$ the maximal solution $r(t, t_0, 0, \mu)$ of

$$v(t) = (1-\alpha)g'(t)\left(h(t, v(t)) + \mu\right) + \frac{\alpha}{\Gamma(\alpha)} \int_{t_0}^{t} g'(\tau)(t-\tau)^{\alpha-1} \left(h(\tau, v(\tau)) + \mu\right) d\tau, \tag{9.427}$$

$$v(t_0) = v_0$$

existing with the compact interval $[t_0, t_0 + a]$ and more importantly satisfies

$$r(t, t_0, 0, \mu) < \varepsilon. \tag{9.428}$$

In addition due to the following condition:

$$\lim_{\eta \to \eta_0} f(t, y, \eta) = f(t, y, \eta_0), \tag{9.429}$$

we have $\forall \mu > 0, \exists \delta(\mu) > 0$ such that the following inequality holds:

$$|f(t, y, \eta) - f(t, y, \eta_0)| < \mu \tag{9.430}$$

when

$$|\eta - \eta_0| < \delta. \tag{9.431}$$

Now letting

$$\Omega(t) = |y(t) - y_0(t)|. \tag{9.432}$$

With $y(t)$ and $y_0(t)$ solutions of our equations

$$y(t) = (1-\alpha)g'(t)f(t, y, \eta_0) + \frac{\alpha}{\Gamma(\alpha)} \int_{t_0}^{t} g'(\tau)(t-\tau)^{\alpha-1} f(\tau, y, \eta_0) d\tau, \tag{9.433}$$

$$y(t_0) = y_0.$$

$$y_0(t) = (1-\alpha)g'(t)f(t, y, \eta) + \frac{\alpha}{\Gamma(\alpha)} \int_{t_0}^{t} g'(\tau)(t-\tau)^{\alpha-1} f(\tau, y, \eta) d\tau, \tag{9.434}$$

$$y(t_0) = y_0,$$

respectively.

$$\begin{aligned} |y(t) - y_0(t)| \leq\ & (1-\alpha)g'(t)\,|f(t, y, \eta_0) - f(t, y, \eta)| \\ & + \frac{\alpha}{\Gamma(\alpha)} \int_{t_0}^{t} g'(\tau)(t-\tau)^{\alpha-1} |f(\tau, y, \eta_0) - f(\tau, y, \eta)|\, d\tau, \\ \leq\ & (1-\alpha)g'(t)\left(h(t, |y(t) - y_0(t)|) + |f(t, y_0(t), \eta) - f(t, y_0(t), \eta_0)|\right) \\ & + \frac{\alpha}{\Gamma(\alpha)} \int_{t_0}^{t} g'(\tau)(t-\tau)^{\alpha-1} \begin{pmatrix} h(\tau, |y(\tau) - y_0(\tau)|) \\ + |f(\tau, y_0(\tau), \eta) - f(\tau, y_0(\tau), \eta_0)| \end{pmatrix} d\tau. \end{aligned} \tag{9.435}$$

But

$$|\eta - \eta_0| < \delta \tag{9.436}$$

then

$$\Omega(t) = |y(t) - y_0(t)| \tag{9.437}$$

we have

$$\begin{aligned}\Omega(t) \leq \quad & (1-\alpha)g'(t)\left(h(t,\Omega(t))+\mu\right) \\ & + \frac{\alpha}{\Gamma(\alpha)}\int_{t_0}^{t} g'(\tau)(t-\tau)^{\alpha-1}\left(h(\tau,\Omega(\tau))+\mu\right)d\tau, \\ \Omega(t_0) = \Omega_0. &\end{aligned} \tag{9.438}$$

Therefore, we have that

$$\Omega(t) \leq r\left(t, t_0, 0, \mu\right), t > t_0 \tag{9.439}$$

and hence, we have that if

$$|\eta - \eta_0| < \delta \tag{9.440}$$

then

$$|y(t) - y_0(t)| < \varepsilon, t > t_0. \tag{9.441}$$

Indeed the parameter δ depends on ε since μ does. The proof is therefore completed.

References

1. Kurpelj, N.S.: Some generalization and modification of Chaplygin's method. In: Mitropolski, Y.A., Luchka, Y.A. (eds.) Priblizhennye i kachestvennye metody teorii differencialniyh i integralnyh uravneni, pp. 51–72. Kiev (1971)
2. Vidossich, G.: Chaplygin's method is Newton's method. J. Math. Anal. Appl. **66**, 188–206 (1978)
3. Tošić, D.: One way of discretization of Chaplygin's method. In: Milovanovi, G. (ed.) Proceedings of the Numerical Methods and Approximations Theory, pp. 149–153. Niš (1984)
4. Lakshmikantham, V., Leela, S.: Differential and Integral Inequalities: Theory and Applications, vol. I. Ordinary Differential Equations, Academic Press (1969)

Chapter 10
Numerical Analysis of IVP with Classical Global Derivative

In previous chapters, we provided various significant theoretical discoveries, including inequalities, existence, and uniqueness theories. The Euler technique is used to solve nonlinear differential equations with many fractional global differential equations in this chapter. The forward Euler method is a first-order numerical algorithm for solving ordinary differential equations with a fixed starting point. It is the simplest Runge–Kutta method and the most fundamental explicit method for numerical integration of ordinary differential equations.

Because the Euler method is of first order, the local error is proportional to the square of the step size, and the global error is proportional to the step size. The Euler technique is frequently used to build more complicated methods. We will also make use of some variant numerical methods based on the Euler-like Heun's, Midpoint, linear approximation [1–12]. In addition, different approaches including the Lagrange and the Newton interpolation polynomial will be used too.

10.1 Applying Euler Method on IVP with Classical Global Derivative

We demonstrate the adaptation of the well-known Euler method for solving differential equations, incorporating the classical global derivative approach. This adaptation enhances the method's ability to handle more complex scenarios, offering a clear and effective strategy for numerical solutions.

$$\begin{cases} {}_{t_0}D_g y(t) = f(t, y(t)) & \text{if } t > 0 \\ y(t_0) = y_0 & \text{if } t = 0 \end{cases}. \tag{10.1}$$

A. Atangana and İ. Koca, *Fractional Differential and Integral Operators with Respect to a Function*, Industrial and Applied Mathematics,
https://doi.org/10.1007/978-981-97-9951-0_10

If $g'(t)$ is not differentiable, we have

$$y(t) = y(t_0) + \int_{t_0}^{t} f(\tau, y(\tau))dg(\tau). \tag{10.2}$$

Then at $t = t_{n+1}$, and $t = t_n$, we have

$$y(t_{n+1}) = y(t_n) + \int_{t_n}^{t_{n+1}} f(\tau, y(\tau))dg(\tau). \tag{10.3}$$

Using Euler approximation, we have

$$\begin{aligned} y_{n+1} &= y_n + \int_{t_n}^{t_{n+1}} f(t_n, y_n)dg(\tau), \\ &= y_n + f(c_n, y_n)\left(g(t_{n+1}) - g(t_n)\right), c_n \in \left[t_n, t_{n+1}\right]. \end{aligned} \tag{10.4}$$

If $g(t)$ is differentiable with $g'(t) > 0$ then

$$y_{n+1} = y_n + \int_{t_n}^{t_{n+1}} f(\tau, y(\tau))g'(\tau)d\tau. \tag{10.5}$$

Two ways to evaluate the first is

$$\begin{aligned} y_{n+1} &= y_n + \int_{t_n}^{t_{n+1}} f(t_n, y_n)g'(t_n)d\tau, \\ &= y_n + f(t_n, y_n)\left(\frac{g(t_{n+1}) - g(t_n)}{h}\right)\int_{t_n}^{t_{n+1}} d\tau, \\ &= y_n + f(t_n, y_n)\left(g(t_{n+1}) - g(t_n)\right). \end{aligned} \tag{10.6}$$

The second way is to approximate

$$f(\tau, y(\tau)) \simeq f(t_n, y_n) \tag{10.7}$$

then

$$\begin{aligned} y_{n+1} &= y_n + \int_{t_n}^{t_{n+1}} f(t_n, y_n) g'(\tau) d\tau, \\ &= y_n + f(t_n, y_n) \int_{t_n}^{t_{n+1}} g'(\tau) d\tau, \\ &= y_n + f(t_n, y_n) \left(g(t_{n+1}) - g(t_n) \right). \end{aligned} \tag{10.8}$$

We obtain the same result. It is very easy to show that

$$\lim_{h \to 0} |y_{n+1} - y(t_{n+1})| = 0 \tag{10.9}$$

and

$$|\widetilde{y}_{n+1}| < c\, |\widetilde{y}_0|. \tag{10.10}$$

This fundamental method can be modified instead of taking $f(t_n, y_n)$.

We take $f(t_{n+1}, y_{n+1})$ as follows:

$$y_{n+1} = y_n + \left(g(t_{n+1}) - g(t_n) \right) f(t_{n+1}, y_{n+1}), \tag{10.11}$$

which leads to implicit method; to make it explicit, we have

$$y_{n+1} = y_n + \left(g(t_{n+1}) - g(t_n) \right) f(t_{n+1}, \overline{y}_{n+1}), \tag{10.12}$$

where

$$\overline{y}_{n+1} = y_n + \left(g(t_{n+1}) - g(t_n) \right) f(t_n, y_n). \tag{10.13}$$

Then the method becomes

$$y_{n+1} = y_n + \left(g(t_{n+1}) - g(t_n) \right) f(t_{n+1}, y_n + \left(g(t_{n+1}) - g(t_n) \right) f(t_n, y_n)). \tag{10.14}$$

Theorem 10.1 *We assume that the function $f(t, y(t))$ satisfies the linear growth that is to say*

$$|f(t, y(t))|^2 < k \left(1 + |y(t)|^2 \right). \tag{10.15}$$

Then $\forall n \geq 0$,

$$|y_{n+1}|^2 < A\, |y_n|^2 + B = A^{n+1}\, |y_0|^2 + \sum_{j=0}^{n} A^j B. \tag{10.16}$$

Proof

$$\begin{aligned}
|y_{n+1}|^2 &= |y_n + (g(t_{n+1}) - g(t_n)) f(t_{n+1}, y_n + (g(t_{n+1}) - g(t_n)) f(t_n, y_n))|, \\
&\le 2|y_n|^2 + 2|g(t_{n+1}) - g(t_n)| |f(t_{n+1}, y_n + (g(t_{n+1}) - g(t_n)) f(t_n, y_n))|, \\
&\le 2|y_n|^2 + 2h^2 |g'(t_{n+1})|^2 |f(t_{n+1}, y_n + (g(t_{n+1}) - g(t_n)) f(t_n, y_n)|^2, \\
&\le 2|y_n|^2 + 2h^2 |g'(t_{n+1})|^2 k \left(1 + |y_n + (g(t_{n+1}) - g(t_n)) f(t_n, y_n)|^2\right), \\
&\le 2|y_n|^2 + 2h^2 \|g'\|_\infty^2 k \left\{1 + 2|y_n|^2 + h^2 \|g'\|_\infty^2 |f(t_n, y_n)|^2\right\}, \\
&\le 2|y_n|^2 + 2h^2 \|g'\|_\infty^2 k \left\{1 + 2|y_n|^2 + h^2 \|g'\|_\infty^2 k \left(1 + |y_n|^2\right)\right\}, \\
&\le |y_n|^2 \left\{2 + 4h^2 \|g'\|_\infty^2 k + 4h^4 \|g'\|_\infty^4 k^2\right\} + \left\{2h^2 \|g'\|_\infty^2 + 2h^4 \|g'\|_\infty^4 k^2\right\}, \\
&\le |y_n|^2 A + B
\end{aligned} \tag{10.17}$$

where

$$\begin{aligned}
A &= 2 + 4h^2 \|g'\|_\infty^2 k + 4h^4 \|g'\|_\infty^4 k^2, \\
B &= 2h^2 \|g'\|_\infty^2 + 2h^4 \|g'\|_\infty^4 k^2.
\end{aligned} \tag{10.18}$$

We can see that when $n = 0$

$$\begin{aligned}
|y_1|^2 &\le A|y_0|^2 + B, \\
|y_2|^2 &\le A|y_1|^2 + B = A\left\{A|y_0|^2 + B\right\} + B, \\
&= A^2|y_0|^2 + AB + B, \\
|y_3|^2 &\le A|y_2|^2 + B = A\left\{A^2|y_0|^2 + AB + B\right\} + B \\
&= A^3|y_0|^2 + A^2B + AB + B, \\
|y_4|^2 &\le A|y_3|^2 + B = A\left\{A^3|y_0|^2 + A^2B + AB + B\right\} + B, \\
&= A^4|y_0|^2 + A^3B + A^2B + AB + B.
\end{aligned} \tag{10.19}$$

We then assume that $\forall n \ge 1$, we have

$$|y_n|^2 \le A^n |y_0|^2 + \sum_{j=0}^{n-1} A^j B, \tag{10.20}$$

$$|y_{n+1}|^2 < A|y_n|^2 + B = A^{n+1}|y_0|^2 + \sum_{j=0}^{n} A^j B$$

which completes the proof.

10.2 Applying Heun's Method on IVP with Classical Global Derivative

In this section, we present the extension of Heun's method for solving numerically this class of differential equations:

$$\begin{cases} {}_{t_0}D_g y(t) = f(t, y(t)) \text{ if } t > 0 \\ y(t_0) = y_0 \qquad\qquad\quad \text{if } t = 0 \end{cases}. \tag{10.21}$$

We assume that $g'(t) > 0$, then we have

$$\begin{aligned} y(t) &= y(t_0) + \int_{t_0}^{t} f(\tau, y(\tau))g'(\tau)d\tau, \\ y(t_0) &= y_0; \end{aligned} \tag{10.22}$$

at $t = t_{n+1}$, we have

$$y(t_{n+1}) = y(t_0) + \int_{t_0}^{t_{n+1}} f(\tau, y(\tau))g'(\tau)d\tau, \tag{10.23}$$

and at $t = t_n$, we have

$$y(t_n) = y(t_0) + \int_{t_0}^{t_n} f(\tau, y(\tau))g'(\tau)d\tau. \tag{10.24}$$

The difference produces

$$y(t_{n+1}) = y(t_n) + \int_{t_n}^{t_{n+1}} f(\tau, y(\tau))g'(\tau)d\tau. \tag{10.25}$$

There are two ways to apply Heun's method here. The first is to simply approximate $f(\tau, y(\tau)$ within $[t_n, t_{n+1}]$ using

$$f(\tau, y(\tau)) \simeq \frac{f(t_{n+1}, y_{n+1}) + f(t_n, y_n)}{2}, \tag{10.26}$$

then we will have

$$\begin{aligned} y_{n+1} &= y_n + \int_{t_n}^{t_{n+1}} \frac{f(t_{n+1}, y_{n+1}) + f(t_n, y_n)}{2} g'(\tau) d\tau, \\ &= y_n + \frac{f(t_{n+1}, y_{n+1}) + f(t_n, y_n)}{2} \left(g(t_{n+1}) - g(t_n)\right). \end{aligned} \tag{10.27}$$

To make the above explicit, we reformulate is as

$$y_{n+1} = y_n + \frac{g(t_{n+1}) - g(t_n)}{2} \left(f(t_{n+1}, \overline{y}_{n+1}) + f(t_n, y_n)\right) \tag{10.28}$$

where

$$\overline{y}_{n+1} = y_n + \frac{g(t_{n+1}) - g(t_n)}{2} f(t_n, y_n). \tag{10.29}$$

The second approach is to approximate $g'(\tau) f(\tau, y(\tau)$ as

$$g'(\tau) f(\tau, y(\tau) \simeq \frac{1}{2} \left(g'(t_{n+1}) f(t_{n+1}, y_{n+1}) + g'(t_n) f(t_n, y_n)\right). \tag{10.30}$$

Then, we have

$$\begin{aligned} y_{n+1} &= y_n + \frac{h}{2} \left(g'(t_{n+1}) f(t_{n+1}, y_{n+1}) + g'(t_n) f(t_n, y_n)\right) \\ &= y_n + \frac{h}{2} \left(\frac{g(t_{n+1}) - g(t_n)}{h} f(t_{n+1}, \overline{y}_{n+1}) + \frac{g(t_n) - g(t_{n-1})}{h} f(t_n, y_n)\right) \\ &= y_n + \frac{1}{2} \left(\left(g(t_{n+1}) - g(t_n)\right) f(t_{n+1}, \overline{y}_{n+1}) + \left(g(t_n) - g(t_{n-1})\right) f(t_n, y_n)\right) \end{aligned} \tag{10.31}$$

where

$$\overline{y}_{n+1} = y_n + \left(g(t_{n+1}) - g(t_n)\right) f(t_n, y_n). \tag{10.32}$$

We note that if $g(t) = t$, we have

$$y_{n+1} = y_n + \frac{h}{2} \left(f(t_{n+1}, \overline{y}_{n+1}) + f(t_n, y_n)\right) \tag{10.33}$$

with

$$\overline{y}_{n+1} = y_n + h f(t_n, y_n). \tag{10.34}$$

We recover Heun's method for the classical case.

Theorem 10.2 *Let $\widetilde{y}_n$ be the perturbed term of y_n, then $\forall n \geq 0$, $\exists c > 0$ such that*

$$|\widetilde{y}_{n+1}| < c\,|\widetilde{y}_0|\,. \tag{10.35}$$

Proof

$$\widetilde{y}_{n+1} + y_{n+1} = y_n + \widetilde{y}_n + \frac{1}{2}\begin{pmatrix} (g(t_{n+1}) - g(t_n))\, f(t_{n+1}, \widetilde{y}_{n+1} + y_{n+1}) \\ + (g(t_n) - g(t_{n-1}))\, f(t_n, y_n + \widetilde{y}_n) \end{pmatrix}. \tag{10.36}$$

Therefore rearranging, we get

$$\widetilde{y}_{n+1} = \widetilde{y}_n + \frac{1}{2}\begin{pmatrix} (g(t_{n+1}) - g(t_n))\,(f(t_{n+1}, \widetilde{y}_{n+1} + y_{n+1}) - f(t_{n+1}, y_{n+1})) \\ + (g(t_n) - g(t_{n-1}))\,((f(t_n, y_n + \widetilde{y}_n) - f(t_n, y_n))) \end{pmatrix}. \tag{10.37}$$

Then applying the norm on both sides yields

$$|\widetilde{y}_{n+1}| \leq |\widetilde{y}_n| + \frac{1}{2}\begin{pmatrix} h\left|g'(t_{n+1})\right|\left|f(t_{n+1}, \widetilde{y}_{n+1} + y_{n+1}) - f(t_{n+1}, y_{n+1})\right| \\ h\left|g'(t_n)\right|\left|f(t_n, y_n + \widetilde{y}_n) - f(t_n, y_n)\right| \end{pmatrix}. \tag{10.38}$$

Using the Lipschitz condition of f with respect to y, we obtain

$$\begin{aligned} |\widetilde{y}_{n+1}| &\leq |\widetilde{y}_n| + \frac{h}{2}\left\|g'\right\|_\infty L\,|\widetilde{y}_{n+1}| + \frac{h}{2}\left\|g'\right\|_\infty L\,|\widetilde{y}_n|\,, \\ &\leq |\widetilde{y}_n| + \frac{h}{2}\left\|g'\right\|_\infty L\,\{|\widetilde{y}_{n+1}| + |\widetilde{y}_n|\}\,. \end{aligned} \tag{10.39}$$

Noting that

$$\begin{aligned} \widetilde{\overline{y}}_{n+1} + \overline{y}_{n+1} &= y_n + \widetilde{y}_n + \big(g(t_{n+1}) - g(t_n)\big)\,((f(t_n, y_n + \widetilde{y}_n) - f(t_n, y_n))) \\ \widetilde{\overline{y}}_{n+1} &= \widetilde{y}_n + \big(g(t_{n+1}) - g(t_n)\big)\,((f(t_n, y_n + \widetilde{y}_n) - f(t_n, y_n))) \\ \left|\widetilde{\overline{y}}_{n+1}\right| &\leq |\widetilde{y}_n| + h\left\|g'\right\|_\infty L\,|\widetilde{y}_n|\,, \end{aligned} \tag{10.40}$$

and replacing in the corrector equation, we get

$$\begin{aligned} |\widetilde{y}_{n+1}| &\leq |\widetilde{y}_n| + \frac{h}{2}\left\|g'\right\|_\infty L\left\{|\widetilde{y}_n| + h\left\|g'\right\|_\infty L\,|\widetilde{y}_n|\right\} \\ &\leq |\widetilde{y}_n|\left\{1 + \frac{h}{2}\left\|g'\right\|_\infty L\left\{1 + h\left\|g'\right\|_\infty L\right\}\right\} \\ &\leq |\widetilde{y}_n|\,\Omega. \end{aligned} \tag{10.41}$$

Then

$$\begin{aligned}|\widetilde{y}_1| &\leq |\widetilde{y}_0|\,\Omega,\\ &\vdots\\ |\widetilde{y}_{n+1}| &\leq |\widetilde{y}_0|\,\Omega^{n+1},\\ &\leq |\widetilde{y}_0|\,c,\end{aligned} \tag{10.42}$$

which complete the proof.

Theorem 10.3 $\forall n \geq 0$, *if* $y(t_{n+1})$ *is the exact solution at* $y(t_{n+1})$ *and* y_{n+1} *approximate solution*

$$\lim_{h\to 0} |y(t_{n+1}) - y_{n+1}| = 0 \tag{10.43}$$

if $f(t, y(t))$ *is differentiable.*

Proof

$$|y(t_{n+1}) - y_{n+1}| = \left| \begin{array}{l} y(t_n) - y_n + \int\limits_{t_n}^{t_{n+1}} f(\tau, y(\tau))g'(\tau)d\tau \\ -\frac{1}{2}\int\limits_{t_n}^{t_{n+1}} \frac{g(t_{n+1})-g(t_n)}{h} f(t_{n+1}, y_{n+1})d\tau \\ -\frac{1}{2}\int\limits_{t_n}^{t_{n+1}} \frac{g(t_n)-g(t_{n-1})}{h} f(t_n, y_n)d\tau \end{array} \right| \tag{10.44}$$

$$\leq |y(t_n) - y_n| + \frac{1}{2}\int\limits_{t_n}^{t_{n+1}} |f(\tau, y(\tau)) - f(t_{n+1}, y_{n+1})| \left|g'(\tau)\right| d\tau$$

$$+\frac{1}{2}\int\limits_{t_n}^{t_{n+1}} |f(\tau, y(\tau)) - f(t_n, y_n)| \left|g'(\tau)\right| d\tau$$

$$\leq |y(t_n) - y_n| + \frac{1}{2}\left\|g'\right\|_\infty \left\|f'(., y(.))\right\|_\infty \int\limits_{t_n}^{t_{n+1}} (t_{n+1} - \tau)\, d\tau$$

$$+\frac{1}{2}\left\|g'\right\|_\infty \left\|f'(., y(.))\right\|_\infty \int\limits_{t_n}^{t_{n+1}} (t_n - \tau)\, d\tau$$

$$\leq |y(t_n) - y_n| + \frac{1}{2}\left\|g'\right\|_\infty \left\|f'(., y(.))\right\|_\infty \frac{h^2}{2}$$

$$+\frac{1}{2}\left\|g'\right\|_\infty \left\|f'(., y(.))\right\|_\infty \frac{h}{2}.$$

$$\lim_{h\to 0} |y(t_{n+1}) - y_{n+1}| \leq \lim_{h\to 0} |y(t_n) - y_n|,\ \forall n \geq 0. \tag{10.45}$$

When $n = 0$, we have

$$\begin{aligned}
\lim_{h\to 0} |y(t_1) - y_1| &\le \lim_{h\to 0} |y(t_0) - y_0| = 0 \\
\lim_{h\to 0} |y(t_2) - y_2| &\le \lim_{h\to 0} |y(t_1) - y_1| = 0 \\
&\vdots \\
\lim_{h\to 0} |y(t_{n+1}) - y_{n+1}| &= 0
\end{aligned} \tag{10.46}$$

which completes the proof.

10.3 Applying Midpoint Scheme Method on IVP with Classical Global Derivative

In this section, we aim to solve the classical global differential equation using the Midpoint method, a well-known numerical technique. The equation we will be addressing is as follows:

$$\begin{cases} {}_{t_0}D_g y(t) = f(t, y(t)) & \text{if } t > 0 \\ y(t_0) = y_0 & \text{if } t = 0 \end{cases}. \tag{10.47}$$

We assume that $g'(t) > 0$, then we have

$$\begin{aligned}
y(t) &= y(t_0) + \int_{t_0}^{t} f(\tau, y(\tau))g'(\tau)d\tau, \\
y(t_0) &= y_0;
\end{aligned} \tag{10.48}$$

at $t = t_{n+1}$, we have

$$y(t_{n+1}) = y(t_0) + \int_{t_0}^{t_{n+1}} f(\tau, y(\tau))g'(\tau)d\tau, \tag{10.49}$$

and at $t = t_n$, we have

$$y(t_n) = y(t_0) + \int_{t_0}^{t_n} f(\tau, y(\tau))g'(\tau)d\tau. \tag{10.50}$$

The difference produces

$$y(t_{n+1}) = y(t_n) + \int_{t_n}^{t_{n+1}} f(\tau, y(\tau))g'(\tau)d\tau. \tag{10.51}$$

We shall make use of two ways to derive this. The first way is to approximate only $f(\tau, y(\tau))$ using the midpoint

$$\begin{aligned} y_{n+1} &= y_n + (g(t_{n+1}) - g(t_n))\, f\left(t_n + \frac{h}{2}, \frac{y_n + \overline{y}_{n+1}}{2}\right), \qquad (10.52)\\ \overline{y}_{n+1} &= y_n + \int_{t_n}^{t_{n+1}} f(\tau, y(\tau))g'(\tau)d\tau, \\ &= y_n + f(t_n, y_n)\left(g\left(t_{n+1}\right) - g(t_n)\right). \end{aligned}$$

Therefore we have

$$y_{n+1} = y_n + (g(t_{n+1}) - g(t_n))\, f\left(t_n + \frac{h}{2}, y_n + \frac{1}{2}\left(g\left(t_{n+1}\right) - g(t_n)\right) f(t_n, y_n)\right). \tag{10.53}$$

Noting that if $g(t) = t$, we have

$$y_{n+1} = y_n + hf\left(t_n + \frac{h}{2}, y_n + \frac{h}{2} f(t_n, y_n)\right) \tag{10.54}$$

which is the classical midpoint.

The second way is to approximate

$$f(\tau, y(\tau))g'(\tau) = F(t, y(t)). \tag{10.55}$$

By the above replacement we have

$$\begin{aligned} y_{n+1} &= y_n + hF\left(t_n + \frac{h}{2}, y_n + \frac{h}{2} F(t_n, y_n)\right), \qquad (10.56)\\ &= y_n + hg'(t_n + \frac{h}{2}) f\left(t_n + \frac{h}{2}, y_n + \frac{h}{2} g'(t_n) f(t_n, y_n)\right), \\ &= y_n + h\left(\frac{g(t_{n+1}) - g(t_n)}{h}\right) f\left(t_n + \frac{h}{2}, y_n + \frac{h}{2}\left(\frac{g(t_{n+1}) - g(t_n)}{h}\right) f(t_n, y_n)\right), \\ &= y_n + (g(t_{n+1}) - g(t_n))\, f\left(t_n + \frac{h}{2}, y_n + \frac{1}{2}\left(g(t_{n+1}) - g(t_n)\right) f(t_n, y_n)\right). \end{aligned}$$

We recover the same formula by considering

$$M = \max\left|g'(t)\right|, \forall t \in [0, T]. \tag{10.57}$$

We can show that the scheme is stable and consistent.

10.4 Applying Linear Approximation Method on IVP with Classical Global Derivative

In this section, we solve the IVP problem with a numerical method based on the linear approximation. We the consider the following general IVP:

$$\begin{cases} {}_{t_0}D_g y(t) = f(t, y(t)) \text{ if } t > 0 \\ y(t_0) = y_0 \qquad\qquad\quad \text{ if } t = 0 \end{cases}. \tag{10.58}$$

We assume that $g'(t) > 0$, then we have

$$\begin{aligned} y(t) &= y(t_0) + \int_{t_0}^{t} f(\tau, y(\tau))g'(\tau)d\tau, \\ y(t_0) &= y_0; \end{aligned} \tag{10.59}$$

at $t = t_{n+1}$, we have

$$y(t_{n+1}) = y(t_0) + \int_{t_0}^{t_{n+1}} f(\tau, y(\tau))g'(\tau)d\tau, \tag{10.60}$$

and at $t = t_n$, we have

$$y(t_n) = y(t_0) + \int_{t_0}^{t_n} f(\tau, y(\tau))g'(\tau)d\tau. \tag{10.61}$$

So we have

$$y(t_{n+1}) = y(t_n) + \int_{t_n}^{t_{n+1}} f(\tau, y(\tau))g'(\tau)d\tau. \tag{10.62}$$

We will apply two approaches: the first will be to approximate only $f(\tau, y(\tau))$ with the linear approximation. The second will be to approximate

$$F(\tau, y(\tau)) = f(\tau, y(\tau))g'(\tau). \tag{10.63}$$

In this first approach, we approximate

$$f(\tau, y(\tau)) \simeq f(t_n, y_n) + \frac{f(t_{n+1}, y_{n+1}) - f(t_n, y_n)}{h}(\tau - t_n). \tag{10.64}$$

Replacing yields

$$\begin{aligned} y_{n+1} &= y_n + \int_{t_n}^{t_{n+1}} \left(f(t_n, y_n) + \frac{f(t_{n+1}, y_{n+1}) - f(t_n, y_n)}{h} (\tau - t_n) \right) g'(\tau) d\tau, \quad (10.65) \\ &= y_n + f(t_n, y_n) \left(g(t_{n+1}) - g(t_n) \right) \\ &+ \left(\frac{f(t_{n+1}, y_{n+1}) - f(t_n, y_n)}{h} \right) \left(g(t_{n+1}) - g(t_n) \right) \int_{t_n}^{t_{n+1}} (\tau - t_n)\, d\tau, \\ &= y_n + f(t_n, y_n) \left(g(t_{n+1}) - g(t_n) \right) \\ &+ \left(\frac{f(t_{n+1}, y_{n+1}) - f(t_n, y_n)}{h} \right) \frac{(t_{n+1} - t_n)^2}{2} \left(g(t_{n+1}) - g(t_n) \right), \\ &= y_n + f(t_n, y_n) \left(g(t_{n+1}) - g(t_n) \right) \\ &+ \left(\frac{f(t_{n+1}, y_{n+1}) - f(t_n, y_n)}{2} \right) h \left(g(t_{n+1}) - g(t_n) \right). \end{aligned}$$

Therefore

$$y_{n+1} = y_n + f(t_n, y_n) \left(g(t_{n+1}) - g(t_n) \right) + \left(\frac{f(t_{n+1}, y_{n+1}) - f(t_n, y_n)}{2} \right) h \left(g(t_{n+1}) - g(t_n) \right). \quad (10.66)$$

The second way: we approximate

$$\begin{aligned} F(\tau, y(\tau)) &= f(\tau, y(\tau)) g'(\tau) \simeq F(t_n, y_n) + \frac{F(t_{n+1}, y_{n+1}) - F(t_n, y_n)}{h} (t_{n+1} - \tau), \quad (10.67) \\ &\simeq f(t_n, y_n) g'(t_n) + \frac{f(t_{n+1}, y_{n+1}) g'(t_{n+1}) - f(t_n, y_n) g'(t_n)}{h} (t_{n+1} - \tau), \end{aligned}$$

and replacing yields

$$\begin{aligned} y_{n+1} &= y_n + f(t_n, y_n) \left(g(t_{n+1}) - g(t_n) \right) \quad (10.68) \\ &+ \left(\frac{g(t_{n+1}) - g(t_n)}{h} f(t_{n+1}, y_{n+1}) - \frac{g(t_{n+1}) - g(t_n)}{h} f(t_n, y_n) \right) \int_{t_n}^{t_{n+1}} \frac{(\tau - t_{n+1})}{h} d\tau, \\ y_{n+1} &= y_n + f(t_n, y_n) \left(g(t_{n+1}) - g(t_n) \right) + \left(\frac{g(t_{n+1}) - g(t_n)}{2} \right) \left(f(t_{n+1}, y_{n+1}) - f(t_n, y_n) \right). \end{aligned}$$

Both yield an implicit scheme; we shall make it explicit:

$$y_{n+1} = y_n + f(t_n, y_n) \left(g(t_{n+1}) - g(t_n) \right) + \frac{g(t_{n+1}) - g(t_n)}{2} \left(f(t_{n+1}, \overline{y}_{n+1}) - f(t_n, y_n) \right), \quad (10.69)$$

where

$$\overline{y}_{n+1} = y_n + f(t_n, y_n) \left(g(t_{n+1}) - g(t_n) \right). \quad (10.70)$$

We shall present in detail the stability, consistency, and convergence analysis.

Theorem 10.4 *Assuming that $\|g\|_\infty < M_1$, then the error is*

$$\left|\overline{R}_T\right| < M_1 \left\|f''(., y(.))\right\|_\infty \frac{h^3}{8} \tag{10.71}$$

if $f(., y(.))$ has a bounded second derivative.

Proof We call R_T the error of the approximation, and it is defined as

$$R_T = f(\tau, y(\tau)) - P_n(\tau) \tag{10.72}$$

where

$$\begin{aligned} P_n &\simeq g'(t_{n+1}) f(t_n, y_n) + g'(t_{n+1}) \left(f(t_{n+1}, y_{n+1}) - f(t_n, y_n)\right)(\tau - t_n), \\ R_T &= \frac{(\tau - t_n)(\tau - t_{n+1})}{2} f''(\xi, y(\xi)), \end{aligned} \tag{10.73}$$

where $\xi \in [t_n, t_{n+1}]$.

$$\begin{aligned} y(t_{n+1}) &= y(t_n) + \int_{t_n}^{t_{n+1}} f(\tau, y(\tau)) g'(\tau) d\tau, \\ y_{n+1} &= y_n + \int_{t_n}^{t_{n+1}} f(\tau, y(\tau)) g'(\tau) d\tau, \\ &= y_n + \int_{t_n}^{t_{n+1}} g'(\tau) \left[R_T(\tau) + P_n(\tau)\right] d\tau, \\ &= y_n + \int_{t_n}^{t_{n+1}} g'(\tau) R_T(\tau)\, d\tau + y_{n+1}, \\ &= \int_{t_n}^{t_{n+1}} g'(\tau) \frac{(\tau - t_n)(\tau - t_{n+1})}{2} f''(\xi, y(\xi)) d\tau. \end{aligned} \tag{10.74}$$

$$\begin{aligned} \left|\overline{R}_T\right| &= \left| \int_{t_n}^{t_{n+1}} g'(\tau) \frac{(\tau - t_n)(\tau - t_{n+1})}{2} f''(\xi, y(\xi)) d\tau \right|, \\ &\leq \int_{t_n}^{t_{n+1}} \left|g'(\tau)\right| \frac{(\tau - t_n)(t_{n+1} - \tau)}{2} \left|f''(\xi, y(\xi))\right| d\tau, \end{aligned} \tag{10.75}$$

$$\leq \int_{t_n}^{t_{n+1}} \sup_{l\in[t_n,t_{n+1}]} \left|g'(l)\right| \frac{(\tau - t_n)(t_{n+1}-\tau)}{2} \left|f''(\xi, y(\xi))\right| d\tau,$$

$$\leq \left\|g'\right\|_\infty \left\|f''(.,y(.))\right\|_\infty \left\{\frac{\tau^2}{2}t_{n+1} - \frac{\tau^3}{3} - t_n t_{n+1}\tau + t_n\frac{\tau^2}{2}\right\}\Bigg|_{t_n}^{t_{n+1}},$$

$$\leq \left\|g'\right\|_\infty \left\|f''(.,y(.))\right\|_\infty \left\{\begin{array}{c} \frac{t_{n+1}^3}{2} - \frac{t_{n+1}^3}{3} - t_n t_{n+1}^2 + t_n\frac{t_{n+1}^2}{2} - t_n^2\frac{t_{n+1}}{2} \\ +\frac{t_n^3}{3} + t_n^2 t_{n+1} - \frac{t_n^3}{2} \end{array}\right\},$$

$$\left|\overline{R}_T\right| \leq \left\|g'\right\|_\infty \left\|f''(.,y(.))\right\|_\infty \frac{h^3}{8},$$

$$\leq M_1 \left\|f''(.,y(.))\right\|_\infty \frac{h^3}{8},$$

which completes the proof.

Theorem 10.5 *Let $y(t_{n+1})$ be the exact solution at $t = t_{n+1}$ and y_{n+1} be the approximate solution, then*

$$\lim_{h\to 0} |y(t_{n+1}) - y_{n+1}| = 0. \tag{10.76}$$

Proof

$$\lim_{h\to 0} |y(t_{n+1}) - y_{n+1}| = \lim_{h\to 0}\left|\begin{array}{c} y(t_n) + \int_{t_n}^{t_{n+1}} f(\tau, y(\tau))g'(\tau)d\tau \\ -y_n \\ -\int_{t_n}^{t_{n+1}} g'(\tau)\left\{f(t_n, y_n) + \frac{(f(t_{n+1},y_{n+1})-f(t_n,y_n))}{h}(\tau - t_n)\right\}d\tau \end{array}\right|, \tag{10.77}$$

$$\leq \lim_{h\to 0}|y(t_n) - y_n| + \lim_{h\to 0}\int_{t_n}^{t_{n+1}} |g'(\tau)|\,|f(\tau, y(\tau)) - f(t_n, y_n)|\,d\tau +$$

$$+\lim_{h\to 0}\int_{t_n}^{t_{n+1}} \left|\frac{(f(t_{n+1}, y_{n+1}) - f(t_n, y_n))}{h}\right|(\tau - t_n)\,d\tau,$$

$$\leq \lim_{h\to 0}|y(t_n) - y_n| + \lim_{h\to 0}|g'|_\infty \int_{t_n}^{t_{n+1}} |f(t_n, y_n) - f(\tau, y(\tau)|\,d\tau$$

$$+\lim_{h\to 0}\int_{t_n}^{t_{n+1}} \left|\frac{(f(t_{n+1}, y_{n+1}) - f(t_n, y_n))}{h}\right|(\tau - t_n)\,d\tau;$$

with the use of the differentiation of the function $f(t, y(t))$ and the Mean value theorem, we have

$$\begin{aligned}
\lim_{h\to 0}\left|y(t_{n+1})-y_{n+1}\right| &\leq \lim_{h\to 0}|y(t_n)-y_n| + \lim_{h\to 0}\left|g'\right|_\infty \frac{h^2}{2}\left|f'(.,y(.))\right|_\infty \\
&+\lim_{h\to 0}\left|g'\right|_\infty \frac{h}{2}\left|f(.,y(.))\right|_\infty, \\
&\leq \lim_{h\to 0}|y(t_n)-y_n|.
\end{aligned} \tag{10.78}$$

Now when $n=0$, we have

$$\begin{aligned}
\lim_{h\to 0}|y(t_1)-y_1| &\leq \lim_{h\to 0}|y(t_0)-y_0|, \\
&\leq 0, \\
\lim_{h\to 0}|y(t_2)-y_2| &\leq \lim_{h\to 0}|y(t_1)-y_1| = 0, \\
&\vdots \\
\lim_{h\to 0}|y(t_{n+1})-y_{n+1}| &\leq 0.
\end{aligned} \tag{10.79}$$

We conclude that

$$\lim_{h\to 0}|y(t_{n+1})-y_{n+1}| = 0 \tag{10.80}$$

which completes the proof.

Theorem 10.6 *If $\widetilde{y}_{n+1}$, $\widetilde{y}_n$ and $\widetilde{y}_0$ are perturbed terms of y_{n+1}, y_n and y_0 respectively, then $\exists c \geq 0$ such that $\forall n > 0$*

$$|\widetilde{y}_{n+1}| < c\,|\widetilde{y}_0|. \tag{10.81}$$

Proof

$$\begin{aligned}
y_{n+1}+\widetilde{y}_{n+1} &= y_n+\widetilde{y}_n+f(t_n,y_n+\widetilde{y}_n)\left(g(t_{n+1})-g(t_n)\right) \\
&+\frac{g(t_{n+1})-g(t_n)}{2}\left(f(t_{n+1},\overline{y}_{n+1}+\widetilde{\overline{y}}_{n+1})-f(t_n,y_n+\widetilde{y}_n)\right),
\end{aligned} \tag{10.82}$$

therefore

$$\begin{aligned}
\widetilde{y}_{n+1} &= \widetilde{y}_n+\left(g(t_{n+1})-g(t_n)\right)\{f(t_n,y_n+\widetilde{y}_n)-f(t_n,y_n)\} \\
&+\frac{g(t_{n+1})-g(t_n)}{2}\left\{f(t_{n+1},\overline{y}_{n+1}+\widetilde{\overline{y}}_{n+1})-f(t_{n+1},\overline{y}_{n+1})\right\} \\
&+\frac{g(t_{n+1})-g(t_n)}{2}\{f(t_n,y_n+\widetilde{y}_n)-f(t_n,y_n)\},
\end{aligned} \tag{10.83}$$

$$
\begin{aligned}
|\widetilde{y}_{n+1}| &\leq |\widetilde{y}_n| + |g(t_{n+1}) - g(t_n)| \, |f(t_n, y_n + \widetilde{y}_n) - f(t_n, y_n)| \\
&+ \left| \frac{g(t_{n+1}) - g(t_n)}{2} \right| \left| f(t_{n+1}, \overline{y}_{n+1} + \widetilde{\overline{y}}_{n+1}) - f(t_{n+1}, \overline{y}_{n+1}) \right| \\
&+ \left| \frac{g(t_{n+1}) - g(t_n)}{2} \right| |f(t_n, y_n + \widetilde{y}_n) - f(t_n, y_n)| \\
&\leq |\widetilde{y}_n| + h \left| g'(t_{n+1}) \right|_\infty L \, |\widetilde{y}_n| + \frac{h}{2} \left| g'(t_n) \right|_\infty L \left| \widetilde{\overline{y}}_{n+1} \right| + h \left| g'(t_n) \right|_\infty |\widetilde{y}_n| .
\end{aligned}
$$

Here we have used the Lipschitz condition of $f(t, y(t))$ with respect to y. Thus

$$
|\widetilde{y}_{n+1}| \leq |\widetilde{y}_n| + \overline{M}_1 \, |\widetilde{y}_n| + \overline{M}_2 \left| \widetilde{\overline{y}}_{n+1} \right| + \overline{M}_3 \, |\widetilde{y}_n| . \tag{10.84}
$$

We now evaluate

$$
\begin{aligned}
|\widetilde{y}_{n+1}| &\leq |\widetilde{y}_n| + h \left| g'(t_{n+1}) \right| L \, |\widetilde{y}_n| \\
&\leq |\widetilde{y}_n| \left(1 + \overline{M}_1\right),
\end{aligned} \tag{10.85}
$$

and replacing yields

$$
\begin{aligned}
|\widetilde{y}_{n+1}| &\leq |\widetilde{y}_n| + \overline{M}_1 \, |\widetilde{y}_n| + \overline{M}_2 \, |\widetilde{y}_n| \left(1 + \overline{M}_1\right) + \overline{M}_3 \, |\widetilde{y}_n| \\
&\leq |\widetilde{y}_n| \left\{1 + \overline{M}_1 + \overline{M}_2 \left(1 + \overline{M}_1\right) + \overline{M}_3\right\} \\
&\leq |\widetilde{y}_n| \, B;
\end{aligned} \tag{10.86}
$$

when $n = 0$, we have

$$
\begin{aligned}
|\widetilde{y}_1| &\leq B \, |\widetilde{y}_0| , \\
|\widetilde{y}_2| &\leq B \, |\widetilde{y}_1| \leq B^2 \, |\widetilde{y}_0| , \\
|\widetilde{y}_3| &\leq B \, |\widetilde{y}_2| \leq B^3 \, |\widetilde{y}_0| , \\
&\vdots \\
|\widetilde{y}_{n+1}| &\leq B^{n+1} \, |\widetilde{y}_0| = c \, |\widetilde{y}_0| ,
\end{aligned} \tag{10.87}
$$

which completes the proof.

10.5 Applying Lagrange Interpolation Method on IVP with Classical Global Derivative

We consider the general IVP with classical global derivative

$$
\begin{cases}
{}_{t_0}D_g y(t) = f(t, y(t)) & \text{if } t > 0 \\
y(t_0) = y_0 & \text{if } t = 0
\end{cases} . \tag{10.88}
$$

We assume that $g'(t) > 0$, then we have

$$\begin{aligned} y(t) &= y(t_0) + \int_{t_0}^{t} f(\tau, y(\tau))g'(\tau)d\tau, \\ y(t_0) &= y_0; \end{aligned} \tag{10.89}$$

at $t = t_{n+1}$, we have

$$y(t_{n+1}) = y(t_0) + \int_{t_0}^{t_{n+1}} f(\tau, y(\tau))g'(\tau)d\tau, \tag{10.90}$$

and at $t = t_n$, we have

$$y(t_n) = y(t_0) + \int_{t_0}^{t_n} f(\tau, y(\tau))g'(\tau)d\tau. \tag{10.91}$$

So we have

$$y(t_{n+1}) = y(t_n) + \int_{t_n}^{t_{n+1}} f(\tau, y(\tau))g'(\tau)d\tau. \tag{10.92}$$

We put

$$F(t, y(t)) = g'(t)f(t, y(t)). \tag{10.93}$$

We approximate $F(\tau, y(\tau))$ within $\left[t_n, t_{n+1}\right]$ by

$$F(\tau, y(\tau)) \simeq \frac{\tau - t_{n-1}}{h} F(t_n, y(t_n)) - \frac{\tau - t_n}{h} F(t_{n-1}, y(t_{n-1})) \tag{10.94}$$

and replacing and integrating we obtain

$$\begin{aligned} y_{n+1} &= y_n + \frac{3}{2} hF(t_n, y_n) - \frac{h}{2} F(t_{n-1}, y_{n-1}), \\ y_{n+1} &= y_n + \frac{3}{2} hg'(t_n)f(t_n, y_n) - \frac{h}{2} g'(t_{n-1})f(t_{n-1}, y_{n-1}), \\ &= y_n + \frac{3}{2}\left(g(t_{n+1}) - g(t_n)\right) f(t_n, y_n) - \frac{1}{2}\left(g(t_n) - g(t_{n-1})\right) f(t_{n-1}, y_{n-1}). \end{aligned} \tag{10.95}$$

Remark 10.1 When $n = 0$, one will notice that on the right-hand side we will have $f(t_{-1}, y_{-1})$ which always caused confusion among young researches. This method is well known to be a two-step approach; therefore, it requires the first and the second step to start the process. The first step is obtained via initial condition $f(t_0, y(t_0))$.

The second is calculated as follows:

$$\begin{aligned}
y_1 &= y(t_0) + \int_{t_0}^{t_1} f(\tau, y(\tau)) g'(t_1) d\tau, \quad (10.96)\\
&= y(t_0) + g'(t_1) f(t_1, \overline{y}_1) h,\\
&= y(t_0) + \frac{g(t_1) - g(t_0)}{h} f(t_1, \overline{y}_1) h,\\
&= y(t_0) + (g(t_1) - g(t_0)) f(t_1, \overline{y}_1),\\
y_1 &= y_0 + (g(t_1) - g(t_0)) f(t_1, \overline{y}_1),
\end{aligned}$$

where

$$\begin{aligned}
\overline{y}_1 &= y_0 + (g(t_1) - g(t_0)) f(t_0, y_0), \quad (10.97)\\
\overline{y}_1 &= y_0 + (g(t_1) - g(t_0)) f(t_1, y_0 + (g(t_1) - g(t_0)) f(t_0, y_0)).
\end{aligned}$$

Therefore, the explicit formula is given as

$$\begin{aligned}
y_{n+1} &= y_n + \frac{3}{2}(g(t_{n+1}) - g(t_n)) f(t_n, y_n) - \frac{1}{2}(g(t_n) - g(t_{n-1})) f(t_{n-1}, y_{n-1}), \text{ if } n > 1 \quad (10.98)\\
y_1 &= y_0 + (g(t_1) - g(t_0)) f(t_1, y_0 + (g(t_1) - g(t_0)) f(t_0, y_0)), \text{ if } n = 1\\
y(t_0) &= y_0;
\end{aligned}$$

when $g(t) = t$ we recover the following:

$$\begin{aligned}
y_{n+1} &= y_n + \frac{3}{2} h f(t_n, y_n) - \frac{h}{2} f(t_{n-1}, y_{n-1}), \text{ if } n > 1 \quad (10.99)\\
y_1 &= y_0 + h f(t_1, y_0 + h f(t_0, y_0)), \text{ if } n = 1\\
y(t_0) &= y_0.
\end{aligned}$$

The accuracy of the proposed or modified version should be tested and compared with the original one. Indeed the main issue with multistep approaches is the initial steps. Now we choose to approximate

$$\begin{aligned}
F(\tau, y(\tau)) &\simeq \frac{\tau - t_n}{h} F(t_{n+1}, y_{n+1}) + \frac{t_{n+1} - \tau}{h} F(t_n, y_n) \quad (10.100)\\
&= \frac{F(t_n, y_n)}{h}(t_{n+1} - \tau) + \frac{F(t_{n+1}, y_{n+1})}{h}(\tau - t_n)
\end{aligned}$$

which is the piecewise linear interpolation. In this case by replacing the approximation into the main equation, we obtain

$$y_{n+1} = y_n + \int_{t_n}^{t_{n+1}} \left(\frac{F(t_n, y_n)}{h}(t_{n+1} - \tau) + \frac{F(t_{n+1}, y_{n+1})}{h}(\tau - t_n) \right) d\tau. \quad (10.101)$$

Noting that

$$\int_{t_n}^{t_{n+1}} (t_{n+1} - \tau)\, d\tau = \frac{h^2}{2}, \tag{10.102}$$

$$\int_{t_n}^{t_{n+1}} (\tau - t_n)\, d\tau = \frac{h^2}{2},$$

and replacing, we obtain

$$y_{n+1} = y_n + \frac{h}{2} F(t_{n+1}, y_{n+1}) + \frac{h}{2} F(t_n, y_n), \tag{10.103}$$

which produces

$$y_{n+1} = y_n + \frac{h}{2} \frac{(g(t_{n+1}) - g(t_n))}{h} f(t_{n+1}, y_{n+1}) + \frac{h}{2} \frac{g(t_n) - g(t_{n-1})}{h} f(t_n, y_n) \tag{10.104}$$

which produces an implicit scheme; to make it explicit, we reformulate

$$\overline{y}_{n+1} = y_n + (g(t_{n+1}) - g(t_n))\, f(t_n, y_n). \tag{10.105}$$

Therefore, the explicit scheme is given as

$$y_{n+1} = y_n + \frac{1}{2} (g(t_{n+1}) - g(t_n)) \left[f(t_{n+1}, y_n + (g(t_{n+1}) - g(t_n))\, f(t_n, y_n)) + f(t_n, y_n) \right]. \tag{10.106}$$

For this case theoretical analysis will be presented.

10.6 Applying Newton Polynomial Method on IVP with Classical Global Derivative

Here we aim to derive a numerical solution of the general Cauchy problem with global derivative using the Newton interpolation:

$$\begin{cases} D_g y(t) = f(t, y(t)), & if\ t > 0 \\ y(0) = y_0. \end{cases} \tag{10.107}$$

We apply the global integral on the above to obtain

$$y(t) = y(0) + \int_0^t f(\tau, y(\tau))g^{'}(\tau)d\tau, \tag{10.108}$$

$$y(t_{n+1}) = y(0) + \int_0^{t_{n+1}} f(\tau, y(\tau))g^{'}(\tau)d\tau,$$

$$y(t_n) = y(0) + \int_0^{t_n} f(\tau, y(\tau))g^{'}(\tau)d\tau.$$

$$y(t_{n+1}) - y(t_n) = \int_{t_n}^{t_{n+1}} f(\tau, y(\tau))g^{'}(\tau)d\tau, \tag{10.109}$$

$$y(t_{n+1}) = y(t_n) + \int_{t_n}^{t_{n+1}} f(\tau, y(\tau))g^{'}(\tau)d\tau.$$

We let

$$F(t, y(t)) = f(\tau, y(\tau))g^{'}(\tau). \tag{10.110}$$

$$y(t_{n+1}) = y(t_n) + \int_{t_n}^{t_{n+1}} F(\tau, y(\tau))d\tau; \tag{10.111}$$

within $[t_n, t_{n+1}]$ we approximate

$$\begin{aligned} F(\tau, y(\tau)) &\simeq P_n(\tau) = F(t_{n-2}, y_{n-2}) + \frac{F(t_{n-1}, y_{n-1}) - F(t_{n-2}, y_{n-2})}{\Delta t}(\tau - t_{n-2}) \\ &+ \frac{F(t_n, y_n) - 2F(t_{n-1}, y_{n-1}) - F(t_{n-2}, y_{n-2})}{2\Delta t^2}(\tau - t_{n-2})(\tau - t_{n-1}). \end{aligned} \tag{10.112}$$

Therefore

$$y(t_{n+1}) \simeq y(t_n) + \int_{t_n}^{t_{n+1}} P_n(\tau)d\tau \tag{10.113}$$

and integrating yields

$$y_{n+1} = y_n + \frac{23}{12}F(t_n, y_n)\Delta t - \frac{4}{3}F(t_{n-1}, y_{n-1})\Delta t + \frac{5}{12}F(t_{n-2}, y_{n-2})\Delta t. \tag{10.114}$$

This is to say

$$\begin{aligned} y_{n+1} &= y_n + \frac{23}{12}\frac{g(t_{n+1}) - g(t_n)}{\Delta t} f(t_n, y_n) \triangle t \\ &-\frac{4}{3}\frac{g(t_n) - g(t_{n-1})}{\Delta t} f(t_{n-1}, y_{n-1}) \triangle t \\ &+\frac{5}{12}\frac{g(t_{n-2}) - g(t_{n-2})}{\Delta t} f(t_{n-2}, y_{n-2}) \triangle t \end{aligned} \tag{10.115}$$

and simplifying

$$\begin{aligned} y_{n+1} &= y_n + \frac{23}{12} g(t_{n+1}) - g(t_n) f(t_n, y_n) \\ &-\frac{4}{3} g(t_n) - g(t_{n-1}) f(t_{n-1}, y_{n-1}) \\ &+\frac{5}{12} g(t_{n-2}) - g(t_{n-2}) f(t_{n-2}, y_{n-2}). \end{aligned} \tag{10.116}$$

Note that if $g(t) = t$, we recover

$$y_{n+1} = y_n + \frac{23}{12} h f(t_n, y_n) - \frac{4}{3} h f(t_{n-1}, y_{n-1}) + \frac{5}{12} h f(t_{n-2}, y_{n-2}). \tag{10.117}$$

The recursive formula presented above will be employed to obtain numerical solutions for the general Cauchy problem, utilizing classical global derivatives. This approach aims to provide a robust method for solving complex differential equations by leveraging recursive techniques to iteratively refine the solution, ensuring accuracy and efficiency in the numerical analysis of the problem at hand. Let us perturb the formula by introducing perturbed terms $\widetilde{y}_0, \widetilde{y}_1, \ldots, \widetilde{y}_{n+1}$ as

$$\begin{aligned} y_{n+1} + \widetilde{y}_{n+1} &= y_n + \widetilde{y}_n + \frac{23}{12} \left(g(t_{n+1}) - g(t_n) \right) f(t_n, y_n + \widetilde{y}_n) \\ &-\frac{4}{3} \left(g(t_n) - g(t_{n-1}) \right) f(t_{n-1}, y_{n-1} + \widetilde{y}_{n-1}) \\ &+\frac{5}{12} \left(g(t_{n-2}) - g(t_{n-2}) \right) f(t_{n-2}, y_{n-2} + \widetilde{y}_{n-2}). \end{aligned} \tag{10.118}$$

$$\begin{aligned} y_{n+1} + \widetilde{y}_{n+1} - y_{n+1} &= \widetilde{y}_{n+1} = \widetilde{y}_n + \frac{23}{12} \left(g(t_{n+1}) - g(t_n) \right) \left(f(t_n, y_n + \widetilde{y}_n) - f(t_n, y_n) \right) \\ &-\frac{4}{3} \left(g(t_n) - g(t_{n-1}) \right) \left(f(t_{n-1}, y_{n-1} + \widetilde{y}_{n-1}) - f(t_{n-1}, y_{n-1}) \right) \\ &+\frac{5}{12} \left(g(t_{n-2}) - g(t_{n-2}) \right) \left(f(t_{n-2}, y_{n-2} + \widetilde{y}_{n-2}) - f(t_{n-2}, y_{n-2}) \right). \end{aligned} \tag{10.119}$$

$$\begin{aligned}|\widetilde{y}_{n+1}| &= |\widetilde{y}_n| + \frac{23}{12}\left(g(t_{n+1}) - g(t_n)\right)|f(t_n, y_n + \widetilde{y}_n) - f(t_n, y_n)| \\ &+ \frac{4}{3}\left(g(t_n) - g(t_{n-1})\right)|f(t_{n-1}, y_{n-1} + \widetilde{y}_{n-1}) - f(t_{n-1}, y_{n-1})| \\ &+ \frac{5}{12}\left(g(t_{n-2}) - g(t_{n-2})\right)|f(t_{n-2}, y_{n-2} + \widetilde{y}_{n-2}) - f(t_{n-2}, y_{n-2})|.\end{aligned} \tag{10.120}$$

Using the Lipschitz condition of $f()$ with respect to the second variable yields

$$\begin{aligned}|\widetilde{y}_{n+1}| &= |\widetilde{y}_n| + \frac{23}{12}\left(g(t_{n+1}) - g(t_n)\right)|\widetilde{y}_n| \\ &+ \frac{4}{3}\left(g(t_n) - g(t_{n-1})\right)|\widetilde{y}_{n-1}| \\ &+ \frac{5}{12}\left(g(t_{n-2}) - g(t_{n-2})\right)|\widetilde{y}_{n-2}|.\end{aligned} \tag{10.121}$$

Let

$$G_n = \max_{1\leq n\leq N}\left\{\left(g(t_{n+1}) - g(t_n)\right), \left(g(t_n) - g(t_{n-1})\right), \left(g(t_{n-2}) - g(t_{n-2})\right)\right\} \tag{10.122}$$

$$|\widetilde{y}_{n+1}| = |\widetilde{y}_n| + G_n\left\{\frac{23}{12}|\widetilde{y}_n| + \frac{5}{12}|\widetilde{y}_{n-2}| + \frac{4}{3}|\widetilde{y}_{n-1}|\right\}. \tag{10.123}$$

We let

$$\widetilde{Z}_n = \max_{1\leq n\leq N}\left\{|\widetilde{y}_n|, |\widetilde{y}_{n-2}|, |\widetilde{y}_{n-1}|\right\} \tag{10.124}$$

$$\begin{aligned}|\widetilde{y}_{n+1}| &< \widetilde{Z}_n\left\{1 + \frac{11}{12}G_n\right\} \\ &< \widetilde{Z}_n\left\{1 + \frac{11}{3}G_n\right\}.\end{aligned} \tag{10.125}$$

If $g(t) = t$, we have

$$|\widetilde{y}_{n+1}| < \widetilde{Z}_n\left\{1 + \frac{11h}{3}\right\}. \tag{10.126}$$

$$\lim_{h\to 0}|\widetilde{y}_{n+1}| < \widetilde{Z}_n\left\{1 + \frac{11h}{3}\right\} = \widetilde{Z}_n. \tag{10.127}$$

This produces a satisfactory result.

We next evaluate

$$|y(t_{n+1}) - y_{n+1}|. \tag{10.128}$$

$$y(t_{n+1}) = y(t_n) + \int_{t_n}^{t_{n+1}} f(\tau, y(\tau))g'(\tau)d\tau, \tag{10.129}$$

$$y(t_{n+1}) = y(t_n) + \int_{t_n}^{t_{n+1}} \left[\begin{array}{c} \frac{g(t_{n-1})-g(t_{n-2})}{h} f(t_{n-2}, y_{n-2}) \\ + \frac{g(t_n)-g(t_{n-1})}{h} \frac{f(t_{n-1},y_{n-1})-f(t_{n-2},y_{n-2})}{h}(\tau - t_{n-2}) \\ + \frac{g(t_{n-1})-g(t_{n-2})}{h} \frac{f(t_n,y_n)-2f(t_{n-1},y_{n-1})+f(t_{n-2},y_{n-2})}{2h^2}(\tau - t_{n-2})(\tau - t_{n-1}) \end{array} \right] d\tau,$$

$$\begin{aligned} |y_{n+1} - y(t_{n+1})| &\leq |y(t_n) - y_n| + \int_{t_n}^{t_{n+1}} \left|g'(\tau)\right| |f(t_{n-2}, y_{n-2}) - f(\tau, y(\tau))| \, d\tau \\ &+ \int_{t_n}^{t_{n+1}} \left|g'(\tau)\right| \frac{f(t_{n-1}, y_{n-1}) - f(t_{n-2}, y_{n-2})}{h}(\tau - t_{n-2})d\tau \\ &+ \int_{t_n}^{t_{n+1}} \left|g'(\tau)\right| \frac{f(t_n, y_n) - f(t_n, y_{n-1})}{2h^2}(\tau - t_{n-2})(\tau - t_{n-1})d\tau \\ &+ \int_{t_n}^{t_{n+1}} \left|g'(\tau)\right| \frac{f(t_{n-1}, y_{n-1}) - f(t_{n-2}, y_{n-2})}{2h^2}(\tau - t_{n-2})(\tau - t_{n-1})d\tau. \end{aligned} \tag{10.130}$$

Here we have that $f(t, y(t))$ is differentiable by the Mean Value theorem; we have

$$\begin{aligned} \left|y_{n+1} - y(t_{n+1})\right| &\leq |y(t_n) - y_n| + \int_{t_n}^{t_{n+1}} g'(\tau) \left|f'(a_1, y(a_1))\right| (\tau - t_{n-2})d\tau \\ &+ \int_{t_n}^{t_{n+1}} g'(\tau) \left|f'(a_2, y(a_2))\right| (\tau - t_{n-2})d\tau \\ &+ \int_{t_n}^{t_{n+1}} \frac{g'(\tau)}{2h} \left|f'(a_3, y(a_3))\right| (\tau - t_{n-1})(\tau - t_{n-2})d\tau \\ &+ \int_{t_n}^{t_{n+1}} \frac{g'(\tau)}{2h} \left|f'(a_4, y(a_4))\right| (\tau - t_{n-1})(\tau - t_{n-2})d\tau \\ &\leq |y_n - y(t_n)| + \left\|g'\right\|_\infty \left|f'(a, y(a))\right| \int_{t_n}^{t_{n+1}} (\tau - t_{n-2})d\tau \\ &+ \left\|g'\right\|_\infty \left|f'(a_1, y(a_1))\right| \int_{t_n}^{t_{n+1}} (\tau - t_{n-2})d\tau \\ &+ \left(\frac{\left\|g'\right\|_\infty}{2h} \left|f'(a_3, y(a_3))\right| + \frac{\left\|g'\right\|_\infty}{2h} \left|f'(a_4, y(a_4))\right| \right) \end{aligned} \tag{10.131}$$

$$\int_{t_n}^{t_{n+1}} (\tau - t_{n-1})(\tau - t_{n-2})d\tau$$

$$\leq |y(t_n) - y_n| + \frac{5h^2}{2} \left\| g' \right\|_\infty \left(\left| f'(a, y(a)) + f'(a_1, y(a_1)) \right| \right)$$
$$+ \frac{23h^2}{6} \left\| g' \right\|_\infty \left(\left| f'(a_3, y(a_3)) + f'(a_4, y(a_4)) \right| \right),$$

$$\lim_{h \to 0} |y(t_{n+1}) - y_{n+1}| < \lim_{h \to 0} |y(t_n) - y_n|, \tag{10.132}$$

$$\lim_{h \to 0} \left| \frac{y(t_{n+1}) - y_{n+1}}{y(t_n) - y_n} \right| < 1. \tag{10.133}$$

$$\lim_{\substack{h \to 0 \\ n \geq 1}} |y_{n+1} - y(t_{n+1})| \leq q^n |y_1 - y(0)|, q < 1 \tag{10.134}$$
$$\leq q^n |hf(t_0, y(0))|, q < 1$$

$$\lim_{\substack{h \to 0 \\ n \to \infty}} |y_{n+1} - y(t_{n+1})| = 0. \tag{10.135}$$

References

1. Ascher, U.M., Petzold, L.R.: Computer Methods for Ordinary Differential Equations and Differential-Algebraic Equations. Society for Industrial and Applied Mathematics, Philadelphia (1998). 978-0-89871-412-8
2. Butcher, J.C.: Numerical Methods for Ordinary Differential Equations. Wiley, New York (2003)978-0-471-96758-3
3. Hairer, E., Nørsett, S.P., Wanner, G.: Solving ordinary differential equations I: Nonstiff problems. Springer, Berlin, New York (1993). 978-3-540-56670-0
4. Stoer, J., Bulirsch, R.: Introduction to Numerical Analysis (3rd ed.). Springer, Berlin, New York (2002). ISBN 978-0-387-95452-3
5. Baleanu, D., Jajarmi, A., Hajipour, M.: On the nonlinear dynamical systems within the generalized fractional derivatives with Mittag–Leffler kernel. Nonlinear Dyn. (2018)
6. Baleanu, Dumitru, Jajarmi, Amin, Hajipour, Mojtaba: On the nonlinear dynamical systems within the generalized fractional derivatives with Mittag-Leffler kernel. Nonlinear Dyn. **94**, 397–414 (2018)
7. Leader, J.J.: Numerical Analysis and Scientific Computation. Addison-Wesley, Boston (2004)0-201-73499-0
8. Griffiths, D.V., Smith, I.M.: Numerical Methods for Engineers: A Programming Approach. CRC Press, Boca Raton, p. 218 (1991). ISBN 0-8493-8610-1
9. Meijering, Erik: A chronology of interpolation: from ancient astronomy to modern signal and image processing. Proc. IEEE **90**(3), 319–342 (2002)

10. Lagrange, J.-L.: Leçon Cinquième. Sur l'usage des courbes dans la solution des problèmes". Leçons Elémentaires sur les Mathématiques (in French). Paris. Republished In: Joseph-Alfred, S. (ed.) (1877). Oeuvres de Lagrange, vol. 7, pp. 271–287. Gauthier-Villars (1795)
11. Toufik, M., Atangana, A.: New numerical approximation of fractional derivative with non-local and non-singular kernel: application to chaotic models. Eur. Phys. J. **132**(10), 444 (2017)
12. Atangana, A., Araz, S.İ.: New Numerical Scheme with Newton Polynomial: Theory, Methods, and Applications. Academic Press (2021)

Chapter 11
Numerical Analysis of IVP with Riemann–Liouville Global Derivative

11.1 Applying Euler Method on IVP with Riemann–Liouville Global Derivative

In this section, we examine and discuss the following initial value problem (IVP), providing a detailed analysis of its formulation and the underlying assumptions. The problem presented will serve as the foundation for the upcoming solutions and insights.

$$\begin{cases} {}_{t_0}^{RL}D_g^{\alpha} y(t) = f(t, y(t)) & \text{if } t > 0 \\ y(t_0) = y_0 & \text{if } t = 0 \end{cases}. \tag{11.1}$$

Then we have

$$\begin{cases} y(t) = \frac{1}{\Gamma(\alpha)} \int\limits_{t_0}^{t} g'(\tau) f(\tau, y(\tau))(t-\tau)^{\alpha-1} d\tau, \\ y(t_0) = y_0. \end{cases} \tag{11.2}$$

At $t = t_{n+1}$, then we have

$$\begin{aligned} y(t_{n+1}) &= \frac{1}{\Gamma(\alpha)} \int\limits_{t_0}^{t_{n+1}} g'(\tau) f(\tau, y(\tau))(t_{n+1}-\tau)^{\alpha-1} d\tau, \\ &= \frac{1}{\Gamma(\alpha)} \sum_{j=0}^{n} \int\limits_{t_j}^{t_{j+1}} g'(\tau) f(\tau, y(\tau))(t_{n+1}-\tau)^{\alpha-1} d\tau; \end{aligned} \tag{11.3}$$

with the Euler approximation of

$$g'(\tau) f(\tau, y(\tau)), \tag{11.4}$$

A. Atangana and İ. Koca, *Fractional Differential and Integral Operators with Respect to a Function*, Industrial and Applied Mathematics,
https://doi.org/10.1007/978-981-97-9951-0_11

within $[t_j, t_{j+1}]$ we have

$$y_{n+1} = \frac{1}{\Gamma(\alpha)} \sum_{j=0}^{n} \int_{t_j}^{t_{j+1}} g'(t_j)(t_{n+1}-\tau)^{\alpha-1} f(t_j, y(t_j)) d\tau, \tag{11.5}$$

$$= \frac{1}{\Gamma(\alpha)} \sum_{j=0}^{n} \frac{g(t_j)-g(t_{j-1})}{h} f(t_j, y(t_j)) \int_{t_j}^{t_{j+1}} (t_{n+1}-\tau)^{\alpha-1} d\tau,$$

$$= \frac{1}{\Gamma(\alpha)} \sum_{j=0}^{n} \frac{g(t_j)-g(t_{j-1})}{h} f(t_j, y(t_j)) \frac{h^{\alpha}}{\alpha} \left((n-j+1)^{\alpha} - (n-j)^{\alpha}\right),$$

$$= \frac{h^{\alpha-1}}{\Gamma(\alpha+1)} \sum_{j=0}^{n} \left(g(t_j)-g(t_{j-1})\right) f(t_j, y(t_j)) \left((n-j+1)^{\alpha} - (n-j)^{\alpha}\right).$$

We consider the second version. We start when $t = t_1$,

$$y_1 = \frac{1}{\Gamma(\alpha)} \int_{t_0}^{t_1} g'(\tau) f(\tau, y(\tau))(t_1-\tau)^{\alpha-1} d\tau, \tag{11.6}$$

$$= \frac{1}{\Gamma(\alpha)} \int_{t_0}^{t_1} \frac{g(t_1)-g(t_0)}{h} f(t_0, y(t_0))(t_1-\tau)^{\alpha-1} d\tau,$$

$$= \frac{1}{\Gamma(\alpha)} \frac{g(t_1)-g(t_0)}{h} f(t_0, y(t_0)) \int_{t_0}^{t_1} (t_1-\tau)^{\alpha-1} d\tau,$$

$$y_1 = \frac{h^{\alpha-1}}{\Gamma(\alpha+1)} f(t_0, y_0) \left(g(t_1)-g(t_0)\right).$$

$\forall n \geq 1$ we have that

$$y(t_{n+1}) = \frac{1}{\Gamma(\alpha)} \int_{t_1}^{t_{n+1}} g'(\tau) f(\tau, y(\tau))(t_{n+1}-\tau)^{\alpha-1} d\tau, \tag{11.7}$$

$$= \frac{1}{\Gamma(\alpha)} \sum_{j=1}^{n} \int_{t_j}^{t_{j+1}} g'(\tau) f(\tau, y(\tau))(t_{n+1}-\tau)^{\alpha-1} d\tau,$$

$$= \frac{1}{\Gamma(\alpha)} \sum_{j=1}^{n} \frac{g(t_{j+1})-g(t_j)}{h} f(t_{j+1}, y(t_{j+1})) \int_{t_j}^{t_{j+1}} (t_{n+1}-\tau)^{\alpha-1} d\tau,$$

$$= \frac{h^{\alpha-1}}{\Gamma(\alpha+1)} \sum_{j=1}^{n} \left(g(t_{j+1})-g(t_j)\right) f(t_{j+1}, y(t_{j+1})) \left((n-j+1)^{\alpha} - (n-j)^{\alpha}\right),$$

$$y_{n+1} = \frac{h^{\alpha-1}}{\Gamma(\alpha+1)} \sum_{j=1}^{n-1} \left(g(t_{j+1}) - g(t_j)\right) f(t_{j+1}, y(t_{j+1})) \left((n-j+1)^{\alpha} - (n-j)^{\alpha}\right)$$
$$+ \frac{h^{\alpha-1}}{\Gamma(\alpha+1)} f(t_{n+1}, \overline{y}_{n+1}),$$

where

$$\overline{y}_{n+1} = \frac{h^{\alpha-1}}{\Gamma(\alpha+1)} \sum_{j=0}^{n} \left(g(t_{j+1}) - g(t_j)\right) f(t_j, y_j) \left((n-j+1)^{\alpha} - (n-j)^{\alpha}\right). \tag{11.8}$$

11.2 Applying Heun's Method on IVP with Riemann–Liouville Global Derivative

In the study, we consider using Heun's numerical method to solve the above-mentioned nonlinear equation involving the Riemann-Liouville global derivative. By limiting ourselves to cases where we can still make relevant approximations with regards to our specific system we can end up with a very efficient process to obtain a solution to our problem. Applying the Riemann-Liouville derivative provides a better description of fractional-order dynamics by reflecting the complexity of the equation's original framework.

$$\begin{cases} {}^{RL}_{t_0}D^{\alpha}_{g} y(t) = f(t, y(t)) & \text{if } t > 0 \\ y(t_0) = y_0 & \text{if } t = 0 \end{cases}. \tag{11.9}$$

Here we assume that $g'(t) > 0$. Also here we will need in addition that $f(t, y(t))$ satisfies the Lipschitz condition and is bounded within $[a, b]$. By the fundamental theorem of calculus, the above IVP can be transformed as

$$y(t) = \frac{1}{\Gamma(\alpha)} \int_{t_0}^{t} g'(\tau)(t-\tau)^{\alpha-1} f(\tau, y(\tau)) d\tau, \tag{11.10}$$
$$y(t_0) = y_0.$$

Here we only consider at $t = t_{n+1}$, then we have

$$y(t_{n+1}) = \frac{1}{\Gamma(\alpha)} \int_{t_0}^{t_{n+1}} g'(\tau)(t_{n+1}-\tau)^{\alpha-1} f(\tau, y(\tau)) d\tau, \tag{11.11}$$

$$= \sum_{j=0}^{n} \int_{t_j}^{t_{j+1}} g'(\tau) \frac{(t_{n+1} - \tau)^{\alpha-1}}{\Gamma(\alpha)} f(\tau, y(\tau)) d\tau.$$

Here we approximate within $[t_j, t_{j+1}]$ using

$$\begin{aligned} g'(\tau) f(\tau, y(\tau)) &\simeq \frac{g'(t_{j+1}) f(t_{j+1}, y_{j+1}) + g'(t_j) f(t_j, y_j)}{2}, \\ &\simeq \frac{g(t_{j+1}) - g(t_j)}{2h} f(t_{j+1}, y_{j+1}) + \frac{g(t_j) - g(t_{j-1})}{2h} f(t_j, y_j), \end{aligned} \tag{11.12}$$

and by replacing we obtain

$$\begin{aligned} y_{n+1} &= y_0 + \frac{1}{2\Gamma(\alpha) h} \sum_{j=1}^{n} \begin{pmatrix} (g(t_{j+1}) - g(t_j)) f(t_{j+1}, y_{j+1}) \\ + (g(t_j) - g(t_{j-1})) f(t_j, y_j) \end{pmatrix} \int_{t_j}^{t_{j+1}} (t_{n+1} - \tau)^{\alpha-1} d\tau, \\ &= y_0 + \frac{1}{2\Gamma(\alpha) h} \sum_{j=1}^{n} \begin{pmatrix} (g(t_{j+1}) - g(t_j)) f(t_{j+1}, y_{j+1}) \\ + (g(t_j) - g(t_{j-1})) f(t_j, y_j) \end{pmatrix} \left(\frac{h^\alpha}{\alpha} ((n - j + 1)^\alpha - (n - j)^\alpha) \right), \\ &= y_0 + \frac{h^{\alpha-1}}{2\Gamma(\alpha+1)} \sum_{j=1}^{n} \begin{pmatrix} (g(t_{j+1}) - g(t_j)) f(t_{j+1}, y_{j+1}) \\ + (g(t_j) - g(t_{j-1})) f(t_j, y_j) \end{pmatrix} ((n - j + 1)^\alpha - (n - j)^\alpha). \end{aligned} \tag{11.13}$$

When $j = n + 1$ we have an implicit function; therefore, we reformulated the formula as

$$\begin{aligned} y_{n+1} &= \frac{h^{\alpha-1}}{2\Gamma(\alpha+1)} \sum_{j=1}^{n-1} \begin{pmatrix} (g(t_{j+1}) - g(t_j)) f(t_{j+1}, y_{j+1}) \\ + (g(t_j) - g(t_{j-1})) f(t_j, y_j) \end{pmatrix} \begin{pmatrix} (n - j + 1)^\alpha \\ -(n - j)^\alpha \end{pmatrix} \\ &+ \frac{h^{\alpha-1}}{2\Gamma(\alpha+1)} \begin{pmatrix} (g(t_{n+1}) - g(t_n)) f(t_{n+1}, \overline{y}_{n+1}) \\ + (g(t_n) - g(t_{n-1})) f(t_n, y_n) \end{pmatrix}, \end{aligned} \tag{11.14}$$

where

$$\overline{y}_{n+1} = \frac{h^{\alpha-1}}{2\Gamma(\alpha+1)} \sum_{j=1}^{n-1} \begin{pmatrix} (g(t_{j+1}) - g(t_j)) f(t_{j+1}, y_{j+1}) \\ + (g(t_j) - g(t_{j-1})) f(t_j, y_j) \end{pmatrix} \begin{pmatrix} (n - j + 1)^\alpha \\ -(n - j)^\alpha \end{pmatrix}. \tag{11.15}$$

Theorem 11.1 *Let $\widetilde{y}_n$ be the perturbed term of y_n, then $\forall n \geq 0$, $\exists c > 0$ such that*

$$|\widetilde{y}_{n+1}| < c\, |\widetilde{y}_0|\,. \tag{11.16}$$

Proof Let us put our iteration equation below

$$y_{n+1} = \frac{h^{\alpha-1}}{2\Gamma(\alpha+1)} \sum_{j=1}^{n-1} \begin{pmatrix} \left(g(t_{j+1}) - g(t_j)\right) f(t_{j+1}, y_{j+1}) \\ + \left(g(t_j) - g(t_{j-1})\right) f(t_j, y_j) \end{pmatrix} \tag{11.17}$$
$$\times \left((n-j+1)^{\alpha} - (n-j)^{\alpha}\right)$$
$$+ \frac{h^{\alpha-1}}{2\Gamma(\alpha+1)} \begin{pmatrix} (g(t_{n+1}) - g(t_n)) f(t_{n+1}, \overline{y}_{n+1}) \\ + (g(t_n) - g(t_{n-1})) f(t_n, y_n) \end{pmatrix};$$

with perturbed term

$$y_{n+1} + \widetilde{y}_{n+1} = \frac{h^{\alpha-1}}{2\Gamma(\alpha+1)} \sum_{j=1}^{n-1} \begin{pmatrix} \left(g(t_{j+1}) - g(t_j)\right) f(t_{j+1}, y_{j+1} + \widetilde{y}_{j+1}) \\ + \left(g(t_j) - g(t_{j-1})\right) f(t_j, y_j + \widetilde{y}_j) \end{pmatrix} \tag{11.18}$$
$$\times \left((n-j+1)^{\alpha} - (n-j)^{\alpha}\right)$$
$$+ \frac{h^{\alpha-1}}{2\Gamma(\alpha+1)} \begin{pmatrix} (g(t_{n+1}) - g(t_n)) f(t_{n+1}, \overline{y}_{n+1} + \widetilde{\overline{y}}_{n+1}) \\ + (g(t_n) - g(t_{n-1})) f(t_n, y_n + \widetilde{y}_n) \end{pmatrix} + \widetilde{y}(t_0) + y(t_0).$$

Thus we have that

$$|\widetilde{y}_{n+1}| = \left| \frac{h^{\alpha-1}}{2\Gamma(\alpha+1)} \left(\sum_{j=1}^{n-1} \begin{matrix} (g(t_{j+1}) - g(t_j)) \left(f(t_{j+1}, y_{j+1} + \widetilde{y}_{j+1}) - f(t_{j+1}, y_{j+1})\right) \\ + (g(t_j) - g(t_{j-1})) \left(f(t_j, y_j + \widetilde{y}_j) - f(t_j, y_j)\right) \end{matrix} \right) \delta_{n,j}^{\alpha} \right| \tag{11.19}$$
$$+ \left| \frac{h^{\alpha-1}}{2\Gamma(\alpha+1)} \begin{pmatrix} (g(t_{n+1}) - g(t_n)) \left(f(t_{n+1}, \overline{y}_{n+1} + \widetilde{\overline{y}}_{n+1}) - f(t_{n+1}, \overline{y}_{n+1})\right) \\ + (g(t_n) - g(t_{n-1})) \left(f(t_n, y_n + \widetilde{y}_n) - f(t_n, y_n)\right) \end{pmatrix} \right|$$
$$+ |\widetilde{y}(t_0)|.$$

Using the properties of inequality, we get

$$|\widetilde{y}_{n+1}| < \frac{h^{\alpha-1}}{2\Gamma(\alpha+1)} \sum_{j=0}^{n-1} \begin{pmatrix} h\left|g'(t_{j+1})\right| \left|f(t_{j+1}, y_{j+1} + \widetilde{y}_{j+1}) - f(t_{j+1}, y_{j+1})\right| \\ + h\left|g'(t_j)\right| \left|f(t_j, y_j + \widetilde{y}_j) - f(t_j, y_j)\right| \end{pmatrix} \delta_{n,j}^{\alpha} \tag{11.20}$$
$$+ \frac{h^{\alpha-1}}{2\Gamma(\alpha+1)} \begin{pmatrix} h\left|g'(t_{n+1})\right| \left|f(t_{n+1}, \overline{y}_{n+1} + \widetilde{\overline{y}}_{n+1}) - f(t_{n+1}, \overline{y}_{n+1})\right| \\ + h\left|g'(t_n)\right| \left|f(t_n, y_n + \widetilde{y}_n) - f(t_n, y_n)\right| \end{pmatrix}$$
$$+ |\widetilde{y}(t_0)|.$$

Using the Lipschitz condition of $f(., y(.))$ with respect to y, we get

$$|\widetilde{y}_{n+1}| < \frac{h^{\alpha-1}}{2\Gamma(\alpha+1)} \sum_{j=0}^{n-1} h \left\|g'\right\|_{\infty} \left(L\left|\widetilde{y}_{j+1}\right| + L\left|\widetilde{y}_j\right|\right) \delta_{n,j}^{\alpha} \tag{11.21}$$
$$+ \frac{h^{\alpha-1}}{2\Gamma(\alpha+1)} h \left\|g'\right\|_{\infty} \left(L\left|\widetilde{\overline{y}}_{n+1}\right| + L\left|\widetilde{y}_n\right|\right) + |\widetilde{y}(t_0)|,$$
$$\leq \frac{h^{\alpha}\left\|g'\right\|_{\infty} L}{2\Gamma(\alpha+1)} \left(\sum_{j=0}^{n-1} \left\{\left|\widetilde{y}_{j+1}\right| + \left|\widetilde{y}_j\right|\right\} \delta_{n,j}^{\alpha} + \left|\widetilde{\overline{y}}_{n+1}\right| + |\widetilde{y}_n| \right) + |\widetilde{y}(t_0)|.$$

We have that

$$\left|\widetilde{y}_{n+1}\right| \leq |\widetilde{y}(t_0)| + \frac{h^\alpha \left\|g'\right\|_\infty L}{\Gamma(\alpha+1)} \sum_{j=0}^{n} \left|\widetilde{y}_j\right| \delta_{n,j}^\alpha. \tag{11.22}$$

Therefore replacing yields

$$\begin{aligned} |\widetilde{y}_{n+1}| \leq\ & \frac{h^\alpha \left\|g'\right\|_\infty L}{2\Gamma(\alpha+1)} \left(\sum_{j=0}^{n-1} \left(\left|\widetilde{y}_{j+1}\right| + \left|\widetilde{y}_j\right|\right) \delta_{n,j}^\alpha \right) \\ & + \frac{h^\alpha \left\|g'\right\|_\infty L}{2\Gamma(\alpha+1)} \left(|\widetilde{y}_0| + \frac{h^\alpha \left\|g'\right\|_\infty L}{\Gamma(\alpha+1)} \sum_{j=0}^{n} \left|\widetilde{y}_j\right| \delta_{n,j}^\alpha \right) \\ & + \frac{h^\alpha \left\|g'\right\|_\infty L}{2\Gamma(\alpha+1)} |\widetilde{y}_n| + |\widetilde{y}_0|. \end{aligned} \tag{11.23}$$

$$\begin{aligned} y(t_1) =\ & y(t_0) + \frac{1}{\Gamma(\alpha)} \int_{t_0}^{t_1} (t_1 - \tau)^{\alpha-1} f(\tau, y(\tau) d\tau, \\ \simeq\ & y(t_0) + \frac{1}{\Gamma(\alpha)} \int_{t_0}^{t_1} (t_1 - \tau)^{\alpha-1} f(t_0, y(t_0) d\tau, \\ \simeq\ & y(t_0) + \frac{f(t_0, y_0)}{\Gamma(\alpha)} \int_{t_0}^{t_1} (t_1 - \tau)^{\alpha-1} d\tau, \\ \simeq\ & y(t_0) + \frac{f(t_0, y_0)}{\Gamma(\alpha)} \frac{(t_1 - t_0)^\alpha}{\alpha}, \\ \simeq\ & y(t_0) + \frac{f(t_0, y_0) h^\alpha}{\Gamma(\alpha+1)}. \end{aligned} \tag{11.24}$$

$$\begin{aligned} y_1 =\ & y_0 + \frac{f(t_0, y_0) h^\alpha}{\Gamma(\alpha+1)}, \\ y_1 + \widetilde{y}_1 =\ & y_0 + \widetilde{y}_0 + \frac{f(t_0, y_0 + \widetilde{y}_0) h^\alpha}{\Gamma(\alpha+1)}, \\ \widetilde{y}_1 =\ & \widetilde{y}_0 + \frac{h^\alpha}{\Gamma(\alpha+1)} \left(f(t_0, y_0 + \widetilde{y}_0) - f(t_0, y_0)\right), \\ |\widetilde{y}_1| \leq\ & |\widetilde{y}_0| + \frac{h^\alpha}{\Gamma(\alpha+1)} L |\widetilde{y}_0| = |\widetilde{y}_0| \left(1 + \frac{h^\alpha}{\Gamma(\alpha+1)} L\right), \\ |\widetilde{y}_2| \leq\ & \frac{h^\alpha \left\|g'\right\|_\infty L}{2\Gamma(\alpha+1)} \sum_{j=0}^{n-1} \left(\left|\widetilde{y}_{j+1}\right| + \left|\widetilde{y}_j\right| \delta_{n,j}^\alpha\right) \end{aligned} \tag{11.25}$$

$$+\frac{h^{\alpha}\left\|g'\right\|_{\infty}L}{2\Gamma\left(\alpha+1\right)}\left(\left|\widetilde{y}_{0}\right|+\frac{h^{\alpha}\left\|g'\right\|_{\infty}L}{\Gamma\left(\alpha+1\right)}\sum_{j=0}^{n}\left|\widetilde{y}_{j}\right|\delta_{n,j}^{\alpha}\right)$$
$$+\frac{h^{\alpha}\left\|g'\right\|_{\infty}L\left|\widetilde{y}_{1}\right|}{2\Gamma\left(\alpha+1\right)}+\left|\widetilde{y}_{0}\right|.$$

$$\begin{aligned}\left|\widetilde{y}_{2}\right| = & \frac{h^{\alpha}\left\|g'\right\|_{\infty}L}{2\Gamma\left(\alpha+1\right)}\left(\left|\widetilde{y}_{1}\right|+\left|\widetilde{y}_{0}\right|\right)\delta_{1,0}^{\alpha} \\ & +\frac{h^{\alpha}\left\|g'\right\|_{\infty}L}{2\Gamma\left(\alpha+1\right)}\left(\left|\widetilde{y}_{0}\right|+\frac{h^{\alpha}\left\|g'\right\|_{\infty}L}{\Gamma\left(\alpha+1\right)}\left(\left|\widetilde{y}_{0}\right|\delta_{1,0}^{\alpha}+\left|\widetilde{y}_{1}\right|\delta_{1,1}^{\alpha}\right)\right) \\ & +\frac{h^{\alpha}\left\|g'\right\|_{\infty}L}{2\Gamma\left(\alpha+1\right)}\left|\widetilde{y}_{1}\right|+\left|\widetilde{y}_{0}\right|,\end{aligned} \tag{11.26}$$

$$\begin{aligned}\left|\widetilde{y}_{2}\right| = & \frac{h^{\alpha}\left\|g'\right\|_{\infty}L}{2\Gamma\left(\alpha+1\right)}\left(\left|\widetilde{y}_{1}\right|+\left|\widetilde{y}_{0}\right|\right)\delta_{1,0}^{\alpha} \\ & +\frac{h^{\alpha}\left\|g'\right\|_{\infty}L}{2\Gamma\left(\alpha+1\right)}\left|\widetilde{y}_{0}\right|+\frac{1}{2}\left(\frac{h^{\alpha}\left\|g'\right\|_{\infty}L}{\Gamma\left(\alpha+1\right)}\right)^{2}\left|\widetilde{y}_{0}\right|\delta_{1,0}^{\alpha} \\ & +\frac{1}{2}\left(\frac{h^{\alpha}\left\|g'\right\|_{\infty}L}{\Gamma\left(\alpha+1\right)}\right)^{2}\left|\widetilde{y}_{1}\right|+\frac{1}{2}\left(\frac{h^{\alpha}\left\|g'\right\|_{\infty}L}{\Gamma\left(\alpha+1\right)}\right)\left|\widetilde{y}_{1}\right|+\left|\widetilde{y}_{0}\right|.\end{aligned} \tag{11.27}$$

If we arrange it, we will obtain

$$\left|\widetilde{y}_{2}\right| \leq \frac{h^{\alpha}\left\|g'\right\|_{\infty}L}{2\Gamma\left(\alpha+1\right)}\left(\begin{array}{c}\left(2+\frac{h^{\alpha}L}{\Gamma(\alpha+1)}\right)\delta_{1,0}^{\alpha}\left|\widetilde{y}_{0}\right| \\ +\left(1+\frac{h^{\alpha}\|g'\|_{\infty}L}{\Gamma(\alpha+1)}\delta_{1,0}^{\alpha}\right)\left|\widetilde{y}_{0}\right| \\ +\left(1+\frac{h^{\alpha}\|g'\|_{\infty}L}{\Gamma(\alpha+1)}\right)\left(1+\frac{h^{\alpha}L}{\Gamma(\alpha+1)}\right)\left|\widetilde{y}_{0}\right|\end{array}\right)+\left|\widetilde{y}_{0}\right| \tag{11.28}$$

$$\left|\widetilde{y}_{2}\right| \leq c_{1}\left|\widetilde{y}_{0}\right|$$

where

$$c_{1} = \frac{h^{\alpha}\left\|g'\right\|_{\infty}L}{2\Gamma\left(\alpha+1\right)}\left(\begin{array}{c}\left(2+\frac{h^{\alpha}L}{\Gamma(\alpha+1)}\right)\delta_{1,0}^{\alpha} \\ +\left(1+\frac{h^{\alpha}\|g'\|_{\infty}L}{\Gamma(\alpha+1)}\delta_{1,0}^{\alpha}\right) \\ +\left(1+\frac{h^{\alpha}\|g'\|_{\infty}L}{\Gamma(\alpha+1)}\right)\left(1+\frac{h^{\alpha}L}{\Gamma(\alpha+1)}\right)\end{array}\right)+1. \tag{11.29}$$

We can now assume that $\forall n \geq 2$

$$\left|\widetilde{y}_{n}\right| < c_{n}\left|\widetilde{y}_{0}\right|, \tag{11.30}$$

and c_n is constant; we want to show that

$$|\widetilde{y}_{n+1}| < \overline{c}_n |\widetilde{y}_0| . \tag{11.31}$$

$$\begin{aligned} |\widetilde{y}_{n+1}| \leq & \frac{h^{\alpha} \|g'\|_{\infty} L}{2\Gamma(\alpha+1)} \left(\sum_{j=0}^{n-1} \{c_n |\widetilde{y}_0| + c_n |\widetilde{y}_0|\} \delta_{n,j}^{\alpha} \right) \\ & + \frac{h^{\alpha} \|g'\|_{\infty} L}{2\Gamma(\alpha+1)} \left(|\widetilde{y}_0| + \frac{h^{\alpha} \|g'\|_{\infty} L}{\Gamma(\alpha+1)} \sum_{j=0}^{n} c_n |\widetilde{y}_0| \delta_{n,j}^{\alpha} \right) \\ & + \frac{h^{\alpha} \|g'\|_{\infty} L}{2\Gamma(\alpha+1)} c_n |\widetilde{y}_0| + |\widetilde{y}_0| . \end{aligned} \tag{11.32}$$

$$\begin{aligned} |\widetilde{y}_{n+1}| \leq & \left(\begin{array}{c} \frac{h^{\alpha}\|g'\|_{\infty}L}{\Gamma(\alpha+1)} c_n \sum_{j=0}^{n-1} \delta_{n,j}^{\alpha} \\ + \frac{h^{\alpha}\|g'\|_{\infty}L}{2\Gamma(\alpha+1)} \left(1 + \frac{h^{\alpha}\|g'\|_{\infty}L}{\Gamma(\alpha+1)} c_n \sum_{j=0}^{n} \delta_{n,j}^{\alpha} \right) \\ + \frac{h^{\alpha}\|g'\|_{\infty}L}{2\Gamma(\alpha+1)} c_n + 1 \end{array} \right) |\widetilde{y}_0| \\ \leq & \; c_{n+1} |\widetilde{y}_0| , \end{aligned}$$

which completes the proof.

Theorem 11.2 *It is assumed that $g'(t)$ is bounded. We assume that $f(., y(.))$ is Lipschitz with respect to y.*

We also assume that $f(., y(.))$ is bounded. $\forall n \geq 0$,

$$\lim_{n \to 0} |y(t_{n+1}) - y_{n+1}| = 0. \tag{11.33}$$

Proof

$$|y(t_{n+1}) - y_{n+1}| = \left| \begin{array}{c} \frac{1}{\Gamma(\alpha)} \int_{t_0}^{t_{n+1}} g'(\tau)(t_{n+1}-\tau)^{\alpha-1} f(\tau, y(\tau)) d\tau \\ - \frac{h^{\alpha-1}}{2\Gamma(\alpha+1)} \sum_{j=1}^{n-1} \left(\begin{array}{c} (g(t_{j+1}) - g(t_j)) f(t_{j+1}, y_{j+1}) \\ + (g(t_j) - g(t_{j-1})) f(t_j, y_j) \end{array} \right) \\ \times ((n-j+1)^{\alpha} - (n-j)^{\alpha}) \\ - \frac{h^{\alpha-1}}{2\Gamma(\alpha+1)} \left(\begin{array}{c} (g(t_{n+1}) - g(t_n)) f(t_{n+1}, \overline{y}_{n+1}) \\ + (g(t_n) - g(t_{n-1})) f(t_n, y_n) \end{array} \right) \end{array} \right| \tag{11.34}$$

$$\leq \frac{1}{2\Gamma(\alpha)} \left| \begin{array}{c} \int_{t_0}^{t_{n+1}} g'(\tau)(t_{n+1}-\tau)^{\alpha-1} f(\tau, y(\tau)) d\tau \\ - \sum_{j=0}^{n-1} \int_{t_j}^{t_{j+1}} (t_{n+1}-\tau)^{\alpha-1} \left(\frac{g(t_{j+1}) - g(t_j)}{h} \right) f(t_{j+1}, y_{j+1}) d\tau \end{array} \right|$$

$$+\frac{1}{2\Gamma(\alpha)}\left|\begin{array}{c}\int\limits_{t_0}^{t_{n+1}} g'(\tau)(t_{n+1}-\tau)^{\alpha-1} f(\tau, y(\tau)d\tau \\ -\sum\limits_{j=0}^{n-1}\int\limits_{t_j}^{t_{j+1}} (t_{n+1}-\tau)^{\alpha-1}\left(\frac{g(t_j)-g(t_{j-1})}{h}\right) f(t_j, y_j)d\tau\end{array}\right|$$

$$+\left|\frac{(g(t_{n+1})-g(t_n))\,h^{\alpha-1}}{2\Gamma(\alpha+1)} f(t_{n+1}, \overline{y}_{n+1}) + \frac{(g(t_n)-g(t_{n-1}))\,h^{\alpha-1}}{2\Gamma(\alpha+1)} f(t_n, y_n)\right|$$

$$\begin{aligned}
\leq\ & \frac{1}{2\Gamma(\alpha)}\left|\sum_{j=0}^{n-1}\int\limits_{t_j}^{t_{j+1}} g'(\tau)\left(f(\tau, y(\tau)-f(t_{j+1}, y_{j+1})\right)(t_{n+1}-\tau)^{\alpha-1}d\tau\right| \qquad (11.35)\\
& +\frac{1}{2\Gamma(\alpha)}\left|\sum_{j=0}^{n-1}\int\limits_{t_j}^{t_{j+1}} g'(\tau)\left(f(\tau, y(\tau)-f(t_j, y_j)\right)(t_{n+1}-\tau)^{\alpha-1}d\tau\right| \\
& +\left|\frac{g'(t_{n+1})h^{\alpha}}{2\Gamma(\alpha+1)} f(t_{n+1}, \overline{y}_{n+1}) + \frac{g'(t_n)h^{\alpha}}{2\Gamma(\alpha+1)} f(t_n, y_n)\right| \\
\leq\ & \frac{1}{2\Gamma(\alpha)}\left\|g'\right\|_\infty \left\|f'(., y(.))\right\|_\infty \sum_{j=0}^{n}\int\limits_{t_j}^{t_{j+1}} (t_{j+1}-\tau)(t_{n+1}-\tau)^{\alpha-1}d\tau \\
& +\frac{1}{2\Gamma(\alpha)}\left\|g'\right\|_\infty \left\|f'(., y(.))\right\|_\infty \sum_{j=0}^{n}\int\limits_{t_j}^{t_{j+1}} (t_{j+1}-\tau)(t_{n+1}-\tau)^{\alpha-1}d\tau \\
\leq\ & \frac{\left\|g'\right\|_\infty \left\|f'(., y(.))\right\|_\infty}{\Gamma(\alpha)} \sum_{j=0}^{n}\int\limits_{t_j}^{t_{j+1}} (t_{j+1}-\tau)(t_{n+1}-\tau)^{\alpha-1}d\tau.
\end{aligned}$$

We note that

$$\int\limits_{t_j}^{t_{j+1}} (t_{j+1}-\tau)(t_{n+1}-\tau)^{\alpha-1}d\tau = t_{j+1}\int\limits_{t_j}^{t_{j+1}} (t_{n+1}-\tau)^{\alpha-1}d\tau - \int\limits_{t_j}^{t_{j+1}} \tau(t_{n+1}-\tau)^{\alpha-1}d\tau, \qquad (11.36)$$

$$\int\limits_{t_j}^{t_{j+1}} (t_{n+1}-\tau)^{\alpha-1}d\tau = \left(\frac{(t_{n+1}-t_j)^{\alpha}}{\alpha} - \frac{(t_{n+1}-t_{j-1})^{\alpha}}{\alpha}\right),$$

$$\int\limits_{t_j}^{t_{j+1}} (t_{n+1}-\tau)^{\alpha-1}\tau d\tau = t_{n+1}^{\alpha+1}\left(B\left(\frac{t_{j+1}}{t_{n+1}}; \alpha, 2\right) - B\left(\frac{t_j}{t_{n+1}}; \alpha, 2\right)\right).$$

By replacing, we get

$$
\begin{aligned}
&\left|y(t_{n+1})-y_{n+1}\right| \\
&= \left\|g'\right\|_{\infty}\left\|f'(.,y(.))\right\|_{\infty}\left\{\begin{array}{c}\frac{(n+1)h^{\alpha+1}}{\alpha}\left((n-j+1)^{\alpha}-(n-j)^{\alpha}\right) \\ -(n+1)^{\alpha+1}\left(B\left(\frac{t_{j+1}}{t_{n+1}},\alpha,2\right)-B\left(\frac{t_j}{t_{n+1}},\alpha,2\right)\right)\end{array}\right\} \\
&\leq \left\|g'\right\|_{\infty}\left\|f'(.,y(.))\right\|_{\infty}(n+1)\frac{h^{\alpha+1}}{\Gamma(\alpha+1)},
\end{aligned}
\tag{11.37}
$$

$$
\begin{aligned}
\lim_{h\to 0}\left|y(t_{n+1})-y_{n+1}\right| &\leq \lim_{h\to 0}\left\|g'\right\|_{\infty}\left\|f'(.,y(.))\right\|_{\infty}(n+1)\frac{h^{\alpha+1}}{\Gamma(\alpha+1)} \\
&= 0.
\end{aligned}
\tag{11.38}
$$

Therefore

$$
\lim_{h\to 0}\left|y(t_{n+1})-y_{n+1}\right| = 0,
\tag{11.39}
$$

which completes the proof.

11.3 Applying Midpoint Scheme Method on IVP with Riemann–Liouville Global Derivative

We present the adaptation of the midpoint method for solving the global Riemann–Liouville differential equation, offering a more efficient approach for addressing the complexities inherent in fractional calculus.

$$
\begin{cases}{}^{RL}_{t_0}D^{\alpha}_{g}y(t) = f(t,y(t)) & \text{if } t>0 \\ y(t_0)=y_0 & \text{if } t=0\end{cases}.
\tag{11.40}
$$

Here we assume that $g'(t)>0$. By the fundamental theorem of calculus, the above IVP can be transformed as

$$
\begin{aligned}
y(t) &= \frac{1}{\Gamma(\alpha)}\int_{t_0}^{t} g'(\tau)(t-\tau)^{\alpha-1}f(\tau,y(\tau))d\tau, \\
y(t_0) &= y_0.
\end{aligned}
\tag{11.41}
$$

Here we only consider at $t=t_{n+1}$, then we have

$$y(t_{n+1}) = \frac{1}{\Gamma(\alpha)} \int_{t_0}^{t_{n+1}} g'(\tau)(t_{n+1}-\tau)^{\alpha-1} f(\tau, y(\tau))d\tau, \tag{11.42}$$

$$= \sum_{j=0}^{n} \int_{t_j}^{t_{j+1}} g'(\tau) \frac{(t_{n+1}-\tau)^{\alpha-1}}{\Gamma(\alpha)} f(\tau, y(\tau))d\tau.$$

$$y_{n+1} = \frac{1}{\Gamma(\alpha)} \sum_{j=0}^{n} \int_{t_j}^{t_{j+1}} g'\left(t_j + \frac{h}{2}\right) f\left(t_j + \frac{h}{2}, \frac{y_j + y_{j+1}}{2}\right) (t_{n+1}-\tau)^{\alpha-1} d\tau, \tag{11.43}$$

$$= \frac{h^\alpha}{\Gamma(\alpha)} \sum_{j=0}^{n} g'\left(t_j + \frac{h}{2}\right) f\left(t_j + \frac{h}{2}, \frac{y_j + y_{j+1}}{2}\right) \left\{ \frac{(n-j+1)^\alpha}{\alpha} - \frac{(n-j)^\alpha}{\alpha} \right\},$$

$$= \frac{h^\alpha}{\Gamma(\alpha+1)} \sum_{j=0}^{n-1} g'\left(t_j + \frac{h}{2}\right) f\left(t_j + \frac{h}{2}, \frac{y_j + y_{j+1}}{2}\right) \delta_{n,j}^{\alpha}$$

$$+ \frac{h^\alpha}{\Gamma(\alpha+1)} g'\left(t_n + \frac{h}{2}\right) f\left(t_n + \frac{h}{2}, \frac{y_n + y_{n+1}}{2}\right),$$

$$= \frac{h^{\alpha-1}}{\Gamma(\alpha+1)} \sum_{j=0}^{n-1} \left(g(t_{j+1}) - g(t_j)\right) f\left(t_j + \frac{h}{2}, \frac{y_j + y_{j+1}}{2}\right) \delta_{n,j}^{\alpha}$$

$$+ \frac{h^{\alpha-1}}{\Gamma(\alpha+1)} \left(g(t_{n+1}) - g(t_n)\right) f\left(t_n + \frac{h}{2}, \frac{y_n + \overline{y}_{n+1}}{2}\right).$$

Here

$$\overline{y}_{n+1} = y_n + \frac{h^{\alpha-1}}{\Gamma(\alpha+1)} \sum_{j=0}^{n} f(t_j, y_j) \left(g(t_{j+1}) - g(t_j)\right) \left((n-j+1)^\alpha - (n-j)^\alpha\right). \tag{11.44}$$

11.4 Applying Linear Approximation Method on IVP with Riemann–Liouville Global Derivative

We consider the following IVP problem:

$$\begin{cases} {}_{t_0}^{RL}D_g^\alpha y(t) = f(t, y(t)) & \text{if } t > 0 \\ y(t_0) = y_0 & \text{if } t = 0 \end{cases}. \tag{11.45}$$

$$y(t) = \frac{1}{\Gamma(\alpha)} \int_{t_0}^{t} g'(\tau) f(\tau, y(\tau))(t-\tau)^{\alpha-1} d\tau, \tag{11.46}$$

$$y(t_0) = y_0.$$

At $t = t_{n+1}$, then we have

$$y(t_{n+1}) = \frac{1}{\Gamma(\alpha)} \int_{t_0}^{t_{n+1}} g'(\tau) f(\tau, y(\tau))(t_{n+1} - \tau)^{\alpha-1} d\tau, \tag{11.47}$$

$$= \frac{1}{\Gamma(\alpha)} \sum_{j=0}^{n} \int_{t_j}^{t_{j+1}} g'(\tau) f(\tau, y(\tau))(t_{n+1} - \tau)^{\alpha-1} d\tau;$$

within $[t_j, t_{j+1}]$, we fix

$$F(t, y(t)) = g'(t) f(t, y(t)) \tag{11.48}$$

and approximate

$$F(\tau, y(\tau)) \simeq F(t_j, y(t_j)) + \left(\frac{F(t_{j+1}, y_{j+1}) - F(t_j, y_j)}{h}\right)\left(\tau - t_j\right). \tag{11.49}$$

We replace it by the original function to have

$$\begin{aligned} F(\tau, y(\tau)) \simeq\ & g'(t_j) f(t_j, y(t_j)) + \left(\frac{g'(t_j) f(t_{j+1}, y_{j+1}) - g'(t_j) f(t_j, y_j)}{h}\right)(\tau - t_j) \\ \simeq\ & \frac{(g(t_{j+1}) - g(t_j))}{h} f(t_j, y(t_j)) + \left(\begin{array}{c} \frac{(g(t_{j+1})-g(t_j))}{h} f(t_{j+1}, y_{j+1}) \\ -\frac{(g(t_{j+1})-g(t_j))}{h} f(t_j, y(t_j)) \end{array}\right) \frac{(\tau - t_j)}{h}. \end{aligned} \tag{11.50}$$

Replacing $f(\tau, y(\tau))$ by the linear polynomial yields

$$\begin{aligned} y_{n+1} =\ & \frac{1}{\Gamma(\alpha)} \sum_{j=0}^{n} \int_{t_j}^{t_{j+1}} \frac{(g(t_{j+1}) - g(t_j))}{h} f(t_j, y(t_j))(t_{n+1} - \tau)^{\alpha-1} d\tau \\ & + \frac{1}{\Gamma(\alpha)} \sum_{j=0}^{n} \int_{t_j}^{t_{j+1}} \frac{(g(t_{j+1}) - g(t_j))}{h} \left(f(t_{j+1}, y_{j+1}) - f(t_j, y(t_j))\right)(t_{n+1} - \tau)^{\alpha-1} d\tau, \\ =\ & \frac{h^{\alpha-1}}{\Gamma(\alpha+1)} \sum_{j=0}^{n} \left(g(t_{j+1}) - g(t_j)\right) f(t_j, y(t_j)) \left\{(n - j + 1)^{\alpha} - (n - j)^{\alpha}\right\} \\ & + \frac{1}{\Gamma(\alpha)} \sum_{j=0}^{n} \int_{t_j}^{t_{j+1}} \frac{(g(t_{j+1}) - g(t_j))}{h} \left(f(t_{j+1}, y_{j+1}) - f(t_j, y(t_j))\right) \int_{t_j}^{t_{j+1}} (t_{n+1} - \tau)^{\alpha-1}\left(\tau - t_j\right) d\tau. \end{aligned} \tag{11.51}$$

We shall expand the integral

$$\int_{t_j}^{t_{j+1}} (t_{n+1} - \tau)^{\alpha-1}\left(\tau - t_j\right) d\tau = \int_{t_j}^{t_{j+1}} (t_{n+1} - \tau)^{\alpha-1} \tau d\tau - t_j \int_{t_j}^{t_{j+1}} (t_{n+1} - \tau)^{\alpha-1} d\tau. \tag{11.52}$$

If we put $t_{n+1}y = \tau$ then we get

$$\int_{t_j}^{t_{j+1}} (t_{n+1} - \tau)^{\alpha-1} (\tau - t_j)\, d\tau = t_{n+1}^{\alpha+1} \int_{\frac{t_j}{t_{n+1}}}^{\frac{t_{j+1}}{t_{n+1}}} y^{2-1}(1-y)dy - t_j \int_{t_j}^{t_{j+1}} (t_{n+1} - \tau)^{\alpha-1} d\tau \tag{11.53}$$

$$= t_{n+1}^{\alpha+1} \left(B\left(\frac{t_{j+1}}{t_{n+1}}, \alpha, 2\right) - B\left(\frac{t_j}{t_{n+1}}, \alpha, 2\right)\right) - t_j \left(\frac{(t_{n+1} - t_j)^{\alpha}}{\alpha} - \frac{(t_{n+1} - t_{j+1})^{\alpha}}{\alpha} \right)$$

where $B(z, \alpha, \beta)$ is the incompleted Beta function defined as

$$B(z, \alpha, \beta) = \int_0^z u^{\alpha-1}(1-u)^{\beta-1} du, \tag{11.54}$$

and replacing yields

$$\begin{aligned} y_{n+1} = & \frac{h^{\alpha-1}}{\Gamma(\alpha+1)} \sum_{j=0}^{n} \left(g(t_{j+1}) - g(t_j)\right) f(t_j, y(t_j)) \left((n-j+1)^{\alpha} - (n-j)^{\alpha}\right) \\ & + \frac{1}{\Gamma(\alpha)} \sum_{j=0}^{n} \frac{\left(g(t_{j+1}) - g(t_j)\right)}{h} \left(f(t_{j+1}, y_{j+1}) - f(t_j, y(t_j))\right) \\ & \times \left(\begin{array}{c} t_{n+1}^{\alpha+1} \left(B\left(\frac{t_{j+1}}{t_{n+1}}, \alpha, 2\right) - B\left(\frac{t_j}{t_{n+1}}, \alpha, 2\right)\right) \\ -t_j \left(\frac{(t_{n+1}-t_j)^{\alpha}}{\alpha} - \frac{(t_{n+1}-t_{j-1})^{\alpha}}{\alpha} \right) \end{array} \right). \end{aligned} \tag{11.55}$$

But the scheme is implicit; to make it explicit, we do the following:

$$\begin{aligned} y_{n+1} = & \frac{h^{\alpha-1}}{\Gamma(\alpha+1)} \sum_{j=0}^{n} \left(g(t_{j+1}) - g(t_j)\right) f(t_j, y(t_j)) \left((n-j+1)^{\alpha} - (n-j)^{\alpha}\right) \\ & + \frac{1}{\Gamma(\alpha)} \sum_{j=0}^{n-1} \frac{\left(g(t_{j+1}) - g(t_j)\right)}{h} \left(f(t_{j+1}, y_{j+1}) - f(t_j, y(t_j))\right) \\ & \times \left(\begin{array}{c} t_{n+1}^{\alpha+1} \left(B\left(\frac{t_{j+1}}{t_{n+1}}, \alpha, 2\right) - B\left(\frac{t_j}{t_{n+1}}, \alpha, 2\right)\right) \\ -t_j \left(\frac{(t_{n+1}-t_j)^{\alpha}}{\alpha} - \frac{(t_{n+1}-t_{j-1})^{\alpha}}{\alpha} \right) \end{array} \right) \\ & + \frac{1}{\Gamma(\alpha)} \frac{\left(g(t_{n+1}) - g(t_n)\right)}{h} \left(f(t_{n+1}, \overline{y}_{n+1}) - f(t_n, y(t_n))\right) \\ & \times \left(t_{n+1}^{\alpha+1} \left(B(1, \alpha, 2) - B\left(\frac{t_n}{t_{n+1}}, \alpha, 2\right)\right) - t_n \frac{h^{\alpha}}{\alpha} \right), \end{aligned} \tag{11.56}$$

where

$$\overline{y}_{n+1} = \frac{h^{\alpha-1}}{\Gamma(\alpha+1)} \sum_{j=0}^{n} \left(g(t_{j+1}) - g(t_j)\right) f(t_j, y(t_j)) \left\{(n-j+1)^{\alpha} - (n-j)^{\alpha}\right\}. \tag{11.57}$$

Theorem 11.3 *Let $y(t_{n+1})$ be the exact solution at the point t_{n+1} and y_{n+1} the approximate solution at the point t_{n+1}. If R_T is the error then*

$$|R_T| < \frac{\|g'\|}{2\Gamma(\alpha)} \left|f''(\xi, y(\xi))\right| h^{\alpha+2} \left(\begin{array}{c} (n+1)^{\alpha+1} \left(B(1,\alpha,2) - B\left(\frac{t_0}{t_{n+1}}, \alpha, 2\right)\right) \\ + \frac{n(n+1)}{\alpha} \left((n+1)^{\alpha} + 1\right) \\ + (n+1)^{\alpha+2} \left(B(1,\alpha,2) - B\left(\frac{t_0}{t_{n+1}}, \alpha, 2\right)\right) \end{array} \right) \tag{11.58}$$

if the second derivative of f is bounded.

Proof

$$y(t_{n+1}) = \frac{1}{\Gamma(\alpha)} \sum_{j=0}^{n} \int_{t_j}^{t_{j+1}} g'(\tau) f(\tau, y(\tau))(t_{n+1}-\tau)^{\alpha-1} d\tau, \tag{11.59}$$

$$= \frac{1}{\Gamma(\alpha)} \sum_{j=0}^{n} \int_{t_j}^{t_{j+1}} g'(\tau)(t_{n+1}-\tau)^{\alpha-1} \left\{P_n(\tau) + R_1(\tau)\right\} d\tau,$$

$$= y_{n+1} + \frac{1}{\Gamma(\alpha)} \sum_{j=0}^{n} \int_{t_j}^{t_{j+1}} g'(\tau)(t_{n+1}-\tau)^{\alpha-1} R_1(\tau)\, d\tau.$$

We know that

$$R_1(\tau) = \frac{(\tau - t_j)(\tau - t_{j+1})}{2!} f''(\xi, y(\xi)),\ \xi \in [t_j, t_{j+1}]. \tag{11.60}$$

$$R_T = \frac{1}{\Gamma(\alpha)} \sum_{j=0}^{n} \int_{t_j}^{t_{j+1}} g'(\tau)(t_{n+1}-\tau)^{\alpha-1} \frac{(\tau - t_j)(\tau - t_{j+1})}{2!} f''(\xi, y(\xi)) d\tau,$$

$$|R_T| \leq \frac{1}{2\Gamma(\alpha)} \sum_{j=0}^{n} \int_{t_j}^{t_{j+1}} \left|g'(\tau)\right| (t_{n+1}-\tau)^{\alpha-1} \left|f''(\xi, y(\xi))\right| (t_{j+1}-\tau)(\tau - t_j)\, d\tau,$$

$$\leq \frac{1}{2\Gamma(\alpha)} \sum_{j=0}^{n} \int_{t_j}^{t_{j+1}} (t_{n+1}-\tau)^{\alpha-1} \max_{l \in [t_j, \tau]} \left|g'(\tau)\right| \left|f''(\xi, y(\xi))\right| (t_{j+1}-\tau)(\tau - t_j)\, d\tau,$$

$$\leq \frac{1}{2\Gamma(\alpha)} \sum_{j=0}^{n} \left\|g'\right\|_{\infty} \left\|f''(\xi, y(\xi))\right\| \int_{t_j}^{t_{j+1}} (t_{n+1}-\tau)^{\alpha-1} (t_{j+1}-\tau)(\tau - t_j)\, d\tau.$$

We shall evaluate

$$\begin{aligned}\int_{t_j}^{t_{j+1}} (t_{n+1}-\tau)^{\alpha-1}(t_{j+1}-\tau)\left(\tau-t_j\right)d\tau &= \int_{t_j}^{t_{j+1}} (t_{n+1}-\tau)^{\alpha-1}\left\{\tau t_{j+1}-\tau^2-t_jt_{j+1}-t_j\tau\right\}d\tau, \quad (11.61)\\ &= \int_{t_j}^{t_{j+1}} (t_{n+1}-\tau)^{\alpha-1}\left\{\left(t_{j+1}-t_j\right)\tau-t_jt_{j+1}-\tau^2\right\}d\tau,\\ &= h\int_{t_j}^{t_{j+1}} (t_{n+1}-\tau)^{\alpha-1}\tau d\tau-h^2(j+1)j\int_{t_j}^{t_{j+1}} (t_{n+1}-\tau)^{\alpha-1}d\tau\\ &\quad-\int_{t_j}^{t_{j+1}} (t_{n+1}-\tau)^{\alpha-1}\tau^2 d\tau,\\ &= h^{\alpha}t_{n+1}^{\alpha+1}\left(B\left(\frac{t_{j+1}}{t_{n+1}},\alpha,2\right)-B\left(\frac{t_j}{t_{n+1}},\alpha,2\right)\right)\\ &\quad-\frac{jh^{\alpha+2}(j+1)}{\alpha}\left\{(n-j+1)^{\alpha}-(n-j)^{\alpha}\right\}\\ &\quad-t_{n+1}^{\alpha+2}\left(B\left(\frac{t_{j+1}}{t_{n+1}},\alpha,3\right)-B\left(\frac{t_j}{t_{n+1}},\alpha,3\right)\right).\end{aligned}$$

By replacing we get

$$\begin{aligned}|R_T| \le\ & \frac{\|g'\|_{\infty}\left\|f''(\xi,y(\xi))\right\|}{2\Gamma(\alpha)}\left(\sum_{j=0}^{n} h^{\alpha}t_{n+1}^{\alpha+1}\left(B\left(\frac{t_{j+1}}{t_{n+1}},\alpha,2\right)-B\left(\frac{t_j}{t_{n+1}},\alpha,2\right)\right)\right) \quad (11.62)\\ &+\sum_{j=0}^{n}\frac{jh^{\alpha+2}(j+1)}{\alpha}\left\{(n-j+1)^{\alpha}-(n-j)^{\alpha}\right\}\\ &+\sum_{j=0}^{n} t_{n+1}^{\alpha+2}\left(B\left(\frac{t_{j+1}}{t_{n+1}},\alpha,3\right)-B\left(\frac{t_j}{t_{n+1}},\alpha,3\right)\right).\end{aligned}$$

We start by

$$\begin{aligned}\sum_{j=0}^{n}\left(B\left(\frac{t_{j+1}}{t_{n+1}},\alpha,2\right)-B\left(\frac{t_j}{t_{n+1}},\alpha,2\right)\right) &= B\left(\frac{t_1}{t_{n+1}},\alpha,2\right)-B\left(\frac{t_0}{t_{n+1}},\alpha,2\right) \quad (11.63)\\ &\quad+B\left(\frac{t_2}{t_{n+1}},\alpha,2\right)-B\left(\frac{t_1}{t_{n+1}},\alpha,2\right)\\ &\quad+\cdots+B\left(\frac{t_{n+1}}{t_{n+1}},\alpha,2\right)-B\left(\frac{t_n}{t_{n+1}},\alpha,2\right)\\ &= B(1,\alpha,2)-B\left(\frac{t_0}{t_{n+1}},\alpha,2\right).\end{aligned}$$

$$\sum_{j=0}^{n} B\left(\frac{t_{j+1}}{t_{n+1}},\alpha,2\right)-B\left(\frac{t_j}{t_{n+1}},\alpha,2\right)=B(1,\alpha,2)-B\left(\frac{t_0}{t_{n+1}},\alpha,2\right),$$

$$\sum_{j=0}^{n} j(j+1)\left\{(n-j+1)^{\alpha}-(n-j)^{\alpha}\right\} \le n(n+1)\sum_{j=0}^{n}\left\{(n-j+1)^{\alpha}-(n-j)^{\alpha}\right\}.$$

But

$$\sum_{j=0}^{n}\{(n-j+1)^{\alpha}-(n-j)^{\alpha}\} = (n+1)^{\alpha}-n^{\alpha}+n^{\alpha} \tag{11.64}$$

$$-(n-1)^{\alpha}+\cdots+(n-n-1)^{\alpha}-(n-n)^{\alpha}$$

$$= (n+1)^{\alpha}+1.$$

Replacing all the values yields

$$|R_T| < \frac{\|g'\|}{2\Gamma(\alpha)}\left|f''(\xi,y(\xi))\right|h^{\alpha+2}\left(\begin{array}{c}(n+1)^{\alpha+1}\left(B(1,\alpha,2)-B\left(\frac{t_0}{t_{n+1}},\alpha,2\right)\right)\\ +\frac{n(n+1)}{\alpha}\left((n+1)^{\alpha}+1\right)\\ +(n+1)^{\alpha+2}\left(B(1,\alpha,2)-B\left(\frac{t_0}{t_{n+1}},\alpha,2\right)\right)\end{array}\right) \tag{11.65}$$

which completes the proof.

Theorem 11.4 *Let $y(t_{n+1})$ be the exact solution at $t=t_{n+1}$ and y_{n+1} be the approximate solution, then*

$$\lim_{h\to 0}|y(t_{n+1})-y_{n+1}|=0. \tag{11.66}$$

Proof

$$\left|y(t_{n+1})-y_{n+1}\right| \tag{11.67}$$

$$=\left|\begin{array}{c}\frac{1}{\Gamma(\alpha)}\sum_{j=0}^{n}\int_{t_j}^{t_{j+1}}g'(\tau)f(\tau,y(\tau))(t_{n+1}-\tau)^{\alpha-1}d\tau\\ -\frac{1}{\Gamma(\alpha)}\sum_{j=0}^{n}\int_{t_j}^{t_{j+1}}g'(t_{j+1})\left(f(t_j,y_j)+\left(\frac{f(t_{j+1},y_{j+1})-f(t_j,y_j)}{h}\right)(\tau-t_j)\right)(t-\tau)^{\alpha-1}d\tau\end{array}\right|$$

$$\leq \frac{\|g'\|_{\infty}}{\Gamma(\alpha)}\sum_{j=0}^{n}\int_{t_j}^{t_{j+1}}(t_{n+1}-\tau)^{\alpha-1}\left(f(\tau,y(\tau))-f(t_j,y_j)\right)d\tau$$

$$+\frac{\|g'\|_{\infty}}{\Gamma(\alpha)}\sum_{j=0}^{n}\int_{t_j}^{t_{j+1}}(t_{n+1}-\tau)^{\alpha-1}\left(\frac{f(t_{j+1},y_{j+1})-f(t_j,y_j)}{h}\right)(\tau-t_j)\,d\tau,$$

$$\leq \frac{\|g'\|_{\infty}}{\Gamma(\alpha)}\sum_{j=0}^{n}\int_{t_j}^{t_{j+1}}\left|f'(\xi,y(\xi))\right|(\tau-t_j)(t_{n+1}-\tau)^{\alpha-1}d\tau$$

$$+\frac{\|g'\|_{\infty}}{\Gamma(\alpha)}\sum_{j=0}^{n}\int_{t_j}^{t_{j+1}}(t_{n+1}-\tau)^{\alpha-1}\left|f'(\xi,y(\xi))\right|(\tau-t_j)\,d\tau.$$

The above is achieved via the differentiability of the function f and the Mean Value theorem:

$$
\begin{aligned}
|y(t_{n+1}) - y_{n+1}| \leq\ & \frac{\|g'\|_\infty \left|f'(\xi, y(\xi))\right|}{\Gamma(\alpha)} \sum_{j=0}^{n} \begin{pmatrix} t_{n+1}^{\alpha+1}\left(B\left(\frac{t_{j+1}}{t_{n+1}}, \alpha, 2\right) - B\left(\frac{t_j}{t_{n+1}}, \alpha, 2\right)\right) \\ -t_j\left\{\frac{(t_{n+1}-t_j)^\alpha}{\alpha} - \frac{(t_{n+1}-t_{j+1})^\alpha}{\alpha}\right\} \end{pmatrix} \\
& + \frac{\|g'\|_\infty \left|f'(\xi, y(\xi))\right|}{\Gamma(\alpha)} \sum_{j=0}^{n} \begin{pmatrix} t_{n+1}^{\alpha+1}\left(B\left(\frac{t_{j+1}}{t_{n+1}}, \alpha, 2\right) - B\left(\frac{t_j}{t_{n+1}}, \alpha, 2\right)\right) \\ -t_j\left\{\frac{(t_{n+1}-t_j)^\alpha}{\alpha} - \frac{(t_{n+1}-t_{j+1})^\alpha}{\alpha}\right\} \end{pmatrix}, \\
\leq\ & \frac{2\|g'\|_\infty \left|f'(\xi, y(\xi))\right|}{\Gamma(\alpha)} \sum_{j=0}^{n} \begin{pmatrix} t_{n+1}^{\alpha+1}\left(B\left(\frac{t_{j+1}}{t_{n+1}}, \alpha, 2\right) - B\left(\frac{t_j}{t_{n+1}}, \alpha, 2\right)\right) \\ -t_j\left\{\frac{(t_{n+1}-t_j)^\alpha}{\alpha} - \frac{(t_{n+1}-t_{j+1})^\alpha}{\alpha}\right\} \end{pmatrix}, \\
\leq\ & 2\|g'\|_\infty \left|f'(\xi, y(\xi))\right| \sum_{j=0}^{n} \begin{pmatrix} t_{n+1}^{\alpha+1}\left(B\left(\frac{t_{j+1}}{t_{n+1}}, \alpha, 2\right) - B\left(\frac{t_j}{t_{n+1}}, \alpha, 2\right)\right) \\ -t_j\left\{\frac{(t_{n+1}-t_j)^\alpha}{\alpha} - \frac{(t_{n+1}-t_{j+1})^\alpha}{\alpha}\right\} \end{pmatrix}, \\
\leq\ & 2\|g'\|_\infty \left|f'(\xi, y(\xi))\right| h^{\alpha+1} \sum_{j=0}^{n+1} \begin{pmatrix} (n+1)^{\alpha+1}\left(B\left(\frac{t_{j+1}}{t_{n+1}}, \alpha, 2\right) - B\left(\frac{t_j}{t_{n+1}}, \alpha, 2\right)\right) \\ +n\left\{\frac{(n-j+1)^\alpha}{\alpha} - \frac{(n-j)^\alpha}{\alpha}\right\} \end{pmatrix}, \\
\leq\ & 2\|g'\|_\infty \left|f'(\xi, y(\xi))\right| h^{\alpha+1} \begin{pmatrix} (n+1)^{\alpha}\left(B(1, \alpha, 2) - B\left(\frac{t_0}{t_{n+1}}, \alpha, 2\right)\right) \\ +n\left((n+1)^\alpha + 1\right) \end{pmatrix}, \\
\leq\ & h^{\alpha+1}\lambda(n, \alpha)
\end{aligned}
\tag{11.68}
$$

where

$$
\lambda(n, \alpha) = 2\|g'\|_\infty \left|f'(\xi, y(\xi))\right| \left\{(n+1)^\alpha \left(B(1, \alpha, 2) - B\left(\frac{t_0}{t_{n+1}}, \alpha, 2\right)\right) + n\left((n+1)^\alpha + 1\right)\right\}. \tag{11.69}
$$

Therefore

$$
\begin{aligned}
\lim_{h\to 0} |y(t_{n+1}) - y_{n+1}| &\leq \lim_{h\to 0} h^{\alpha+1}\lambda(n, \alpha), \\
&\leq 0.
\end{aligned}
\tag{11.70}
$$

Since $\lambda(n, \alpha) < \infty$, thus

$$
\lim_{h\to 0} |y(t_{n+1}) - y_{n+1}| = 0, \tag{11.71}
$$

which completes the proof.

Theorem 11.5 *If the function $f(t, y(t))$ satisfies the global Lipschitz condition, we have*

$$
|y_{n+1}| < a_{n+1} + b_{n+1} |y_0| . \tag{11.72}
$$

Proof

$$|y_{n+1}| = \left| \begin{array}{c} \frac{h^{\alpha-1}}{\Gamma(\alpha+1)} \sum_{j=0}^{n} \left(g(t_{j+1}) - g(t_j)\right) f(t_j, y(t_j)) \left((n-j+1)^{\alpha} - (n-j)^{\alpha}\right) \\ + \frac{1}{\Gamma(\alpha)} \sum_{j=0}^{n-1} \frac{\left(g(t_{j+1}) - g(t_j)\right)}{h} \left(f(t_{j+1}, y_{j+1}) - f(t_j, y(t_j))\right) \\ \times \left(\begin{array}{c} t_{n+1}^{\alpha+1} \left(B\left(\frac{t_{j+1}}{t_{n+1}}, \alpha, 2\right) - B\left(\frac{t_j}{t_{n+1}}, \alpha, 2\right)\right) \\ -t_j \left(\frac{(t_{n+1}-t_j)^{\alpha}}{\alpha} - \frac{(t_{n+1}-t_{j+1})^{\alpha}}{\alpha}\right) \end{array} \right) \\ + \frac{1}{h\Gamma(\alpha)} \left(g(t_{n+1}) - g(t_n)\right) \left(f\left(t_{n+1}, \overline{y}_{n+1}\right) - f(t_n, y_n)\right) \\ \times \left(t_{n+1}^{\alpha+1} \left(B(1, \alpha, 2) - B\left(\frac{t_n}{t_{n+1}}, \alpha, 2\right)\right) - t_n \frac{h^{\alpha}}{\alpha}\right) \end{array} \right|. \tag{11.73}$$

For simplicity, we let

$$\begin{aligned} \beta_n^{\alpha,j} &= \left(g(t_{j+1}) - g(t_j)\right) \left((n-j+1)^{\alpha} - (n-j)^{\alpha}\right), \qquad (11.74) \\ \beta_{n,1}^{\alpha,j} &= \frac{\left(g(t_{j+1}) - g(t_j)\right)}{h} \left(\begin{array}{c} t_{n+1}^{\alpha+1} \left(B\left(\frac{t_{j+1}}{t_{n+1}}, \alpha, 2\right) - B\left(\frac{t_j}{t_{n+1}}, \alpha, 2\right)\right) \\ -t_j \left(\frac{(t_{n+1}-t_j)^{\alpha}}{\alpha} - \frac{(t_{n+1}-t_{j+1})^{\alpha}}{\alpha}\right) \end{array} \right), \\ \beta_n^{\alpha} &= \frac{1}{h\Gamma(\alpha)} \left(g(t_{n+1}) - g(t_n)\right) \\ &\quad \times \left(t_{n+1}^{\alpha+1} \left(B(1, \alpha, 2) - B\left(\frac{t_n}{t_{n+1}}, \alpha, 2\right)\right) - t_n \frac{h^{\alpha}}{\alpha}\right). \end{aligned}$$

Then, the equation becomes

$$\begin{aligned} |y_{n+1}| &= \left| \begin{array}{c} \frac{h^{\alpha-1}}{\Gamma(\alpha+1)} \sum_{j=0}^{n} \beta_n^{\alpha,j} f(t_j, y(t_j)) + \frac{1}{\Gamma(\alpha)} \sum_{j=0}^{n-1} \beta_{n,1}^{\alpha,j} \left(f(t_{j+1}, y_{j+1}) - f(t_j, y(t_j))\right) \\ + \beta_n^{\alpha} \left(f\left(t_{n+1}, \overline{y}_{n+1}\right) - f(t_n, y_n)\right) \end{array} \right|, \qquad (11.75) \\ &\leq \frac{h^{\alpha-1}}{\Gamma(\alpha+1)} \sum_{j=0}^{n} \left|\beta_n^{\alpha,j}\right| \left|f(t_j, y(t_j))\right| \\ &\quad + \frac{1}{\Gamma(\alpha)} \sum_{j=0}^{n-1} \left|\beta_{n,1}^{\alpha,j}\right| \left|\left(f(t_{j+1}, y_{j+1}) - f(t_j, y(t_j))\right)\right| \\ &\quad + \left|\beta_n^{\alpha}\right| \left|f\left(t_{n+1}, \overline{y}_{n+1}\right) - f(t_n, y_n)\right|. \end{aligned}$$

Using the global Lipschitz condition of f yields

$$\begin{aligned} \left|f(t_{j+1}, y_{j+1})\right| &< c\left(1 + \left|y_{j+1}\right|\right), \qquad (11.76) \\ \left|f(t_j, y_j)\right| &< c\left(1 + \left|y_j\right|\right), \\ \left|f\left(t_{n+1}, \overline{y}_{n+1}\right)\right| &< c\left(1 + \left|\overline{y}_{n+1}\right|\right), \\ \left|f(t_n, y_n)\right| &< c\left(1 + |y_n|\right), \end{aligned}$$

where indeed

$$\left|\overline{y}_{n+1}\right| \leq \frac{h^{\alpha-1}}{\Gamma(\alpha+1)} \sum_{j=0}^{n} \left|\beta_n^{\alpha,j}\right| \left(c\left(1+\left|y_j\right|\right)\right). \tag{11.77}$$

Therefore

$$\begin{aligned} \left|f\left(t_{n+1}, \overline{y}_{n+1}\right)\right| &< c\left(1+\frac{h^{\alpha-1}}{\Gamma(\alpha+1)} \sum_{j=0}^{n} \left|\beta_n^{\alpha,j}\right| \left(c\left(1+\left|y_j\right|\right)\right)\right) \\ &< c+\frac{h^{\alpha-1}}{\Gamma(\alpha+1)} h\left|g'(t_{n+1})\right| c^2\left((n+1)^\alpha+1\right) \\ &\quad +\frac{c^2 h^{\alpha-1}}{\Gamma(\alpha+1)} \sum_{j=0}^{n} \left|\beta_n^{\alpha,j}\right| \left|y_j\right|. \end{aligned} \tag{11.78}$$

By replacing, we obtain

$$\begin{aligned} \left|y_{n+1}\right| = &\ \frac{h^{\alpha-1}}{\Gamma(\alpha+1)} \sum_{j=0}^{n} \left|\beta_n^{\alpha,j}\right| \left(c\left(1+\left|y_j\right|\right)\right) \\ &+\frac{1}{\Gamma(\alpha)} \sum_{j=0}^{n-1} \left|\beta_{n,1}^{\alpha,j}\right| \left(c\left(1+\left|y_{j+1}\right|\right)+c\left(1+\left|y_j\right|\right)\right) \\ &+\left|\beta_n^\alpha\right| \left(\begin{array}{c} c+\frac{h^\alpha}{\Gamma(\alpha+1)}\left|g'(t_{n+1})\right| c^2\left((n+1)^\alpha+1\right) \\ +\frac{c^2 h^{\alpha-1}}{\Gamma(\alpha+1)} \sum_{j=0}^{n} \left|\beta_n^{\alpha,j}\right| \left|y_j\right| + c\left(1+\left|y_n\right|\right) \end{array}\right). \end{aligned} \tag{11.79}$$

For $n = 0$,

$$\begin{aligned} \left|y_1\right| < &\ \frac{h^{\alpha-1}}{\Gamma(\alpha+1)} \left|\beta_0^\alpha\right| \left(c\left(1+\left|y_0\right|\right)\right) \\ &+\left|\beta_0^\alpha\right| \left(\begin{array}{c} c+\frac{2h^\alpha}{\Gamma(\alpha+1)}\left|g'(t_{n+1})\right| c^2 \\ +\frac{c^2 h^{\alpha-1}}{\Gamma(\alpha+1)}\left|\beta_0^\alpha\right| \left|y_0\right| + c + c\left|y_0\right| \end{array}\right), \\ < &\ a_1 + b_1\left|y_0\right|, \end{aligned} \tag{11.80}$$

where

$$\begin{aligned} a_1 &= \frac{h^{\alpha-1}}{\Gamma(\alpha+1)} \left|\beta_0^\alpha\right| c + \left|\beta_0^\alpha\right| \left(c+\frac{2h^\alpha}{\Gamma(\alpha+1)} + \left|g'(t_{n+1})\right| c^2 + c\right), \\ b_1 &= \frac{h^{\alpha-1}}{\Gamma(\alpha+1)} \left|\beta_0^\alpha\right| c + \left|\beta_0^\alpha\right| \frac{c^2 h^{\alpha-1}}{\Gamma(\alpha+1)} + c. \end{aligned} \tag{11.81}$$

For $n = 1$,

$$|y_2| < \frac{h^{\alpha-1}}{\Gamma(\alpha+1)} \sum_{j=0}^{n} \left|\beta_1^{\alpha,j}\right| \left(c\left(1+|y_j|\right)\right) + \frac{1}{\Gamma(\alpha)} \left|\beta_{1,1}^{\alpha,0}\right| \{2c + c|y_1| + c|y_0|\} + \left|\beta_1^{\alpha}\right| \begin{pmatrix} c + \frac{h^{\alpha}}{\Gamma(\alpha+1)} \left|g'(t_2)\right| c^2 \left(2^{\alpha}+1\right) \\ + \frac{c^2 h^{\alpha-1}}{\Gamma(\alpha+1)} \sum_{j=0}^{1} \left|\beta_1^{\alpha,j}\right| |y_j| + c\left(1+|y_1|\right) \end{pmatrix}, \tag{11.82}$$

and replacing

$$\begin{aligned} |y_1| &< a_1 + b_1 |y_0|, \\ |y_1| &< a_2 + b_2 |y_0|. \end{aligned} \tag{11.83}$$

We assume $\forall n > 1$

$$|y_n| < a_n + b_n |y_0|. \tag{11.84}$$

Then

$$|y_{n+1}| = \frac{h^{\alpha-1}}{\Gamma(\alpha+1)} \sum_{j=0}^{n} \left|\beta_n^{\alpha,j}\right| \left(c\left(1+|y_j|\right)\right) + \frac{1}{\Gamma(\alpha)} \sum_{j=0}^{n-1} \left|\beta_{n,1}^{\alpha,j}\right| \left(2c + |y_{j+1}| + |y_j|\right) + \left|\beta_n^{\alpha}\right| \begin{Bmatrix} c + \frac{h^{\alpha}}{\Gamma(\alpha+1)} \left|g'(t_{n+1})\right| c^2 \left((n+1)^{\alpha}+1\right) \\ + \frac{c^2 h^{\alpha-1}}{\Gamma(\alpha+1)} \sum_{j=0}^{n} \left|\beta_n^{\alpha,j}\right| |y_j| + c\left(1+|y_n|\right) \end{Bmatrix}. \tag{11.85}$$

By inductive hypothesis, we have that

$$\begin{aligned} |y_j| &< a_j + b_j |y_0|, \\ |y_{j+1}| &< a_{j+1} + b_j |y_0|, \\ |y_n| &< a_n + b_n |y_0|. \end{aligned} \tag{11.86}$$

Let

$$\begin{aligned} \overline{a} &= \max_{1 \le j \le n} \left\{a_j, a_{j+1}, a_n\right\}, \\ \overline{b} &= \max_{1 \le j \le n} \left\{b_j, b_{j+1}, b_n\right\}, \end{aligned} \tag{11.87}$$

then

$$\begin{aligned} \left|y_j\right| &< \overline{a}+\overline{b}\left|y_0\right|, \\ \left|y_{j+1}\right| &< \overline{a}+\overline{b}\left|y_0\right|, \\ \left|y_n\right| &< \overline{a}+\overline{b}\left|y_0\right|. \end{aligned} \tag{11.88}$$

By replacing we obtain

$$\begin{aligned} \left|y_{n+1}\right| = & \frac{h^{\alpha-1}}{\Gamma(\alpha+1)} \sum_{j=0}^{n}\left|\beta_n^{\alpha,j}\right|\left(c\left(1+\overline{a}+\overline{b}\left|y_0\right|\right)\right) \\ & +\frac{1}{\Gamma(\alpha)} \sum_{j=0}^{n-1}\left|\beta_{n,1}^{\alpha,j}\right|\left(2c+2\overline{a}+2\overline{b}\left|y_0\right|\right) \\ & +\left|\beta_n^{\alpha}\right|\left\{\begin{array}{c} c+\frac{h^{\alpha}}{\Gamma(\alpha+1)}\left|g'(t_{n+1})\right| c^2\left((n+1)^{\alpha}+1\right) \\ +\frac{c^2 h^{\alpha-1}}{\Gamma(\alpha+1)} \sum_{j=0}^{n}\left|\beta_n^{\alpha,j}\right|\left(\overline{a}+\overline{b}\left|y_0\right|\right) \\ +c\left(1+\overline{a}+\overline{b}\left|y_0\right|\right) \end{array}\right\} \\ & < a_{n+1}+b_{n+1}\left|y_0\right|, \end{aligned} \tag{11.89}$$

where

$$\begin{aligned} a_{n+1} = & \frac{h^{\alpha-1}}{\Gamma(\alpha+1)} \sum_{j=0}^{n}\left|\beta_n^{\alpha,j}\right| c(1+\overline{a})+\frac{1}{\Gamma(\alpha)} \sum_{j=0}^{n-1}\left|\beta_{n,1}^{\alpha,j}\right|(2c+2\overline{a}) \\ & +\left|\beta_n^{\alpha}\right|\left\{\begin{array}{c} c+\frac{h^{\alpha}}{\Gamma(\alpha+1)}\left|g'(t_{n+1})\right| c^2\left((n+1)^{\alpha}+1\right) \\ +\frac{c^2 h^{\alpha-1}}{\Gamma(\alpha+1)} \sum_{j=0}^{n}\left|\beta_n^{\alpha,j}\right| \overline{a}+c(1+\overline{a}) \end{array}\right\}, \\ b_{n+1} = & \frac{h^{\alpha-1}}{\Gamma(\alpha+1)} \sum_{j=0}^{n}\left|\beta_n^{\alpha,j}\right| c\overline{b}+\frac{1}{\Gamma(\alpha)} \sum_{j=0}^{n-1}\left|\beta_{n,1}^{\alpha,j}\right| 2\overline{b} \\ & +\frac{h^{\alpha-1}}{\Gamma(\alpha+1)} \sum_{j=0}^{n}\left|\beta_n^{\alpha,j}\right| \overline{b}+c\overline{b} \end{aligned} \tag{11.90}$$

which completes the proof.

Theorem 11.6 *If $\widetilde{y}_{n+1}$, $\widetilde{y}_n$, $\widetilde{y}_j$, $\widetilde{y}_{j+1}$, and $\widetilde{y}_0$ are perturbed terms of y_{n+1}, y_n, y_j, y_{j+1}, and y_0 respectively, then if L is Lipschitz and then $\exists c \geq 0$ such that $\forall n > 0$*

$$\left|\widetilde{y}_{n+1}\right| < \Omega\left|\widetilde{y}_0\right|. \tag{11.91}$$

Proof

$$
\begin{aligned}
y_{n+1}+\widetilde{y}_{n+1} &= \tfrac{h^{\alpha-1}}{\Gamma(\alpha+1)}\sum_{j=0}^{n}\beta_n^{\alpha,j} f(t_j, y_j+\widetilde{y}_j)\\
&+\tfrac{1}{\Gamma(\alpha)}\sum_{j=0}^{n-1}\beta_{n,1}^{\alpha,j}\left(f(t_{j+1}, y_{j+1}+\widetilde{y}_{j+1})-f(t_j, y_j+\widetilde{y}_j)\right)\\
&+\beta_n^{\alpha}\left(f\left(t_{n+1}, \overline{y}_{n+1}+\widetilde{\overline{y}}_{n+1}\right)-f(t_n, y_n+\widetilde{y}_n)\right),\\
\widetilde{y}_{n+1} &= \tfrac{h^{\alpha-1}}{\Gamma(\alpha+1)}\sum_{j=0}^{n}\beta_n^{\alpha,j}\left(f(t_j, y_j+\widetilde{y}_j)-f(t_j, y_j)\right)\\
&+\tfrac{1}{\Gamma(\alpha)}\sum_{j=0}^{n-1}\beta_{n,1}^{\alpha,j}\begin{pmatrix}\left(f(t_{j+1}, y_{j+1}+\widetilde{y}_{j+1})-f(t_{j+1}, y_{j+1})\right)\\ -\left(f(t_j, y_j+\widetilde{y}_j)-f(t_j, y_j)\right)\end{pmatrix}\\
&+\beta_n^{\alpha}\begin{pmatrix}\left(f\left(t_{n+1}, \overline{y}_{n+1}+\widetilde{\overline{y}}_{n+1}\right)-f\left(t_{n+1}, \overline{y}_{n+1}\right)\right)\\ -\left(f(t_n, y_n+\widetilde{y}_n)-f(t_n, y_n)\right)\end{pmatrix},\\
|\widetilde{y}_{n+1}| &\le \tfrac{h^{\alpha-1}}{\Gamma(\alpha+1)}\sum_{j=0}^{n}\left|\beta_n^{\alpha,j}\right| L\left|\widetilde{y}_j\right|\\
&+\tfrac{1}{\Gamma(\alpha)}\sum_{j=0}^{n-1}\left|\beta_{n,1}^{\alpha,j}\right|\left(L\left|\widetilde{y}_{j+1}\right|+L\left|\widetilde{y}_j\right|\right)\\
&+\beta_n^{\alpha}\left(L\left|\widetilde{\overline{y}}_{n+1}\right|+L\left|\widetilde{y}_n\right|\right),
\end{aligned}
\tag{11.92}
$$

where

$$
\left|\widetilde{\overline{y}}_{n+1}\right| \le \frac{h^{\alpha-1}}{\Gamma(\alpha+1)}\sum_{j=0}^{n}\left|\beta_n^{\alpha,j}\right| L\left|\widetilde{y}_j\right|. \tag{11.93}
$$

$$
\begin{aligned}
|\widetilde{y}_{n+1}| \le\ & \frac{h^{\alpha-1}}{\Gamma(\alpha+1)}\sum_{j=0}^{n}\left|\beta_n^{\alpha,j}\right| L\left|\widetilde{y}_j\right|\\
&+\frac{1}{\Gamma(\alpha)}\sum_{j=0}^{n-1}\left|\beta_{n,1}^{\alpha,j}\right|\left(L\left|\widetilde{y}_{j+1}\right|+L\left|\widetilde{y}_j\right|\right)\\
&+\left|\beta_n^{\alpha}\right| L\left(\frac{h^{\alpha-1}}{\Gamma(\alpha+1)}\sum_{j=0}^{n}\left|\beta_n^{\alpha,j}\right| L\left|\widetilde{y}_j\right|+\left|\widetilde{y}_n\right|\right).
\end{aligned}
$$

When $n=0$, we have

$$
\begin{aligned}
|\widetilde{y}_1| &\le \frac{h^{\alpha-1}}{\Gamma(\alpha+1)}\left|\beta_0^{\alpha,j}\right| L\,|\widetilde{y}_0| + \frac{\left|\beta_0^{\alpha}\right| L^2 h^{\alpha-1}\left|\beta_0^{\alpha}\right|}{\Gamma(\alpha+1)}|\widetilde{y}_0| + \left|\beta_0^{\alpha}\right||\widetilde{y}_0|, \qquad (11.94)\\
&\le \left(\frac{h^{\alpha-1}}{\Gamma(\alpha+1)}\left|\beta_0^{\alpha,j}\right| L + \frac{\left|\beta_0^{\alpha}\right|^2 L^2 h^{\alpha-1}}{\Gamma(\alpha+1)} + \left|\beta_0^{\alpha}\right|\right)|\widetilde{y}_0|,\\
&\le c\,|\widetilde{y}_0|.\\
|\widetilde{y}_2| &\le \frac{h^{\alpha-1}}{\Gamma(\alpha+1)}\sum_{j=0}^{2}\left|\beta_2^{\alpha,j}\right| L\left|\widetilde{y}_j\right| + \frac{L}{\Gamma(\alpha)}\left|\beta_{1,1}^{\alpha,j}\right|(|\widetilde{y}_1|+|\widetilde{y}_0|)\\
&\quad + L\left|\beta_1^{\alpha}\right|\left(\frac{Lh^{\alpha-1}}{\Gamma(\alpha+1)}\sum_{j=0}^{1}\left|\beta_1^{\alpha,j}\right| L\left|\widetilde{y}_j\right| + |\widetilde{y}_1|\right),\\
|\widetilde{y}_2| &\le \frac{h^{\alpha-1}}{\Gamma(\alpha+1)}\left(\left|\beta_2^{\alpha}\right| L\,|\widetilde{y}_0| + \left|\beta_2^{\alpha}\right| Lc\,|\widetilde{y}_0|\right) + \frac{L}{\Gamma(\alpha)}\left|\beta_{1,1}^{\alpha}\right|(c\,|\widetilde{y}_0|+|\widetilde{y}_0|)\\
&\quad + L\left|\beta_2^{\alpha}\right|\left(\frac{Lh^{\alpha-1}}{\Gamma(\alpha+1)}\left|\beta_1^{\alpha}\right||\widetilde{y}_0| + \frac{Lh^{\alpha-1}}{\Gamma(\alpha+1)}\left|\beta_1^{\alpha,1}\right||\widetilde{y}_0|\,c + |\widetilde{y}_0|\,c\right)\\
&\le |\widetilde{y}_0|\begin{pmatrix} \frac{h^{\alpha-1}}{\Gamma(\alpha+1)}\left|\beta_2^{\alpha}\right| L + \left|\beta_2^{\alpha}\right| Lc + \frac{L}{\Gamma(\alpha)}\left|\beta_{1,1}^{\alpha}\right|(c+1)\\ +L^2\left|\beta_2^{\alpha}\right|\frac{h^{\alpha-1}}{\Gamma(\alpha+1)}\left|\beta_1^{\alpha}\right| + \frac{Lh^{\alpha-1}}{\Gamma(\alpha+1)}\left|\beta_1^{\alpha,1}\right| c + c\end{pmatrix},\\
|\widetilde{y}_2| &\le c_n\,|\widetilde{y}_0|.
\end{aligned}
$$

We assume that $\forall n \ge 2$

$$
|\widetilde{y}_n| \le c_n\,|\widetilde{y}_0|. \qquad (11.95)
$$

$$
\begin{aligned}
|\widetilde{y}_{n+1}| &\le \left[\begin{array}{c} \frac{h^{\alpha-1}}{\Gamma(\alpha+1)}\sum_{j=0}^{n}\left|\beta_n^{\alpha,j}\right| Lc_j + \frac{L}{\Gamma(\alpha)}\sum_{j=0}^{n-1}\left|\beta_{n,1}^{\alpha,j}\right|(c_{j+1}+c_j)\\ +\left|\beta_n^{\alpha}\right| L\left\{\frac{Lh^{\alpha-1}}{\Gamma(\alpha+1)}\sum_{j=0}^{n}\left|\beta_n^{\alpha,j}\right| c_j + c_n\right\}\end{array}\right]|\widetilde{y}_0|, \qquad (11.96)\\
&\le c_{n+1}\,|\widetilde{y}_0| = \Omega\,|\widetilde{y}_0|,
\end{aligned}
$$

which completes the proof.

11.5 Applying Linear Piecewise Interpolation Method on IVP with Riemann–Liouville Global Derivative

In this section, we consider the following IVP with Riemann–Liouville global derivative:

$$
\begin{cases} {}_{t_0}^{RL}D_g^{\alpha} y(t) = f(t, y(t)) & \text{if } t > 0\\ y(t_0) = y_0 & \text{if } t = 0 \end{cases}. \qquad (11.97)
$$

$$y(t) = \frac{1}{\Gamma(\alpha)} \int_{t_0}^{t} g'(\tau) f(\tau, y(\tau))(t-\tau)^{\alpha-1} d\tau, \tag{11.98}$$
$$y(t_0) = y_0.$$

At $t = t_{n+1}$, then we have

$$\begin{aligned} y(t_{n+1}) &= \frac{1}{\Gamma(\alpha)} \int_{t_0}^{t_{n+1}} g'(\tau) f(\tau, y(\tau))(t_{n+1}-\tau)^{\alpha-1} d\tau, \\ &= \frac{1}{\Gamma(\alpha)} \sum_{j=0}^{n} \int_{t_j}^{t_{j+1}} g'(\tau) f(\tau, y(\tau))(t_{n+1}-\tau)^{\alpha-1} d\tau, \\ &= \frac{1}{\Gamma(\alpha)} \sum_{j=0}^{n} \int_{t_j}^{t_{j+1}} \frac{g(t_{j+1}) - g(t_j)}{h} \begin{pmatrix} \frac{\tau - t_{j-1}}{h} f(t_j, y_j) \\ -\frac{\tau - t_j}{h} f(t_{j-1}, y_{j-1}) \end{pmatrix} (t_{n+1}-\tau)^{\alpha-1} d\tau. \end{aligned} \tag{11.99}$$

We shall calculate each interval separately

$$\begin{aligned} \int_{t_j}^{t_{j+1}} (\tau - t_{j-1})(t_{n+1}-\tau)^{\alpha-1} d\tau &= \int_{t_j}^{t_{j+1}} \tau (t_{n+1}-\tau)^{\alpha-1} d\tau - t_{j-1} \int_{t_j}^{t_{j+1}} (t_{n+1}-\tau)^{\alpha-1} d\tau \\ &= t_{n+1} \left(B\left(\frac{t_{j+1}}{t_{n+1}}, 2, \alpha\right) - B\left(\frac{t_j}{t_{n+1}}, 2, \alpha\right) \right) \\ &\quad - t_{j-1} \left(\frac{(t_{n+1} - t_j)^{\alpha}}{\alpha} - \frac{(t_{n+1} - t_{j-1})^{\alpha}}{\alpha} \right), \end{aligned} \tag{11.100}$$

$$\begin{aligned} \int_{t_j}^{t_{j+1}} (\tau - t_j)(t_{n+1}-\tau)^{\alpha-1} d\tau &= t_{n+1} \left(B\left(\frac{t_{j+1}}{t_{n+1}}, 2, \alpha\right) - B\left(\frac{t_j}{t_{n+1}}, 2, \alpha\right) \right) \\ &\quad - t_j \left(\frac{(t_{n+1} - t_j)^{\alpha}}{\alpha} - \frac{(t_{n+1} - t_{j-1})^{\alpha}}{\alpha} \right). \end{aligned}$$

By replacing we obtain

$$y_{n+1} = \sum_{j=0}^{n} \begin{pmatrix} \frac{g(t_{j+1}) - g(t_j)}{h^2} f(t_j, y_j) \begin{pmatrix} t_{n+1} \left(B\left(\frac{t_{j+1}}{t_{n+1}}, 2, \alpha\right) - B\left(\frac{t_j}{t_{n+1}}, 2, \alpha\right) \right) \\ -t_{j-1} \left(\frac{(t_{n+1}-t_j)^{\alpha}}{\alpha} - \frac{(t_{n+1}-t_{j-1})^{\alpha}}{\alpha} \right) \end{pmatrix} \\ -\frac{g(t_j) - g(t_{j-1})}{h^2} f(t_{j-1}, y_{j-1}) \begin{pmatrix} t_{n+1} \left(B\left(\frac{t_{j+1}}{t_{n+1}}, 2, \alpha\right) - B\left(\frac{t_j}{t_{n+1}}, 2, \alpha\right) \right) \\ -t_j \left(\frac{(t_{n+1}-t_j)^{\alpha}}{\alpha} - \frac{(t_{n+1}-t_{j-1})^{\alpha}}{\alpha} \right) \end{pmatrix} \end{pmatrix}. \tag{11.101}$$

But

$$y_1 = \frac{1}{\Gamma(\alpha)} \int_{t_0}^{t_1} g'(\tau) f(\tau, y(\tau))(t_1 - \tau)^{\alpha-1} d\tau, \tag{11.102}$$

$$= \frac{1}{\Gamma(\alpha)} \int_{t_0}^{t_1} g'(t_1) f(t_1, y_1)(t_1 - \tau)^{\alpha-1} d\tau,$$

$$= \frac{1}{\Gamma(\alpha+1)} \frac{g(t_1) - g(t_0)}{h} f(t_1, \overline{y}_1)(t_1 - t_0)^{\alpha}$$

where

$$\overline{y}_1 = \frac{1}{\Gamma(\alpha+1)} \frac{g(t_1) - g(t_0)}{h} f(t_0, y_0)(t_1 - t_0)^{\alpha}. \tag{11.103}$$

Thus

$$y_1 = \frac{1}{\Gamma(\alpha+1)} \frac{g(t_1) - g(t_0)}{h} f(t_1, \frac{1}{\Gamma(\alpha+1)} \frac{g(t_1) - g(t_0)}{h} f(t_0, y_0)(t_1 - t_0)^{\alpha})(t_1 - t_0)^{\alpha}. \tag{11.104}$$

Therefore the numerical scheme obtained is given as

$$\begin{aligned} y_{n+1} &= \sum_{j=0}^{n} \left(\begin{array}{c} \frac{g(t_{j+1})-g(t_j)}{h^2} f(t_j, y_j) \left(\begin{array}{c} t_{n+1}\left(B\left(\frac{t_{j+1}}{t_{n+1}}, 2, \alpha\right) - B\left(\frac{t_j}{t_{n+1}}, 2, \alpha\right)\right) \\ -t_{j-1}\left(\frac{(t_{n+1}-t_j)^{\alpha}}{\alpha} - \frac{(t_{n+1}-t_{j-1})^{\alpha}}{\alpha}\right) \end{array} \right) \\ -\frac{g(t_j)-g(t_{j-1})}{h^2} f(t_{j-1}, y_{j-1}) \left(\begin{array}{c} t_{n+1}\left(B\left(\frac{t_{j+1}}{t_{n+1}}, 2, \alpha\right) - B\left(\frac{t_j}{t_{n+1}}, 2, \alpha\right)\right) \\ -t_j\left(\frac{(t_{n+1}-t_j)^{\alpha}}{\alpha} - \frac{(t_{n+1}-t_{j-1})^{\alpha}}{\alpha}\right) \end{array} \right) \end{array} \right) & if\ n > 1 \\ y_1 &= \frac{1}{\Gamma(\alpha+1)} \frac{g(t_1)-g(t_0)}{h} f(t_1, \frac{1}{\Gamma(\alpha+1)} \frac{g(t_1)-g(t_0)}{h} f(t_0, y_0)(t_1 - t_0)^{\alpha})(t_1 - t_0)^{\alpha} & if\ n = 1. \end{aligned} \tag{11.105}$$

References

1. Ascher, U.M., Petzold, L.R.: Computer Methods for Ordinary Differential Equations and Differential-Algebraic Equations. Society for Industrial and Applied Mathematics, Philadelphia (1998). 978-0-89871-412-8
2. Butcher, J.C.: Numerical Methods for Ordinary Differential Equations. Wiley, New York (2003). 978-0-471-96758-3
3. Hairer, E., Nørsett, S.P., Wanner, G.: Solving Ordinary Differential Equations I: Nonstiff Problems. Springer, Berlin, New York (1993). 978-3-540-56670-0
4. Stoer, J., Bulirsch, R.: Introduction to Numerical Analysis (3rd ed.). Springer, Berlin, New York (2002). ISBN 978-0-387-95452-3
5. Baleanu, D., Jajarmi, A., Hajipour, M.: On the nonlinear dynamical systems within the generalized fractional derivatives with Mittag–Leffler kernel. Nonlinear Dyn (2018)

6. Baleanu, D., Jajarmi, A., Hajipour, M.: On the nonlinear dynamical systems within the generalized fractional derivatives with Mittag–Leffler kernel. Nonlinear Dyn. **94**, 397–414 (2018)
7. Leader, J.J.: Numerical Analysis and Scientific Computation. Addison-Wesley, Boston (2004)0-201-73499-0
8. Griffiths, D.V., Smith, I.M.: Numerical Methods for Engineers: A Programming Approach, p. 218. CRC Press, Boca Raton (1991). ISBN 0-8493-8610-1
9. Meijering, E.: A chronology of interpolation: from ancient astronomy to modern signal and image processing. Proc. IEEE **90**(3), 319–342 (2002)
10. Lagrange, J.-L., Leçon C.: Sur l'usage des courbes dans la solution des problèmes. Leçons Elémentaires sur les Mathématiques (in French). Paris. Republished In: Serret, J.-A. (ed.) (1877). Oeuvres de Lagrange, vol. 7, pp. 271–287. Gauthier-Villars (1795)
11. Toufik, M., Atangana, A.: New numerical approximation of fractional derivative with non-local and non-singular kernel: application to chaotic models. Eur. Phys. J. Plus **132**(10), 444 (2017)
12. Atangana, A., Araz, S.İ.: New Numerical Scheme with Newton Polynomial: Theory, Methods, and Applications. Academic Press (2021)

Chapter 12
Numerical Analysis of IVP with Caputo–Fabrizio Global Derivative

12.1 Applying Euler Method on IVP with Caputo–Fabrizio Global Derivative

We provide the finite differences representation of the Euler method for our problems and the subsequent changes made to the algorithm to fit this specific class of ODEs. In this direction several significant research works [1–6] have been developed to numerically solve these complex nonlinear differential equations showing up in several areas of science and technology. The endeavour aims to improve the accuracy, stability, and computational efficiency of numerical methods, with a focus on equations containing complex nonlinear terms, variable coefficients, and complex boundary conditions. As a result of these advancements, researchers have tackled a range of real-world issues, including fluid dynamics, chemical reaction modeling, population dynamics, and heat transfer in heterogeneous media. For example, the following references specify various approaches that may be used for such problems to improve solutions to nonlinear differential equations [7–12], which prospective readers can aim to go through.

$$\begin{cases} {}_{t_0}^{CF}D_g^{\alpha} y(t) = f(t, y(t)) & \text{if } t > 0 \\ y(t_0) = y_0 & \text{if } t = 0 \end{cases}. \tag{12.1}$$

Here we shall rely on the differentiation of the function $g(t)$:

$$\begin{cases} {}_{t_0}^{CF}D_t^{\alpha} y(t) = g'(t) f(t, y(t)) & \text{if } t > 0 \\ y(t_0) = y_0 & \text{if } t = 0 \end{cases}. \tag{12.2}$$

$$y(t) = \ y(t_0) + (1-\alpha) f(t, y(t)) g'(t) + \alpha \int_{t_0}^{t} f(\tau, y(\tau) g'(\tau) d\tau, \tag{12.3}$$

A. Atangana and İ. Koca, *Fractional Differential and Integral Operators with Respect to a Function*, Industrial and Applied Mathematics,
https://doi.org/10.1007/978-981-97-9951-0_12

$$y(t_0) = y_0.$$

At $t = t_{n+1} = (n+1)h$ and $t = t_n = nh$, we have

$$y(t_{n+1}) = y(t_0) + (1-\alpha) f(t_{n+1}, y(t_{n+1}))g'(t_{n+1}) + \alpha \int_{t_0}^{t_{n+1}} f(\tau, y(\tau)g'(\tau)d\tau, \tag{12.4}$$

and

$$y(t_n) = y(t_0) + (1-\alpha) f(t_n, y(t_n))g'(t_n) + \alpha \int_{t_0}^{t_n} f(\tau, y(\tau)g'(\tau)d\tau. \tag{12.5}$$

Subtracting yields

$$\begin{aligned} y(t_{n+1}) =\ & y(t_n) + (1-\alpha) f(t_{n+1}, y(t_{n+1}))g'(t_{n+1}) \\ & -(1-\alpha) f(t_n, y(t_n))g'(t_n) + \alpha \int_{t_n}^{t_{n+1}} f(\tau, y(\tau)g'(\tau)d\tau. \end{aligned} \tag{12.6}$$

Two results could be obtained: the first is

$$\begin{aligned} y_{n+1} =\ & y_n + (1-\alpha) f(t_{n+1}, \overline{y}_{n+1}) \left(\frac{g(t_{n+1}) - g(t_n)}{h} \right) \\ & -(1-\alpha) f(t_n, y_n) \left(\frac{g(t_n) - g(t_{n-1})}{h} \right) \\ & +\alpha \left(g(t_{n+1}) - g(t_n) \right) f(t_n, y_n), \end{aligned} \tag{12.7}$$

where

$$\begin{aligned} \overline{y}_{n+1} =\ & y_n + (1-\alpha) f(t_n, y_n) \left(\frac{g(t_{n+1}) - g(t_n)}{h} \right) \\ & -(1-\alpha) f(t_{n-1}, y_{n-1}) \left(\frac{g(t_n) - g(t_{n-1})}{h} \right) \\ & +\alpha \left(g(t_{n+1}) - g(t_n) \right) f(t_n, y_n). \end{aligned} \tag{12.8}$$

The second result is

$$\begin{aligned} y_{n+1} =\ & y_n + (1-\alpha) f(t_{n+1}, \overline{y}_{n+1}) \left(\frac{g(t_{n+1}) - g(t_n)}{h} \right) \\ & -(1-\alpha) f(t_n, y_n) \left(\frac{g(t_n) - g(t_{n-1})}{h} \right) \end{aligned} \tag{12.9}$$

$$+\alpha\left(g(t_{n+1})-g(t_n)\right)f(t_{n+1},\overline{y}_{n+1}),$$

where $\overline{y}_{n+1}$ is the same as above . Indeed this version is different from the classical case; thus we shall show some analysis.

Theorem 12.1 *If the function $f(t, y(t))$ is Lipschitz with respect to y then, if $t_{n+1} - t_n = h$, $y(t_{n+1})$ is the exact solution at $t = t_{n+1}$ and y_{n+1} is the approximate solution, we have*

$$\lim_{h\to 0}|y(t_{n+1})-y_{n+1}|=0. \tag{12.10}$$

Proof

$$\begin{aligned}
&|y(t_{n+1})-y_{n+1}| \\
&= \left|\begin{array}{c}
y(t_0)-y_0+(1-\alpha)f(t_{n+1},y(t_{n+1}))g'(t_{n+1}) \\
-(1-\alpha)f(t_n,y(t_n))g'(t_n) \\
-(1-\alpha)f(t_{n+1},y_{n+1})\left(\frac{g(t_{n+1})-g(t_n)}{h}\right) \\
+(1-\alpha)f(t_n,y_n)\left(\frac{g(t_n)-g(t_{n-1})}{h}\right) \\
+\alpha\int_{t_n}^{t_{n+1}} f(\tau,y(\tau))g'(\tau)d\tau-\alpha\int_{t_n}^{t_{n+1}} f(t_n,y_n)g'(\tau)d\tau
\end{array}\right| \\
&\le |y(t_0)-y_0|+(1-\alpha)\left\|g'\right\|_\infty|f(t_{n+1},y(t_{n+1}))-f(t_{n+1},y_{n+1})| \\
&\quad+(1-\alpha)\left\|g'\right\|_\infty|f(t_n,y(t_n))-f(t_n,y_n)| \\
&\quad+\left\|g'\right\|_\infty\int_{t_n}^{t_{n+1}}|f(\tau,y(\tau)-f(t_n,y_n)|\,d\tau.
\end{aligned} \tag{12.11}$$

We use the Lipschitz condition and the differentiability of $f(t, y(t))$ with respect to t to have

$$\begin{aligned}
|y(t_{n+1})-y_{n+1}| \le\ & (1-\alpha)\left\|g'\right\|_\infty L\,|y(t_{n+1})-y_{n+1}| \\
&+(1-\alpha)\left\|g'\right\|_\infty\left\|f'(.,y(.))\right\|_\infty h \\
&+\left\|g'\right\|_\infty\alpha\frac{h^2}{2}\left\|f'(.,y(.))\right\|_\infty,
\end{aligned} \tag{12.12}$$

$$\begin{aligned}
\lim_{h\to 0}|y(t_{n+1})-y_{n+1}| \le\ & \lim_{h\to 0}\frac{(1-\alpha)\left\|g'\right\|_\infty\left\|f'(.,y(.))\right\|_\infty h}{\left(1-(1-\alpha)\|g'\|_\infty L\right)} \\
&+\lim_{h\to 0}\frac{\left\|g'\right\|_\infty\alpha\frac{h^2}{2}\left\|f'(.,y(.))\right\|_\infty}{\left(1-(1-\alpha)\|g'\|_\infty L\right)} \\
=\ & 0.
\end{aligned}$$

Therefore

$$\lim_{h \to 0} |y(t_{n+1}) - y_{n+1}| = 0. \tag{12.13}$$

12.2 Applying Heun's Method on IVP with Caputo–Fabrizio Global Derivative

In the current section we introduce an extension of Heun's method applied to the numerical solution of the Caputo-Fabrizio global initial value problem (IVP). The Caputo-Fabrizio fractional derivative has emerged as a versatile and remarkable approach in the mathematical modeling of systems exhibiting memory effects and anomalous diffusion, garnering considerable interest in recent years owing to its potential to provide more accurate representations of physical phenomena. This paper utilizes Heun's method, a second-order numerical method, which can be applied for the solution of the ordinary differential equations and its extension to the corresponding fractional case would yield an efficacious algorithm for solving the Caputo-Fabrizio fractional IVP.

Since the Caputo-Fabrizio derivative is a global operator, initial conditions and fractional order need to be addressed simultaneously, which adds complexity as compared to normal differential equations. To overcome these difficulties with conventional methods, we extend Heun's method as we will incorporate the fractional derivative into the iterative process of Heun's method which can solve with high accuracy on a large time interval. This technique consists of a forecasting phase followed by a correction phase, and it has improved considerably the convergence speed and accuracy of the numerical resolution compared to single-step methods.

In this way, the fractional derivative is approximated by a series of steps in a way that keeps the memory effects of the process, so that we offer a very efficient algorithm for the numerical solving of processes of the anomalous diffusion type, as well as of processes of relaxation and other kind of phenomena described by the Caputo-Fabrizio derivative. This is a useful extension of the third-order colony predictor corrector which is applicable to global initial value problems in fractional-order differential equations; suitable for a number of scientific and engineering applications.

$$\begin{cases} {}^{CF}_{t_0}D^{\alpha}_{g} y(t) = f(t, y(t)) & \text{if } t > 0 \\ y(t_0) = y_0 & \text{if } t = 0 \end{cases}. \tag{12.14}$$

Here we assume that $g'(t) > 0$. Also here we will need in addition that $f(t, y(t))$ satisfies the Lipschitz condition and is bounded within $[a, b]$. By the fundamental theorem of calculus, the above IVP can be transformed as

$$
\begin{aligned}
y(t) = {} & y(t_0) + (1-\alpha) f(t, y(t)) g'(t) + \alpha \int_{t_0}^{t} f(\tau, y(\tau) g'(\tau) d\tau, && (12.15) \\
y(t_0) = {} & y_0.
\end{aligned}
$$

At $t = t_{n+1}$, we have

$$
\begin{aligned}
y(t_{n+1}) = {} & y(t_0) + (1-\alpha) f(t_{n+1}, y(t_{n+1})) g'(t_{n+1}) \\
& + \alpha \int_{t_0}^{t_{n+1}} f(\tau, y(\tau) g'(\tau) d\tau,
\end{aligned}
\tag{12.16}
$$

at $t = t_n$, we have

$$
\begin{aligned}
y(t_n) = {} & y(t_0) + (1-\alpha) f(t_n, y(t_n)) g'(t_n) \\
& + \alpha \int_{t_0}^{t_n} f(\tau, y(\tau) g'(\tau) d\tau.
\end{aligned}
\tag{12.17}
$$

If we subtract the last two equations we get

$$
\begin{aligned}
y(t_{n+1}) - y(t_n) = {} & (1-\alpha) f(t_{n+1}, y(t_{n+1})) g'(t_{n+1}) \\
& - (1-\alpha) f(t_n, y(t_n)) g'(t_n) + \alpha \int_{t_n}^{t_{n+1}} f(\tau, y(\tau) g'(\tau) d\tau.
\end{aligned}
\tag{12.18}
$$

We approximate $f(\tau, y(\tau)$ within $[t_n, t_{n+1}]$ using

$$
f(\tau, y(\tau)) \simeq \frac{f(t_{n+1}, y_{n+1}) + f(t_n, y_n)}{2} \tag{12.19}
$$

Then we will get

$$
\begin{aligned}
y(t_{n+1}) - y(t_n) = {} & (1-\alpha) f(t_{n+1}, y(t_{n+1})) g'(t_{n+1}) \\
& - (1-\alpha) f(t_n, y(t_n)) g'(t_n) \\
& + \alpha \int_{t_n}^{t_{n+1}} \frac{f(t_{n+1}, y_{n+1}) + f(t_n, y_n)}{2} g'(\tau) d\tau.
\end{aligned}
\tag{12.20}
$$

So we have

$$
\begin{aligned}
y(t_{n+1}) = {} & y(t_n) + (1-\alpha) f(t_{n+1}, y(t_{n+1})) g'(t_{n+1}) \\
& - (1-\alpha) f(t_n, y(t_n)) g'(t_n)
\end{aligned}
\tag{12.21}
$$

$$+\alpha\left(\frac{f(t_{n+1},y_{n+1})+f(t_n,y_n)}{2}\right)(g(t_{n+1})-g(t_n)).$$

Then we have

$$\begin{aligned}y_{n+1}= & \; y_n+(1-\alpha)f(t_{n+1},\overline{y}_{n+1})\left(\frac{g(t_{n+1})-g(t_n)}{h}\right)\\ & -(1-\alpha)f(t_n,y_n)\left(\frac{g(t_n)-g(t_{n-1})}{h}\right)\\ & +\alpha\left(\frac{f(t_{n+1},\overline{y}_{n+1})+f(t_n,y_n)}{2}\right)(g(t_{n+1})-g(t_n)),\end{aligned} \tag{12.22}$$

where

$$\begin{aligned}\overline{y}_{n+1}= & \; y_n+(1-\alpha)f(t_n,y_n)\left(\frac{g(t_{n+1})-g(t_n)}{h}\right)\\ & -(1-\alpha)f(t_{n-1},y_{n-1})\left(\frac{g(t_n)-g(t_{n-1})}{h}\right)\\ & +\alpha f(t_n,y_n)(g(t_{n+1})-g(t_n)).\end{aligned} \tag{12.23}$$

The second approach is to approximate $g'(\tau)f(\tau,y(\tau)$ as

$$g'(\tau)f(\tau,y(\tau)\simeq\frac{1}{2}\left(g'(t_{n+1})f(t_{n+1},y_{n+1})+g'(t_n)f(t_n,y_n)\right). \tag{12.24}$$

Then we have

$$\begin{aligned}y_{n+1}= & \; y_n+(1-\alpha)f(t_{n+1},\overline{y}_{n+1})\left(\frac{g(t_{n+1})-g(t_n)}{h}\right)\\ & -(1-\alpha)f(t_n,y_n)\left(\frac{g(t_n)-g(t_{n-1})}{h}\right)\\ & +\frac{\alpha}{2}\left(\frac{g(t_{n+1})-g(t_n)}{h}\right)f(t_{n+1},\overline{y}_{n+1})\\ & +\frac{\alpha}{2}(g(t_n)-g(t_{n-1}))f(t_n,y_n),\end{aligned} \tag{12.25}$$

where

$$\begin{aligned}\overline{y}_{n+1}= & \; y_n+(1-\alpha)f(t_n,y_n)\left(\frac{g(t_{n+1})-g(t_n)}{h}\right)\\ & -(1-\alpha)f(t_{n-1},y_{n-1})\left(\frac{g(t_n)-g(t_{n-1})}{h}\right)\\ & +\alpha f(t_n,y_n)(g(t_n)-g(t_{n-1})).\end{aligned} \tag{12.26}$$

Theorem 12.2 *Let $\overline{y}_n$, be the perturbed term of y_n, then $\forall n \geq 0$, $\exists c > 0$ such that*

$$|\widetilde{y}_{n+1}| < c\,|\widetilde{y}_0|\,. \tag{12.27}$$

Proof Let us put our iteration equation below

$$\begin{aligned} y_{n+1} = {} & y_n + (1-\alpha) f(t_{n+1}, y_{n+1}) \left(\frac{g(t_{n+1}) - g(t_n)}{h}\right) \\ & -(1-\alpha) f(t_n, y_n) \left(\frac{g(t_n) - g(t_{n-1})}{h}\right) \\ & +\frac{\alpha}{2}\left(\frac{g(t_{n+1}) - g(t_n)}{h}\right) f(t_{n+1}, y_{n+1}) \\ & +\frac{\alpha}{2}\left(g(t_n) - g(t_{n-1})\right) f(t_n, y_n), \end{aligned} \tag{12.28}$$

with perturbed term

$$\begin{aligned} \widetilde{y}_{n+1} + y_{n+1} = {} & y_n + \widetilde{y}_n \\ & +(1-\alpha) f(t_{n+1}, \widetilde{y}_{n+1} + y_{n+1}) \left(\frac{g(t_{n+1}) - g(t_n)}{h}\right) \\ & -(1-\alpha) f(t_n, y_n + \widetilde{y}_n) \left(\frac{g(t_n) - g(t_{n-1})}{h}\right) \\ & +\frac{\alpha}{2} f(t_{n+1}, \widetilde{y}_{n+1} + y_{n+1}) \left(\frac{g(t_{n+1}) - g(t_n)}{h}\right) \\ & +\frac{\alpha}{2} f(t_n, y_n + \widetilde{y}_n) \left(g(t_n) - g(t_{n-1})\right). \end{aligned} \tag{12.29}$$

If we rearrange the above, we get

$$\begin{aligned} \widetilde{y}_{n+1} = {} & \widetilde{y}_n + (1-\alpha)\left(f(t_{n+1}, \widetilde{y}_{n+1} + y_{n+1})\right. \\ & \left.- f(t_{n+1}, y_{n+1})\right) \left(\frac{g(t_{n+1}) - g(t_n)}{h}\right) \\ & -(1-\alpha)\left(f(t_n, y_n + \widetilde{y}_n) - f(t_n, y_n)\right) \left(\frac{g(t_n) - g(t_{n-1})}{h}\right) \\ & +\frac{\alpha}{2}\left(f(t_{n+1}, \widetilde{y}_{n+1} + y_{n+1}) - f(t_{n+1}, y_{n+1})\right) \left(\frac{g(t_{n+1}) - g(t_n)}{h}\right) \\ & +\frac{\alpha}{2}\left(f(t_n, y_n + \widetilde{y}_n) - f(t_n, y_n)\right)\left(g(t_n) - g(t_{n-1})\right). \end{aligned} \tag{12.30}$$

Then applying the norm on both sides yields

$$\begin{aligned} |\widetilde{y}_{n+1}| \leq {} & |\widetilde{y}_n| + (1-\alpha)\left|f(t_{n+1}, \widetilde{y}_{n+1} + y_{n+1}) - f(t_{n+1}, y_{n+1})\right| \left|g'(t_{n+1})\right| \\ & -(1-\alpha)\left|f(t_n, y_n + \widetilde{y}_n) - f(t_n, y_n)\right| \left|g'(t_n)\right| \end{aligned} \tag{12.31}$$

$$+\frac{\alpha}{2}\left|f(t_{n+1},\widetilde{y}_{n+1}+y_{n+1})-f(t_{n+1},y_{n+1})\right|\left|g'(t_{n+1})\right|$$
$$+\frac{\alpha h}{2}\left|f(t_n,y_n+\widetilde{y}_n)-f(t_n,y_n)\right|\left|g'(t_n)\right|.$$

Now using the Lipschitz condition of f with respect to y we obtain

$$\begin{aligned}|\widetilde{y}_{n+1}| \leq\ & |\widetilde{y}_n|+(1-\alpha)\left\|g'\right\|_\infty L\left|\widetilde{y}_{n+1}\right| \\ & -(1-\alpha)\left\|g'\right\|_\infty L\left|\widetilde{y}_n\right|+\frac{\alpha}{2}\left\|g'\right\|_\infty L\left|\widetilde{y}_{n+1}\right|+\frac{\alpha h}{2}\left\|g'\right\|_\infty L\left|\widetilde{y}_n\right| \\ \leq\ & |\widetilde{y}_n|+(1-\alpha)\left\|g'\right\|_\infty L\left\{\left|\widetilde{y}_{n+1}\right|-\left|\widetilde{y}_n\right|\right\}+\frac{\alpha}{2}\left\|g'\right\|_\infty L\left\{\left|\widetilde{y}_{n+1}\right|+h\left|\widetilde{y}_n\right|\right\}.\end{aligned} \tag{12.32}$$

On the other side

$$\begin{aligned}\overline{y}_{n+1} =\ & y_n+(1-\alpha)f(t_n,y_n)g'(t_{n+1}) \\ & -(1-\alpha)f(t_{n-1},y_{n-1})g'(t_n)+\alpha hf\left(t_n,y_n\right)g'(t_n).\end{aligned} \tag{12.33}$$

$$\begin{aligned}\widetilde{\overline{y}}_{n+1}+\overline{y}_{n+1} =\ & y_n+\widetilde{y}_n+(1-\alpha)\left(f(t_n,y_n+\widetilde{y}_n)\right. \\ & \left.-f(t_n,y_n)\right)g'(t_{n+1}) \\ & -(1-\alpha)\left(f(t_{n-1},y_{n-1}+\widetilde{y}_{n-1})-f(t_{n-1},y_{n-1})\right)g'(t_n) \\ & +\alpha h\left(f\left(t_n,y_n+\widetilde{y}_n\right)-f\left(t_n,y_n\right)\right)g'(t_n).\end{aligned} \tag{12.34}$$

Now using the Lipschitz condition of f with respect to y and applying the norm on both sides yields

$$\begin{aligned}\left|\widetilde{\overline{y}}_{n+1}\right| \leq\ & |\widetilde{y}_n|+(1-\alpha)\left\|g'\right\|_\infty L\left|\widetilde{y}_n\right| \\ & -(1-\alpha)\left\|g'\right\|_\infty L\left|\widetilde{y}_{n-1}\right|+\alpha h\left\|g'\right\|_\infty L\left|\widetilde{y}_n\right| \\ \leq\ & |\widetilde{y}_n|+(1-\alpha)\left\|g'\right\|_\infty L\left|\widetilde{y}_n\right|+\alpha h\left\|g'\right\|_\infty L\left|\widetilde{y}_n\right|.\end{aligned} \tag{12.35}$$

If we put this in the below equation

$$|\widetilde{y}_{n+1}|\leq|\widetilde{y}_n|+(1-\alpha)\left\|g'\right\|_\infty L\left\{\left|\widetilde{y}_{n+1}\right|-\left|\widetilde{y}_n\right|\right\}+\frac{\alpha}{2}\left\|g'\right\|_\infty L\left\{\left|\widetilde{y}_{n+1}\right|+h\left|\widetilde{y}_n\right|\right\} \tag{12.36}$$

$$\begin{aligned}|\widetilde{y}_{n+1}|\leq\ & |\widetilde{y}_n|+(1-\alpha)\left\|g'\right\|_\infty L\left\{\begin{array}{c}|\widetilde{y}_n|+(1-\alpha)\left\|g'\right\|_\infty L\left|\widetilde{y}_n\right| \\ +\alpha h\left\|g'\right\|_\infty L\left|\widetilde{y}_n\right|-\left|\widetilde{y}_n\right|\end{array}\right\} \\ & +\frac{\alpha}{2}\left\|g'\right\|_\infty L\left\{\begin{array}{c}|\widetilde{y}_n|+(1-\alpha)\left\|g'\right\|_\infty L\left|\widetilde{y}_n\right| \\ +\alpha h\left\|g'\right\|_\infty L\left|\widetilde{y}_n\right|+h\left|\widetilde{y}_n\right|\end{array}\right\}.\end{aligned} \tag{12.37}$$

If we rearrange

$$\begin{aligned} |\widetilde{y}_{n+1}| &\leq |\widetilde{y}_n| \left\{ \begin{array}{c} 1+(1-\alpha)^2\left(\|g'\|_\infty\right)^2 L^2+(1-\alpha)\left(\|g'\|_\infty\right)^2 h\alpha L \\ +\frac{\alpha}{2}\|g'\|_\infty L+\frac{(1-\alpha)\alpha}{2}\left(\|g'\|_\infty\right)^2 L \\ +\frac{\alpha^2 h}{2}\left(\|g'\|_\infty\right)^2 L+\frac{\alpha h}{2}\|g'\|_\infty L \end{array} \right\} \\ &\leq |\widetilde{y}_n|\,\Omega \end{aligned} \tag{12.38}$$

where

$$\Omega=\left\{ \begin{array}{c} 1+(1-\alpha)^2\left(\|g'\|_\infty\right)^2 L^2+(1-\alpha)\left(\|g'\|_\infty\right)^2 h\alpha L \\ +\frac{\alpha}{2}\|g'\|_\infty L+\frac{(1-\alpha)\alpha}{2}\left(\|g'\|_\infty\right)^2 L \\ +\frac{\alpha^2 h}{2}\left(\|g'\|_\infty\right)^2 L+\frac{\alpha h}{2}\|g'\|_\infty L \end{array} \right\}. \tag{12.39}$$

Then

$$\begin{aligned} |\widetilde{y}_1| &\leq |\widetilde{y}_0|\,\Omega, \\ &\vdots \\ |\widetilde{y}_{n+1}| &\leq |\widetilde{y}_0|\,\Omega^{n+1}, \\ &\leq |\widetilde{y}_0|\,c, \end{aligned} \tag{12.40}$$

which completes the proof.

Theorem 12.3 *$\forall n \geq 0$, if $y(t_{n+1})$ is the exact solution at $y(t_{n+1})$ and y_{n+1} approximate solution*

$$\lim_{h\to 0}|y(t_{n+1})-y_{n+1}|=0 \tag{12.41}$$

if $f(t, y(t))$ is differentiable.

Proof

$$\begin{aligned} &|y(t_{n+1})-y_{n+1}| \\ &= \left| \begin{array}{c} y(t_n)-y_n+(1-\alpha)f(t_{n+1},y(t_{n+1}))g'(t_{n+1}) \\ -(1-\alpha)f(t_n,y(t_n))g'(t_n)+\alpha\displaystyle\int_{t_n}^{t_{n+1}} f(\tau,y(\tau))g'(\tau)d\tau \\ -(1-\alpha)f(t_{n+1},y_{n+1})g'(t_{n+1}) \\ +(1-\alpha)f(t_n,y_n)g'(t_n) \\ -\frac{\alpha}{2}\left(\frac{g(t_{n+1})-g(t_n)}{h}\right)f(t_{n+1},y_{n+1}) \\ -\frac{\alpha}{2}\left(g(t_n)-g(t_{n-1})\right)f(t_n,y_n), \end{array} \right| \end{aligned} \tag{12.42}$$

$$= \left| \begin{array}{l} y(t_n) - y_n + \alpha \int\limits_{t_n}^{t_{n+1}} f(\tau, y(\tau))g'(\tau)d\tau \\ -\alpha \int\limits_{t_n}^{t_{n+1}} \frac{f(t_{n+1}, y_{n+1}) + f(t_n, y_n)}{2} g'(\tau)d\tau, \end{array} \right|$$

$$\begin{aligned} &\leq |y(t_n) - y_n| \\ &+\frac{\alpha}{2}\left| \int\limits_{t_n}^{t_{n+1}} f(\tau, y(\tau))g'(\tau)d\tau - \int\limits_{t_n}^{t_{n+1}} f(t_{n+1}, y_{n+1})g'(\tau)d\tau \right| \\ &\frac{\alpha}{2}\left| \int\limits_{t_n}^{t_{n+1}} f(\tau, y(\tau))g'(\tau)d\tau - \int\limits_{t_n}^{t_{n+1}} f(t_n, y_n)g'(\tau)d\tau \right| \end{aligned} \quad (12.43)$$

$$\begin{aligned} &\leq |y(t_n) - y_n| + \frac{\alpha}{2} \int\limits_{t_n}^{t_{n+1}} |f(\tau, y(\tau)) - f(t_{n+1}, y_{n+1})| \left|g'(\tau)\right| d\tau \\ &+\frac{\alpha}{2} \int\limits_{t_n}^{t_{n+1}} |f(\tau, y(\tau)) - f(t_n, y_n)| \left|g'(\tau)\right| d\tau \\ &\leq |y(t_n) - y_n| + \frac{\alpha}{2} \left\|g'\right\|_\infty \left\|f'(., y(.))\right\|_\infty \int\limits_{t_n}^{t_{n+1}} (t_{n+1} - \tau)\, d\tau \\ &+\frac{\alpha}{2} \left\|g'\right\|_\infty \left\|f'(., y(.))\right\|_\infty \int\limits_{t_n}^{t_{n+1}} (t_n - \tau)\, d\tau \\ &\leq |y(t_n) - y_n| + \frac{\alpha}{2} \left\|g'\right\|_\infty \left\|f'(., y(.))\right\|_\infty \frac{h^2}{2} + \frac{\alpha}{2} \left\|g'\right\|_\infty \left\|f'(., y(.))\right\|_\infty \frac{h}{2} \end{aligned}$$

$$\lim_{h\to 0} |y(t_{n+1}) - y_{n+1}| \leq \lim_{h\to 0} |y(t_n) - y_n|, \forall n \geq 0. \quad (12.44)$$

When $n = 0$, we have

$$\begin{aligned} \lim_{h\to 0} |y(t_1) - y_1| &\leq \lim_{h\to 0} |y(t_0) - y_0| = 0 \\ \lim_{h\to 0} |y(t_2) - y_2| &\leq \lim_{h\to 0} |y(t_1) - y_1| = 0 \\ &\vdots \\ \lim_{h\to 0} |y(t_{n+1}) - y_{n+1}| &= 0 \end{aligned} \quad (12.45)$$

which completes the proof.

12.3 Applying Midpoint Method on IVP with Caputo–Fabrizio Global Derivative

In this section, we solve the classical global differential equation using the Midpoint approach. We consider the following equation:

$$\begin{cases} {}_{t_0}^{CF}D_g y(t) = f(t, y(t)) \text{ if } t > 0 \\ y(t_0) = y_0 \qquad\qquad \text{ if } t = 0 \end{cases}. \tag{12.46}$$

We assume that $g'(t) > 0$, then we have

$$\begin{aligned} y(t) = & \; y(t_0) + (1-\alpha) f(t, y(t)) g'(t) + \alpha \int_{t_0}^{t} f(\tau, y(\tau) g'(\tau) d\tau, \\ y(t_0) = & \; y_0. \end{aligned} \tag{12.47}$$

At $t = t_{n+1} = (n+1)h$ and $t = t_n = nh$, we have

$$y(t_{n+1}) = y(t_0) + (1-\alpha) f(t_{n+1}, y(t_{n+1})) g'(t_{n+1}) + \alpha \int_{t_0}^{t_{n+1}} f(\tau, y(\tau) g'(\tau) d\tau, \tag{12.48}$$

and

$$y(t_n) = y(t_0) + (1-\alpha) f(t_n, y(t_n)) g'(t_n) + \alpha \int_{t_0}^{t_n} f(\tau, y(\tau) g'(\tau) d\tau. \tag{12.49}$$

Subtracting yields

$$\begin{aligned} y(t_{n+1}) = & \; y(t_n) + (1-\alpha) f(t_{n+1}, y(t_{n+1})) g'(t_{n+1}) \\ & -(1-\alpha) f(t_n, y(t_n)) g'(t_n) \\ & +\alpha \int_{t_n}^{t_{n+1}} f(\tau, y(\tau) g'(\tau) d\tau. \end{aligned} \tag{12.50}$$

We shall make use of two ways to derive this. The first way is to approximate only $f(\tau, y(\tau))$ using the midpoint

$$\begin{aligned} y_{n+1} = & \; y_n + (1-\alpha) f(t_{n+1}, \bar{y}_{n+1}) \left(\frac{g(t_{n+1}) - g(t_n)}{h} \right) \\ & -(1-\alpha) f(t_n, y_n) \left(\frac{g(t_n) - g(t_{n-1})}{h} \right) \end{aligned}$$

$$
\begin{aligned}
&+\alpha\left(g(t_{n+1})-g(t_n)\right)f\left(t_n+\frac{h}{2},\frac{y_n+\overline{y}_{n+1}}{2}\right),\\
\overline{y}_{n+1} = {}& y_n+(1-\alpha)f(t_n,y_n)\left(\frac{g(t_{n+1})-g(t_n)}{h}\right)\\
&-(1-\alpha)f(t_{n-1},y_{n-1})\left(\frac{g(t_n)-g(t_{n-1})}{h}\right)+\alpha f(t_n,y_n)\left(g\left(t_{n+1}\right)-g(t_n)\right).
\end{aligned}
$$

The second way is to approximate

$$
f(\tau,y(\tau))g'(\tau)=F(t,y(t)). \tag{12.51}
$$

By the above replacement we have

$$
\begin{aligned}
y_{n+1} &= y_n+(1-\alpha)f(t_{n+1},y_{n+1})\left(\frac{g(t_{n+1})-g(t_n)}{h}\right) \\
&\quad -(1-\alpha)f(t_n,y_n)\left(\frac{g(t_n)-g(t_{n-1})}{h}\right)\\
&\quad +h\alpha F\left(t_n+\frac{h}{2},y_n+\frac{h}{2}F(t_n,y_n)\right),\\
&= y_n+(1-\alpha)f(t_{n+1},y_{n+1})g'(t_{n+1})-(1-\alpha)f(t_n,y_n)g'(t_n)\\
&\quad +h\alpha g'(t_n+\frac{h}{2})f\left(t_n+\frac{h}{2},y_n+\frac{h}{2}g'(t_n)f(t_n,y_n)\right),\\
&= y_n+(1-\alpha)f(t_{n+1},y_{n+1})g'(t_{n+1})-(1-\alpha)f(t_n,y_n)g'(t_n)\\
&\quad +h\alpha\left(\frac{g(t_{n+1})-g(t_n)}{h}\right)f\left(t_n+\frac{h}{2},y_n+\frac{h}{2}\left(\frac{g(t_{n+1})-g(t_n)}{h}\right)f(t_n,y_n)\right),\\
&= y_n+(1-\alpha)f(t_{n+1},y_{n+1})g'(t_{n+1})-(1-\alpha)f(t_n,y_n)g'(t_n)\\
&\quad +\alpha\left(g(t_{n+1})-g(t_n)\right)f\left(t_n+\frac{h}{2},y_n+\frac{1}{2}\left(g(t_{n+1})-g(t_n)\right)f(t_n,y_n)\right).
\end{aligned} \tag{12.52}
$$

We recover the same formula. By considering

$$
M=\max\left|g'(t)\right|,\ \forall t\in[0,T], \tag{12.53}
$$

we can show that the scheme is stable and consistent.

12.4 Applying Linear Approximation Method on IVP with Caputo–Fabrizio Global Derivative

In this section, we solve the IVP problem with a numerical method based on the linear approximation. We the consider the following general IVP:

$$\begin{cases} {}_{t_0}^{CF}D_g y(t) = f(t, y(t)) & \text{if } t > 0 \\ y(t_0) = y_0 & \text{if } t = 0 \end{cases}. \tag{12.54}$$

We assume that $g'(t) > 0$, then we have

$$\begin{aligned} y(t) = & \; y(t_0) + (1-\alpha) f(t, y(t)) g'(t) + \alpha \int_{t_0}^{t} f(\tau, y(\tau) g'(\tau) d\tau, \quad (12.55) \\ y(t_0) = & \; y_0. \end{aligned}$$

At $t = t_{n+1} = (n+1)h$ and $t = t_n = nh$, we have

$$y(t_{n+1}) = y(t_0) + (1-\alpha) f(t_{n+1}, y(t_{n+1})) g'(t_{n+1}) + \alpha \int_{t_0}^{t_{n+1}} f(\tau, y(\tau) g'(\tau) d\tau, \tag{12.56}$$

and

$$y(t_n) = y(t_0) + (1-\alpha) f(t_n, y(t_n)) g'(t_n) + \alpha \int_{t_0}^{t_n} f(\tau, y(\tau) g'(\tau) d\tau. \tag{12.57}$$

Subtracting yields

$$\begin{aligned} y(t_{n+1}) = & \; y(t_n) + (1-\alpha) f(t_{n+1}, y(t_{n+1})) g'(t_{n+1}) \\ & -(1-\alpha) f(t_n, y(t_n)) g'(t_n) \\ & +\alpha \int_{t_n}^{t_{n+1}} f(\tau, y(\tau) g'(\tau) d\tau. \end{aligned} \tag{12.58}$$

We will apply two approaches: the first will be to approximate only $f(\tau, y(\tau))$ with the linear approximation. The second will be to approximate

$$F(\tau, y(\tau)) = f(\tau, y(\tau)) g'(\tau). \tag{12.59}$$

In this first approach, we approximate

$$f(\tau, y(\tau)) \simeq f(t_n, y_n) + \frac{f(t_{n+1}, y_{n+1}) - f(t_n, y_n)}{h} (\tau - t_n). \tag{12.60}$$

Replacing yields

$$\begin{aligned} y_{n+1} & = y_n + (1-\alpha) f(t_{n+1}, y(t_{n+1})) g'(t_{n+1}) \\ & \quad -(1-\alpha) f(t_n, y(t_n)) g'(t_n) \end{aligned} \tag{12.61}$$

$$
\begin{aligned}
&+\alpha \int_{t_n}^{t_{n+1}} \left(f(t_n, y_n) + \frac{f(t_{n+1}, y_{n+1}) - f(t_n, y_n)}{h} (\tau - t_n) \right) g'(\tau) d\tau, \\
&= y_n + (1-\alpha) f(t_{n+1}, y(t_{n+1})) g'(t_{n+1}) \\
&\quad + \alpha f(t_n, y_n) \left(g(t_{n+1}) - g(t_n) \right) - (1-\alpha) f(t_n, y(t_n)) g'(t_n) \\
&\quad + \alpha \left(\frac{f(t_{n+1}, y_{n+1}) - f(t_n, y_n)}{h} \right) \left(g(t_{n+1}) - g(t_n) \right) \int_{t_n}^{t_{n+1}} (\tau - t_n) \, d\tau, \\
&= y_n + (1-\alpha) f(t_{n+1}, y(t_{n+1})) \left(\frac{g(t_{n+1}) - g(t_n)}{h} \right) \\
&\quad - (1-\alpha) f(t_n, y(t_n)) \left(\frac{g(t_n) - g(t_{n-1})}{h} \right) \\
&\quad + \alpha f(t_n, y_n) \left(g(t_{n+1}) - g(t_n) \right) \\
&\quad + \alpha \left(\frac{f(t_{n+1}, y_{n+1}) - f(t_n, y_n)}{h} \right) \frac{(t_{n+1} - t_n)^2}{2} \left(g(t_{n+1}) - g(t_n) \right), \\
&= y_n + (1-\alpha) f(t_{n+1}, y(t_{n+1})) \left(\frac{g(t_{n+1}) - g(t_n)}{h} \right) \\
&\quad - (1-\alpha) f(t_n, y(t_n)) \left(\frac{g(t_n) - g(t_{n-1})}{h} \right) \\
&\quad + \alpha f(t_n, y_n) \left(g(t_{n+1}) - g(t_n) \right) \\
&\quad + \alpha \left(\frac{f(t_{n+1}, y_{n+1}) - f(t_n, y_n)}{2} \right) h \left(g(t_{n+1}) - g(t_n) \right).
\end{aligned}
$$

Therefore

$$
\begin{aligned}
y_{n+1} = \; & y_n + (1-\alpha) f(t_{n+1}, y(t_{n+1})) h^{-1} \left(g(t_{n+1}) - g(t_n) \right) \\
& - (1-\alpha) f(t_n, y(t_n)) h^{-1} \left(g(t_n) - g(t_{n-1}) \right) \\
& + \alpha f(t_n, y_n) \left(g(t_{n+1}) - g(t_n) \right) \\
& + \alpha \left(\frac{f(t_{n+1}, y_{n+1}) - f(t_n, y_n)}{2} \right) h \left(g(t_{n+1}) - g(t_n) \right).
\end{aligned} \tag{12.62}
$$

The second way: we approximate

$$
\begin{aligned}
F(\tau, y(\tau)) = \; & f(\tau, y(\tau)) g'(\tau) \simeq F(t_n, y_n) \\
& + \frac{F(t_{n+1}, y_{n+1}) - F(t_n, y_n)}{h} (t_{n+1} - \tau), \\
\simeq \; & f(t_n, y_n) g'(t_n) \\
& + \frac{f(t_{n+1}, y_{n+1}) g'(t_{n+1}) - f(t_n, y_n) g'(t_n)}{h} (t_{n+1} - \tau),
\end{aligned} \tag{12.63}
$$

and replacing yields

$$
\begin{aligned}
y_{n+1} = {} & y_n + (1-\alpha) f(t_{n+1}, y(t_{n+1})) h^{-1} \left(g(t_{n+1}) - g(t_n)\right) \\
& -(1-\alpha) f(t_n, y(t_n)) h^{-1} \left(g(t_n) - g(t_{n-1})\right) \\
& +\alpha f(t_n, y_n) \left(g(t_{n+1}) - g(t_n)\right) \\
& +\alpha \begin{pmatrix} \frac{g(t_{n+1})-g(t_n)}{h} f(t_{n+1}, y_{n+1}) \\ -\frac{g(t_{n+1})-g(t_n)}{h} f(t_n, y_n) \end{pmatrix} \int_{t_n}^{t_{n+1}} \frac{(\tau - t_{n+1})}{h} d\tau,
\end{aligned} \tag{12.64}
$$

$$
\begin{aligned}
y_{n+1} = {} & y_n + (1-\alpha) f(t_{n+1}, y(t_{n+1})) h^{-1} \left(g(t_{n+1}) - g(t_n)\right) \\
& -(1-\alpha) f(t_n, y(t_n)) h^{-1} \left(g(t_n) - g(t_{n-1})\right) \\
& +\alpha f(t_n, y_n) \left(g(t_{n+1}) - g(t_n)\right) \\
& +\alpha \left(\frac{g(t_{n+1}) - g(t_n)}{2}\right) \left(f(t_{n+1}, y_{n+1}) - f(t_n, y_n)\right).
\end{aligned}
$$

Both yield an implicit scheme; we shall make it explicit:

$$
\begin{aligned}
y_{n+1} = {} & y_n + (1-\alpha) f(t_{n+1}, \overline{y}(t_{n+1})) h^{-1} \left(g(t_{n+1}) - g(t_n)\right) \\
& -(1-\alpha) f(t_n, y(t_n)) h^{-1} \left(g(t_n) - g(t_{n-1})\right) \\
& +\alpha f(t_n, y_n) \left(g(t_{n+1}) - g(t_n)\right) \\
& +\alpha \frac{g(t_{n+1}) - g(t_n)}{2} \left(f(t_{n+1}, \overline{y}_{n+1}) - f(t_n, y_n)\right),
\end{aligned} \tag{12.65}
$$

where

$$
\begin{aligned}
\overline{y}_{n+1} = {} & y_n + (1-\alpha) f(t_n, y_n) h^{-1} \left(g(t_{n+1}) - g(t_n)\right) \\
& -(1-\alpha) f(t_{n-1}, y_{n-1}) h^{-1} \left(g(t_n) - g(t_{n-1})\right) \\
& +\alpha f(t_n, y_n) \left(g(t_{n+1}) - g(t_n)\right).
\end{aligned} \tag{12.66}
$$

We shall present in detail the stability, consistency, and convergence analysis.

Theorem 12.4 *Assuming that $\|g\|_\infty < M_1$, then the error is*

$$
\left|\overline{R}_T\right| < \alpha M_1 \left\| f''(., y(.)) \right\|_\infty \frac{h^3}{8} \tag{12.67}
$$

if $f(., y(.))$ has a bounded second derivative.

Proof We call R_T the error of the approximation and is defined as

$$
R_T = f(\tau, y(\tau)) - P_n(\tau) \tag{12.68}
$$

where

$$
\begin{aligned}
P_n \simeq {} & (1-\alpha) f(t_{n+1}, y(t_{n+1})) g'(t_{n+1}) - (1-\alpha) f(t_n, y(t_n)) g'(t_n) \\
& +\alpha g'(t_{n+1}) f(t_n, y_n) + \alpha g'(t_{n+1}) \left(f(t_{n+1}, y_{n+1}) - f(t_n, y_n)\right) (\tau - t_n),
\end{aligned} \tag{12.69}
$$

$$R_T = \frac{(\tau - t_n)(\tau - t_{n+1})}{2} f''(\xi, y(\xi)),$$

where $\xi \in [t_n, t_{n+1}]$,

$$\begin{aligned} y(t_{n+1}) = & \; y(t_n) + (1-\alpha) f(t_{n+1}, y(t_{n+1})) g'(t_{n+1}) \\ & -(1-\alpha) f(t_n, y(t_n)) g'(t_n) \\ & +\alpha \int_{t_n}^{t_{n+1}} f(\tau, y(\tau)) g'(\tau) d\tau. \end{aligned} \tag{12.70}$$

$$\begin{aligned} y_{n+1} = & \; y_n + (1-\alpha) f(t_{n+1}, y(t_{n+1})) g'(t_{n+1}) - (1-\alpha) f(t_n, y(t_n)) g'(t_n) \\ & +\alpha \int_{t_n}^{t_{n+1}} f(\tau, y(\tau)) g'(\tau) d\tau. \\ = & \; y_n + (1-\alpha) f(t_{n+1}, y(t_{n+1})) g'(t_{n+1}) - (1-\alpha) f(t_n, y(t_n)) g'(t_n) \\ & +\alpha \int_{t_n}^{t_{n+1}} g'(\tau) \left[R_T(\tau) + P_n(\tau) \right] d\tau, \\ = & \; y_n + \alpha \int_{t_n}^{t_{n+1}} g'(\tau) R_T(\tau) \, d\tau + y_{n+1}, \\ = & \; \alpha \int_{t_n}^{t_{n+1}} g'(\tau) \frac{(\tau - t_n)(\tau - t_{n+1})}{2} f''(\xi, y(\xi)) d\tau. \end{aligned}$$

$$\begin{aligned} \left| \overline{R}_T \right| = & \left| \alpha \int_{t_n}^{t_{n+1}} g'(\tau) \frac{(\tau - t_n)(\tau - t_{n+1})}{2} f''(\xi, y(\xi)) d\tau \right|, \\ \leq & \; \alpha \int_{t_n}^{t_{n+1}} \left| g'(\tau) \right| \frac{(\tau - t_n)(t_{n+1} - \tau)}{2} \left| f''(\xi, y(\xi)) \right| d\tau, \\ \leq & \; \alpha \int_{t_n}^{t_{n+1}} \sup_{l \in [t_n, t_{n+1}]} \left| g'(l) \right| \frac{(\tau - t_n)(t_{n+1} - \tau)}{2} \left| f''(\xi, y(\xi)) \right| d\tau, \\ \leq & \; \alpha \left\| g' \right\|_\infty \left\| f''(., y(.)) \right\|_\infty \left\{ \frac{\tau^2}{2} t_{n+1} - \frac{\tau^3}{3} - t_n t_{n+1} \tau + t_n \frac{\tau^2}{2} \right\} \Bigg|_{t_n}^{t_{n+1}}, \\ \leq & \; \alpha \left\| g' \right\|_\infty \left\| f''(., y(.)) \right\|_\infty \left\{ \begin{array}{c} \frac{t_{n+1}^3}{2} - \frac{t_{n+1}^3}{3} - t_n t_{n+1}^2 + t_n \frac{t_{n+1}^2}{2} - t_n^2 \frac{t_{n+1}}{2} \\ + \frac{t_n^3}{3} + t_n^2 t_{n+1} - \frac{t_n^3}{2} \end{array} \right\}, \end{aligned} \tag{12.71}$$

$$\begin{aligned}\left|\overline{R}_T\right| &\le \alpha \left\|g'\right\|_\infty \left\|f''(., y(.))\right\|_\infty \frac{h^3}{8},\\ &\le \alpha M_1 \left\|f''(., y(.))\right\|_\infty \frac{h^3}{8},\end{aligned}$$

which completes the proof.

Theorem 12.5 *Let $y(t_{n+1})$ be the exact solution at $t = t_{n+1}$ and y_{n+1} be the approximate solution, then*

$$\lim_{h\to 0} |y(t_{n+1}) - y_{n+1}| = 0. \tag{12.72}$$

Proof

$$\lim_{h\to 0} |y(t_{n+1}) - y_{n+1}| =$$

$$\lim_{h\to 0}\left|\begin{array}{c} y(t_n) + (1-\alpha) f(t_{n+1}, y(t_{n+1}))g'(t_{n+1}) \\ -(1-\alpha) f(t_n, y(t_n))g'(t_n) \\ +\alpha \int\limits_{t_n}^{t_{n+1}} f(\tau, y(\tau))g'(\tau)d\tau - y_n \\ -(1-\alpha) f(t_{n+1}, y(t_{n+1}))g'(t_{n+1}) \\ +(1-\alpha) f(t_n, y(t_n))g'(t_n) \\ -\alpha \int\limits_{t_n}^{t_{n+1}} g'(\tau)\left\{f(t_n, y_n) + \frac{(f(t_{n+1}, y_{n+1}) - f(t_n, y_n))}{h}(\tau - t_n)\right\} d\tau \end{array}\right|, \tag{12.73}$$

$$= \lim_{h\to 0}\left|\begin{array}{c} y(t_n) + \alpha \int\limits_{t_n}^{t_{n+1}} f(\tau, y(\tau))g'(\tau)d\tau \\ -y_n - \alpha \int\limits_{t_n}^{t_{n+1}} g'(\tau)\left\{f(t_n, y_n) + \frac{(f(t_{n+1}, y_{n+1}) - f(t_n, y_n))}{h}(\tau - t_n)\right\} d\tau \end{array}\right|,$$

$$\le \lim_{h\to 0} |y(t_n) - y_n| + \lim_{h\to 0} \alpha \int\limits_{t_n}^{t_{n+1}} \left|g'(\tau)\right| |f(\tau, y(\tau)) - f(t_n, y_n)|\, d\tau +$$

$$+ \lim_{h\to 0} \alpha \int\limits_{t_n}^{t_{n+1}} \left|\frac{(f(t_{n+1}, y_{n+1}) - f(t_n, y_n))}{h}\right| (\tau - t_n)\, d\tau,$$

$$\le \lim_{h\to 0} |y(t_n) - y_n| + \lim_{h\to 0} \left\|g'\right\|_\infty \alpha \int\limits_{t_n}^{t_{n+1}} |f(t_n, y_n) - f(\tau, y(\tau)|\, d\tau$$

$$+ \lim_{h\to 0} \alpha \int\limits_{t_n}^{t_{n+1}} \left|\frac{(f(t_{n+1}, y_{n+1}) - f(t_n, y_n))}{h}\right| (\tau - t_n)\, d\tau.$$

By systematically applying the differentiation of the function $f(t, y(t))$ with respect to its variables and incorporating the principles of the Mean Value Theorem, we are able to derive a more comprehensive understanding of the behavior of the function, leading to the following result.

$$\begin{aligned}\lim_{h\to 0}|y(t_{n+1})-y_{n+1}| \leq & \lim_{h\to 0}|y(t_n)-y_n| \\ & +\lim_{h\to 0}\left\|g'\right\|_\infty \frac{\alpha h^2}{2}\left\|f'(.,y(.))\right\|_\infty \\ & +\lim_{h\to 0}\left\|g'\right\|_\infty \frac{\alpha h}{2}\left\|f(.,y(.))\right\|_\infty, \\ \leq & \lim_{h\to 0}|y(t_n)-y_n|.\end{aligned} \tag{12.74}$$

Now when $n = 0$, we have

$$\begin{aligned}\lim_{h\to 0}|y(t_1)-y_1| \leq & \lim_{h\to 0}|y(t_0)-y_0|, \\ \leq & \ 0, \\ \lim_{h\to 0}|y(t_2)-y_2| \leq & \lim_{h\to 0}|y(t_1)-y_1| = 0, \\ & \vdots \\ \lim_{h\to 0}|y(t_{n+1})-y_{n+1}| \leq & \ 0.\end{aligned} \tag{12.75}$$

We conclude that

$$\lim_{h\to 0}|y(t_{n+1})-y_{n+1}| = 0 \tag{12.76}$$

which completes the proof.

Theorem 12.6 *If $\widetilde{y}_{n+1}$, $\widetilde{y}_n$, and $\widetilde{y}_0$ are perturbed terms of y_{n+1}, y_n, and y_0 respectively, the $\exists c \geq 0$ such that $\forall n > 0$*

$$|\widetilde{y}_{n+1}| < c\,|\widetilde{y}_0|. \tag{12.77}$$

Proof

$$\begin{aligned}y_{n+1}+\widetilde{y}_{n+1} = & \ y_n+\widetilde{y}_n \\ & +(1-\alpha)f(t_{n+1},\overline{y}_{n+1}+\widetilde{\overline{y}}_{n+1})h^{-1}\left(g(t_{n+1})-g(t_n)\right) \\ & -(1-\alpha)f(t_n,y_n+\widetilde{y}_n)h^{-1}\left(g(t_n)-g(t_{n-1})\right) \\ & +\alpha f(t_n,y_n+\widetilde{y}_n)\left(g(t_{n+1})-g(t_n)\right) \\ & +\alpha\frac{g(t_{n+1})-g(t_n)}{2}\left(f(t_{n+1},\overline{y}_{n+1}+\widetilde{\overline{y}}_{n+1})-f(t_n,y_n+\widetilde{y}_n)\right),\end{aligned} \tag{12.78}$$

therefore

$$\begin{aligned}
\widetilde{y}_{n+1} = &\ \widetilde{y}_n + (1-\alpha)\left(f(t_{n+1}, \overline{y}_{n+1} + \widetilde{\overline{y}}_{n+1})\right. \\
&\left. - f(t_{n+1}, \overline{y}_{n+1})\right) h^{-1} (g(t_{n+1}) - g(t_n)) \\
&- (1-\alpha)\left(f(t_n, y_n + \widetilde{y}_n) - f(t_n, y_n)\right) h^{-1} (g(t_n) - g(t_{n-1})) \\
&+ \alpha \left(f(t_n, y_n + \widetilde{y}_n) - f(t_n, y_n)\right)(g(t_{n+1}) - g(t_n)) \\
&+ \alpha \frac{g(t_{n+1}) - g(t_n)}{2} \left(f(t_{n+1}, \overline{y}_{n+1} + \widetilde{\overline{y}}_{n+1}) - f(t_{n+1}, \overline{y}_{n+1})\right) \\
&+ \alpha \frac{g(t_{n+1}) - g(t_n)}{2} \left(f(t_n, y_n + \widetilde{y}_n) - f(t_n, y_n)\right) \\
|\widetilde{y}_{n+1}| \leq &\ |\widetilde{y}_n| + (1-\alpha)\left|f(t_{n+1}, \overline{y}_{n+1} + \widetilde{\overline{y}}_{n+1}) - f(t_{n+1}, \overline{y}_{n+1})\right| h^{-1} |g(t_{n+1}) - g(t_n)| \\
&+ (1-\alpha)\left|f(t_n, y_n + \widetilde{y}_n) - f(t_n, y_n)\right| h^{-1} |g(t_n) - g(t_{n-1})| \\
&+ \alpha \left|f(t_n, y_n + \widetilde{y}_n) - f(t_n, y_n)\right| |g(t_{n+1}) - g(t_n)| \\
&+ \frac{\alpha}{2} |g(t_{n+1}) - g(t_n)| \left|f(t_{n+1}, \overline{y}_{n+1} + \widetilde{\overline{y}}_{n+1}) - f(t_{n+1}, \overline{y}_{n+1})\right| \\
&+ \frac{\alpha}{2} |g(t_{n+1}) - g(t_n)| \left|f(t_n, y_n + \widetilde{y}_n) - f(t_n, y_n)\right| \\
\leq &\ |\widetilde{y}_n| + (1-\alpha) L \left|g'(t_{n+1})\right|_\infty \left|\widetilde{\overline{y}}_{n+1}\right| + (1-\alpha) L \left|g'(t_n)\right|_\infty |\widetilde{y}_n| \\
&+ \alpha h L \left|g'(t_{n+1})\right|_\infty |\widetilde{y}_n| + \frac{\alpha h}{2} \left|g'(t_{n+1})\right|_\infty L \left|\widetilde{\overline{y}}_{n+1}\right| + \frac{\alpha h}{2} \left|g'(t_{n+1})\right|_\infty L |\widetilde{y}_n| .
\end{aligned} \tag{12.79}$$

Here we have used the Lipschitz condition of $f(t, y(t))$ with respect to y. Thus

$$|\widetilde{y}_{n+1}| \leq |\widetilde{y}_n| + \overline{M}_1 \left|\widetilde{\overline{y}}_{n+1}\right| + \overline{M}_2 |\widetilde{y}_n| + \overline{M}_3 |\widetilde{y}_n| + \overline{M}_4 \left|\widetilde{\overline{y}}_{n+1}\right| + \overline{M}_5 |\widetilde{y}_n| . \tag{12.80}$$

We now evaluate

$$\begin{aligned}
|\widetilde{\overline{y}}_{n+1}| \leq &\ |\widetilde{y}_n| + h \left|g'(t_{n+1})\right| L |\widetilde{y}_n| \\
\leq &\ |\widetilde{y}_n| \left(1 + \overline{M}_1\right),
\end{aligned} \tag{12.81}$$

and replacing yields

$$\begin{aligned}
|\widetilde{y}_{n+1}| \leq &\ |\widetilde{y}_n| + \overline{M}_1 \left(|\widetilde{y}_n| \left(1 + \overline{M}_1\right)\right) + \overline{M}_2 |\widetilde{y}_n| \\
&+ \overline{M}_3 |\widetilde{y}_n| + \overline{M}_4 \left(|\widetilde{y}_n| \left(1 + \overline{M}_1\right)\right) + \overline{M}_5 |\widetilde{y}_n| , \\
|\widetilde{y}_{n+1}| \leq &\ |\widetilde{y}_n| \left\{1 + \overline{M}_1 + \overline{M}_1^2 + \overline{M}_2 + \overline{M}_3 + \overline{M}_4 + \overline{M}_4 \overline{M}_1 + \overline{M}_5\right\}, \\
\leq &\ |\widetilde{y}_n| B.
\end{aligned} \tag{12.82}$$

When $n = 0$, we have

$$\begin{aligned}
|\widetilde{y}_1| \leq &\ B |\widetilde{y}_0| , \\
|\widetilde{y}_2| \leq &\ B |\widetilde{y}_1| \leq B^2 |\widetilde{y}_0| , \\
|\widetilde{y}_3| \leq &\ B |\widetilde{y}_2| \leq B^3 |\widetilde{y}_0| , \\
&\ \vdots
\end{aligned} \tag{12.83}$$

$$|\widetilde{y}_{n+1}| \leq B^{n+1} |\widetilde{y}_0| = c |\widetilde{y}_0| ,$$

which completes the proof.

12.5 Applying Lagrange Interpolation Method on IVP with Caputo–Fabrizio Global Derivative

Here we consider the following IVP with the Caputo–Fabrizio global derivative:

$$\begin{cases} {}^{CF}_{t_0}D_g y(t) = f(t, y(t)) \text{ if } t > 0 \\ y(t_0) = y_0 \qquad\qquad \text{if } t = 0 \end{cases}. \tag{12.84}$$

We assume that $g'(t) > 0$, then we have

$$y(t) = y(t_0) + (1-\alpha) f(t, y(t)) g'(t) + \alpha \int_{t_0}^{t} f(\tau, y(\tau) g'(\tau) d\tau, \tag{12.85}$$

$$y(t_0) = y_0.$$

At $t = t_{n+1} = (n+1)h$ and $t = t_n = nh$, we have

$$y(t_{n+1}) = y(t_0) + (1-\alpha) f(t_{n+1}, y(t_{n+1})) g'(t_{n+1}) + \alpha \int_{t_0}^{t_{n+1}} f(\tau, y(\tau) g'(\tau) d\tau, \tag{12.86}$$

and

$$y(t_n) = y(t_0) + (1-\alpha) f(t_n, y(t_n)) g'(t_n) + \alpha \int_{t_0}^{t_n} f(\tau, y(\tau) g'(\tau) d\tau. \tag{12.87}$$

Subtracting yields

$$\begin{aligned} y(t_{n+1}) = \;& y(t_n) + (1-\alpha) f(t_{n+1}, y_{n+1}) g'(t_{n+1}) \\ & -(1-\alpha) f(t_n, y_n) g'(t_n) \\ & +\alpha \int_{t_n}^{t_{n+1}} f(\tau, y(\tau) g'(\tau) d\tau. \end{aligned} \tag{12.88}$$

Approximating and integrating yields

$$\begin{aligned} y_{n+1} = & \ y_n + \frac{g(t_{n+1}) - g(t_n)}{h}(1-\alpha) f(t_{n+1}, y_{n+1}) \\ & -\frac{g(t_n) - g(t_{n-1})}{h}(1-\alpha) f(t_n, y_n) \\ & +\alpha \left\{ \frac{3}{2} (g(t_{n+1}) - g(t_n)) f(t_n, y_n) - (g(t_n) - g(t_{n-1})) f(t_{n-1}, y_{n-1}) \right\}. \end{aligned} \tag{12.89}$$

We note that

$$\begin{aligned} y_1 = & \ y_0 + (1-\alpha) f(t_1, y_1) g'(t_1) + \alpha \int_{t_0}^{t_1} f(\tau, y(\tau) g'(\tau) d\tau \\ = & \ y_0 + (1-\alpha) f(t_1, \overline{y}_1) \frac{g(t_1) - g(t_0)}{h} + \alpha (g(t_1) - g(t_0)) f(t_1, \overline{y}_1). \end{aligned} \tag{12.90}$$

Here

$$\overline{y}_1 = y_0 + (1-\alpha) f(t_0, y_0) \frac{g(t_1) - g(t_0)}{h} + \alpha (g(t_1) - g(t_0)) f(t_0, y_0). \tag{12.91}$$

Also

$$\begin{aligned} \overline{y}_{n+1} = & \ y_n + (1-\alpha) f(t_n, y_n) \frac{g(t_{n+1}) - g(t_n)}{h} \\ & -(1-\alpha) f(t_{n-1}, y_{n-1}) \frac{g(t_n) - g(t_{n-1})}{h} \\ & +\alpha (g(t_{n+1}) - g(t_n)) f(t_n, y_n). \end{aligned} \tag{12.92}$$

Therefore the explicit scheme is given by

$$\begin{aligned} y_{n+1} = & \ y_n + (1-\alpha) \frac{g(t_{n+1}) - g(t_n)}{h} f(t_{n+1}, \overline{y}_{n+1}) \quad \text{if } n > 1 \\ & -(1-\alpha) \frac{g(t_n) - g(t_{n-1})}{h} f(t_n, y_n) \\ & +\frac{3\alpha}{2} (g(t_{n+1}) - g(t_n)) f(t_n, y_n) - \frac{\alpha}{2} (g(t_n) - g(t_{n-1})) f(t_{n-1}, y_{n-1}), \\ \overline{y}_{n+1} = & \ y_n + (1-\alpha) f(t_n, y_n) \frac{g(t_{n+1}) - g(t_n)}{h} \\ & -(1-\alpha) f(t_{n-1}, y_{n-1}) \frac{g(t_n) - g(t_{n-1})}{h} + \alpha (g(t_{n+1}) - g(t_n)) f(t_n, y_n) \\ y_1 = & \ y_0 + (1-\alpha) f(t_1, \overline{y}_1) \frac{g(t_1) - g(t_0)}{h} + \alpha (g(t_1) - g(t_0)) f(t_1, y_1), \\ \overline{y}_1 = & \ y_0 + (1-\alpha) f(t_0, y_0) \frac{g(t_1) - g(t_0)}{h} + \alpha (g(t_1) - g(t_0)) f(t_0, y_0). \end{aligned} \tag{12.93}$$

If, instead, we opt to approximate the values using linear interpolation, we can derive the following result, which provides a simpler yet effective estimation of the function's behavior.

$$
\begin{aligned}
y_{n+1} = \; & y_n + (1-\alpha)\frac{g(t_{n+1}) - g(t_n)}{h} f(t_{n+1}, \overline{y}_{n+1}) \\
& - (1-\alpha)\frac{g(t_n) - g(t_{n-1})}{h} f(t_n, y_n) \\
& + \frac{\alpha h}{2}\left\{ \begin{array}{c} (g(t_{n+1}) - g(t_n))\, f(t_{n+1}, y_{n+1}) \\ + (g(t_n) - g(t_{n-1}))\, f(t_n, y_n) \end{array} \right\}.
\end{aligned}
\tag{12.94}
$$

The $\overline{y}_{n+1}$ is the same here to be in the case of Lagrange.

References

1. Ascher, U.M., Petzold, L.R.: Computer Methods for Ordinary Differential Equations and Differential-Algebraic Equations. Society for Industrial and Applied Mathematics, Philadelphia (1998). ISBN 978-0-89871-412-8
2. Butcher, J.C.: Numerical Methods for Ordinary Differential Equations. WileySons, New York (2003). ISBN 978-0-471-96758-3
3. Hairer, E., Nørsett, S.P., Wanner, G.: Solving Ordinary Differential Equations I: Nonstiff Problems. Springer, Berlin, New York (1993). ISBN 978-3-540-56670-0
4. Stoer, J., Bulirsch, R.: Introduction to Numerical Analysis (3rd ed.). Springer, Berlin, New York (2002). ISBN 978-0-387-95452-3
5. Baleanu, D., Jajarmi, A., Hajipour, M.: On the nonlinear dynamical systems within the generalized fractional derivatives with Mittag–Leffler kernel. Nonlinear Dyn. (2018)
6. Dumitru, B., Amin, J., Mojtaba, H.: On the nonlinear dynamical systems within the generalized fractional derivatives with Mittag–Leffler kernel. Nonlinear Dyn. **94**, 397–414 (2018)
7. Leader, J.J.: Numerical Analysis and Scientific Computation. Addison-Wesley, Boston (2004). ISBN 0-201-73499-0
8. Griffiths, D.V., Smith, I.M.: Numerical Methods for Engineers: A Programming Approach. CRC Press, Boca Raton, p. 218 (1991). ISBN 0-8493-8610-1
9. Meijering, Erik: A chronology of interpolation: from ancient astronomy to modern signal and image processing. Proc. IEEE **90**(3), 319–342 (2002)
10. Lagrange, J.-L.: "Leçon Cinquième. Sur l'usage des courbes dans la solution des problèmes". Leçons Elémentaires sur les Mathématiques (in French). Paris. Republished in Serret, Joseph-Alfred, ed. (1877). Oeuvres de Lagrange, vol. 7. Gauthier-Villars, pp. 271–287 (1795)
11. Toufik, M., Atangana, A.: New numerical approximation of fractional derivative with non-local and non-singular kernel: application to chaotic models. Eur. Phys. J. Plus **132**(10), 444 (2017)
12. Atangana, A., Araz, S.İ.: New Numerical Scheme with Newton Polynomial: Theory, Methods, and Applications. Academic Press (2021)

Chapter 13
Numerical Analysis of IVP with Atangana–Baleanu Global Derivative

13.1 Applying Euler Method on IVP with Atangana–Baleanu Global Derivative

In the fascinating world of fractional calculus, we explore the compelling application of the Euler Method to solve an Initial Value Problem (IVP) [1–6] governed by the Atangana-Baleanu Global Derivative. This approach brings forth a novel dimension to traditional problem-solving, where the influence of past states is encoded through the memory-dependent nature of fractional derivatives. The Atangana-Baleanu derivative, which captures both local and non-local dynamics, allows us to model complex systems that are governed by histories and long-range interactions, providing a deeper understanding of phenomena in various fields, from physics to biology. By employing the Euler Method, a well-known numerical technique, we can efficiently approximate the solution to this intricate IVP, offering valuable insights into real-world systems where classical derivatives fail to capture essential behaviors [7–12]. This powerful combination of methods opens up exciting possibilities in the modeling of processes with memory effects, making it an indispensable tool for tackling advanced problems in mathematical modeling, engineering [7–12], and beyond.

$$\begin{cases} {}_{t_0}^{ABR}D_g^{\alpha} y(t) = f(t, y(t)) & \text{if } t > 0 \\ y(t_0) = y_0 & \text{if } t = 0 \end{cases}. \tag{13.1}$$

If we write its integral form, then we get

$$\begin{aligned} y(t) = {} & (1-\alpha)g'(t)f(t, y(t)) + \frac{\alpha}{\Gamma(\alpha)} \int_{t_0}^{t} g'(\tau)(t-\tau)^{\alpha-1} f(\tau, y(\tau)d\tau, \\ y(t_0) = {} & y_0. \end{aligned} \tag{13.2}$$

A. Atangana and İ. Koca, *Fractional Differential and Integral Operators with Respect to a Function*, Industrial and Applied Mathematics,
https://doi.org/10.1007/978-981-97-9951-0_13

At $t = t_{n+1}$, we have

$$\begin{aligned} y(t_{n+1}) = & \ (1-\alpha)g'(t_{n+1})f(t_{n+1}, y(t_{n+1})) \\ & + \frac{\alpha}{\Gamma(\alpha)} \int_{t_0}^{t_{n+1}} g'(\tau)f(\tau, y(\tau))(t_{n+1}-\tau)^{\alpha-1}d\tau, \\ = & \ (1-\alpha)g'(t_{n+1})f(t_{n+1}, y(t_{n+1})) \\ & + \frac{\alpha}{\Gamma(\alpha)} \sum_{j=0}^{n} \int_{t_j}^{t_{j+1}} g'(\tau)f(\tau, y(\tau))(t_{n+1}-\tau)^{\alpha-1}d\tau. \end{aligned} \tag{13.3}$$

With the Euler approximation of $g'(\tau)f(\tau, y(\tau)$ within $\left[t_j, t_{j+1}\right]$ we have

$$\begin{aligned} y_{n+1} = & \ (1-\alpha)g'(t_{n+1})f(t_{n+1}, y(t_{n+1})) \\ & + \frac{\alpha}{\Gamma(\alpha)} \sum_{j=0}^{n} \int_{t_j}^{t_{j+1}} g'(t_j)(t_{n+1}-\tau)^{\alpha-1} f(t_j, y(t_j))d\tau, \\ = & \ (1-\alpha)g'(t_{n+1})f(t_{n+1}, y(t_{n+1})) \\ & + \frac{\alpha}{\Gamma(\alpha)} \sum_{j=0}^{n} \frac{g(t_j)-g(t_{j-1})}{h} f(t_j, y(t_j)) \int_{t_j}^{t_{j+1}} (t_{n+1}-\tau)^{\alpha-1}d\tau, \\ = & \ (1-\alpha)g'(t_{n+1})f(t_{n+1}, y(t_{n+1})) \\ & + \frac{h^{\alpha}}{\Gamma(\alpha)} \sum_{j=0}^{n} \frac{g(t_j)-g(t_{j-1})}{h} f(t_j, y(t_j)) \left((n-j+1)^{\alpha} - (n-j)^{\alpha}\right), \\ = & \ (1-\alpha)g'(t_{n+1})f(t_{n+1}, y(t_{n+1})) \\ & + \frac{h^{\alpha-1}}{\Gamma(\alpha)} \sum_{j=0}^{n} \left(g(t_j)-g(t_{j-1})\right) f(t_j, y(t_j)) \left((n-j+1)^{\alpha} - (n-j)^{\alpha}\right). \end{aligned} \tag{13.4}$$

We consider the second version. We start when $t = t_1$

$$\begin{aligned} y_1 = & \ (1-\alpha)g'(t_1)f(t_1, y(t_1)) + \frac{\alpha}{\Gamma(\alpha)} \int_{t_0}^{t_1} g'(\tau)f(\tau, y(\tau))(t_1-\tau)^{\alpha-1}d\tau, \\ = & \ (1-\alpha)\frac{g(t_1)-g(t_0)}{h} f(t_1, y(t_1)) \\ & + \frac{\alpha}{\Gamma(\alpha)} \int_{t_0}^{t_1} \frac{g(t_1)-g(t_0)}{h} f(t_0, y(t_0))(t_1-\tau)^{\alpha-1}d\tau, \\ = & \ (1-\alpha)\frac{g(t_1)-g(t_0)}{h} f(t_1, y(t_1)) + \end{aligned} \tag{13.5}$$

$$
\begin{aligned}
& \frac{\alpha}{\Gamma(\alpha)} \frac{g(t_1)-g(t_0)}{h} f(t_0, y(t_0)) \int_{t_0}^{t_1} (t_1-\tau)^{\alpha-1} d\tau, \\
= \; & (1-\alpha)h^{-1} f(t_1, y(t_1)) \left(g(t_1)-g(t_0)\right) \\
& + \frac{h^{\alpha-1}}{\Gamma(\alpha)} f(t_0, y_0) \left(g(t_1)-g(t_0)\right), \\
y_1 = \; & \left\{(1-\alpha)h^{-1} f(t_1, y(t_1)) + \frac{h^{\alpha-1}}{\Gamma(\alpha)} f(t_0, y_0)\right\} \left(g(t_1)-g(t_0)\right).
\end{aligned}
$$

$\forall n \geq 1$ we have that

$$
\begin{aligned}
y(t_{n+1}) = \; & (1-\alpha) g'(t_{n+1}) f(t_{n+1}, y(t_{n+1})) \\
& + \frac{\alpha}{\Gamma(\alpha)} \int_{t_1}^{t_{n+1}} g'(\tau) f(\tau, y(\tau)) (t_{n+1}-\tau)^{\alpha-1} d\tau, \\
= \; & (1-\alpha) g'(t_{n+1}) f(t_{n+1}, y(t_{n+1})) \\
& + \frac{\alpha}{\Gamma(\alpha)} \sum_{j=1}^{n} \int_{t_j}^{t_{j+1}} g'(\tau) f(\tau, y(\tau)) (t_{n+1}-\tau)^{\alpha-1} d\tau, \\
= \; & (1-\alpha) \left(\frac{g(t_{n+1})-g(t_n)}{h}\right) f(t_{n+1}, y(t_{n+1})) \\
& + \frac{\alpha}{\Gamma(\alpha)} \sum_{j=1}^{n} \frac{g(t_{j+1})-g(t_j)}{h} f(t_{j+1}, y(t_{j+1})) \int_{t_j}^{t_{j+1}} (t_{n+1}-\tau)^{\alpha-1} d\tau, \\
= \; & (1-\alpha) \left(\frac{g(t_{n+1})-g(t_n)}{h}\right) f(t_{n+1}, y(t_{n+1})) \\
& + \frac{h^{\alpha-1}}{\Gamma(\alpha)} \sum_{j=1}^{n} \left(g(t_{j+1})-g(t_j)\right) f(t_{j+1}, y(t_{j+1})) \left((n-j+1)^{\alpha}-(n-j)^{\alpha}\right), \\
y_{n+1} = \; & (1-\alpha)h^{-1} \left(g(t_{n+1})-g(t_n)\right) f(t_{n+1}, \overline{y}(t_{n+1})) \\
& + \frac{h^{\alpha-1}}{\Gamma(\alpha)} \sum_{j=1}^{n-1} \left(g(t_{j+1})-g(t_j)\right) f(t_{j+1}, y(t_{j+1})) \left((n-j+1)^{\alpha}-(n-j)^{\alpha}\right) \\
& + \frac{h^{\alpha-1}}{\Gamma(\alpha)} f(t_{n+1}, \overline{y}_{n+1}),
\end{aligned}
\tag{13.6}
$$

where

$$
\begin{aligned}
\overline{y}_{n+1} = \; & (1-\alpha)h^{-1} \left(g(t_{n+1})-g(t_n)\right) f(t_n, y(t_n)) \\
& + \frac{h^{\alpha-1}}{\Gamma(\alpha+1)} \sum_{j=0}^{n} \left(g(t_{j+1})-g(t_j)\right) f(t_j, y_j) \left((n-j+1)^{\alpha}-(n-j)^{\alpha}\right).
\end{aligned}
\tag{13.7}
$$

13.2 Applying Heun's Method on IVP with Atangana–Baleanu Global Derivative

We utilize the Heun's numerical method to solve the below nonlinear equation which is defined in a Atangana-Baleanu global derivative. The approach is used to obtain the solution of the equation according to the properties and complexity of the fractional derivative on the context of this problem. This is with the goal of obtaining a feasible numerical solution that accurately represents the dynamics of the system, with its fractional order and the distinctive properties of the Atangana-Baleanu derivative taken into consideration.

$$\begin{cases} {}_{t_0}^{ABR}D_g^{\alpha} y(t) = f(t, y(t)) \text{ if } t > t_0 \\ y(t_0) = y_0 \end{cases}. \tag{13.8}$$

Here we assume that $g'(t) > 0$. Also here we will need in addition that $f(t, y(t))$ satisfies the Lipschitz condition and is bounded within $[a, b]$. By the fundamental theorem of calculus, the above IVP can be transformed with Atangana–Baleanu integral

$$\begin{aligned} y(t) = &\ (1-\alpha)g'(t)f(t, y(t)) + \frac{\alpha}{\Gamma(\alpha)} \int_{t_0}^{t} g'(\tau)(t-\tau)^{\alpha-1} f(\tau, y(\tau)d\tau, \qquad (13.9) \\ y(t_0) = &\ y_0. \end{aligned}$$

Here we only consider at $t = t_{n+1}$, then we have

$$\begin{aligned} y(t_{n+1}) = &\ (1-\alpha)g'(t_{n+1})f(t_{n+1}, y(t_{n+1})) \qquad (13.10) \\ &+ \frac{\alpha}{\Gamma(\alpha)} \int_{t_0}^{t_{n+1}} g'(\tau)(t_{n+1}-\tau)^{\alpha-1} f(\tau, y(\tau)d\tau, \\ = &\ (1-\alpha)g'(t_{n+1})f(t_{n+1}, y(t_{n+1})) \\ &+ \sum_{j=0}^{n} \int_{t_j}^{t_{j+1}} g'(\tau) \frac{\alpha(t_{n+1}-\tau)^{\alpha-1}}{\Gamma(\alpha)} f(\tau, y(\tau)d\tau. \end{aligned}$$

Here we approximate within $[t_j, t_{j+1}]$ using

$$\begin{aligned} g'(\tau)f(\tau, y(\tau)) \simeq &\ \frac{g'(t_{j+1})f(t_{j+1}, y_{j+1}) + g'(t_j)f(t_j, y_j)}{2}, \qquad (13.11) \\ \simeq &\ \frac{g(t_{j+1}) - g(t_j)}{2h} f(t_{j+1}, y_{j+1}) + \frac{g(t_j) - g(t_{j-1})}{2h} f(t_j, y_j), \end{aligned}$$

and replacing, we obtain

$$
\begin{aligned}
y_{n+1} &= (1-\alpha)\left(\frac{g(t_{n+1})-g(t_n)}{h}\right) f(t_{n+1}, y(t_{n+1})) \\
&+\frac{\alpha}{2\Gamma(\alpha)h}\sum_{j=1}^{n}\left\{\begin{array}{c}\left(g(t_{j+1})-g(t_j)\right) f(t_{j+1}, y_{j+1}) \\ +\left(g(t_j)-g(t_{j-1})\right) f(t_j, y_j)\end{array}\right\}\int_{t_j}^{t_{j+1}}(t_{n+1}-\tau)^{\alpha-1}d\tau, \\
&= (1-\alpha)\left(\frac{g(t_{n+1})-g(t_n)}{h}\right) f(t_{n+1}, y(t_{n+1})) \\
&+\frac{\alpha}{2\Gamma(\alpha)h}\sum_{j=1}^{n}\left\{\begin{array}{c}\left(g(t_{j+1})-g(t_j)\right) f(t_{j+1}, y_{j+1}) \\ +\left(g(t_j)-g(t_{j-1})\right) f(t_j, y_j)\end{array}\right\} \\
&\quad\left\{\frac{h^\alpha}{\alpha}\left((n-j+1)^\alpha-(n-j)^\alpha\right)\right\}, \\
&= (1-\alpha)\left(\frac{g(t_{n+1})-g(t_n)}{h}\right) f(t_{n+1}, y(t_{n+1})) \\
&+\frac{\alpha h^{\alpha-1}}{2\Gamma(\alpha+1)}\sum_{j=1}^{n}\left\{\begin{array}{c}\left(g(t_{j+1})-g(t_j)\right) f(t_{j+1}, y_{j+1}) \\ +\left(g(t_j)-g(t_{j-1})\right) f(t_j, y_j)\end{array}\right\} \\
&\quad\left\{\left((n-j+1)^\alpha-(n-j)^\alpha\right)\right\}.
\end{aligned} \tag{13.12}
$$

When $j=n+1$ we have an implicit function; therefore, we reformulated the formula as

$$
\begin{aligned}
y_{n+1} &= (1-\alpha)\left(\frac{g(t_{n+1})-g(t_n)}{h}\right) f(t_{n+1}, \overline{y}(t_{n+1})) \\
&+\frac{\alpha h^{\alpha-1}}{2\Gamma(\alpha+1)}\sum_{j=1}^{n-1}\left\{\begin{array}{c}\left(g(t_{j+1})-g(t_j)\right) f(t_{j+1}, y_{j+1}) \\ +\left(g(t_j)-g(t_{j-1})\right) f(t_j, y_j)\end{array}\right\} \\
&\quad\left\{\left((n-j+1)^\alpha-(n-j)^\alpha\right)\right\} \\
&+\frac{\alpha h^{\alpha-1}}{2\Gamma(\alpha+1)}\left\{\left(g(t_{n+1})-g(t_n)\right) f(t_{n+1}, \overline{y}_{n+1})\right. \\
&\left.+\left(g(t_n)-g(t_{n-1})\right) f(t_n, y_n)\right\},
\end{aligned} \tag{13.13}
$$

where

$$
\begin{aligned}
\overline{y}_{n+1} &= (1-\alpha)\left(\frac{g(t_{n+1})-g(t_n)}{h}\right) f(t_n, y(t_n)) \\
&+\frac{\alpha h^{\alpha-1}}{2\Gamma(\alpha+1)}\sum_{j=1}^{n-1}\left\{\begin{array}{c}\left(g(t_{j+1})-g(t_j)\right) f(t_{j+1}, y_{j+1}) \\ +\left(g(t_j)-g(t_{j-1})\right) f(t_j, y_j)\end{array}\right\} \\
&\quad\left\{\left((n-j+1)^\alpha-(n-j)^\alpha\right)\right\}.
\end{aligned} \tag{13.14}
$$

Theorem 13.1 *Let $\overline{y}_n$ be the perturbed term of y_n, then $\forall n \geq 0$, $\exists c > 0$ such that*

$$|\widetilde{y}_{n+1}| < c\,|\widetilde{y}_0|\,. \tag{13.15}$$

Proof Let us put our iteration below

$$\begin{aligned} y_{n+1} = {} & (1-\alpha)\left(\frac{g(t_{n+1})-g(t_n)}{h}\right) f(t_{n+1},\overline{y}_{n+1}) \\ & +\frac{\alpha h^{\alpha-1}}{2\Gamma(\alpha+1)}\sum_{j=1}^{n-1}\left\{\begin{array}{c}\left(g(t_{j+1})-g(t_j)\right) f(t_{j+1},y_{j+1}) \\ +\left(g(t_j)-g(t_{j-1})\right) f(t_j,y_j)\end{array}\right\} \\ & \{((n-j+1)^{\alpha}-(n-j)^{\alpha})\} \\ & +\frac{\alpha h^{\alpha-1}}{2\Gamma(\alpha+1)}\left\{(g(t_{n+1})-g(t_n))\, f(t_{n+1},\overline{y}_{n+1})\right. \\ & \left.+(g(t_n)-g(t_{n-1}))\, f(t_n,y_n)\right\}. \end{aligned} \tag{13.16}$$

With the perturbed term

$$\begin{aligned} y_{n+1}+\widetilde{y}_{n+1} = {} & (1-\alpha)\left(\frac{g(t_{n+1})-g(t_n)}{h}\right) f(t_{n+1},\overline{y}_{n+1}+\widetilde{\overline{y}}_{n+1}) \\ & +\frac{\alpha h^{\alpha-1}}{2\Gamma(\alpha+1)}\sum_{j=1}^{n-1}\left\{\begin{array}{c}\left(g(t_{j+1})-g(t_j)\right) f(t_{j+1},y_{j+1}+\widetilde{y}_{j+1}) \\ +\left(g(t_j)-g(t_{j-1})\right) f(t_j,y_j+\widetilde{y}_j)\end{array}\right\} \\ & \{((n-j+1)^{\alpha}-(n-j)^{\alpha})\} \\ & +\frac{\alpha h^{\alpha-1}}{2\Gamma(\alpha+1)}\left\{\begin{array}{c}(g(t_{n+1})-g(t_n))\, f(t_{n+1},\overline{y}_{n+1}+\widetilde{\overline{y}}_{n+1}) \\ +(g(t_n)-g(t_{n-1}))\, f(t_n,y_n+\widetilde{y}_n)\end{array}\right\} \\ & +\widetilde{y}(t_0)+y(t_0). \end{aligned}$$

Thus we have that

$$\begin{aligned} & |\widetilde{y}_{n+1}| \\ & = \left|\begin{array}{c} (1-\alpha)\left(\frac{g(t_{n+1})-g(t_n)}{h}\right)\left(f(t_{n+1},\overline{y}_{n+1}+\widetilde{\overline{y}}_{n+1})-f(t_{n+1},\overline{y}_{n+1})\right) \\ +\frac{\alpha h^{\alpha-1}}{2\Gamma(\alpha+1)}\sum_{j=1}^{n-1}\left\{\begin{array}{c}\left(g(t_{j+1})-g(t_j)\right)\left(f(t_{j+1},y_{j+1}+\widetilde{y}_{j+1})-f(t_{j+1},y_{j+1})\right) \\ +\left(g(t_j)-g(t_{j-1})\right)\left(f(t_j,y_j+\widetilde{y}_j)-f(t_j,y_j)\right)\end{array}\right\}\delta_{n,j}^{\alpha} \\ +\frac{\alpha h^{\alpha-1}}{2\Gamma(\alpha+1)}\left\{\begin{array}{c}(g(t_{n+1})-g(t_n))\left(f(t_{n+1},\overline{y}_{n+1}+\widetilde{\overline{y}}_{n+1})-f(t_{n+1},\overline{y}_{n+1})\right) \\ +(g(t_n)-g(t_{n-1}))\left(f(t_n,y_n+\widetilde{y}_n)-f(t_n,y_n)\right)\end{array}\right\}\end{array}\right| \\ & +|\widetilde{y}(t_0)|\,. \end{aligned} \tag{13.17}$$

Using the properties of inequality, we get

$$\begin{aligned} |\widetilde{y}_{n+1}| < {} & (1-\alpha)g'(t_{n+1})\left|f(t_{n+1},\overline{y}_{n+1}+\widetilde{\overline{y}}_{n+1})-f(t_{n+1},\overline{y}_{n+1})\right| \\ & +\frac{\alpha h^{\alpha-1}}{2\Gamma(\alpha+1)}\sum_{j=0}^{n-1}\left(\begin{array}{c}h\left|g'(t_{j+1})\right|\left|f(t_{j+1},y_{j+1}+\widetilde{y}_{j+1})-f(t_{j+1},y_{j+1})\right| \\ +h\left|g'(t_j)\right|\left|f(t_j,y_j+\widetilde{y}_j)-f(t_j,y_j)\right|\end{array}\right)\delta_{n,j}^{\alpha} \end{aligned} \tag{13.18}$$

$$+\frac{\alpha h^{\alpha-1}}{2\Gamma(\alpha+1)}\begin{pmatrix} h\left|g'(t_{n+1})\right|\left|f(t_{n+1},\overline{y}_{n+1}+\widetilde{\overline{y}}_{n+1})-f(t_{n+1},\overline{y}_{n+1})\right| \\ +h\left|g'(t_n)\right|\left|f(t_n,y_n+\widetilde{y}_n)-f(t_n,y_n)\right| \end{pmatrix}$$
$$+\left|\widetilde{y}(t_0)\right|.$$

Using the Lipschitz condition of $f(.,y(.))$ with respect to y, we get

$$\begin{aligned}\left|\widetilde{y}_{n+1}\right| &< (1-\alpha)\left\|g'\right\|_{\infty} L\left|\widetilde{\overline{y}}_{n+1}\right| \\ &+\frac{\alpha h^{\alpha-1}}{2\Gamma(\alpha+1)}\sum_{j=0}^{n-1} h\left\|g'\right\|_{\infty}\left\{L\left|\widetilde{y}_{j+1}\right|+L\left|\widetilde{y}_{j+1}\right|\right\}\delta_{n,j}^{\alpha} \\ &+\frac{\alpha h^{\alpha-1}}{2\Gamma(\alpha+1)} h\left\|g'\right\|_{\infty}\left\{L\left|\widetilde{\overline{y}}_{n+1}\right|+L\left|\widetilde{y}_n\right|\right\}+\left|\widetilde{y}(t_0)\right|.\end{aligned} \tag{13.19}$$

We have that

$$\left|\widetilde{\overline{y}}_{n+1}\right| \leq \left|\widetilde{y}(t_0)\right|+\frac{\alpha h^{\alpha}\left\|g'\right\|_{\infty} L}{\Gamma(\alpha+1)}\sum_{j=0}^{n}\left|\widetilde{y}_j\right|\delta_{n,j}^{\alpha}. \tag{13.20}$$

Therefore replacing yields

$$\begin{aligned}\left|\widetilde{y}_{n+1}\right| \leq\ & (1-\alpha)\left\|g'\right\|_{\infty} L\left|\widetilde{y}_0\right|+\frac{(1-\alpha)h^{\alpha}\left\|g'\right\|_{\infty}^{2} L^{2}}{\Gamma(\alpha)}\sum_{j=0}^{n}\left|\widetilde{y}_j\right|\delta_{n,j}^{\alpha} \\ &+\frac{\alpha h^{\alpha}\left\|g'\right\|_{\infty} L}{2\Gamma(\alpha+1)}\left\{\sum_{j=0}^{n-1}\left\{\left|\widetilde{y}_{j+1}\right|+\left|\widetilde{y}_j\right|\right\}\delta_{n,j}^{\alpha}\right\} \\ &+\frac{\alpha h^{\alpha}\left\|g'\right\|_{\infty} L}{2\Gamma(\alpha+1)}\left\{\left|\widetilde{y}_0\right|+\frac{h^{\alpha}\left\|g'\right\|_{\infty} L}{\Gamma(\alpha+1)}\sum_{j=0}^{n}\left|\widetilde{y}_j\right|\delta_{n,j}^{\alpha}\right\} \\ &+\frac{\alpha h^{\alpha}\left\|g'\right\|_{\infty} L}{2\Gamma(\alpha+1)}\left|\widetilde{y}_n\right|+\left|\widetilde{y}_0\right|.\end{aligned} \tag{13.21}$$

$$y(t_1)=(1-\alpha)f(t_0,y(t_0))+\frac{\alpha}{\Gamma(\alpha)}\int_{t_0}^{t_1}(t_1-\tau)^{\alpha-1}f(\tau,y(\tau))d\tau, \tag{13.22}$$

$$\simeq (1-\alpha)f(t_0,y(t_0))+\frac{\alpha}{\Gamma(\alpha)}\int_{t_0}^{t_1}(t_1-\tau)^{\alpha-1}f(t_0,y(t_0))d\tau$$

$$\simeq (1-\alpha)f(t_0,y(t_0))+\frac{\alpha f(t_0,y(t_0))}{\Gamma(\alpha)}\int_{t_0}^{t_1}(t_1-\tau)^{\alpha-1}d\tau$$

$$\simeq (1-\alpha)f(t_0,y(t_0))+\frac{\alpha f(t_0,y(t_0))}{\Gamma(\alpha)}\frac{(t_1-t_0)^{\alpha}}{\alpha},$$

$$\simeq (1-\alpha) f(t_0, y(t_0)) + \frac{\alpha f(t_0, y(t_0))}{\Gamma(\alpha)} \frac{h^\alpha}{\alpha},$$
$$\simeq \left((1-\alpha) + \frac{h^\alpha}{\Gamma(\alpha)}\right) f(t_0, y(t_0)).$$

$$\begin{aligned} y_1 &= \left((1-\alpha) + \frac{h^\alpha}{\Gamma(\alpha)}\right) f(t_0, y_0), \\ y_1 + \widetilde{y}_1 &= \left((1-\alpha) + \frac{h^\alpha}{\Gamma(\alpha)}\right) f(t_0, y_0 + \widetilde{y}_0), \\ \widetilde{y}_1 &= \left((1-\alpha) + \frac{h^\alpha}{\Gamma(\alpha)}\right) (f(t_0, y_0 + \widetilde{y}_0) - f(t_0, y_0)) \\ |\widetilde{y}_1| &\le \left((1-\alpha) + \frac{h^\alpha}{\Gamma(\alpha)}\right) L |\widetilde{y}_0| \end{aligned} \tag{13.23}$$

$$\begin{aligned} |\widetilde{y}_2| \le\ & (1-\alpha) \left\|g'\right\|_\infty L |\widetilde{y}_0| + \frac{(1-\alpha)h^\alpha \left\|g'\right\|_\infty^2 L^2}{\Gamma(\alpha)} \sum_{j=0}^{1} \left|\widetilde{y}_j\right| \delta_{1,j}^\alpha \\ & + \frac{\alpha h^\alpha \left\|g'\right\|_\infty L}{2\Gamma(\alpha+1)} \left\{ \sum_{j=0}^{0} \left\{ \left|\widetilde{y}_{j+1}\right| + \left|\widetilde{y}_j\right| \right\} \delta_{1,j}^\alpha \right\} \\ & + \frac{\alpha h^\alpha \left\|g'\right\|_\infty L}{2\Gamma(\alpha+1)} \left\{ |\widetilde{y}_0| + \frac{h^\alpha \left\|g'\right\|_\infty L}{\Gamma(\alpha+1)} \sum_{j=0}^{1} \left|\widetilde{y}_j\right| \delta_{1,j}^\alpha \right\} \\ & + \frac{\alpha h^\alpha \left\|g'\right\|_\infty L}{2\Gamma(\alpha+1)} |\widetilde{y}_1| + |\widetilde{y}_0|. \end{aligned}$$

$$\begin{aligned} |\widetilde{y}_2| =\ & (1-\alpha) \left\|g'\right\|_\infty L |\widetilde{y}_0| + \frac{(1-\alpha)h^\alpha \left\|g'\right\|_\infty^2 L^2}{\Gamma(\alpha)} \left\{ |\widetilde{y}_0| \delta_{1,0}^\alpha + |\widetilde{y}_1| \right\} \\ & + \frac{h^\alpha \left\|g'\right\|_\infty L}{2\Gamma(\alpha+1)} \left\{ |\widetilde{y}_1| + |\widetilde{y}_0| \right\} \delta_{1,0}^\alpha \\ & + \frac{h^\alpha \left\|g'\right\|_\infty L}{2\Gamma(\alpha+1)} |\widetilde{y}_0| + \frac{1}{2} \left(\frac{h^\alpha \left\|g'\right\|_\infty L}{\Gamma(\alpha+1)} \right)^2 |\widetilde{y}_0| \delta_{1,0}^\alpha \\ & + \frac{1}{2} \left(\frac{h^\alpha \left\|g'\right\|_\infty L}{\Gamma(\alpha+1)} \right)^2 |\widetilde{y}_1| + \frac{1}{2} \left(\frac{h^\alpha \left\|g'\right\|_\infty L}{\Gamma(\alpha+1)} \right) |\widetilde{y}_1| + |\widetilde{y}_0|. \end{aligned}$$

If we arrange it, we will obtain

$$|\widetilde{y}_2| \le \left\{ (1-\alpha) \left\|g'\right\|_\infty L + \left(\frac{(1-\alpha)h^\alpha \left\|g'\right\|_\infty^2 L^2}{\Gamma(\alpha)} \right) \delta_{1,0}^\alpha \right.$$

$$+\left((1-\alpha)+\frac{h^{\alpha}}{\Gamma(\alpha)}\right)L\Bigg\}\left|\widetilde{y}_0\right| \tag{13.24}$$
$$+\frac{\alpha h^{\alpha}\left\|g'\right\|_{\infty}L}{2\Gamma(\alpha+1)}\left\{\begin{array}{c}\left(2+\frac{h^{\alpha}L}{\Gamma(\alpha+1)}\right)\delta_{1,0}^{\alpha}\left|\widetilde{y}_0\right|+\left(1+\frac{h^{\alpha}\|g'\|_{\infty}L}{\Gamma(\alpha+1)}\delta_{1,0}^{\alpha}\right)\left|\widetilde{y}_0\right| \\ +\left(1+\frac{h^{\alpha}\|g'\|_{\infty}L}{\Gamma(\alpha+1)}\right)\left(1+\frac{h^{\alpha}L}{\Gamma(\alpha+1)}\right)\left|\widetilde{y}_0\right|\end{array}\right\}$$
$$+\left|\widetilde{y}_0\right|,$$
$$\left|\widetilde{y}_2\right| \le c_1\left|\widetilde{y}_0\right|,$$

where

$$c_1=\left\{(1-\alpha)\left\|g'\right\|_{\infty}L+\left(\frac{(1-\alpha)h^{\alpha}\left\|g'\right\|_{\infty}^{2}L^{2}}{\Gamma(\alpha)}\right)\delta_{1,0}^{\alpha}\right.$$
$$\left.+\left((1-\alpha)+\frac{h^{\alpha}}{\Gamma(\alpha)}\right)L\right\} \tag{13.25}$$
$$+\frac{\alpha h^{\alpha}\left\|g'\right\|_{\infty}L}{2\Gamma(\alpha+1)}\left\{\begin{array}{c}\left(2+\frac{h^{\alpha}L}{\Gamma(\alpha+1)}\right)\delta_{1,0}^{\alpha}+\left(1+\frac{h^{\alpha}\|g'\|_{\infty}L}{\Gamma(\alpha+1)}\delta_{1,0}^{\alpha}\right) \\ +\left(1+\frac{h^{\alpha}\|g'\|_{\infty}L}{\Gamma(\alpha+1)}\right)\left(1+\frac{h^{\alpha}L}{\Gamma(\alpha+1)}\right)\end{array}\right\}+1.$$

We can now assume that $\forall n \ge 2$

$$\left|\widetilde{y}_n\right| < c_n\left|\widetilde{y}_0\right|, \tag{13.26}$$

c_n is constant; we want to show that

$$\left|\widetilde{y}_{n+1}\right| < \overline{c}_n\left|\widetilde{y}_0\right|. \tag{13.27}$$

$$\left|\widetilde{y}_{n+1}\right| \le (1-\alpha)\left\|g'\right\|_{\infty}L\left|\widetilde{y}_0\right|+\frac{(1-\alpha)h^{\alpha}\left\|g'\right\|_{\infty}^{2}L^{2}}{\Gamma(\alpha)}\sum_{j=0}^{n}c_n\left|\widetilde{y}_0\right|\delta_{n,j}^{\alpha} \tag{13.28}$$
$$+\frac{\alpha h^{\alpha}\left\|g'\right\|_{\infty}L}{2\Gamma(\alpha+1)}\left\{\sum_{j=0}^{n-1}\left\{c_n\left|\widetilde{y}_0\right|+c_n\left|\widetilde{y}_0\right|\right\}\delta_{n,j}^{\alpha}\right\}$$
$$+\frac{\alpha h^{\alpha}\left\|g'\right\|_{\infty}L}{2\Gamma(\alpha+1)}\left\{\left|\widetilde{y}_0\right|+\frac{h^{\alpha}\left\|g'\right\|_{\infty}L}{\Gamma(\alpha+1)}\sum_{j=0}^{n}c_n\left|\widetilde{y}_0\right|\delta_{n,j}^{\alpha}\right\}$$
$$+\frac{\alpha h^{\alpha}\left\|g'\right\|_{\infty}L}{2\Gamma(\alpha+1)}c_n\left|\widetilde{y}_0\right|+\left|\widetilde{y}_0\right|.$$

$$|\widetilde{y}_{n+1}| \leq \left\{ \begin{array}{c} (1-\alpha)\left\|g'\right\|_\infty L + \frac{(1-\alpha)h^\alpha \|g'\|_\infty^2 L^2}{\Gamma(\alpha)} c_n \left\{ \sum_{j=0}^{n} \delta_{n,j}^\alpha \right\} \\ + \frac{\alpha h^\alpha \|g'\|_\infty L}{\Gamma(\alpha+1)} c_n \left\{ \sum_{j=0}^{n-1} \delta_{n,j}^\alpha \right\} \\ + \frac{\alpha h^\alpha \|g'\|_\infty L}{2\Gamma(\alpha+1)} \left\{ 1 + \frac{h^\alpha \|g'\|_\infty L}{\Gamma(\alpha+1)} c_n \sum_{j=0}^{n} \delta_{n,j}^\alpha \right\} \\ + \frac{\alpha h^\alpha \|g'\|_\infty L}{2\Gamma(\alpha+1)} c_n + 1 \end{array} \right\} |\widetilde{y}_0|$$

$$\leq c_{n+1} |\widetilde{y}_0|,$$

which completes the proof.

Theorem 13.2 *$\forall n \geq 0$, if $y(t_{n+1})$ is the exact solution at $y(t_{n+1})$ and y_{n+1} approximate solution*

$$\lim_{h\to 0} |y(t_{n+1}) - y_{n+1}| = 0 \tag{13.29}$$

if $f(t, y(t))$ is differentiable.

Proof

$$\left|y(t_{n+1}) - y_{n+1}\right| = \left| \begin{array}{c} (1-\alpha)g'(t_{n+1})f(t_{n+1}, y(t_{n+1})) \\ + \frac{\alpha}{\Gamma(\alpha)} \int_{t_0}^{t_{n+1}} g'(\tau)(t_{n+1}-\tau)^{\alpha-1} f(\tau, y(\tau)d\tau \\ -(1-\alpha)g'(t_{n+1})f(t_{n+1}, y(t_{n+1})) \\ - \frac{h^{\alpha-1}}{2\Gamma(\alpha+1)} \sum_{j=1}^{n-1} \left\{ \begin{array}{c} (g(t_{j+1}) - g(t_j))\, f(t_{j+1}, y_{j+1}) \\ + (g(t_j) - g(t_{j-1}))\, f(t_j, y_j) \end{array} \right\} \{((n-j+1)^\alpha - (n-j)^\alpha)\} \\ - \frac{h^{\alpha-1}}{2\Gamma(\alpha+1)} \{(g(t_{n+1}) - g(t_n))\, f(t_{n+1}, \overline{y}_{n+1}) + (g(t_n) - g(t_{n-1}))\, f(t_n, y_n)\} \end{array} \right|, \tag{13.30}$$

$$\left|y(t_{n+1}) - y_{n+1}\right| = \left| \begin{array}{c} \frac{\alpha}{\Gamma(\alpha)} \int_{t_0}^{t_{n+1}} g'(\tau)(t_{n+1}-\tau)^{\alpha-1} f(\tau, y(\tau)d\tau \\ - \frac{\alpha h^{\alpha-1}}{2\Gamma(\alpha+1)} \sum_{j=1}^{n-1} \left\{ \begin{array}{c} (g(t_{j+1}) - g(t_j))\, f(t_{j+1}, y_{j+1}) \\ + (g(t_j) - g(t_{j-1}))\, f(t_j, y_j) \end{array} \right\} \{((n-j+1)^\alpha - (n-j)^\alpha)\} \\ - \frac{\alpha h^{\alpha-1}}{2\Gamma(\alpha+1)} \left\{ \begin{array}{c} (g(t_{n+1}) - g(t_n))\, f(t_{n+1}, \overline{y}_{n+1}) \\ + (g(t_n) - g(t_{n-1}))\, f(t_n, y_n) \end{array} \right\} \end{array} \right|, \tag{13.31}$$

$$\leq \frac{\alpha}{2\Gamma(\alpha)} \left| \begin{array}{c} \int_{t_0}^{t_{n+1}} g'(\tau)(t_{n+1}-\tau)^{\alpha-1} f(\tau, y(\tau)d\tau \\ - \sum_{j=0}^{n-1} \int_{t_j}^{t_{j+1}} (t_{n+1}-\tau)^{\alpha-1} \left(\frac{g(t_{j+1}) - g(t_j)}{h} \right) f(t_{j+1}, y_{j+1})d\tau \end{array} \right|$$

$$+\frac{\alpha}{2\Gamma(\alpha)}\left|\begin{array}{l}\int_{t_0}^{t_{n+1}} g'(\tau)(t_{n+1}-\tau)^{\alpha-1} f(\tau, y(\tau)d\tau \\ -\sum_{j=0}^{n-1}\int_{t_j}^{t_{j+1}} (t_{n+1}-\tau)^{\alpha-1}\left(\frac{g(t_j)-g(t_{j-1})}{h}\right) f(t_j, y_j)d\tau\end{array}\right|,$$

$$+\left|\frac{\alpha\left(g(t_{n+1})-g(t_n)\right)h^{\alpha-1}}{2\Gamma(\alpha+1)} f(t_{n+1}, y_{n+1})+\frac{\alpha\left(g(t_n)-g(t_{n-1})\right)h^{\alpha-1}}{2\Gamma(\alpha+1)} f(t_n, y_n)\right|$$

$$\begin{aligned}
&\leq \frac{\alpha}{2\Gamma(\alpha)}\left|\sum_{j=0}^{n-1}\int_{t_j}^{t_{j+1}} g'(\tau)\left(f(\tau, y(\tau)-f(t_{j+1}, y_{j+1})\right)(t_{n+1}-\tau)^{\alpha-1}d\tau\right| \\
&+\frac{\alpha}{2\Gamma(\alpha)}\left|\sum_{j=0}^{n-1}\int_{t_j}^{t_{j+1}} g'(\tau)\left(f(\tau, y(\tau)-f(t_j, y_j)\right)(t_{n+1}-\tau)^{\alpha-1}d\tau\right| \\
&+\left|\frac{\alpha g'(t_{n+1})h^{\alpha}}{2\Gamma(\alpha+1)} f(t_{n+1}, \overline{y}_{n+1})+\frac{\alpha g'(t_n)h^{\alpha}}{2\Gamma(\alpha+1)} f(t_n, y_n)\right| \\
&\leq \frac{\alpha}{2\Gamma(\alpha)}\left\|g'\right\|_\infty \left\|f'(., y(.))\right\|_\infty \sum_{j=0}^{n}\int_{t_j}^{t_{j+1}} (t_{j+1}-\tau)(t_{n+1}-\tau)^{\alpha-1}d\tau \\
&+\frac{\alpha}{2\Gamma(\alpha)}\left\|g'\right\|_\infty \left\|f'(., y(.))\right\|_\infty \sum_{j=0}^{n}\int_{t_j}^{t_{j+1}} (t_{j+1}-\tau)(t_{n+1}-\tau)^{\alpha-1}d\tau \\
&\leq \frac{\alpha\left\|g'\right\|_\infty \left\|f'(., y(.))\right\|_\infty}{\Gamma(\alpha)} \sum_{j=0}^{n}\int_{t_j}^{t_{j+1}} (t_{j+1}-\tau)(t_{n+1}-\tau)^{\alpha-1}d\tau
\end{aligned} \tag{13.32}$$

We note that

$$\begin{aligned}
\int_{t_j}^{t_{j+1}} (t_{j+1}-\tau)(t_{n+1}-\tau)^{\alpha-1}d\tau &= t_{j+1}\int_{t_j}^{t_{j+1}} (t_{n+1}-\tau)^{\alpha-1}d\tau \\
&\quad -\int_{t_j}^{t_{j+1}} \tau(t_{n+1}-\tau)^{\alpha-1}d\tau,
\end{aligned} \tag{13.33}$$

$$\int_{t_j}^{t_{j+1}} (t_{n+1}-\tau)^{\alpha-1}d\tau = \left\{\frac{(t_{n+1}-t_j)^{\alpha}}{\alpha}-\frac{(t_{n+1}-t_{j-1})^{\alpha}}{\alpha}\right\},$$

$$\int_{t_j}^{t_{j+1}} (t_{n+1}-\tau)^{\alpha-1}\tau d\tau = t_{n+1}^{\alpha+1}\left\{B\left(\frac{t_{j+1}}{t_{n+1}}; \alpha, 2\right)-B\left(\frac{t_j}{t_{n+1}}; \alpha, 2\right)\right\}.$$

Replacing, we get

$$
\begin{aligned}
& |y(t_{n+1}) - y_{n+1}| \\
&= \left\|g'\right\|_\infty \left\|f'(., y(.))\right\|_\infty \left\{ \begin{array}{c} (n+1)h^{\alpha+1}\{((n-j+1)^\alpha - (n-j)^\alpha)\} \\ -\alpha(n+1)^{\alpha+1}\left\{B\left(\frac{t_{j+1}}{t_{n+1}}, \alpha, 2\right) - B\left(\frac{t_j}{t_{n+1}}, \alpha, 2\right)\right\} \end{array} \right\} \\
&\leq \left\|g'\right\|_\infty \left\|f'(., y(.))\right\|_\infty (n+1)\frac{\alpha h^{\alpha+1}}{\Gamma(\alpha+1)},
\end{aligned} \tag{13.34}
$$

$$
\begin{aligned}
\lim_{h\to 0} |y(t_{n+1}) - y_{n+1}| &\leq \lim_{h\to 0} \left\|g'\right\|_\infty \left\|f'(., y(.))\right\|_\infty (n+1)\frac{\alpha h^{\alpha+1}}{\Gamma(\alpha+1)} \\
&= 0.
\end{aligned} \tag{13.35}
$$

Therefore

$$
\lim_{h\to 0} |y(t_{n+1}) - y_{n+1}| = 0, \tag{13.36}
$$

which completes the proof.

13.3 Applying Midpoint Scheme Method on IVP with Atangana–Baleanu Global Derivative

In this section we consider the following IVP problem:

$$
\begin{cases} {}^{ABR}_{t_0}D^\alpha_g y(t) = f(t, y(t)) \text{ if } t > t_0 \\ y(t_0) = y_0 \end{cases}. \tag{13.37}
$$

$$
\begin{aligned}
y(t) &= (1-\alpha)g'(t)f(t, y(t)) + \frac{\alpha}{\Gamma(\alpha)} \int_{t_0}^{t} g'(\tau)(t-\tau)^{\alpha-1} f(\tau, y(\tau)d\tau, \\
y(t_0) &= y_0.
\end{aligned} \tag{13.38}
$$

At $t = t_{n+1}$, we have

$$
\begin{aligned}
y(t_{n+1}) &= (1-\alpha)g'(t_{n+1})f(t_{n+1}, y(t_{n+1})) \\
&\quad + \frac{\alpha}{\Gamma(\alpha)} \int_{t_0}^{t_{n+1}} g'(\tau)f(\tau, y(\tau))(t_{n+1}-\tau)^{\alpha-1}d\tau, \\
&= (1-\alpha)g'(t_{n+1})f(t_{n+1}, y(t_{n+1})) \\
&\quad + \frac{\alpha}{\Gamma(\alpha)} \sum_{j=0}^{n} \int_{t_j}^{t_{j+1}} g'(\tau)f(\tau, y(\tau))(t_{n+1}-\tau)^{\alpha-1}d\tau,
\end{aligned} \tag{13.39}
$$

$$
\begin{aligned}
&= (1-\alpha)g'(t_{n+1})f(t_{n+1}, y(t_{n+1})) \\
&\quad + \frac{\alpha}{\Gamma(\alpha)} \sum_{j=0}^{n} \int_{t_j}^{t_{j+1}} g'\left(t_j + \frac{h}{2}\right) f\left(t_j + \frac{h}{2}, \frac{y_j + y_{j+1}}{2}\right) (t_{n+1} - \tau)^{\alpha-1} d\tau, \\
&= (1-\alpha)g'(t_{n+1})f(t_{n+1}, y(t_{n+1})) \\
&\quad + \frac{\alpha h^\alpha}{\Gamma(\alpha)} \sum_{j=0}^{n} g'\left(t_j + \frac{h}{2}\right) f\left(t_j + \frac{h}{2}, \frac{y_j + y_{j+1}}{2}\right) \\
&\quad \left\{ \frac{(n-j+1)^\alpha}{\alpha} - \frac{(n-j)^\alpha}{\alpha} \right\} \\
&= (1-\alpha)\frac{g(t_{n+1}) - g(t_n)}{h} f(t_{n+1}, y(t_{n+1})) \\
&\quad + \frac{h^\alpha}{\Gamma(\alpha)} \sum_{j=0}^{n-1} g'\left(t_j + \frac{h}{2}\right) f\left(t_j + \frac{h}{2}, \frac{y_j + y_{j+1}}{2}\right) \delta_{n,j}^\alpha \\
&\quad + \frac{h^\alpha}{\Gamma(\alpha)} g'\left(t_n + \frac{h}{2}\right) f\left(t_n + \frac{h}{2}, \frac{y_n + y_{n+1}}{2}\right), \\
&= (1-\alpha)h^{-1}\left(g(t_{n+1}) - g(t_n)\right) f(t_{n+1}, \overline{y}(t_{n+1})) \\
&\quad + \frac{h^{\alpha-1}}{\Gamma(\alpha)} \sum_{j=0}^{n-1} \left(g(t_{j+1}) - g(t_j)\right) f\left(t_j + \frac{h}{2}, \frac{y_j + y_{j+1}}{2}\right) \delta_{n,j}^\alpha \\
&\quad + \frac{h^{\alpha-1}}{\Gamma(\alpha)} \left(g(t_{n+1}) - g(t_n)\right) f\left(t_n + \frac{h}{2}, \frac{y_n + \overline{y}_{n+1}}{2}\right).
\end{aligned}
$$

Here

$$
\begin{aligned}
\overline{y}_{n+1} &= (1-\alpha)h^{-1}\left(g(t_{n+1}) - g(t_n)\right) f(t_n, y_n) \\
&\quad + \frac{h^{\alpha-1}}{\Gamma(\alpha)} \sum_{j=0}^{n} f\left(t_j, y_j\right) \left(g(t_{j+1}) - g(t_j)\right) \delta_{n,j}^\alpha + y_n.
\end{aligned} \tag{13.40}
$$

13.4 Applying Linear Approximation Method on IVP with Atangana–Baleanu Global Derivative

$$
\begin{cases} {}_{t_0}^{ABR}D_g^\alpha y(t) = f(t, y(t)) \text{ if } t > t_0 \\ y(t_0) = y_0 \end{cases}. \tag{13.41}
$$

Here we assume that $g'(t) > 0$. Also here we will need in addition that $f(t, y(t))$ satisfies the Lipschitz condition and is bounded within $[a, b]$. By the fundamental theorem of calculus, the above IVP can be transformed with Atangana–Baleanu integral

$$
\begin{aligned}
y(t) = & \ (1-\alpha)g'(t)f(t, y(t)) + \frac{\alpha}{\Gamma(\alpha)} \int_{t_0}^{t} g'(\tau)(t-\tau)^{\alpha-1} f(\tau, y(\tau)d\tau, \quad (13.42) \\
y(t_0) = & \ y_0.
\end{aligned}
$$

Here we only consider at $t = t_{n+1}$, then we have

$$
\begin{aligned}
y(t_{n+1}) = & \ (1-\alpha)g'(t_{n+1})f(t_{n+1}, y(t_{n+1})) \\
& + \frac{\alpha}{\Gamma(\alpha)} \int_{t_0}^{t_{n+1}} g'(\tau)(t_{n+1}-\tau)^{\alpha-1} f(\tau, y(\tau)d\tau, \quad (13.43) \\
= & \ (1-\alpha)g'(t_{n+1})f(t_{n+1}, y(t_{n+1})) \\
& + \sum_{j=0}^{n} \int_{t_j}^{t_{j+1}} g'(\tau)\frac{\alpha(t_{n+1}-\tau)^{\alpha-1}}{\Gamma(\alpha)} f(\tau, y(\tau)d\tau.
\end{aligned}
$$

Within $[t_j, t_{j+1}]$, we fix

$$
F(t, y(t)) = g'(t)f(t, y(t)) \quad (13.44)
$$

and approximate

$$
F(\tau, y(\tau)) \simeq F(t_j, y(t_j)) + \left(\frac{F(t_{j+1}, y_{j+1}) - F(t_j, y_j)}{h}\right)(\tau - t_j). \quad (13.45)
$$

We replace by the original function to have

$$
\begin{aligned}
F(\tau, y(\tau)) \simeq & \ \frac{\alpha}{\Gamma(\alpha)} g'(t_j)f(t_j, y(t_j)) \\
& + \frac{\alpha}{\Gamma(\alpha)} \left(\frac{g'(t_j)f(t_{j+1}, y_{j+1}) - g'(t_j)f(t_j, y_j)}{h}\right)(\tau - t_j) \\
\simeq & \ \frac{\alpha}{\Gamma(\alpha)} \frac{(g(t_{j+1}) - g(t_j))}{h} f(t_j, y(t_j)) \\
& + \frac{\alpha}{\Gamma(\alpha)} \left(\begin{array}{l} \frac{(g(t_{j+1})-g(t_j))}{h} f(t_{j+1}, y_{j+1}) \\ -\frac{(g(t_{j+1})-g(t_j))}{h} f(t_j, y(t_j)) \end{array}\right) \frac{(\tau - t_j)}{h}.
\end{aligned}
$$

Replacing $f(\tau, y(\tau))$ by the linear polynomial yields

$$
y_{n+1} = \ (1-\alpha)g'(t_{n+1})f(t_{n+1}, y(t_{n+1})) \quad (13.46)
$$

$$
\begin{aligned}
&+\frac{\alpha}{\Gamma(\alpha)}\sum_{j=0}^{n}\int_{t_j}^{t_{j+1}}\frac{\left(g(t_{j+1})-g(t_j)\right)}{h}f(t_j,y(t_j))(t_{n+1}-\tau)^{\alpha-1}d\tau\\
=&\ (1-\alpha)g'(t_{n+1})f(t_{n+1},y(t_{n+1}))\\
&+\frac{\alpha}{\Gamma(\alpha)}\sum_{j=0}^{n}\int_{t_j}^{t_{j+1}}\frac{\left(g(t_{j+1})-g(t_j)\right)}{h}\\
&\quad\left(f(t_{j+1},y_{j+1})-f(t_j,y(t_j))\right)(t_{n+1}-\tau)^{\alpha-1}d\tau,\\
=&\ (1-\alpha)g'(t_{n+1})f(t_{n+1},y(t_{n+1}))\\
&+\frac{\alpha h^{\alpha-1}}{\Gamma(\alpha+1)}\sum_{j=0}^{n}\left(g(t_{j+1})-g(t_j)\right)f(t_j,y(t_j))\left\{(n-j+1)^{\alpha}-(n-j)^{\alpha}\right\}\\
&+\frac{\alpha}{\Gamma(\alpha)}\sum_{j=0}^{n}\int_{t_j}^{t_{j+1}}\frac{\left(g(t_{j+1})-g(t_j)\right)}{h}\\
&\quad\left(f(t_{j+1},y_{j+1})-f(t_j,y(t_j))\right)\int_{t_j}^{t_{j+1}}(t_{n+1}-\tau)^{\alpha-1}\left(\tau-t_j\right)d\tau.
\end{aligned}
$$

We shall expand the integral

$$
\int_{t_j}^{t_{j+1}}(t_{n+1}-\tau)^{\alpha-1}\left(\tau-t_j\right)d\tau=\int_{t_j}^{t_{j+1}}(t_{n+1}-\tau)^{\alpha-1}\tau d\tau-t_j\int_{t_j}^{t_{j+1}}(t_{n+1}-\tau)^{\alpha-1}d\tau. \tag{13.47}
$$

If we put $t_{n+1}y=\tau$ then we get

$$
\begin{aligned}
\int_{t_j}^{t_{j+1}}(t_{n+1}-\tau)^{\alpha-1}\left(\tau-t_j\right)d\tau=&\ t_{n+1}^{\alpha+1}\int_{\frac{t_j}{t_{n+1}}}^{\frac{t_{j+1}}{t_{n+1}}}y^{2-1}(1-y)dy \qquad (13.48)\\
&-t_j\int_{t_j}^{t_{j+1}}(t_{n+1}-\tau)^{\alpha-1}d\tau\\
=&\ t_{n+1}^{\alpha+1}\left(B\left(\frac{t_{j+1}}{t_{n+1}},\alpha,2\right)-B\left(\frac{t_j}{t_{n+1}},\alpha,2\right)\right)\\
&-t_j\left(\frac{\left(t_{n+1}-t_j\right)^{\alpha}}{\alpha}-\frac{\left(t_{n+1}-t_{j+1}\right)^{\alpha}}{\alpha}\right)
\end{aligned}
$$

where $B(z,\alpha,\beta)$ is the incompleted Beta function defined as

$$B(z,\alpha,\beta)=\int_0^z u^{\alpha-1}(1-u)^{\beta-1}du. \tag{13.49}$$

Replacing yields

$$\begin{aligned}
y_{n+1} = & (1-\alpha)g'(t_{n+1})f(t_{n+1},y(t_{n+1})) \\
&+\frac{\alpha h^{\alpha-1}}{\Gamma(\alpha+1)}\sum_{j=0}^{n}\left(g(t_{j+1})-g(t_j)\right)f(t_j,y(t_j))\{(n-j+1)^\alpha-(n-j)^\alpha\} \\
&+\frac{\alpha}{\Gamma(\alpha)}\sum_{j=0}^{n}\frac{\left(g(t_{j+1})-g(t_j)\right)}{h}\left(f(t_{j+1},y_{j+1})-f(t_j,y(t_j))\right) \\
&\times\left\{t_{n+1}^{\alpha+1}\left(B\left(\frac{t_{j+1}}{t_{n+1}},\alpha,2\right)-B\left(\frac{t_j}{t_{n+1}},\alpha,2\right)\right)\right. \\
&\left.-t_j\left(\frac{\left(t_{n+1}-t_j\right)^\alpha}{\alpha}-\frac{\left(t_{n+1}-t_{j-1}\right)^\alpha}{\alpha}\right)\right\}.
\end{aligned} \tag{13.50}$$

But the scheme is implicit; to make it explicit we do the following:

$$\begin{aligned}
y_{n+1} = & (1-\alpha)g'(t_{n+1})f(t_{n+1},\overline{y}(t_{n+1})) \\
&+\frac{\alpha h^{\alpha-1}}{\Gamma(\alpha+1)}\sum_{j=0}^{n}\left(g(t_{j+1})-g(t_j)\right)f(t_j,y(t_j))\{(n-j+1)^\alpha-(n-j)^\alpha\} \\
&+\frac{\alpha}{\Gamma(\alpha)}\sum_{j=0}^{n-1}\frac{\left(g(t_{j+1})-g(t_j)\right)}{h}\left(f(t_{j+1},y_{j+1})-f(t_j,y(t_j))\right) \\
&\times\left\{t_{n+1}^{\alpha+1}\left(B\left(\frac{t_{j+1}}{t_{n+1}},\alpha,2\right)-B\left(\frac{t_j}{t_{n+1}},\alpha,2\right)\right)\right. \\
&\left.-t_j\left(\frac{\left(t_{n+1}-t_j\right)^\alpha}{\alpha}-\frac{\left(t_{n+1}-t_{j-1}\right)^\alpha}{\alpha}\right)\right\} \\
&+\frac{\alpha}{\Gamma(\alpha)}\frac{\left(g(t_{n+1})-g(t_n)\right)}{h}\left(f(t_{n+1},\overline{y}_{n+1})-f(t_n,y(t_n))\right) \\
&\times\left\{t_{n+1}^{\alpha+1}\left(B(1,\alpha,2)-B\left(\frac{t_n}{t_{n+1}},\alpha,2\right)\right)-t_n\frac{h^\alpha}{\alpha}\right\},
\end{aligned} \tag{13.51}$$

where

$$\begin{aligned}
\overline{y}_{n+1} = & (1-\alpha)g'(t_{n+1})f(t_n,y_n) \\
&+\frac{\alpha h^{\alpha-1}}{\Gamma(\alpha+1)}\sum_{j=0}^{n}\left(g(t_{j+1})-g(t_j)\right)f(t_j,y(t_j))\{(n-j+1)^\alpha-(n-j)^\alpha\}.
\end{aligned} \tag{13.52}$$

Theorem 13.3 *Let $y(t_{n+1})$ be the exact solution at the point t_{n+1} and y_{n+1} the approximate solution at the point t_{n+1}. If R_T is the error then*

$$|R_T| < \frac{\|g'\|\,\alpha}{2\Gamma(\alpha)}\left|f''(\xi, y(\xi))\right| h^{\alpha+2}\left[\begin{array}{c}(n+1)^{\alpha+1}\left(B\left(1,\alpha,2\right)-B\left(\frac{t_0}{t_{n+1}},\alpha,2\right)\right)\\ +\frac{n(n+1)}{\alpha}\left((n+1)^{\alpha}+1\right)\\ +(n+1)^{\alpha+2}\left(B\left(1,\alpha,3\right)-B\left(\frac{t_0}{t_{n+1}},\alpha,3\right)\right)\end{array}\right] \tag{13.53}$$

if the second derivative of f is bounded.

Proof

$$\begin{aligned} y(t_{n+1}) = {} & (1-\alpha)g'(t_{n+1})f(t_{n+1}, y(t_{n+1})) \\ & +\frac{\alpha}{\Gamma(\alpha)}\sum_{j=0}^{n}\int_{t_j}^{t_{j+1}} g'(\tau)f(\tau, y(\tau))(t_{n+1}-\tau)^{\alpha-1}d\tau, \\ = {} & (1-\alpha)g'(t_{n+1})f(t_{n+1}, y(t_{n+1})) \\ & +\frac{\alpha}{\Gamma(\alpha)}\sum_{j=0}^{n}\int_{t_j}^{t_{j+1}} g'(\tau)(t_{n+1}-\tau)^{\alpha-1}\left\{P_n(\tau)+R_1(\tau)\right\}d\tau, \\ = {} & y_{n+1}+\frac{\alpha}{\Gamma(\alpha)}\sum_{j=0}^{n}\int_{t_j}^{t_{j+1}} g'(\tau)(t_{n+1}-\tau)^{\alpha-1}R_1(\tau)\,d\tau. \end{aligned} \tag{13.54}$$

We know that

$$R_1(\tau) = \frac{(\tau-t_j)(\tau-t_{j+1})}{2!}f''(\xi, y(\xi)),\ \xi\in[t_j, t_{j+1}] \tag{13.55}$$

$$R_T = \frac{\alpha}{\Gamma(\alpha)}\sum_{j=0}^{n}\int_{t_j}^{t_{j+1}} g'(\tau)(t_{n+1}-\tau)^{\alpha-1}\frac{(\tau-t_j)(\tau-t_{j+1})}{2!}f''(\xi, y(\xi))d\tau,$$

$$\begin{aligned} |R_T| \le {} & \frac{\alpha}{2\Gamma(\alpha)}\sum_{j=0}^{n}\int_{t_j}^{t_{j+1}} \left|g'(\tau)\right|(t_{n+1}-\tau)^{\alpha-1}\left|f''(\xi, y(\xi))\right|(t_{j+1}-\tau)(\tau-t_j)\,d\tau, \\ \le {} & \frac{\alpha}{2\Gamma(\alpha)}\sum_{j=0}^{n}\int_{t_j}^{t_{j+1}} (t_{n+1}-\tau)^{\alpha-1}\sup_{l\in[t_j,\tau]}\left|g'(\tau)\right|\left|f''(\xi, y(\xi))\right| \\ & (t_{j+1}-\tau)(\tau-t_j)\,d\tau, \\ \le {} & \frac{\alpha}{2\Gamma(\alpha)}\sum_{j=0}^{n}\left\|g'\right\|_{\infty}\left\|f''(\xi, y(\xi))\right\|\int_{t_j}^{t_{j+1}} (t_{n+1}-\tau)^{\alpha-1}(t_{j+1}-\tau)(\tau-t_j)\,d\tau. \end{aligned}$$

We shall evaluate

$$\int_{t_j}^{t_{j+1}} (t_{n+1}-\tau)^{\alpha-1}(t_{j+1}-\tau)\left(\tau-t_j\right)d\tau \tag{13.56}$$

$$= \int_{t_j}^{t_{j+1}} (t_{n+1}-\tau)^{\alpha-1}\left\{\tau t_{j+1}-\tau^2-t_jt_{j+1}-t_j\tau\right\}d\tau,$$

$$= \int_{t_j}^{t_{j+1}} (t_{n+1}-\tau)^{\alpha-1}\left\{\left(t_{j+1}-t_j\right)\tau-t_jt_{j+1}-\tau^2\right\}d\tau,$$

$$= h\int_{t_j}^{t_{j+1}} (t_{n+1}-\tau)^{\alpha-1}\tau d\tau - h^2(j+1)j\int_{t_j}^{t_{j+1}} (t_{n+1}-\tau)^{\alpha-1}d\tau$$

$$- \int_{t_j}^{t_{j+1}} (t_{n+1}-\tau)^{\alpha-1}\tau^2 d\tau,$$

$$= h^\alpha t_{n+1}^{\alpha+1}\left(B\left(\frac{t_{j+1}}{t_{n+1}},\alpha,2\right)-B\left(\frac{t_j}{t_{n+1}},\alpha,2\right)\right)$$

$$-\frac{jh^{\alpha+2}(j+1)}{\alpha}\left\{(n-j+1)^\alpha-(n-j)^\alpha\right\}$$

$$-t_{n+1}^{\alpha+2}\left(B\left(\frac{t_{j+1}}{t_{n+1}},\alpha,3\right)-B\left(\frac{t_j}{t_{n+1}},\alpha,3\right)\right).$$

On replacing, we get

$$|R_T| \le \frac{\alpha\left\|g'\right\|_\infty\left\|f''(\xi,y(\xi))\right\|}{2\Gamma(\alpha)} \tag{13.57}$$

$$\left(\sum_{j=0}^{n} h^\alpha t_{n+1}^{\alpha+1}\left(B\left(\frac{t_{j+1}}{t_{n+1}},\alpha,2\right)-B\left(\frac{t_j}{t_{n+1}},\alpha,2\right)\right)\right)$$

$$+\sum_{j=0}^{n}\frac{jh^{\alpha+2}(j+1)}{\alpha}\left\{(n-j+1)^\alpha-(n-j)^\alpha\right\}$$

$$+\sum_{j=0}^{n} t_{n+1}^{\alpha+2}\left(B\left(\frac{t_{j+1}}{t_{n+1}},\alpha,3\right)-B\left(\frac{t_j}{t_{n+1}},\alpha,3\right)\right).$$

We start by

$$\begin{aligned}\sum_{j=0}^{n}\left(B\left(\frac{t_{j+1}}{t_{n+1}},\alpha,2\right)-B\left(\frac{t_j}{t_{n+1}},\alpha,2\right)\right) &= B\left(\frac{t_1}{t_{n+1}},\alpha,2\right)-B\left(\frac{t_0}{t_{n+1}},\alpha,2\right) \\ &\quad +B\left(\frac{t_2}{t_{n+1}},\alpha,2\right)-B\left(\frac{t_1}{t_{n+1}},\alpha,2\right) \\ &\quad +\cdots+B\left(\frac{t_{n+1}}{t_{n+1}},\alpha,2\right)-B\left(\frac{t_n}{t_{n+1}},\alpha,2\right) \\ &= B(1,\alpha,2)-B\left(\frac{t_0}{t_{n+1}},\alpha,2\right). \end{aligned} \tag{13.58}$$

$$\sum_{j=0}^{n}B\left(\frac{t_{j+1}}{t_{n+1}},\alpha,3\right)-B\left(\frac{t_j}{t_{n+1}},\alpha,3\right) = B(1,\alpha,3)-B\left(\frac{t_0}{t_{n+1}},\alpha,3\right),$$

$$\sum_{j=0}^{n}j(j+1)\left\{(n-j+1)^\alpha-(n-j)^\alpha\right\} \le n(n+1)\sum_{j=0}^{n}\left\{(n-j+1)^\alpha-(n-j)^\alpha\right\}.$$

But

$$\begin{aligned}\sum_{j=0}^{n}\{(n-j+1)^\alpha-(n-j)^\alpha\} &= (n+1)^\alpha-n^\alpha+n^\alpha \\ &\quad -(n-1)^\alpha+\ldots+(n-n-1)^\alpha-(n-n)^\alpha \\ &= (n+1)^\alpha+1. \end{aligned} \tag{13.59}$$

Replacing all values yields

$$|R_T| < \frac{\alpha\left\|g'\right\|}{2\Gamma(\alpha)}\left|f''(\xi,y(\xi))\right|h^{\alpha+2}\left[\begin{array}{c}(n+1)^{\alpha+1}\left(B(1,\alpha,2)-B\left(\frac{t_0}{t_{n+1}},\alpha,2\right)\right) \\ +\frac{n(n+1)}{\alpha}\left((n+1)^\alpha+1\right) \\ +(n+1)^{\alpha+2}\left(B(1,\alpha,3)-B\left(\frac{t_0}{t_{n+1}},\alpha,3\right)\right)\end{array}\right] \tag{13.60}$$

which completes the proof.

Theorem 13.4 *Let $y(t_{n+1})$ be the exact solution at $t=t_{n+1}$ and y_{n+1} be the approximate solution, then*

$$\lim_{h\to 0}|y(t_{n+1})-y_{n+1}|=0. \tag{13.61}$$

Proof

$$\begin{aligned}&|y(t_{n+1})-y_{n+1}| \\ &= \left|\begin{array}{c}(1-\alpha)g'(t_{n+1})f(t_{n+1},y(t_{n+1})) \\ +\frac{\alpha}{\Gamma(\alpha)}\sum_{j=0}^{n}\int_{t_j}^{t_{j+1}}g'(\tau)f(\tau,y(\tau))(t_{n+1}-\tau)^{\alpha-1}d\tau \\ -(1-\alpha)g'(t_{n+1})f(t_{n+1},y(t_{n+1})) \\ +\frac{\alpha}{\Gamma(\alpha)}\sum_{j=0}^{n}\int_{t_j}^{t_{j+1}}g'(t_{j+1})\left(f(t_j,y_j)+\left(\frac{f(t_{j+1},y_{j+1})-f(t_j,y_j)}{h}\right)(\tau-t_j)\right)(t-\tau)^{\alpha-1}d\tau\end{array}\right|, \end{aligned} \tag{13.62}$$

$$\leq \frac{\alpha \|g'\|_\infty}{\Gamma(\alpha)} \sum_{j=0}^{n} \int_{t_j}^{t_{j+1}} (t_{n+1}-\tau)^{\alpha-1} \left(f(\tau, y(\tau)) - f(t_j, y_j)\right) d\tau$$
$$+\frac{\alpha \|g'\|_\infty}{\Gamma(\alpha)} \sum_{j=0}^{n} \int_{t_j}^{t_{j+1}} (t_{n+1}-\tau)^{\alpha-1} \left(\frac{f(t_{j+1}, y_{j+1}) - f(t_j, y_j)}{h}\right) (\tau - t_j)\, d\tau,$$
$$\leq \frac{\alpha \|g'\|_\infty}{\Gamma(\alpha)} \sum_{j=0}^{n} \int_{t_j}^{t_{j+1}} \left|f'(\xi, y(\xi))\right| (\tau - t_j)(t_{n+1}-\tau)^{\alpha-1} d\tau$$
$$+\frac{\alpha \|g'\|_\infty}{\Gamma(\alpha)} \sum_{j=0}^{n} \int_{t_j}^{t_{j+1}} (t_{n+1}-\tau)^{\alpha-1} \left|f'(\xi, y(\xi))\right| (\tau - t_j)\, d\tau.$$

The above is achieved via the differentiability of the function f and the Mean Value theorem:

$$\begin{aligned}
\left|y(t_{n+1}) - y_{n+1}\right| \leq\ & \frac{\alpha \|g'\|_\infty \left|f'(\xi, y(\xi))\right|}{\Gamma(\alpha)} \\
& \sum_{j=0}^{n} \left\{ \begin{array}{c} t_{n+1}^{\alpha+1}\left(B\left(\frac{t_{j+1}}{t_{n+1}}, \alpha, 2\right) - B\left(\frac{t_j}{t_{n+1}}, \alpha, 2\right)\right) \\ -t_j \left\{\frac{(t_{n+1}-t_j)^\alpha}{\alpha} - \frac{(t_{n+1}-t_{j+1})^\alpha}{\alpha}\right\} \end{array} \right\} \\
& + \frac{\alpha \|g'\|_\infty \left|f'(\xi, y(\xi))\right|}{\Gamma(\alpha)} \\
& \sum_{j=0}^{n} \left\{ \begin{array}{c} t_{n+1}^{\alpha+1}\left(B\left(\frac{t_{j+1}}{t_{n+1}}, \alpha, 2\right) - B\left(\frac{t_j}{t_{n+1}}, \alpha, 2\right)\right) \\ -t_j \left\{\frac{(t_{n+1}-t_j)^\alpha}{\alpha} - \frac{(t_{n+1}-t_{j+1})^\alpha}{\alpha}\right\} \end{array} \right\}, \\
\leq\ & \frac{2\alpha \|g'\|_\infty \left|f'(\xi, y(\xi))\right|}{\Gamma(\alpha)} \\
& \sum_{j=0}^{n} \left\{ \begin{array}{c} t_{n+1}^{\alpha+1}\left(B\left(\frac{t_{j+1}}{t_{n+1}}, \alpha, 2\right) - B\left(\frac{t_j}{t_{n+1}}, \alpha, 2\right)\right) \\ -t_j \left\{\frac{(t_{n+1}-t_j)^\alpha}{\alpha} - \frac{(t_{n+1}-t_{j+1})^\alpha}{\alpha}\right\} \end{array} \right\}, \\
\leq\ & 2\alpha \|g'\|_\infty \left|f'(\xi, y(\xi))\right| \\
& \sum_{j=0}^{n} \left\{ \begin{array}{c} t_{n+1}^{\alpha+1}\left(B\left(\frac{t_{j+1}}{t_{n+1}}, \alpha, 2\right) - B\left(\frac{t_j}{t_{n+1}}, \alpha, 2\right)\right) \\ -t_j \left\{\frac{(t_{n+1}-t_j)^\alpha}{\alpha} - \frac{(t_{n+1}-t_{j+1})^\alpha}{\alpha}\right\} \end{array} \right\}, \\
\leq\ & 2\alpha \|g'\|_\infty \left|f'(\xi, y(\xi))\right| h^{\alpha+1} \\
& \sum_{j=0}^{n+1} \left\{ \begin{array}{c} (n+1)^{\alpha+1}\left(B\left(\frac{t_{j+1}}{t_{n+1}}, \alpha, 2\right) - B\left(\frac{t_j}{t_{n+1}}, \alpha, 2\right)\right) \\ +n \left\{\frac{(n-j+1)^\alpha}{\alpha} - \frac{(n-j)^\alpha}{\alpha}\right\} \end{array} \right\}, \\
\leq\ & 2\alpha \|g'\|_\infty \left|f'(\xi, y(\xi))\right| h^{\alpha+1} \left\{ \begin{array}{c} (n+1)^{\alpha}\left(B(1, \alpha, 2) - B\left(\frac{t_0}{t_{n+1}}, \alpha, 2\right)\right) \\ +n\left((n+1)^\alpha + 1\right) \end{array} \right\}, \\
\leq\ & h^{\alpha+1}\lambda(n, \alpha)
\end{aligned} \tag{13.63}$$

where

$$\lambda(n,\alpha) = 2\alpha \left\|g'\right\|_{\infty} \left|f'(\xi, y(\xi))\right| \left\{ \begin{array}{c} (n+1)^{\alpha}\left(B(1,\alpha,2) - B\left(\frac{t_0}{t_{n+1}},\alpha,2\right)\right) \\ +n\left((n+1)^{\alpha}+1\right) \end{array} \right\}. \tag{13.64}$$

Therefore

$$\begin{aligned} \lim_{h\to 0} \left|y(t_{n+1}) - y_{n+1}\right| \leq & \lim_{h\to 0} h^{\alpha+1}\lambda(n,\alpha), \\ \leq & \ 0. \end{aligned} \tag{13.65}$$

Since $\lambda(n,\alpha) < \infty$; thus

$$\lim_{h\to 0} \left|y(t_{n+1}) - y_{n+1}\right| = 0, \tag{13.66}$$

which completes the proof.

Theorem 13.5 *If the function $f(t, y(t))$ satisfies the global Lipschitz condition, we have*

$$\left|y_{n+1}\right| < a_{n+1} + b_{n+1}\left|y_0\right|. \tag{13.67}$$

Proof

$$\left|y_{n+1}\right| = \left| \begin{array}{c} (1-\alpha)g'(t_{n+1})f(t_{n+1},\overline{y}(t_{n+1})) \\ +\frac{\alpha h^{\alpha-1}}{\Gamma(\alpha+1)}\sum_{j=0}^{n}\left(g(t_{j+1})-g(t_j)\right)f(t_j,y(t_j))\left\{(n-j+1)^{\alpha}-(n-j)^{\alpha}\right\} \\ +\frac{\alpha}{\Gamma(\alpha)}\sum_{j=0}^{n-1}\frac{\left(g(t_{j+1})-g(t_j)\right)}{h}\left(f(t_{j+1},y_{j+1})-f(t_j,y(t_j))\right) \\ \times\left\{ \begin{array}{c} t_{n+1}^{\alpha+1}\left(B\left(\frac{t_{j+1}}{t_{n+1}},\alpha,2\right)-B\left(\frac{t_j}{t_{n+1}},\alpha,2\right)\right) \\ -t_j\left\{\frac{(t_{n+1}-t_j)^{\alpha}}{\alpha}-\frac{(t_{n+1}-t_{j+1})^{\alpha}}{\alpha}\right\} \end{array} \right\} \\ +\frac{\alpha}{h\Gamma(\alpha)}\left(g(t_{n+1})-g(t_n)\right)\left(f\left(t_{n+1},\overline{y}_{n+1}\right)-f(t_n,y_n)\right) \\ \times\left\{t_{n+1}^{\alpha+1}\left(B(1,\alpha,2)-B\left(\frac{t_n}{t_{n+1}},\alpha,2\right)\right)-t_n\frac{h^{\alpha}}{\alpha}\right\} \end{array} \right|. \tag{13.68}$$

For simplicity, we let

$$\begin{aligned} \beta_n^{\alpha,j} &= \left(g(t_{j+1})-g(t_j)\right)\left\{(n-j+1)^{\alpha}-(n-j)^{\alpha}\right\}, \\ \beta_{n,1}^{\alpha,j} &= \frac{\left(g(t_{j+1})-g(t_j)\right)}{h}\left\{ \begin{array}{c} t_{n+1}^{\alpha+1}\left(B\left(\frac{t_{j+1}}{t_{n+1}},\alpha,2\right)-B\left(\frac{t_j}{t_{n+1}},\alpha,2\right)\right) \\ -t_j\left\{\frac{(t_{n+1}-t_j)^{\alpha}}{\alpha}-\frac{(t_{n+1}-t_{j+1})^{\alpha}}{\alpha}\right\} \end{array} \right\}, \\ \beta_n^{\alpha} &= \frac{\alpha}{h\Gamma(\alpha)}\left(g(t_{n+1})-g(t_n)\right) \\ &\quad \times\left\{t_{n+1}^{\alpha+1}\left(B(1,\alpha,2)-B\left(\frac{t_n}{t_{n+1}},\alpha,2\right)\right)-t_n\frac{h^{\alpha}}{\alpha}\right\}. \end{aligned} \tag{13.69}$$

Then, the equation becomes

$$
\begin{aligned}
&|y_{n+1}| \qquad (13.70)\\
&= \left| \begin{array}{c} (1-\alpha)g'(t_{n+1})f(t_{n+1},\overline{y}(t_{n+1})) \\ +\frac{\alpha h^{\alpha-1}}{\Gamma(\alpha+1)}\sum_{j=0}^{n}\beta_n^{\alpha,j} f(t_j,y(t_j)) \\ +\frac{\alpha}{\Gamma(\alpha)}\sum_{j=0}^{n-1}\beta_{n,1}^{\alpha,j}\left(f(t_{j+1},y_{j+1})-f(t_j,y(t_j))\right) \\ +\beta_n^{\alpha}\left(f\left(t_{n+1},\overline{y}_{n+1}\right)-f(t_n,y_n)\right) \end{array} \right|, \\
&\le (1-\alpha)\left|g'(t_{n+1})\right| \left|f(t_{n+1},\overline{y}(t_{n+1}))\right| \\
&+\frac{\alpha h^{\alpha-1}}{\Gamma(\alpha+1)}\sum_{j=0}^{n}\left|\beta_n^{\alpha,j}\right| \left|f(t_j,y(t_j))\right| \\
&+\frac{\alpha}{\Gamma(\alpha)}\sum_{j=0}^{n-1}\left|\beta_{n,1}^{\alpha,j}\right| \left|\left(f(t_{j+1},y_{j+1})-f(t_j,y(t_j))\right)\right| \\
&+\left|\beta_n^{\alpha}\right| \left|f\left(t_{n+1},\overline{y}_{n+1}\right)-f(t_n,y_n)\right|.
\end{aligned}
$$

Using the global Lipschitz condition of f yields

$$
\begin{aligned}
\left|f(t_{j+1},y_{j+1})\right| &< c\left(1+\left|y_{j+1}\right|\right), \qquad (13.71)\\
\left|f(t_j,y_j)\right| &< c\left(1+\left|y_j\right|\right),\\
\left|f\left(t_{n+1},\overline{y}_{n+1}\right)\right| &< c\left(1+\left|\overline{y}_{n+1}\right|\right),\\
\left|f(t_n,y_n)\right| &< c\left(1+\left|y_n\right|\right),
\end{aligned}
$$

where indeed

$$
\left|\overline{y}_{n+1}\right| \le \frac{\alpha h^{\alpha-1}}{\Gamma(\alpha+1)}\sum_{j=0}^{n}\left|\beta_n^{\alpha,j}\right|\left(c\left(1+\left|y_j\right|\right)\right). \qquad (13.72)
$$

Therefore

$$
\begin{aligned}
\left|f\left(t_{n+1},\overline{y}_{n+1}\right)\right| &< c\left(1+\frac{\alpha h^{\alpha-1}}{\Gamma(\alpha+1)}\sum_{j=0}^{n}\left|\beta_n^{\alpha,j}\right|\left(c\left(1+\left|y_j\right|\right)\right)\right) \qquad (13.73)\\
&< c+\frac{\alpha h^{\alpha-1}}{\Gamma(\alpha+1)}h\left|g'(t_{n+1})\right|c^2\left((n+1)^{\alpha}+1\right)\\
&+\frac{\alpha c^2 h^{\alpha-1}}{\Gamma(\alpha+1)}\sum_{j=0}^{n}\left|\beta_n^{\alpha,j}\right|\left|y_j\right|.
\end{aligned}
$$

Replacing, we obtain

$$|y_{n+1}| = (1-\alpha)g'(t_{n+1})\begin{pmatrix} c+\frac{\alpha h^{\alpha-1}}{\Gamma(\alpha+1)}h\left|g'(t_{n+1})\right|c^2\left((n+1)^\alpha+1\right) \\ +\frac{\alpha c^2 h^{\alpha-1}}{\Gamma(\alpha+1)}\sum_{j=0}^{n}\left|\beta_n^{\alpha,j}\right|\left|y_j\right| \end{pmatrix} \tag{13.74}$$

$$+\frac{\alpha c^2 h^{\alpha-1}}{\Gamma(\alpha+1)}\sum_{j=0}^{n}\left|\beta_n^{\alpha,j}\right|\left|y_j\right| + \frac{\alpha h^{\alpha-1}}{\Gamma(\alpha+1)}\sum_{j=0}^{n}\left|\beta_n^{\alpha,j}\right|\left(c\left(1+\left|y_j\right|\right)\right)$$

$$+\frac{\alpha}{\Gamma(\alpha)}\sum_{j=0}^{n-1}\left|\beta_{n,1}^{\alpha,j}\right|\left(c\left(1+\left|y_{j+1}\right|\right)+c\left(1+\left|y_j\right|\right)\right)$$

$$+\left|\beta_n^\alpha\right|\begin{Bmatrix} c+\frac{\alpha h^\alpha}{\Gamma(\alpha+1)}\left|g'(t_{n+1})\right|c^2\left((n+1)^\alpha+1\right) \\ +\frac{\alpha c^2 h^{\alpha-1}}{\Gamma(\alpha+1)}\sum_{j=0}^{n}\left|\beta_n^{\alpha,j}\right|\left|y_j\right|+c\left(1+\left|y_n\right|\right) \end{Bmatrix}.$$

For $n = 0$,

$$\begin{aligned} |y_1| &< (1-\alpha)g'(t_1)\left(c+\frac{\alpha h^{\alpha-1}}{\Gamma(\alpha+1)}h\left|g'(t_1)\right|c^2+\frac{\alpha c^2 h^{\alpha-1}}{\Gamma(\alpha+1)}\left|\beta_0^{\alpha,0}\right|\left|y_0\right|\right) \\ &+\frac{\alpha h^{\alpha-1}}{\Gamma(\alpha+1)}\left|\beta_0^{\alpha,0}\right|\left(c\left(1+|y_0|\right)\right)+\left|\beta_0^{\alpha,0}\right|\begin{Bmatrix} c+\frac{2\alpha h^\alpha}{\Gamma(\alpha+1)}\left|g'(t_1)\right|c^2 \\ +\frac{c^2\alpha h^{\alpha-1}}{\Gamma(\alpha+1)}\left|\beta_0^{\alpha,0}\right|\left|y_0\right|+c+c\left|y_0\right| \end{Bmatrix}, \\ &< a_1+b_1\left|y_0\right|, \end{aligned} \tag{13.75}$$

$$\begin{aligned} a_1 &= (1-\alpha)\left|g'(t_1)\right|\left(c+\frac{h^\alpha}{\Gamma(\alpha)}\left|g'(t_1)\right|c^2\right)+\frac{ch^{\alpha-1}}{\Gamma(\alpha+1)}\left|\beta_0^\alpha\right| \\ &+\left|\beta_0^\alpha\right|\left(2c+\frac{2h^{\alpha-1}}{\Gamma(\alpha)}\left|g'(t_1)\right|c^2\right), \\ b_1 &= (1-\alpha)\left|g'(t_1)\right|\frac{c^2 h^{\alpha-1}}{\Gamma(\alpha+1)}\left|\beta_0^\alpha\right|+\frac{h^{\alpha-1}}{\Gamma(\alpha)}\left|\beta_0^\alpha\right|c+\left|\beta_0^\alpha\right|\left(c+\frac{c^2 h^{\alpha-1}}{\Gamma(\alpha)}\left|\beta_0^\alpha\right|\right). \end{aligned} \tag{13.76}$$

We assume $\forall n \geq 1$ that

$$|y_n| < a_n + b_n\left|y_0\right|. \tag{13.77}$$

$$\begin{aligned} |y_{n+1}| \leq\ & \overline{\lambda}_0+\overline{\lambda}_1\sum_{j=0}^{n}\left|\beta_n^{\alpha,j}\right|\left|y_j\right|+\overline{\lambda}_2\sum_{j=0}^{n-1}\left|\beta_{n,1}^{\alpha,j}\right|\left|y_{j+1}\right| \\ & +\overline{\lambda}_3\left|y_n\right|+\overline{\lambda}_4\sum_{j=0}^{n-1}\left|\beta_{n,1}^{\alpha,j}\right|\left|y_j\right|, \end{aligned} \tag{13.78}$$

where

$$\begin{aligned}
\overline{\lambda}_0 = & \ (1-\alpha)\left|g'(t_{n+1})\right|\left(c+\frac{h^\alpha}{\Gamma(\alpha)}\left|g'(t_{n+1})\right|c^2\left((n+1)^\alpha+1\right)\right) \quad (13.79)\\
& +\frac{h^\alpha}{\Gamma(\alpha)}\sum_{j=0}^{n}\left|\beta_n^{\alpha,j}\right|c+\frac{2\alpha c}{\Gamma(\alpha)}\sum_{j=0}^{n-1}\left|\beta_{n,j}^{\alpha,j}\right|\\
& +\left|\beta_n^\alpha\right|\left\{c+\frac{h^\alpha}{\Gamma(\alpha)}\left|g'(t_{n+1})\right|c^2\left((n+1)^\alpha+1\right)\right\}+\left|\beta_n^\alpha\right|c,\\
\overline{\lambda}_1 = & \ (1-\alpha)\left|g'(t_{n+1})\right|\frac{c^2h^\alpha}{\Gamma(\alpha)}+\frac{c^2h^{\alpha-1}}{\Gamma(\alpha)}+\frac{ch^{\alpha-1}}{\Gamma(\alpha)},\\
\overline{\lambda}_2 = & \ \left|\beta_n^\alpha\right|\frac{c^2h^{\alpha-1}}{\Gamma(\alpha)},\\
\overline{\lambda}_3 = & \ \left|\beta_n^\alpha\right|c,\\
\overline{\lambda}_4 = & \ \frac{c\alpha}{\Gamma(\alpha)}.
\end{aligned}$$

By inductive hypothesis, we have

$$\begin{aligned}
\left|y_{n+1}\right| \leq & \ \overline{\lambda}_0+\overline{\lambda}_1\sum_{j=0}^{n}\left|\beta_n^{\alpha,j}\right|\left(a_j+b_j\left|y_0\right|\right)+\overline{\lambda}_2\sum_{j=0}^{n-1}\left|\beta_{n,j}^{\alpha,j}\right|\left(a_{j+1}+b_{j+1}\left|y_0\right|\right) \quad (13.80)\\
& +\overline{\lambda}_3\left(a_n+b_n\left|y_0\right|\right)+\overline{\lambda}_4\sum_{j=0}^{n-1}\left|\beta_{n,j}^{\alpha,j}\right|\left(a_j+b_j\left|y_0\right|\right),\\
\leq & \ \overline{\lambda}_0+\overline{\lambda}_1\sum_{j=0}^{n}\left|\beta_n^{\alpha,j}\right|a_j+\overline{\lambda}_2\sum_{j=0}^{n-1}\left|\beta_{n,1}^{\alpha,j}\right|a_{j+1}+\overline{\lambda}_3a_n+\overline{\lambda}_4\sum_{j=0}^{n-1}\left|\beta_{n,j}^{\alpha,j}\right|a_j\\
& +\left(\overline{\lambda}_1\sum_{j=0}^{n}\left|\beta_n^{\alpha,j}\right|b_j+\overline{\lambda}_2\sum_{j=0}^{n-1}\left|\beta_{n,j}^{\alpha,j}\right|b_{j+1}+\overline{\lambda}_3b_n+\overline{\lambda}_4\sum_{j=0}^{n-1}\left|\beta_{n,j}^{\alpha,j}\right|b_j\right)\left|y_0\right|,\\
\leq & \ a_{n+1}+b_{n+1}\left|y_0\right|,
\end{aligned}$$

which concludes the proof.

Theorem 13.6 *If $\widetilde{y}_{n+1}$, $\widetilde{y}_n$, $\widetilde{y}_j$, $\widetilde{y}_{j+1}$, and $\widetilde{y}_0$ are perturbed terms of y_{n+1}, y_n, y_j, y_{j+1}, and y_0 respectively, then if L is Lipschitz and the $\exists c \geq 0$ such that $\forall n > 0$*

$$\left|\widetilde{y}_{n+1}\right| < \Omega\left|\widetilde{y}_0\right|. \quad (13.81)$$

Proof

$$\begin{aligned}
y_{n+1}+\widetilde{y}_{n+1} = & \ (1-\alpha)g'(t_{n+1})\left(f(t_{n+1},\overline{y}_{n+1})-f(t_{n+1},\overline{y}_{n+1}+\widetilde{\overline{y}}_{n+1})\right) \quad (13.82)\\
& +\frac{\alpha h^{\alpha-1}}{\Gamma(\alpha+1)}\sum_{j=0}^{n}\beta_n^{\alpha,j}\left(f(t_j,y_j+\widetilde{y}_j)-f(t_j,y_j)\right)
\end{aligned}$$

$$+\frac{\alpha}{\Gamma(\alpha)}\sum_{j=0}^{n-1}\beta_{n,1}^{\alpha,j}\left(f(t_{j+1},y_{j+1}+\widetilde{y}_{j+1})-f(t_j,y_j+\widetilde{y}_j)\right)$$
$$+\beta_n^{\alpha}\left(f\left(t_{n+1},\overline{y}_{n+1}+\widetilde{\overline{y}}_{n+1}\right)-f(t_n,y_n+\widetilde{y}_n)\right),$$

$$\begin{aligned}\widetilde{y}_{n+1}=&\ (1-\alpha)g'(t_{n+1})\left(f(t_{n+1},\overline{y}_{n+1})-f(t_{n+1},\overline{y}_{n+1}+\widetilde{\overline{y}}_{n+1})\right)\\&+\frac{\alpha h^{\alpha-1}}{\Gamma(\alpha+1)}\sum_{j=0}^{n}\beta_n^{\alpha,j}\left(f(t_j,y_j+\widetilde{y}_j)-f(t_j,y_j)\right)\\&+\frac{\alpha}{\Gamma(\alpha)}\sum_{j=0}^{n-1}\beta_{n,1}^{\alpha,j}\begin{pmatrix}\left(f(t_{j+1},y_{j+1}+\widetilde{y}_{j+1})-f(t_{j+1},y_{j+1})\right)\\-\left(f(t_j,y_j+\widetilde{y}_j)-f(t_j,y_j)\right)\end{pmatrix}\\&+\beta_n^{\alpha}\begin{pmatrix}\left(f\left(t_{n+1},\overline{y}_{n+1}+\widetilde{\overline{y}}_{n+1}\right)-f\left(t_{n+1},\overline{y}_{n+1}\right)\right)\\-\left(f(t_n,y_n+\widetilde{y}_n)-f(t_n,y_n)\right)\end{pmatrix},\end{aligned}\tag{13.83}$$

$$\begin{aligned}\left|\widetilde{y}_{n+1}\right|\leq&\ (1-\alpha)\left|g'\right|_{\infty}L\left|\widetilde{\overline{y}}_{n+1}\right|+\frac{\alpha h^{\alpha-1}}{\Gamma(\alpha+1)}\sum_{j=0}^{n}\left|\beta_n^{\alpha,j}\right|L\left|\widetilde{y}_j\right|\\&+\frac{\alpha}{\Gamma(\alpha)}\sum_{j=0}^{n-1}\left|\beta_{n,1}^{\alpha,j}\right|\left(L\left|\widetilde{y}_{j+1}\right|+L\left|\widetilde{y}_j\right|\right)+\beta_n^{\alpha}\left(L\left|\widetilde{\overline{y}}_{n+1}\right|+L\left|\widetilde{y}_n\right|\right)\end{aligned}\tag{13.84}$$

where

$$\left|\widetilde{\overline{y}}_{n+1}\right|\leq\frac{\alpha h^{\alpha-1}}{\Gamma(\alpha+1)}\sum_{j=0}^{n}\left|\beta_n^{\alpha,j}\right|L\left|\widetilde{y}_j\right|.\tag{13.85}$$

$$\begin{aligned}\left|\widetilde{y}_{n+1}\right|\leq&\ (1-\alpha)\left|g'\right|_{\infty}L\left(\frac{\alpha h^{\alpha-1}}{\Gamma(\alpha+1)}\sum_{j=0}^{n}\left|\beta_n^{\alpha,j}\right|L\left|\widetilde{y}_j\right|\right)\\&+\frac{\alpha h^{\alpha-1}}{\Gamma(\alpha+1)}\sum_{j=0}^{n}\left|\beta_n^{\alpha,j}\right|L\left|\widetilde{y}_j\right|\\&+\frac{\alpha}{\Gamma(\alpha)}\sum_{j=0}^{n-1}\left|\beta_{n,1}^{\alpha,j}\right|\left(L\left|\widetilde{y}_{j+1}\right|+L\left|\widetilde{y}_j\right|\right)\\&+\left|\beta_n^{\alpha}\right|L\left\{\frac{\alpha h^{\alpha-1}}{\Gamma(\alpha+1)}\sum_{j=0}^{n}\left|\beta_n^{\alpha,j}\right|L\left|\widetilde{y}_j\right|+\left|\widetilde{y}_n\right|\right\}.\end{aligned}$$

When $n=0$, we have

$$\left|\widetilde{y}_1\right|\leq(1-\alpha)\left|g'\right|_{\infty}L\frac{\alpha h^{\alpha-1}}{\Gamma(\alpha+1)}\left|\beta_0^{\alpha,j}\right|L\left|\widetilde{y}_0\right|+\frac{\alpha h^{\alpha-1}}{\Gamma(\alpha+1)}\left|\beta_0^{\alpha,j}\right|L\left|\widetilde{y}_0\right|\tag{13.86}$$

$$
\begin{aligned}
&+\frac{\alpha\left|\beta_0^\alpha\right| L^2 h^{\alpha-1}\left|\beta_0^\alpha\right|}{\Gamma(\alpha+1)}\left|\widetilde{y}_0\right|+\left|\beta_0^\alpha\right|\left|\widetilde{y}_0\right| \\
&\leq \left(\frac{\alpha h^{\alpha-1}}{\Gamma(\alpha+1)}\left|\beta_0^{\alpha,j}\right| L+\frac{\alpha\left|\beta_0^\alpha\right|^2 L^2 h^{\alpha-1}}{\Gamma(\alpha+1)}+\left|\beta_0^\alpha\right|\right)\left|\widetilde{y}_0\right| \\
&\leq c\left|\widetilde{y}_0\right| .
\end{aligned}
$$

$$
\begin{aligned}
\left|\widetilde{y}_2\right| \leq & (1-\alpha)\left|g'\right|_{\infty} L\left(\frac{\alpha h^{\alpha-1}}{\Gamma(\alpha+1)} \sum_{j=0}^{1}\left|\beta_1^{\alpha,j}\right| L\left|\widetilde{y}_j\right|\right) \\
&+\frac{\alpha h^{\alpha-1}}{\Gamma(\alpha+1)} \sum_{j=0}^{1}\left|\beta_1^{\alpha,j}\right| L\left|\widetilde{y}_j\right|+\frac{L}{\Gamma(\alpha)}\left|\beta_{1,1}^{\alpha,j}\right|\left(\left|\widetilde{y}_1\right|+\left|\widetilde{y}_0\right|\right) \\
&+L\left|\beta_1^\alpha\right|\left(\frac{\alpha L h^{\alpha-1}}{\Gamma(\alpha+1)} \sum_{j=0}^{1}\left|\beta_1^{\alpha,j}\right| L\left|\widetilde{y}_j\right|+\left|\widetilde{y}_1\right|\right),
\end{aligned}
$$

$$
\begin{aligned}
\left|\widetilde{y}_2\right| \leq & (1-\alpha)\left|g'\right|_{\infty} L \frac{\alpha h^{\alpha-1}}{\Gamma(\alpha+1)}\left(\left|\beta_1^{\alpha,0}\right| L\left|\widetilde{y}_0\right|+\left|\beta_1^{\alpha,1}\right| L\left|\widetilde{y}_1\right|\right) \\
&+\frac{\alpha h^{\alpha-1}}{\Gamma(\alpha+1)}\left(\left|\beta_2^\alpha\right| L\left|\widetilde{y}_0\right|+\left|\beta_2^\alpha\right| L c\left|\widetilde{y}_0\right|\right)+\frac{\alpha L}{\Gamma(\alpha)}\left|\beta_{1,1}^\alpha\right|\left(c\left|\widetilde{y}_0\right|+\left|\widetilde{y}_0\right|\right) \\
&+L\left|\beta_2^\alpha\right|\left\{\frac{\alpha L h^{\alpha-1}}{\Gamma(\alpha+1)}\left|\beta_1^\alpha\right|\left|\widetilde{y}_0\right|+\frac{L \alpha h^{\alpha-1}}{\Gamma(\alpha+1)}\left|\beta_1^{\alpha,1}\right|\left|\widetilde{y}_0\right| c+\left|\widetilde{y}_0\right| c\right\} \\
\leq & (1-\alpha)\left|g'\right|_{\infty} L \frac{\alpha h^{\alpha-1}}{\Gamma(\alpha+1)}\left(\left|\beta_1^{\alpha,0}\right| L\left|\widetilde{y}_0\right|+\left|\beta_1^{\alpha,1}\right| L c\left|\widetilde{y}_0\right|\right) \\
&+\frac{\alpha h^{\alpha-1}}{\Gamma(\alpha+1)}\left(\left|\beta_2^\alpha\right| L\left|\widetilde{y}_0\right|+\left|\beta_2^\alpha\right| L c\left|\widetilde{y}_0\right|\right)+\frac{\alpha L}{\Gamma(\alpha)}\left|\beta_{1,1}^\alpha\right|\left(c\left|\widetilde{y}_0\right|+\left|\widetilde{y}_0\right|\right) \\
&+L\left|\beta_2^\alpha\right|\left\{\frac{\alpha L h^{\alpha-1}}{\Gamma(\alpha+1)}\left|\beta_1^\alpha\right|\left|\widetilde{y}_0\right|+\frac{\alpha L h^{\alpha-1}}{\Gamma(\alpha+1)}\left|\beta_1^{\alpha,1}\right|\left|\widetilde{y}_0\right| c+\left|\widetilde{y}_0\right| c\right\} \\
\leq & \left|\widetilde{y}_0\right|\left\{\begin{array}{c}
(1-\alpha)\left|g'\right|_{\infty} L^2 \frac{\alpha h^{\alpha-1}}{\Gamma(\alpha+1)}\left(\left|\beta_1^{\alpha,0}\right|+\left|\beta_1^{\alpha,1}\right| c\right) \\
+\frac{\alpha h^{\alpha-1}}{\Gamma(\alpha+1)}\left|\beta_2^\alpha\right| L+\left|\beta_2^\alpha\right| L c+\frac{\alpha L}{\Gamma(\alpha)}\left|\beta_{1,1}^\alpha\right|(c+1) \\
+L^2\left|\beta_2^\alpha\right| \frac{\alpha h^{\alpha-1}}{\Gamma(\alpha+1)}\left|\beta_1^\alpha\right|+\frac{\alpha L h^{\alpha-1}}{\Gamma(\alpha+1)}\left|\beta_1^{\alpha,1}\right| c+c
\end{array}\right\}, \\
\left|\widetilde{y}_2\right| \leq & c_n\left|\widetilde{y}_0\right| .
\end{aligned} \tag{13.87}
$$

We assume that $\forall n \geq 2$

$$
\left|\widetilde{y}_n\right| \leq c_n\left|\widetilde{y}_0\right| . \tag{13.88}
$$

$$|\widetilde{y}_{n+1}| \leq \left(\begin{array}{c} (1-\alpha)\left|g'\right|_{\infty} L\left(\frac{\alpha h^{\alpha-1}}{\Gamma(\alpha+1)}\sum_{j=0}^{n}\left|\beta_n^{\alpha,j}\right| Lc_j\right) \\ +\frac{\alpha h^{\alpha-1}}{\Gamma(\alpha+1)}\sum_{j=0}^{n}\left|\beta_n^{\alpha,j}\right| Lc_j + \frac{L\alpha}{\Gamma(\alpha)}\sum_{j=0}^{n-1}\left|\beta_{n,1}^{\alpha,j}\right|\left(c_{j+1}+c_j\right) \\ +\left|\beta_n^{\alpha}\right| L\left\{\frac{\alpha L h^{\alpha-1}}{\Gamma(\alpha+1)}\sum_{j=0}^{n}\left|\beta_n^{\alpha,j}\right| c_j + c_n\right\} \end{array}\right)|\widetilde{y}_0|, \quad (13.89)$$

$$\leq c_{n+1}|\widetilde{y}_0|$$
$$= \Omega|\widetilde{y}_0|,$$

which completes the proof.

13.5 Applying Lagrange Interpolation Method on IVP with Atangana–Baleanu Global Derivative

In this section, we consider the following IVP with Riemann–Liouville global derivative

$$\begin{cases} {}^{ABR}_{t_0}D_g^{\alpha} y(t) = f(t, y(t)) \text{ if } t > t_0 \\ y(t_0) = y_0 \end{cases}. \quad (13.90)$$

$$y(t) = (1-\alpha)g'(t)f(t, y(t)) + \frac{\alpha}{\Gamma(\alpha)}\int_{t_0}^{t} g'(\tau)f(\tau, y(\tau))(t-\tau)^{\alpha-1}d\tau \quad (13.91)$$

$$y(t_0) = y_0.$$

At $t = t_{n+1}$, we have

$$\begin{aligned} y(t_{n+1}) = {} & (1-\alpha)g'(t_{n+1})f(t_{n+1}, y_{n+1}) \quad (13.92) \\ & + \frac{\alpha}{\Gamma(\alpha)}\int_{t_0}^{t_{n+1}} g'(\tau)f(\tau, y(\tau))(t_{n+1}-\tau)^{\alpha-1}d\tau, \\ = {} & (1-\alpha)g'(t_{n+1})f(t_{n+1}, \overline{y}_{n+1}) \\ & + \frac{\alpha}{\Gamma(\alpha)}\sum_{j=0}^{n}\int_{t_j}^{t_{j+1}} g'(\tau)f(\tau, y(\tau))(t_{n+1}-\tau)^{\alpha-1}d\tau, \\ = {} & (1-\alpha)g'(t_{n+1})f(t_{n+1}, \overline{y}_{n+1}) \\ & + \frac{\alpha}{\Gamma(\alpha)}\sum_{j=0}^{n}\int_{t_j}^{t_{j+1}}\left[\begin{array}{c} \frac{g(t_{j+1})-g(t_j)}{h^2}\left(\tau-t_{j-1}\right)f(t_j, y_j) \\ -\frac{g(t_j)-g(t_{j-1})}{h^2}\left(\tau-t_j\right)f(t_{j-1}, y_{j-1}) \end{array}\right](t_{n+1}-\tau)^{\alpha-1}d\tau. \end{aligned}$$

As presented before for the integral part

$$
\begin{aligned}
&\int_{t_j}^{t_{j+1}} \left(\tau - t_{j-1}\right)\left(t_{n+1}-\tau\right)^{\alpha-1} d\tau \qquad (13.93)\\
&= \int_{t_j}^{t_{j+1}} \tau (t_{n+1}-\tau)^{\alpha-1} d\tau - t_{j-1}\int_{t_j}^{t_{j+1}} (t_{n+1}-\tau)^{\alpha-1} d\tau \\
&= t_{n+1}\left(B\left(\frac{t_{j+1}}{t_{n+1}},2,\alpha\right) - B\left(\frac{t_j}{t_{n+1}},2,\alpha\right)\right) \\
&\quad -t_{j-1}\left(\frac{\left(t_{n+1}-t_j\right)^{\alpha}}{\alpha} - \frac{\left(t_{n+1}-t_{j-1}\right)^{\alpha}}{\alpha}\right),
\end{aligned}
$$

$$
\begin{aligned}
\int_{t_j}^{t_{j+1}} \left(\tau - t_j\right)\left(t_{n+1}-\tau\right)^{\alpha-1} d\tau &= t_{n+1}\left(B\left(\frac{t_{j+1}}{t_{n+1}},2,\alpha\right) - B\left(\frac{t_j}{t_{n+1}},2,\alpha\right)\right) \\
&\quad -t_j\left(\frac{\left(t_{n+1}-t_j\right)^{\alpha}}{\alpha} - \frac{\left(t_{n+1}-t_{j-1}\right)^{\alpha}}{\alpha}\right).
\end{aligned}
$$

Replacing in the original equation, we obtain the following:

$$
\begin{aligned}
y_{n+1} &= (1-\alpha)\frac{g(t_{n+1})-g(t_n)}{h} f(t_{n+1},\overline{y}_{n+1}) \qquad (13.94)\\
&+\frac{\alpha}{\Gamma(\alpha)}\sum_{j=0}^{n}\left[\begin{array}{c}
\left(\frac{g(t_{j+1})-g(t_j)}{h^2} f(t_j,y_j)\right)\left(\begin{array}{c} t_{n+1}\left(B\left(\frac{t_{j+1}}{t_{n+1}},2,\alpha\right)-B\left(\frac{t_j}{t_{n+1}},2,\alpha\right)\right)\\ -t_{j-1}\left(\frac{(t_{n+1}-t_j)^{\alpha}}{\alpha}-\frac{(t_{n+1}-t_{j-1})^{\alpha}}{\alpha}\right)\end{array}\right)\\
-\left(\frac{g(t_j)-g(t_{j-1})}{h^2} f(t_{j-1},y_{j-1})\right)\left(\begin{array}{c} t_{n+1}\left(B\left(\frac{t_{j+1}}{t_{n+1}},2,\alpha\right)-B\left(\frac{t_j}{t_{n+1}},2,\alpha\right)\right)\\ -t_j\left(\frac{(t_{n+1}-t_j)^{\alpha}}{\alpha}-\frac{(t_{n+1}-t_{j-1})^{\alpha}}{\alpha}\right)\end{array}\right)
\end{array}\right].
\end{aligned}
$$

We note that the formula is implicit since we have $\overline{y}_{n+1}$ on the right-hand side; therefore to make it explicit, we replace

$$
\begin{aligned}
\overline{y}_{n+1} &= (1-\alpha)\frac{g(t_{n+1})-g(t_n)}{h} f(t_n,y_n) \qquad (13.95)\\
&+\frac{\alpha}{\Gamma(\alpha+1)}\sum_{j=0}^{n}\frac{g\left(t_{j+1}\right)-g\left(t_j\right)}{h^{1-\alpha}} f(t_j,y_j)\left\{(n-j+1)^{\alpha}-(n-j)^{\alpha}\right\},\\
&= (1-\alpha)\frac{g(t_{n+1})-g(t_n)}{h} f(t_n,y_n)\\
&+\frac{\alpha h^{\alpha-1}}{\Gamma(\alpha+1)}\sum_{j=0}^{n}\left(g\left(t_{j+1}\right)-g\left(t_j\right)\right) f(t_j,y_j)\left\{(n-j+1)^{\alpha}-(n-j)^{\alpha}\right\},
\end{aligned}
$$

$$\begin{aligned} y_1 = & (1-\alpha)\frac{g(t_1)-g(t_0)}{h}f(t_1,\overline{y}_1) \\ & +\frac{\alpha}{\Gamma(\alpha)}\int_{t_0}^{t_1} g'(\tau)f(\tau,y(\tau))(t_1-\tau)^{\alpha-1}d\tau, \\ = & (1-\alpha)\frac{g(t_1)-g(t_0)}{h}f(t_1,\overline{y}_1)+\frac{\alpha}{\Gamma(\alpha)}\frac{g(t_1)-g(t_0)}{h}f(t_1,\overline{y}_1)\frac{(t_1-t_0)^{\alpha}}{\alpha}, \end{aligned} \tag{13.96}$$

where

$$\overline{y}_1 = (1-\alpha)\frac{g(t_1)-g(t_0)}{h}f(t_0,y_0)+\frac{\alpha}{\Gamma(\alpha)}\frac{g(t_1)-g(t_0)}{h}f(t_0,y_0)\frac{(t_1-t_0)^{\alpha}}{\alpha}, \tag{13.97}$$

therefore

$$y_1 = (1-\alpha)\frac{g(t_1)-g(t_0)}{h}f(t_1,y_1)+\frac{h^{\alpha-1}}{\Gamma(\alpha)}f(t_1,\overline{y}_1) \tag{13.98}$$

where

$$\overline{y}_1 = (1-\alpha)\frac{g(t_1)-g(t_0)}{h}f(t_0,y_0)+\frac{h^{\alpha-1}}{\Gamma(\alpha)}f(t_0,y_0). \tag{13.99}$$

13.6 Applying Linear Piecewise Interpolation Method on IVP with Atangana–Baleanu Global Derivative

In this section, we consider the following IVP with Riemann–Liouville global derivative:

$$\begin{cases} {}_{t_0}^{ABR}D_g^{\alpha}y(t) = f(t,y(t)) \text{ if } t > t_0 \\ y(t_0) = y_0 \end{cases}. \tag{13.100}$$

$$\begin{cases} y(t) = (1-\alpha)g'(t)f(t,y(t))+\frac{\alpha}{\Gamma(\alpha)}\int_{t_0}^{t} g'(\tau)f(\tau,y(\tau))(t-\tau)^{\alpha-1}d\tau, \\ y(t_0) = y_0. \end{cases} \tag{13.101}$$

At $t = t_{n+1}$, we have

$$\begin{aligned} y(t_{n+1}) = & (1-\alpha)g'(t_{n+1})f(t_{n+1},y_{n+1}) \\ & +\frac{\alpha}{\Gamma(\alpha)}\int_{t_0}^{t_{n+1}} g'(\tau)f(\tau,y(\tau))(t_{n+1}-\tau)^{\alpha-1}d\tau, \\ = & (1-\alpha)g'(t_{n+1})f(t_{n+1},y_{n+1}) \end{aligned} \tag{13.102}$$

$$+\frac{\alpha}{\Gamma(\alpha)}\sum_{j=0}^{n}\int_{t_j}^{t_{j+1}} g'(\tau)f(\tau,y(\tau))(t_{n+1}-\tau)^{\alpha-1}d\tau,$$

$$= (1-\alpha)g'(t_{n+1})f(t_{n+1},y_{n+1})$$

$$+\frac{\alpha}{\Gamma(\alpha)}\sum_{j=0}^{n}\int_{t_j}^{t_{j+1}} \frac{g\left(t_{j+1}\right)-g\left(t_j\right)}{h}$$

$$\left[\frac{\tau-t_{j-1}}{h}f(t_j,y_j)-\frac{\tau-t_j}{h}f(t_{j-1},y_{j-1})\right](t_{n+1}-\tau)^{\alpha-1}d\tau.$$

We shall calculate each interval separately

$$\int_{t_j}^{t_{j+1}} \left(\tau-t_{j-1}\right)\left(t_{n+1}-\tau\right)^{\alpha-1}d\tau \tag{13.103}$$

$$=\int_{t_j}^{t_{j+1}} \tau(t_{n+1}-\tau)^{\alpha-1}d\tau - t_{j-1}\int_{t_j}^{t_{j+1}}(t_{n+1}-\tau)^{\alpha-1}d\tau$$

$$=t_{n+1}\left(B\left(\frac{t_{j+1}}{t_{n+1}},2,\alpha\right)-B\left(\frac{t_j}{t_{n+1}},2,\alpha\right)\right)$$

$$-t_{j-1}\left(\frac{\left(t_{n+1}-t_j\right)^{\alpha}}{\alpha}-\frac{\left(t_{n+1}-t_{j-1}\right)^{\alpha}}{\alpha}\right),$$

$$\int_{t_j}^{t_{j+1}} \left(\tau-t_j\right)\left(t_{n+1}-\tau\right)^{\alpha-1}d\tau = t_{n+1}\left(B\left(\frac{t_{j+1}}{t_{n+1}},2,\alpha\right)-B\left(\frac{t_j}{t_{n+1}},2,\alpha\right)\right)$$

$$-t_j\left(\frac{\left(t_{n+1}-t_j\right)^{\alpha}}{\alpha}-\frac{\left(t_{n+1}-t_{j-1}\right)^{\alpha}}{\alpha}\right).$$

Replacing we obtain

$$y_{n+1} = (1-\alpha)g'(t_{n+1})f(t_{n+1},y_{n+1}) \tag{13.104}$$
$$+\sum_{j=0}^{n}\left[\begin{array}{c}\frac{\alpha(g(t_{j+1})-g(t_j))}{h^2}f(t_j,y_j)\left\{\begin{array}{c}t_{n+1}\left(B\left(\frac{t_{j+1}}{t_{n+1}},2,\alpha\right)-B\left(\frac{t_j}{t_{n+1}},2,\alpha\right)\right)\\ -t_{j-1}\left(\frac{\left(t_{n+1}-t_j\right)^{\alpha}}{\alpha}-\frac{\left(t_{n+1}-t_{j-1}\right)^{\alpha}}{\alpha}\right)\end{array}\right\}\\ -\frac{\alpha(g(t_j)-g(t_{j-1}))}{h^2}f(t_{j-1},y_{j-1})\left\{\begin{array}{c}t_{n+1}\left(B\left(\frac{t_{j+1}}{t_{n+1}},2,\alpha\right)-B\left(\frac{t_j}{t_{n+1}},2,\alpha\right)\right)\\ -t_j\left(\frac{\left(t_{n+1}-t_j\right)^{\alpha}}{\alpha}-\frac{\left(t_{n+1}-t_{j-1}\right)^{\alpha}}{\alpha}\right)\end{array}\right\}\end{array}\right].$$

But

$$y_1 = (1-\alpha)g'(t_1)f(t_1,y_1) + \frac{\alpha}{\Gamma(\alpha)}\int_{t_0}^{t_1} g'(\tau)f(\tau,y(\tau))(t_1-\tau)^{\alpha-1}d\tau, \quad (13.105)$$

$$= (1-\alpha)g'(t_1)f(t_1,y_1) + \frac{\alpha}{\Gamma(\alpha)}\int_{t_0}^{t_1} g'(t_1)f(t_1,y_1)(t_1-\tau)^{\alpha-1}d\tau,$$

$$= (1-\alpha)\frac{g(t_1)-g(t_0)}{h}f(t_1,\overline{y}_1) + \frac{\alpha}{\Gamma(\alpha+1)}\frac{g(t_1)-g(t_0)}{h}f(t_1,\overline{y}_1)(t_1-t_0)^{\alpha}$$

where

$$\overline{y}_1 = (1-\alpha)\frac{g(t_1)-g(t_0)}{h}f(t_0,y_0) + \frac{1}{\Gamma(\alpha+1)}\frac{g(t_1)-g(t_0)}{h}f(t_0,y_0)(t_1-t_0)^{\alpha}. \quad (13.106)$$

Thus

$$\begin{aligned} y_1 = {} & (1-\alpha)\frac{g(t_1)-g(t_0)}{h}f\left(t_1,\left(\begin{array}{c}(1-\alpha)\frac{g(t_1)-g(t_0)}{h}f(t_0,y_0)\\ +\frac{1}{\Gamma(\alpha+1)}\frac{g(t_1)-g(t_0)}{h}f(t_0,y_0)(t_1-t_0)^{\alpha}\end{array}\right)\right) \\ & +\frac{\alpha(t_1-t_0)^{\alpha}}{\Gamma(\alpha+1)}\frac{g(t_1)-g(t_0)}{h}f\left(t_1,\left(\begin{array}{c}(1-\alpha)\frac{g(t_1)-g(t_0)}{h}f(t_0,y_0)\\ +\frac{1}{\Gamma(\alpha+1)}\frac{g(t_1)-g(t_0)}{h}f(t_0,y_0)(t_1-t_0)^{\alpha}\end{array}\right)\right). \end{aligned} \quad (13.107)$$

Therefore the numerical scheme obtained is given as

$$\begin{aligned} y_{n+1} = {} & (1-\alpha)g'(t_{n+1})f(t_{n+1},y_{n+1}) \\ & +\sum_{j=0}^{n}\left(\begin{array}{c} \frac{\alpha(g(t_{j+1})-g(t_j))}{h^2}f(t_j,y_j)\left\{\begin{array}{c} t_{n+1}\left(\begin{array}{c}B\left(\frac{t_{j+1}}{t_{n+1}},2,\alpha\right)\\ -B\left(\frac{t_j}{t_{n+1}},2,\alpha\right)\end{array}\right)\\ -t_{j-1}\left(\frac{(t_{n+1}-t_j)^{\alpha}}{\alpha}-\frac{(t_{n+1}-t_{j-1})^{\alpha}}{\alpha}\right)\end{array}\right\} \\ -\frac{\alpha(g(t_j)-g(t_{j-1}))}{h^2}f(t_{j-1},y_{j-1})\left\{\begin{array}{c} t_{n+1}\left(\begin{array}{c}B\left(\frac{t_{j+1}}{t_{n+1}},2,\alpha\right)\\ -B\left(\frac{t_j}{t_{n+1}},2,\alpha\right)\end{array}\right)\\ -t_j\left(\frac{(t_{n+1}-t_j)^{\alpha}}{\alpha}-\frac{(t_{n+1}-t_{j-1})^{\alpha}}{\alpha}\right)\end{array}\right\}\end{array}\right) \end{aligned} \quad (13.108)$$

$if\ n > 1.$

$$\begin{aligned} y_1 = {} & (1-\alpha)\frac{g(t_1)-g(t_0)}{h}f\left(t_1,\left(\begin{array}{c}(1-\alpha)\frac{g(t_1)-g(t_0)}{h}f(t_0,y_0)\\ +\frac{1}{\Gamma(\alpha+1)}\frac{g(t_1)-g(t_0)}{h}f(t_0,y_0)(t_1-t_0)^{\alpha}\end{array}\right)\right) \\ & +\frac{\alpha(t_1-t_0)^{\alpha}}{\Gamma(\alpha+1)}\frac{g(t_1)-g(t_0)}{h}f\left(t_1,\left(\begin{array}{c}(1-\alpha)\frac{g(t_1)-g(t_0)}{h}f(t_0,y_0)\\ +\frac{1}{\Gamma(\alpha+1)}\frac{g(t_1)-g(t_0)}{h}f(t_0,y_0)(t_1-t_0)^{\alpha}\end{array}\right)\right) \end{aligned} \quad (13.109)$$

$if\ n = 1.$

Remark 13.1 We recall that

$$g'(t_n) = \frac{g(t_{n+1}) - g(t_n)}{h} + O(h) \tag{13.110}$$

is said to be first order approximation. Thus, to maintain the order or to increase the order, we can use $g'(t_n)$ instead of

$$\frac{g(t_{n+1}) - g(t_n)}{h} \tag{13.111}$$

while implementing the code to solve differential equations considered in this book.

References

1. Ascher, U.M., Petzold, L.R.: Computer Methods for Ordinary Differential Equations and Differential-Algebraic Equations. Society for Industrial and Applied Mathematics, Philadelphia (1998). ISBN 978-0-89871-412-8
2. Butcher, J.C.: Numerical Methods for Ordinary Differential Equations. Wiley, New York (2003). ISBN 978-0-471-96758-3
3. Hairer, E., Nørsett, S.P., Wanner, G.: Solving Ordinary Differential Equations I: Nonstiff Problems. Springer, Berlin, New York (1993). ISBN 978-3-540-56670-0
4. Stoer, J., Bulirsch, R.: Introduction to Numerical Analysis (3rd ed.). : Springer, Berlin, New York (2002). ISBN 978-0-387-95452-3
5. Baleanu, D., Jajarmi, A., Hajipour, M.: On the nonlinear dynamical systems within the generalized fractional derivatives with Mittag–Leffler kernel. Nonlinear Dyn. (2018)
6. Dumitru, B., Amin, J., Mojtaba, H.: On the nonlinear dynamical systems within the generalized fractional derivatives with Mittag–Leffler kernel. Nonlinear Dyn. **94**, 397–414 (2018)
7. Leader, J.J.: Numerical Analysis and Scientific Computation. Addison-Wesley, Boston (2004). ISBN 0-201-73499-0
8. Griffiths, D.V., Smith, I.M.: Numerical Methods for Engineers: A Programming Approach. CRC Press, Boca Raton, p. 218 (1991). ISBN 0-8493-8610-1
9. Meijering, Erik: A chronology of interpolation: from ancient astronomy to modern signal and image processing. Proc. IEEE **90**(3), 319–342 (2002)
10. Lagrange, J.-L.: "Leçon Cinquième. Sur l'usage des courbes dans la solution des problèmes". Leçons Elémentaires sur les Mathématiques (in French). Paris. Republished in Serret, Joseph-Alfred, ed. (1877). Oeuvres de Lagrange, vol. 7. Gauthier-Villars, pp. 271–287 (1795)
11. Toufik, M., Atangana, A.: New numerical approximation of fractional derivative with non-local and non-singular kernel: application to chaotic models. Eur. Phys. J. Plus **132**(10), 444 (2017)
12. Atangana, A., Araz, S.İ.: New Numerical Scheme with Newton Polynomial: Theory, Methods, and Applications. Academic Press (2021)

Chapter 14
Examples and Applications of Global Fractional Differential Equations

This chapter will offer a large number of examples of ordinary differential equations and solution techniques so that the breadth of the application of these techniques may be illustrated. Aside from being one of the pillars of mathematical modeling, these equations are useful for almost everything around us [1–8]. We will also explore their applications in various fields, from modeling disease spread in epidemiology to studying the chaotic systems of chaos theory. By exploring these varied applications, we will illustrate how ordinary differential equations serve as a foundation for studying complex systems and provide insights into a wide array of phenomena, ranging from population dynamics and ecological modeling to the behavior of physical systems and the prediction of chaotic behavior. It acts as both an entry point into the material and a comprehensive overview of the practical implications of these mathematical models, with real-world applications for scientific and engineering challenges that make up the bulk of the remaining chapters.

14.1 Examples of Linear and Nonlinear Ordinary Differential Equations

In this section, we shall present linear and nonlinear ordinary differential equations. We start with the following.

Example 14.1

$$\begin{cases} D_g y(t) = -\lambda y(t) & \text{if } t \geq 0 \\ y(0) = y_0 & \text{if } t = 0 \end{cases}, \tag{14.1}$$

where $g(t)$ satisfies the conditions described in the book. Thus

A. Atangana and İ. Koca, *Fractional Differential and Integral Operators with Respect to a Function*, Industrial and Applied Mathematics,
https://doi.org/10.1007/978-981-97-9951-0_14

$$y'(t) = -\lambda g'(t)y(t). \tag{14.2}$$

The solution is given by

$$y(t) = y(0)\exp\left[-\lambda\left(g(t) - g(0)\right)\right]. \tag{14.3}$$

We present in the below figure the solution for a given $g(t)$.

Example 14.2 We consider the following equation:

$$\begin{cases} D_g y(t) = a \text{ if } t \geq 0 \\ y(0) = y_0 \quad \text{if } t = 0 \end{cases}. \tag{14.4}$$

The solution is given by

$$y(t) = y(0) + a\left(g(t) - g(0)\right). \tag{14.5}$$

Example 14.3 Let f be a function on $C\,[0, T]$, such that $f'(t)$ exists

$$\begin{cases} D_g y(t) = f'(t) \text{ if } t \geq 0 \\ y(t_0) = y_0 \qquad \text{if } t = 0 \end{cases} \tag{14.6}$$

The solution is given by

$$\begin{aligned} y(t) = \; & y(0) + \int_0^t g'(\tau)f(\tau)\,d\tau \\ = \; & y(0) + g(t)f(t) - \int_0^t f'(\tau)g(\tau)\,d\tau. \end{aligned} \tag{14.7}$$

Several linear equations can be determined and solved, but we are more interested in nonlinear equations.

Example 14.4 In this example we shall consider an important nonlinear equation that was introduced to explain long-run economic growth by looking at capital accumulation, labor, and increases in productivity longly driven by technological progress. The model is known as the Solow–Swan model

$$D_g k(t) = ak^{\alpha}(t) - (n + g + \delta)\,k(t) \tag{14.8}$$

and manipulating yields

$$\begin{aligned} k'(t) = \; & g'(t)ak^{\alpha}(t) - g'(t)\,(n + g + \delta)\,k(t), \\ = \; & \gamma_1(t)k^{\alpha}(t) - \gamma_2(t)k(t), \end{aligned} \tag{14.9}$$

which becomes a nonlinear ordinary differential equation with variable order. The steady state is given as

$$k^* = \left(\frac{s}{n+g+\delta}\right)^{\frac{n}{1-\alpha}}. \tag{14.10}$$

If $\alpha \neq 1$,

$$k'(t) = \gamma_1(t)k^\alpha(t) - \gamma_2(t)k(t) = f(t, k(t)). \tag{14.11}$$

The above equation can be solved numerically for example by Heun's method:

$$\begin{cases} k_{n+1} = k_n + \frac{h}{2}\left[f(t_n, k_n) + f\left(t_{n+1}, \widetilde{k}_{n+1}\right)\right], \\ \widetilde{k}_{n+1} = k_n + hf(t_n, k_n) \end{cases} \tag{14.12}$$

where

$$f(t_n, k_n) = \gamma_1(t_n)k_n^\alpha - \gamma_2(t_n)k_n. \tag{14.13}$$

So we get

$$k_{n+1} = k_n + \frac{h}{2}\left[\begin{array}{c} \gamma_1(t_n)k_n^\alpha - \gamma_2(t_n)k_n + \gamma_1(t_{n+1})\left(k_n + h\left(\gamma_1(t_n)k_n^\alpha - \gamma_2(t_n)k_n\right)\right)^\alpha \\ -\gamma_2(t_{n+1})\left(k_n + h\left(\gamma_1(t_n)k_n^\alpha - \gamma_2(t_n)k_n\right)\right) \end{array}\right]. \tag{14.14}$$

The numerical solutions are presented in figures below for different $g(t)$ and α.

We replace classical derivative type with the Caputo–Fabrizio type to have

$$\begin{cases} {}_0^{CF}D_g^\alpha k(t) = f(t, k(t)), \\ k(t_0) = k_0, \end{cases} \tag{14.15}$$

$$\begin{cases} {}_0^{CF}D_t^\alpha k(t) = g'(t)f(t, k(t)), \\ k(t_0) = k_0, \end{cases}$$

$$\begin{cases} k(t) = (1-\alpha)g'(t)f(t, k(t)) + \alpha \int\limits_{t_0}^{t} g'(\tau)f(\tau, k(\tau))\, d\tau, \\ k(t_0) = k_0. \end{cases} \tag{14.16}$$

When $t = t_{n+1}$, we have

$$k(t_{n+1}) = (1-\alpha)g'(t_{n+1})f(t_{n+1}, k(t_{n+1})) + \alpha \int\limits_{t_0}^{t_{n+1}} g'(\tau)f(\tau, k(\tau))\, d\tau, \tag{14.17}$$

$$k(t_n) = (1-\alpha)g'(t_n)f(t_n, k(t_n)) + \alpha \int\limits_{t_0}^{t_n} g'(\tau)f(\tau, k(\tau))\, d\tau,$$

$$k_{n+1} - k_n = (1-\alpha)\left[g'(t_{n+1})f(t_{n+1}, \widetilde{k}_{n+1}) - g'(t_n)f(t_n, k_n)\right]$$

$$+\frac{h}{2}\left[f(t_n,k_n)+f(t_{n+1},\widetilde{k}_{n+1})\right]$$

where

$$\widetilde{k}_{n+1}=(1-\alpha)g(t_{n+1})f(t_n,k_n)+\alpha\sum_{j=0}^{n}f\left(t_j,k_j\right)\Delta t. \tag{14.18}$$

We shall now present the numerical solution of the above equation using the Euler method. From the result obtained in Chap. 12, we have

$$\begin{aligned}k_{n+1}=\;&k_n+(1-\alpha)f(t_{n+1},\overline{k}_{n+1})\left(\frac{g(t_{n+1})-g(t_n)}{h}\right)\\&-(1-\alpha)f(t_n,k_n)\left(\frac{g(t_n)-g(t_{n-1})}{h}\right)\\&+\alpha\left(g(t_{n+1})-g(t_n)\right)f(t_n,k_n).\end{aligned} \tag{14.19}$$

For the first case, now replacing $f(t_n,k_n)$ by its value yields

$$\begin{aligned}k_{n+1}=\;&k_n+(1-\alpha)\left[\gamma_1(t_{n+1})\overline{k}_{n+1}^{\alpha}-\gamma_2(t_{n+1})\overline{k}_{n+1}\right]g'(t_{n+1}) \qquad (14.20)\\&-(1-\alpha)\left[\gamma_1(t_n)\overline{k}_n^{\alpha}-\gamma_2(t_n)\overline{k}_n\right]g'(t_n)+\alpha\left(g(t_{n+1})-g(t_n)\right)f(t_n,k_n).\end{aligned}$$

For the second case we will have

$$\begin{aligned}k_{n+1}=\;&k_n+(1-\alpha)f(t_{n+1},\overline{k}_{n+1})\left(\frac{g(t_{n+1})-g(t_n)}{h}\right)\\&-(1-\alpha)f(t_n,k_n)\left(\frac{g(t_n)-g(t_{n-1})}{h}\right)\\&+\alpha\left(g(t_{n+1})-g(t_n)\right)f(t_{n+1},\overline{k}_{n+1}).\end{aligned} \tag{14.21}$$

In both cases, we have

$$\begin{aligned}\overline{k}_{n+1}=\;&k_n+(1-\alpha)\left[\gamma_n(t_n)k_n^{\alpha}-\gamma_2(t_n)k_n\right]\left(\frac{g(t_{n+1})-g(t_n)}{h}\right) \qquad (14.22)\\&-(1-\alpha)\left[\gamma_1(t_{n-1})k_{n-1}^{\alpha}-\gamma_2(t_{n-1})k_{n-1}\right]\left(\frac{g(t_n)-g(t_{n-1})}{h}\right)\\&+\alpha\left(g(t_{n+1})-g(t_n)\right)\left[\gamma_1(t_n)k_n^{\alpha}-\gamma_2(t_n)k_n\right].\end{aligned}$$

Some numerical simulation is performed for different $g(t)$. We consider the model when the differential operator is that of fractional power law kernel

$$\begin{cases}{}_{t_0}^{RL}D_g^{\alpha}k(t)=\gamma_1(t)k^{\alpha}(t)-\gamma_2(t)k(t),\\ k(t_0)=k_0.\end{cases} \tag{14.23}$$

To solve this numerically, we transform the above first to

$$\begin{cases} {}_{t_0}^{RL}D_t^{\alpha}k(t) = g'(t)\left(\gamma_1(t)k^{\alpha}(t) - \gamma_2(t)k(t)\right), \\ k(t_0) = k_0. \end{cases} \tag{14.24}$$

Further, we apply the Riemann–Liouville integral on both sides to have

$$k(t) = \frac{1}{\Gamma(\alpha)} \int_{t_0}^{t} g'(\tau)\left(\gamma_1(\tau)k^{\alpha}(\tau) - \gamma_2(\tau)k(\tau)\right)(t-\tau)^{\alpha-1}\, d\tau \tag{14.25}$$

such that at $t = t_{n+1}$ we will have

$$k(t_{n+1}) = \frac{1}{\Gamma(\alpha)} \int_{t_0}^{t_{n+1}} g'(\tau)\left(\gamma_1(\tau)k^{\alpha}(\tau) - \gamma_2(\tau)k(\tau)\right)(t_{n+1}-\tau)^{\alpha-1}\, d\tau. \tag{14.26}$$

Now using the Midpoint approach, we have

$$\begin{aligned} k_{n+1} = {} & \frac{h^{\alpha-1}}{\Gamma(\alpha+1)} \sum_{j=0}^{n-1} \left(g\left(t_{j+1}\right) - g\left(t_j\right)\right) \\ & \left[\gamma_1(t_j + \frac{h}{2})\frac{k_j^{\alpha} + k_{j+1}^{\alpha}}{2} - \gamma_2\left(t_j + \frac{h}{2}\right)\frac{k_j + k_{j+1}}{2}\right]\delta_{n,j}^{\alpha} \\ & + \frac{h^{\alpha-1}}{\Gamma(\alpha+1)}\left(g(t_{n+1}) - g(t_n)\right) \\ & + \frac{h^{\alpha-1}}{\Gamma(\alpha+1)}\left(g(t_{n+1}) - g(t_n)\right) \\ & \left[\gamma_1(t_n + \frac{h}{2})\frac{k_n^{\alpha} + \overline{k}_{n+1}^{\alpha}}{2} - \gamma_2(t_n + \frac{h}{2})\frac{k_n + \overline{k}_{n+1}}{2}\right] \end{aligned} \tag{14.27}$$

where

$$\overline{k}_{n+1} = k_n + \frac{h^{\alpha-1}}{\Gamma(\alpha+1)} \sum_{j=0}^{n} \left[\left(\gamma_1(t_j)k_j^{\alpha} - \gamma_2(t_j)k_j\right)\right]\left(g\left(t_{j+1}\right) - g\left(t_j\right)\right)\delta_{n,j}^{\alpha}. \tag{14.28}$$

Using the Linear Piecewise interpolation approach, we get

$$k_{n+1} = \sum_{j=1}^{n} \left\{ \begin{array}{c} g'(t_j)\left[\gamma_1(t_j)k_j^{\alpha} - \gamma_2(t_j)k_j\right] \left[\begin{array}{c} t_{n+1}\left(B\left(\frac{t_{j+1}}{t_{n+1}}, 2, \alpha\right) - B\left(\frac{t_j}{t_{n+1}}, 2, \alpha\right)\right) \\ -t_{j-1}\left(\frac{(t_{n+1}-t_j)^{\alpha}}{\alpha}\right) - \left(\frac{(t_{n+1}-t_{j-1})^{\alpha}}{\alpha}\right) \end{array} \right] \\ -g'(t_j)\left[\gamma_1(t_{j-1})k_{j-1}^{\alpha} - \gamma_2(t_{j-1})k_{j-1}\right] \left[\begin{array}{c} t_{n+1}\left(B\left(\frac{t_{j+1}}{t_{n+1}}, 2, \alpha\right) - B\left(\frac{t_j}{t_{n+1}}, 2, \alpha\right)\right) \\ -t_j\left(\frac{(t_{n+1}-t_j)^{\alpha}}{\alpha}\right) - \left(\frac{(t_{n+1}-t_{j-1})^{\alpha}}{\alpha}\right) \end{array} \right] \end{array} \right\} \text{if } n > 1 \tag{14.29}$$

$$k_1 = \frac{g(t_1) - g(t_0)}{h\Gamma(\alpha+1)} \left[\gamma_1(t_1)\overline{k}_1^{\alpha} - \gamma_2(t_1)\overline{k}_1\right], \tag{14.30}$$
$$\overline{k}_1 = \frac{g(t_1) - g(t_0)}{h\Gamma(\alpha+1)} \left[\gamma_1(t_0)k_0^{\alpha} - \gamma_2(t_0)k_0\right](t_1 - t_0)^{\alpha}.$$

Noting that $(t_1 - t_0) = h$, therefore,

$$\overline{k}_1 = \frac{g(t_1) - g(t_0)}{\Gamma(\alpha+1)} h^{\alpha-1} \left[\gamma_1(t_0)k_0^{\alpha} - \gamma_2(t_0)k_0\right]. \tag{14.31}$$

14.2 SIR Model with Global Derivative

In this chapter, we analyze the SIR model with global derivative from classical to fractional framework. We shall recall that SIR stands for susceptible, infected, and removed. One can find several versions of this model in the literature, but in our investigation, we shall consider the simplest version

$$\begin{aligned} D_g S(t) &= -\frac{\beta I S}{N}, \\ D_g I(t) &= \frac{\beta I S}{N} - \gamma I, \\ D_g R(t) &= \gamma I, \end{aligned} \tag{14.32}$$

where D_g will be any fractional global derivative. We shall present the analysis for each case. We shall start with the classical case.

14.3 SIR Model with Classical Global Derivative

We consider in this section SIR model with the classical global derivative as follows:

$$\begin{aligned} {}_0D_g S(t) &= -\frac{\beta I S}{N}, \\ {}_0D_g I(t) &= \frac{\beta I S}{N} - \gamma I, \\ {}_0D_g R(t) &= \gamma I. \end{aligned} \tag{14.33}$$

The initial conditions are the same like in the normal model

$$S(0) = S_0,\ I(0) = I_0,\ R(0) = R_0. \tag{14.34}$$

Using the properties of the global derivative when $g'(t) > 0$, we get

$$\begin{aligned}
\frac{dS(t)}{dt} &= -\beta g'(t)I(t)\frac{S(t)}{N}, \\
\frac{dI(t)}{dt} &= \beta g'(t)I(t)\frac{S(t)}{N} - \gamma I(t)g'(t), \\
\frac{dR(t)}{dt} &= \gamma I(t)g'(t).
\end{aligned} \tag{14.35}$$

Routine analysis will not be presented here. However where there is an extra.

Irrespective to the sign of $\beta - \gamma$, we have that $g(t)$ is bounded, therefore $\forall t \in [t_0, T]$, $I(t)$ is bounded there exists $M_I > 0$ such that

$$\sup_{\forall t \in [t_0, T]} |I(t)| \leq M_I, \tag{14.36}$$

therefore

$$\begin{aligned}
\frac{dR(t)}{dt} &= \gamma I(t)g'(t) \leq \gamma M_I g'(t), \\
R(t) &\leq \gamma M_I \int_{t_0}^{t} g'(\tau)d\tau + R(0), \\
R(t) &\leq \gamma M_I \left(g(t) - g(t_0)\right) + R(0), \\
&\leq \gamma M_I \left(g(T) - g(t_0)\right) + R(0),
\end{aligned} \tag{14.37}$$

which shows that $\forall t \in [t_0, T]$, $R(t)$ is bounded too. Let

$$\overline{m} = \max\left\{ \sup_{\forall t \in [t_0, T]} |S(t)|,\ \sup_{\forall t \in [t_0, T]} |I(t)|,\ \sup_{\forall t \in [t_0, T]} |R(t)| \right\} \tag{14.38}$$

then $\forall t \in [t_0, T]$

$$(S(t), I(t), R(t)) < \overline{m}. \tag{14.39}$$

We shall now present the analysis of the class $S(t)$ and $R(t)$ using the fact that since

$$\frac{dS(t)}{dt} + \frac{dI(t)}{dt} + \frac{dR(t)}{dt} = 0, \tag{14.40}$$

we have that

$$S(t) + I(t) + R(t) = N. \tag{14.41}$$

We can eliminate in this case the component $I(t)$ to have

$$I(t) = N - S(t) - R(t). \tag{14.42}$$

Replacing in the initial equation yields

$$\begin{aligned}\frac{dS(t)}{dt} &= -\beta(N - S(t) - R(t))\frac{S(t)g'(t)}{N}, \\ \frac{dR(t)}{dt} &= \gamma(N - S(t) - R(t))\,g'(t).\end{aligned} \tag{14.43}$$

Note that $(N - S(t) - R(t))$ is positive or zero $\forall t \in [t_0, T]$.

We can see that the function $S(t)$ is decreasing, because $g'(t) > 0\ \forall t \in [t_0, T]$ and $\beta > 0$. However $\frac{dR(t)}{dt} > 0\ \forall t \in [t_0, T]$; this implies $R(t)$ is increasing. We can now name S_∞, R_∞ the limit of S and R when $t \to \infty$ respectively. Assuming the eradication of the disease, we have that $t \to \infty$ $S_\infty + R_\infty = N$ since

$$\lim_{t\to\infty} I(t) = 0. \tag{14.44}$$

$$\frac{\frac{dS(t)}{dt}}{\frac{dR(t)}{dt}} = \frac{-\beta}{\gamma}S. \tag{14.45}$$

Note the reproductive number is

$$R_0 \simeq \frac{-\beta}{\gamma}. \tag{14.46}$$

Therefore

$$\begin{aligned}\frac{dS(t)}{dR(t)} &= -R_0 S(t) \Longrightarrow \\ \frac{dS(t)}{S(t)} &= -R_0 dR(t).\end{aligned} \tag{14.47}$$

Integrating on both sides yields

$$S(t) = S(0)e^{-R_0 R(t)}. \tag{14.48}$$

To find $R(t)$, we apply on both sides to have after manipulation

$$R(t) = \frac{1}{R_0}\ln\frac{s_0}{s(t)}. \tag{14.49}$$

We can now replace $S(t)$ and $R(t)$ in the system to have

$$\frac{dR(t)}{dt} = \gamma\left(N - S_0 e^{-R_0 R(t)} - R(t)\right)g'(t). \tag{14.50}$$

We recall that when $t \to \infty$ $R_\infty + S_\infty = N$, thus,

$$R_\infty(t) = N - S_\infty(t). \tag{14.51}$$

$$S_\infty(t) = S_0 \exp[-R_0(N - S_\infty(t))]. \tag{14.52}$$

If we put $S_\infty(t) = X$ then we shall have

$$X = S_0 \exp[-R_0(N - X)] \Rightarrow S_0 \exp[-R_0(N - X)] - X = 0. \tag{14.53}$$

We put

$$\begin{aligned} f(x) &= S_0 \exp[-R_0(N - X)] - X, \\ f(x) &= 0. \end{aligned} \tag{14.54}$$

We can find the solution using the Newton method

$$X_{n+1} = X_n - \frac{f(X_n)}{f'(X_n)}, \tag{14.55}$$

since f is differentiable.

In a large population, the number of those infected is really very small at beginning of the outbreak; thus we can assume that at $t = 0$,

$S_0 = N$, thus

$$S_\infty(t) = N \exp[-R_0(N - S_\infty(t))]. \tag{14.56}$$

This equation can now be solved to get an idea of accurate approximation of population proportion which indeed stays non-infected at the end. On the other hand we have that

$$\frac{dI(t)}{dt} = \beta\left(S - \frac{1}{R_0}\right) I(t)g'(t). \tag{14.57}$$

We know that the function $S(t)$ decreases in $[t_0, T]$ but for the epidemic model we have $R_0 > 1$. To solve the problem, we divide our interests in 3 stages according to the sign of

$$S - \frac{1}{R_0}. \tag{14.58}$$

Noting that

$$\begin{aligned} \frac{dI(t)}{dt} &= (\beta I(t)S(t) - \gamma I(t))\, g'(t), \\ &= g'(t)I(t)\,(\beta S(t) - \gamma)\,, \\ &= g'(t)\beta I(t)\left(S(t) - \frac{\gamma}{\beta}\right), \end{aligned} \tag{14.59}$$

$$= g'(t)\beta I(t)\left(S(t) - \frac{1}{R_0}\right).$$

If $S(t) - \frac{1}{R_0} > 0$, then we have that $I(t)$ increase which corresponds to the outbreak. If $S(t) - \frac{1}{R_0} = 0$, t^* is the peak which correspond to the turning point where the spread less if strength.

If $S(t) - \frac{1}{R_0} < 0$, then we have the completion of the wave, then the number of those infected goes to zero. We now try to solve

$$\frac{dR(t)}{dt} = g'(t)\gamma\left(N - S_0 e^{-R_0 R(t)} - R(t)\right). \tag{14.60}$$

We cannot solve the above analytically. Therefore we shall provide either approximate solution or numerical solution:

$$\begin{aligned} \frac{dR(t)}{dt} &= g'(t)\gamma\left(N - S_0 e^{-R_0 R(t)} - R(t)\right), \\ &= f(t, R(t)). \end{aligned} \tag{14.61}$$

We integrate both sides to have

$$R(t) = R(0) + \int_{t_0}^{t} f(\tau, R(\tau))d\tau. \tag{14.62}$$

We introduce the Picard iteration

$$R_n(t) = R(0) + \int_{t_0}^{t} f(\tau, R_{n-1}(\tau))d\tau. \tag{14.63}$$

We want to show that $f(t, R(t))$ is bounded $\forall t \in [t_0, T]$ and that $\forall t \in [t_0, T]$ $f(t, R(t))$ satisfies the Lipschitz condition. If these conditions are verified then the

$$R_n(t) = R(0) + \int_{t_0}^{t} f(\tau, R_{n-1}(\tau))d\tau \tag{14.64}$$

converges.

By Peano–Cauchy theorem, it can be converged to a unique solution since the Lipschitz condition is satisfied:

$$f(t, R(t)) = g'(t)\gamma\left(N - S_0 e^{-R_0 R(t)} - R(t)\right) + R(0) \tag{14.65}$$

generally at $R(0) = 0$. Since no one has recovered from the disease:

$$f(t, R(t)) = g'(t)\gamma \left(N - S_0 e^{-R_0 R(t)} - R(t)\right), \tag{14.66}$$

$R(t) \geq 0, \forall t \in [t_0, T]$, $R_0 > 0$, $g'(t) > 0$ therefore

$$|f(t, R(t))| = \left|g'(t)\right| \gamma \left|N - S_0 e^{-R_0 R(t)} - R(t)\right|. \tag{14.67}$$

We show that $R(t)$ was bounded; therefore

$$\begin{aligned} |f(t, R(t))| \leq & \left\|g'\right\|_\infty \gamma \left\{N + S_0 e^{-R_0 R(t)} + R(t)\right\}, \\ \leq & \left\|g'\right\|_\infty \gamma \left\{N + 1 + \sup_{\forall t \in [t_0, T]} |R(t)|\right\}, \\ \leq & \left\|g'\right\|_\infty \gamma \left\{N + 1 + \|R\|_\infty\right\} < \infty \end{aligned} \tag{14.68}$$

which proves that $f(t, R(t))$ is bounded:

$$\begin{aligned} & |f(t, R_1(t)) - f(t, R_2(t))| \\ = & \left|g'(t)\right| \gamma \left|-e^{-R_0 R_1(t)} - R_1(t) + e^{-R_0 R_2(t)} + R_2(t)\right| \\ \leq & \left\|g'\right\|_\infty \gamma \left|R_1(t) - R_2(t)\right| \overline{\alpha} \end{aligned} \tag{14.69}$$

since $e^{-R_0 R_1(t)}$ and $e^{-R_0 R_2(t)}$ could be neglected:

$$|f(t, R_1(t)) - f(t, R_2(t))| \leq K \left|R_1(t) - R_2(t)\right| \tag{14.70}$$

where

$$K = \left\|g'\right\|_\infty \gamma \overline{\alpha}. \tag{14.71}$$

For $\frac{dI(t)}{dt}$, we have

$$\frac{dI(t)}{dt} = f_I(t, I(t)) = g'(t)\beta I(t) \left(\frac{S(t)}{N} - \frac{1}{R_0}\right) \forall t \in [t_0, T], \tag{14.72}$$

$$\begin{aligned} |f_I(t, I(t))| = & \left|g'(t)\right| \beta \left|I(t)\right| \left|\frac{S(t)}{N} - \frac{1}{R_0}\right| \forall t \in [t_0, T], \\ \leq & \sup_{t \in [t_0, T]} \left|g'(t)\right| \sup_{t \in [t_0, T]} |I(t)| \left|1 - \frac{1}{R_0}\right|. \end{aligned} \tag{14.73}$$

Since $\forall t \in [t_0, T]$, $\frac{S(t)}{N} < 1$, therefore,

$$|f_I(t, I(t))| \leq \left\|g'\right\|_\infty \|I\|_\infty \left|1 - \frac{1}{R_0}\right| \tag{14.74}$$

Thus $\forall t \in [t_0, T]$, f_I is bounded:

$$\begin{aligned}
&|f_I(t, I_1(t)) - f_I(t, I_2(t))| \\
&= \left|g'(t)\right| \beta \left|I_1(t) - I_2(t)\right| \left|\frac{S(t)}{N} - \frac{1}{R_0}\right| \quad \forall t \in [t_0, T], \\
&\leq \left\|g'\right\|_\infty \beta \left|1 - \frac{1}{R_0}\right| |I_1(t) - I_2(t)| \\
&\leq K_I |I_1(t) - I_2(t)|
\end{aligned} \tag{14.75}$$

where

$$K_I = \left\|g'\right\|_\infty \beta \left|1 - \frac{1}{R_0}\right|. \tag{14.76}$$

Therefore f_I is a Lipschitz condition. Therefore the Picard sequence

$$I_n(t) = I(0) + \int_{t_0}^{t} f_I(\tau, I_{n-1}(\tau))d\tau \tag{14.77}$$

guarantees the existence and uniqueness.

Via the Cauchy–Peano theorem and Lipschitz condition of $f_I(t, I(t))$, since $I(t)$ is bounded, the function

$$f_S(t, S(t)) = \frac{-\beta I(t)S(t)}{N} \tag{14.78}$$

is also bounded. More importantly

$$\begin{aligned}
|f_S(t, S_1(t)) - f_S(t, S_2(t))| &= \frac{\beta}{N} |S_1(t) - S_2(t)| \, |I(t)| \\
&\leq \beta |S_1(t) - S_2(t)|
\end{aligned} \tag{14.79}$$

which shows that f_S is Lipschitz. Take $\overline{k} = \max\{k_S, k_I, k_R\}$ and $\overline{M} = \max\left\{\|f_S\|_\infty, \|f_I\|_\infty, \|f_R\|_\infty\right\}$, then we can conclude that the system is bounded and satisfies the Lipschitz condition. Now applying the Picard iteration of the system

$$S_n(t) = S(t_0) + \int_{t_0}^{t} f_S(\tau, S_{n-1}(\tau))d\tau \tag{14.80}$$

$$I_n(t) = I(t_0) + \int_{t_0}^{t} f_I(\tau, I_{n-1}(\tau))d\tau \tag{14.81}$$

$$R_n(t) = R(t_0) + \int_{t_0}^{t} f_R(\tau, R_{n-1}(\tau))d\tau. \tag{14.82}$$

We have that the above converges toward a unique solution.

14.4 Exponential Perturbation

In this section, we introduce an exponential perturbation to the modified SIR model. To achieve this, we assume that

$$\begin{aligned} S(t) =& \ S_0(t) + \varepsilon S_1(t)e^{ilt}, \\ I(t) =& \ I_0(t) + \varepsilon I_1(t)e^{ilt}, \\ R(t) =& \ R_0(t) + \varepsilon R_1(t)e^{ilt}. \end{aligned} \tag{14.83}$$

The derivatives are given as

$$\begin{aligned} S^{'}(t) =& \ S_0^{'}(t) + \varepsilon S_1'(t)e^{ilt} + \varepsilon il S_1(t)e^{ilt}, \\ I'(t) =& \ I_0'(t) + \varepsilon I_1'(t)e^{ilt} + \varepsilon il I_1(t)e^{ilt}, \\ R'(t) =& \ R_0'(t) + \varepsilon R_1'(t)e^{ilt} + \varepsilon il R_1(t)e^{ilt}. \end{aligned} \tag{14.84}$$

The right-hand side is given as

$$S_0^{'}(t) + \varepsilon S_1'(t)e^{ilt} + \varepsilon il S_1(t)e^{ilt} = \frac{-\beta}{N}\left(S_0(t) + \varepsilon S_1(t)e^{ilt}\right)\left(I_0(t) + \varepsilon i I_1(t)e^{ilt}\right), \tag{14.85}$$

$$\begin{aligned} & I_0'(t) + \varepsilon I_1'(t)e^{ilt} + \varepsilon il I_1(t)e^{ilt} \\ & = \frac{\beta}{N}\left(S_0(t) + \varepsilon S_1(t)e^{ilt}\right)\left(I_0(t) + \varepsilon i I_1(t)e^{ilt}\right) \\ & - \gamma\left(I_0(t) + \varepsilon i I_1(t)e^{ilt}\right), \end{aligned} \tag{14.86}$$

$$R_0'(t) + \varepsilon R_1'(t)e^{ilt} + \varepsilon il R_1(t)e^{ilt} = \gamma\left(I_0(t) + \varepsilon i I_1(t)e^{ilt}\right). \tag{14.87}$$

Indeed we have that

$$\begin{aligned} S_0^{'}(t) =& \ -g'(t)I_0(t)S_0(t)\frac{\beta}{N} \\ I_0^{'}(t) =& \ g'(t)I_0(t)S_0(t)\frac{\beta}{N} - \gamma I_0(t)g'(t) \\ R_0'(t) =& \ \gamma I_0(t)g'(t) \end{aligned} \tag{14.88}$$

which is the system with perturbation. The system becomes

$$\begin{aligned}
\varepsilon S_1'(t) + \varepsilon i l S_1(t) &= -\frac{\beta}{N} g'(t) \left(S_0(t) \varepsilon i I_1(t) + \varepsilon S_1(t) I_0(t) \right), \\
I_1'(t) \varepsilon i - \varepsilon l I_1(t) &= \frac{\beta}{N} g'(t) \left(S_0(t) \varepsilon i I_1(t) + \varepsilon S_1(t) I_0(t) \right) - \gamma \varepsilon i I_1(t) \\
\varepsilon R_1'(t) + \varepsilon i l R_1(t) &= \gamma g'(t) \varepsilon i I_1(t).
\end{aligned} \tag{14.89}$$

Note that we have removed the term with ε^2 since ε is small. Thus $\varepsilon^2 \to 0$.

We now have a system of complex functions; thus the real point gives

$$\begin{aligned}
S_1'(t) &= -\frac{\beta}{N} g'(t) \left(I_0(t) S_1(t) \right), \\
-I_0'(t) &= \frac{\beta}{N} g'(t) \left(I_0(t) S_1(t) \right), \\
R_0'(t) &= 0.
\end{aligned} \tag{14.90}$$

Imaginary part

$$\begin{aligned}
l S_1(t) &= -\frac{\beta}{N} g'(t) I_1(t) S_0(t), \\
I_1'(t) &= \frac{\beta}{N} g'(t) I_1(t) S_0(t), \\
l R_1(t) &= \gamma g'(t) I_1(t).
\end{aligned} \tag{14.91}$$

The real part will be dismissed since we have a negative $I(t)$; therefore we consider only the imaginary part. Notice that $S_0(t)$ appears here which is the solution of nonperturbed syste; therefore the system is evaluated as

$$\begin{aligned}
S_0'(t) &= -g'(t) \frac{\beta}{N} I_0(t) S_0(t), \\
I_0'(t) &= g'(t) \frac{\beta}{N} I_0(t) S_0(t) - \gamma I_0(t), \\
R_0'(t) &= \gamma I_0(t) g'(t), \\
I_1'(t) &= \frac{\beta}{N} g'(t) I_1(t) S_0(t), \\
l R_1(t) &= \gamma g'(t) I_1(t).
\end{aligned} \tag{14.92}$$

We can quickly check that $\forall t \in [t_0, T]$

$$\begin{aligned}
&S_0'(t) + I_0'(t) + R_0'(t) + I_1'(t) \neq 0, \\
&S_0(t) + I_0(t) + R_0(t) + I_1(t) = \overline{N}(t).
\end{aligned} \tag{14.93}$$

The above system will be analyzed next as it gives information in $I_1(t)$. But to have information about $S_1(t)$ and $R_1(t)$, we consider

$$
\begin{aligned}
S_0'(t) &= -g'(t)\frac{\beta}{N}I_0(t)S_0(t), \\
I_0'(t) &= g'(t)\frac{\beta}{N}I_0(t)S_0(t) - \gamma I_0(t)g'(t), \\
R_0'(t) &= \gamma I_0(t)g'(t), \\
I_1'(t) &= \frac{-\beta}{N}g'(t)I_0(t)S_1(t), \\
R_1(t) &= cte.
\end{aligned}
\tag{14.94}
$$

The above system will be investigated.

We start the analysis of the following perturbed system:

$$
\begin{aligned}
S_0'(t) &= -g'(t)\frac{\beta}{N}I_0(t)S_0(t), \\
I_0'(t) &= g'(t)\frac{\beta}{N}I_0(t)S_0(t) - \gamma I_0(t), \\
R_0'(t) &= \gamma I_0(t)g'(t), \\
I_1'(t) &= \frac{-\beta}{N}g'(t)I_1(t)S_0(t), \\
R_1(t) &= \frac{\gamma}{l}g'(t)I_1(t).
\end{aligned}
\tag{14.95}
$$

We have shown that the first set of equations

$$
\begin{aligned}
S_0'(t) &= f_{S_0}(t, S_0(t), I_0(t)), \\
I_0'(t) &= f_{I_0}(t, S_0(t), I_0(t)), \\
R_0'(t) &= f_{R_0}(t, I_0(t)), \\
I_1'(t) &= f_{I_1}(t, I_1(t), S_0(t)), \\
R_1(t) &= \frac{\gamma}{l}g'(t)I_1(t).
\end{aligned}
\tag{14.96}
$$

We have shown that $f_{S_0}(t, S, I_0)$, $f_{I_0}(t, S, I_0)$, $f_{R_0}(t, I_0)$ are bounded and verify the Lipschitz conditions. We shall now evaluate

$$
\begin{aligned}
\left|f_{I_1}(t, I_1, S_0)\right| &= \frac{\beta}{N}\left|g'(t)\right| |I_1(t)| \, |S_0(t)|, \\
&\le \frac{\beta}{N}\sup_{t\in[t_0,T]}\left|g'(t)\right| \sup_{t\in[t_0,T]}|S_0(t)| \, |I_1(t)|, \\
&\le \frac{\beta}{N}\left\|g'\right\|_\infty \|S_0\|_\infty |I_1(t)|, \\
&\le \beta\left\|g'\right\|_\infty \|S_0\|_\infty .
\end{aligned}
\tag{14.97}
$$

Since $\frac{I_1(t)}{N} < 1 \; \forall t \in [t_0, T]$, therefore,

$$\left| f_{I_1}(t, I_1, S_0) \right| \leq \beta \left\| g' \right\|_\infty \| S_0 \|_\infty < \infty. \tag{14.98}$$

Therefore the function is bounded:

$$\begin{aligned} \left| f_{I_1}(t, I_1, S_0) - f_{I_1}(t, \overline{I}_1, S_0) \right| &= \frac{\beta}{N} \left| g'(t) \right| |S_0(t)| \left| I_1 - \overline{I}_1 \right| \\ &\leq \frac{\beta}{N} \sup_{t \in [t_0, T]} \left| g'(t) \right| |S_0(t)| \left| I_1 - \overline{I}_1 \right| \\ &\leq \frac{\beta}{N} \left\| g' \right\|_\infty |S_0(t)| \left| I_1 - \overline{I}_1 \right| \\ &\leq \beta \left\| g' \right\|_\infty \left| I_1 - \overline{I}_1 \right| \end{aligned}$$

Since $\frac{S_0(t)}{N} < 1 \; \forall t \in [t_0, T]$.

Therefore taking

$$K = \beta \left\| g' \right\|_\infty \tag{14.99}$$

we have

$$\left| f_{I_1}(t, I_1, S_0) - f_{I_1}(t, \overline{I}_1, S_0) \right| \leq K \left| I_1 - \overline{I}_1 \right| \tag{14.100}$$

Thus f_{I_1} verifies the Lipschitz condition.

Taking the max of all K and all M_f, we conclude that the perturbed system is bounded and satisfies the Lipschitz condition; therefore by considering the Picard iteration

$$\begin{aligned} S_0^{n+1}(t) &= S_0(t_0) + \int_{t_0}^{t} f_{S_0}(\tau, S_0^n(\tau), I_0^n(\tau)) d\tau, \\ I_0^{n+1}(t) &= I_0(t_0) + \int_{t_0}^{t} f_{I_0}(\tau, I_0^n(\tau), S_0^n(\tau)) d\tau, \\ R_0^{n+1}(t) &= R_0(t_0) + \int_{t_0}^{t} f_{R_0}(\tau, I_0^n(\tau)) d\tau, \\ I_1^{n+1}(t) &= I_1(t_0) + \int_{t_0}^{t} f_{I_1}(\tau, I_1^n(\tau), S_0^n(\tau)) d\tau, \\ R_1^n(t) &= \frac{\gamma}{l} g'(t) I_1^n(t), \end{aligned} \tag{14.101}$$

it converges to the exact solution according the Cauchy–Peano theorem. We shall now show the numerical solution of the above system. We shall present, first, the numerical scheme to be used.

We consider a general IVP

$$y'(t) = f(t, y(t)). \tag{14.102}$$

We apply the integral on both sides to get

$$y(t) = y(t_0) + \int_{t_0}^{t} f(\tau, y(\tau))d\tau. \tag{14.103}$$

At $t = t_{n+1}$, we have

$$y(t_{n+1}) = y(t_0) + \int_{t_0}^{t_{n+1}} f(\tau, y(\tau))d\tau \tag{14.104}$$

at $t = t_n$, we have

$$y(t_n) = y(t_0) + \int_{t_0}^{t_n} f(\tau, y(\tau))d\tau. \tag{14.105}$$

Then we have

$$y(t_{n+1}) - y(t_n) = \int_{t_n}^{t_{n+1}} f(\tau, y(\tau))d\tau. \tag{14.106}$$

Within $[t_n, t_{n+1}]$ we approximate

$$f(\tau, y(\tau)) = \frac{f(t_n, y(t_n)) + f(t_{n+1}, y(t_{n+1}))}{2}. \tag{14.107}$$

$$y_{n+1} = y_n + \frac{h}{2} f(t_n, y(t_n)) + f(t_{n+1}, y(t_{n+1})). \tag{14.108}$$

The sequence is now implicit; therefore we can on the right-hand side replace y_{n+1} by $\overline{y}_{n+1}$

$$\begin{aligned} y_{n+1} &= y_n + \frac{h}{2}\left(f(t_n, y_n) + f(t_{n+1}, \overline{y}_{n+1})\right) \\ \overline{y}_{n+1} &= y_n + hf(t_n, y(t_n)). \end{aligned} \tag{14.109}$$

Thus the method is known as Predictor–Corrector or Heun's method. In applying it to own system, we get

$$S_0^{n+1} = S_0^n + \frac{h}{2}\left[f_{S_0}(t_n, S_0^n, I_0^n) + f_{S_0}(t_{n+1}, \overline{S}_0^{n+1}, \overline{I}_0^{n+1})\right], \tag{14.110}$$

$$
\begin{aligned}
I_0^{n+1} &= I_0^n + \frac{h}{2}\left[f_{I_0}(t_n, S_0^n, I_0^n) + f_{I_0}(t_{n+1}, \overline{S}_0^{n+1}, \overline{I}_0^{n+1}) \right], \\
R_0^{n+1} &= R_0^n + \frac{h}{2}\left[f_{R_0}(t_n, I_0^n) + f_{R_0}(t_{n+1}, \overline{I}_0^{n+1}) \right], \\
I_1^{n+1} &= I_1^n + \frac{h}{2}\left[f_{I_1}(t_n, S_0^n, I_1^n) + f_{I_1}(t_{n+1}, \overline{S}_0^{n+1}, \overline{I}_1^{n+1}) \right], \\
R_1^{n+1}(t) &= \frac{\gamma}{l} g'(t) I_1^{n+1}(t)
\end{aligned}
$$

where

$$
\begin{aligned}
\overline{S}_0^{n+1} &= S_0^n + h f_{S_0}(t_n, S_0^n, I_0^n), \\
\overline{I}_0^{n+1} &= I_0^n + h f_{I_0}(t_n, S_0^n, I_0^n), \\
\overline{R}_0^{n+1} &= R_0^n + h f_{R_0}(t_n, I_0^n), \\
\overline{I}_1^{n+1} &= I_1^n + h f_{I_1}(t_n, S_0^n, I_1^n).
\end{aligned} \tag{14.111}
$$

In the next section, we present the same analysis when the derivative is the Caputo–Fabrizio derivative.

By replacing the classical derivative by the Caputo–Fabrizio type, we obtain

$$
\begin{aligned}
{}_{t_0}^{CF}D_g^{\alpha} S_0(t) &= -\frac{\beta}{N} I_0(t) S_0(t), \\
{}_{t_0}^{CF}D_g^{\alpha} I_0(t) &= \frac{\beta}{N} I_0(t) S_0(t) - \gamma I_0(t), \\
{}_{t_0}^{CF}D_g^{\alpha} R_0(t) &= \gamma I_0(t), \\
{}_{t_0}^{CF}D_g^{\alpha} I_1(t) &= \frac{-\beta}{N} I_1(t) S_0(t), \\
R_1(t) &= \frac{\gamma}{l} I_1(t).
\end{aligned} \tag{14.112}
$$

Using the fact that $g'(t) > 0$

$$
\begin{aligned}
{}_{t_0}^{CF}D_t^{\alpha} S_0(t) &= -g'(t)\frac{\beta}{N} I_0(t) S_0(t), \\
{}_{t_0}^{CF}D_t^{\alpha} I_0(t) &= g'(t)\frac{\beta}{N} I_0(t) S_0(t) - \gamma g'(t) I_0(t), \\
{}_{t_0}^{CF}D_t^{\alpha} R_0(t) &= \gamma I_0(t) g'(t), \\
{}_{t_0}^{CF}D_t^{\alpha} I_1(t) &= \frac{-\beta}{N} g'(t) I_1(t) S_0(t), \\
R_1(t) &= \frac{\gamma}{l} g'(t) I_1(t).
\end{aligned} \tag{14.113}
$$

Applying the integral on both sides yields

$$\begin{aligned} S_0(t) = \ & S_0(t_0) + (1-\alpha)g'(t)f_{S_0}(t, I_0, S_0) \\ & +\alpha \int_{t_0}^{t} g'(\tau)f_{S_0}(\tau, I_0, S_0)d\tau, \end{aligned} \tag{14.114}$$

$$I_0(t) = I_0(t_0) + (1-\alpha)g'(t)f_{I_0}(t, I_0, S_0) + \alpha \int_{t_0}^{t} g'(\tau)f_{I_0}(\tau, I_0, S_0)d\tau,$$

$$R_0(t) = R_0(t_0) + (1-\alpha)g'(t)f_{R_0}(t, I_0) + \alpha \int_{t_0}^{t} g'(\tau)f_{R_0}(\tau, I_0)d\tau,$$

$$I_1(t) = I_1(t_0) + (1-\alpha)g'(t)f_{I_1}(t, I_1, S_0) + \alpha \int_{t_0}^{t} g'(\tau)f_{I_1}(\tau, I_1, S_0)d\tau,$$

$$R_1(t) = \frac{\gamma}{l}g'(t)I_1(t).$$

At $t = t_{n+1}$and $t = t_n$ we have

$$\begin{aligned} S_0(t_{n+1}) - S_0(t_n) = \ & (1-\alpha)g'(t_{n+1})f_{S_0}(t_{n+1}, I_0\left(t_{n+1}\right), S_0\left(t_{n+1}\right)) \\ & +\alpha \int_{t_n}^{t_{n+1}} g'(\tau)f_{S_0}(\tau, I_0\left(\tau\right), S_0\left(\tau\right))d\tau \\ & -(1-\alpha)g'(t_n)f_{S_0}(t_n, I_0\left(t_n\right), S_0\left(t_n\right)). \end{aligned} \tag{14.115}$$

Within $[t_n, t_{n+1}]$, we approximate

$$f_{S_0}(t, I_0, S_0) \simeq \frac{f(t_n, I_0^n, S_0^n) + f(t_{n+1}, I_0^{n+1}, S_0^{n+1})}{2}. \tag{14.116}$$

Replacing it yields

$$\begin{aligned} S_0^{n+1} = \ & S_0^n + (1-\alpha)\frac{g(t_{n+1}) - g(t_n)}{h}f(t_{n+1}, \overline{I_0}^{n+1}, \overline{S_0}^{n+1}) \\ & -(1-\alpha)\frac{g(t_n) - g(t_{n-1})}{h}f(t_n, I_0^n, S_0^n) \\ & +\frac{g(t_{n+1}) - g(t_n)}{2}\alpha\left(f(t_n, I_0^n, S_0^n) + f(t_{n+1}, \overline{I_0}^{n+1}, \overline{S_0}^{n+1})\right), \end{aligned} \tag{14.117}$$

where

$$\overline{S}_0^{n+1} = S_0^n + (1-\alpha)\frac{g(t_{n+1})-g(t_n)}{h}f(t_n, I_0^n, S_0^n) \\ -\frac{g(t_n)-g(t_{n-1})}{h}f(t_{n-1}, I_0^{n-1}, S_0^{n-1}) \\ +\alpha\frac{g(t_{n+1})-g(t_n)}{h}f(t_n, I_0^n, S_0^n). \tag{14.118}$$

Therefore,

$$S_0^{n+1} = S_0^n + (1-\alpha)\frac{g(t_{n+1})-g(t_n)}{h}f(t_{n+1}, \overline{I}_0^{n+1}, \overline{S}_0^{n+1}) \\ +\frac{g(t_{n+1})-g(t_n)}{2}\alpha\left(f(t_n, I_0^n, S_0^n) + f(t_{n+1}, \overline{I}_0^{n+1}, \overline{S}_0^{n+1})\right) \\ -(1-\alpha)\frac{g(t_n)-g(t_{n-1})}{h}f(t_n, I_0^n, S_0^n), \tag{14.119}$$

$$\overline{S}_0^{n+1} = S_0^n + (1-\alpha)\frac{g(t_{n+1})-g(t_n)}{h}f(t_n, I_0^n, S_0^n) \\ -\frac{g(t_n)-g(t_{n-1})}{h}f(t_{n-1}, I_0^{n-1}, S_0^{n-1}) \\ +\alpha\frac{g(t_{n+1})-g(t_n)}{h}f(t_n, I_0^n, S_0^n). \tag{14.120}$$

Similarly as $S_0(t)$, we will get

$$I_0^{n+1} = I_0^n + (1-\alpha)\frac{g(t_{n+1})-g(t_n)}{h}f(t_{n+1}, \overline{I}_0^{n+1}, \overline{S}_0^{n+1}) \\ +\frac{g(t_{n+1})-g(t_n)}{2}\alpha\left(f(t_n, I_0^n, S_0^n) + f(t_{n+1}, \overline{I}_0^{n+1}, \overline{S}_0^{n+1})\right) \\ -(1-\alpha)\frac{g(t_n)-g(t_{n-1})}{h}f(t_n, I_0^n, S_0^n), \tag{14.121}$$

$$\overline{I}_0^{n+1} = I_0^n + (1-\alpha)\frac{g(t_{n+1})-g(t_n)}{h}f(t_n, I_0^n, S_0^n) \\ -\frac{g(t_n)-g(t_{n-1})}{h}f(t_{n-1}, I_0^{n-1}, S_0^{n-1}) \\ +\alpha\frac{g(t_{n+1})-g(t_n)}{h}f(t_n, I_0^n, S_0^n). \tag{14.122}$$

$$\begin{aligned} R_0^{n+1} = & \ R_0^n + (1-\alpha)\frac{g(t_{n+1}) - g(t_n)}{h} f(t_{n+1}, \overline{I}_0^{n+1}) \\ & + \frac{g(t_{n+1}) - g(t_n)}{2}\alpha\left(f(t_n, I_0^n) + f(t_{n+1}, \overline{I}_0^{n+1})\right) \\ & -(1-\alpha)\frac{g(t_n) - g(t_{n-1})}{h} f(t_n, I_0^n), \end{aligned} \tag{14.123}$$

$$\begin{aligned} \overline{R}_0^{n+1} = & \ R_0^n + (1-\alpha)\frac{g(t_{n+1}) - g(t_n)}{h} f(t_n, I_0^n) \\ & -\frac{g(t_n) - g(t_{n-1})}{h} f(t_{n-1}, I_0^{n-1}) \\ & +\alpha\frac{g(t_{n+1}) - g(t_n)}{h} f(t_n, I_0^n). \end{aligned} \tag{14.124}$$

$$\begin{aligned} I_1^{n+1} = & \ I_1^n + (1-\alpha)\frac{g(t_{n+1}) - g(t_n)}{h} f(t_{n+1}, \overline{I}_1^{n+1}, \overline{S}_0^{n+1}) \\ & + \frac{g(t_{n+1}) - g(t_n)}{2}\alpha\left(f(t_n, I_1^n, S_0^n) + f(t_{n+1}, \overline{I}_1^{n+1}, \overline{S}_0^{n+1})\right) \\ & -(1-\alpha)\frac{g(t_n) - g(t_{n-1})}{h} f(t_n, I_1^n, S_0^n), \end{aligned} \tag{14.125}$$

$$\begin{aligned} \overline{I}_1^{n+1} = & \ I_1^n + (1-\alpha)\frac{g(t_{n+1}) - g(t_n)}{h} f(t_n, I_1^n, S_0^n) \\ & -\frac{g(t_n) - g(t_{n-1})}{h} f(t_{n-1}, I_1^{n-1}, S_0^{n-1}) \\ & +\alpha\frac{g(t_{n+1}) - g(t_n)}{h} f(t_n, I_1^n, S_0^n). \end{aligned} \tag{14.126}$$

And

$$R_1^n(t) = \frac{\gamma}{l} g'(t) I_1^n(t). \tag{14.127}$$

We now replace the classical derivative with the Caputo global derivative based on the power law:

$$\begin{aligned} {}_{t_0}^{RL}D_g^\alpha S_0(t) = & -\frac{\beta}{N} I_0(t) S_0(t), \\ {}_{t_0}^{RL}D_g^\alpha I_0(t) = & \ \frac{\beta}{N} I_0(t) S_0(t) - \gamma I_0(t), \\ {}_{t_0}^{RL}D_g^\alpha R_0(t) = & \ \gamma I_0(t), \\ {}_{t_0}^{RL}D_g^\alpha I_1(t) = & \ \frac{-\beta}{N} I_1(t) S_0(t), \\ R_1(t) = & \ \frac{\gamma}{l} I_1(t). \end{aligned} \tag{14.128}$$

We use the properties of ${}_{t_0}^{RL}D_g^{\alpha}$ when g is differentiable; we obtain

$$\begin{aligned}
{}_{t_0}^{RL}D_t^{\alpha} S_0(t) &= -g'(t)\frac{\beta}{N} I_0(t) S_0(t), \\
{}_{t_0}^{RL}D_t^{\alpha} I_0(t) &= g'(t)\frac{\beta}{N} I_0(t) S_0(t) - \gamma g'(t) I_0(t), \\
{}_{t_0}^{RL}D_t^{\alpha} R_0(t) &= \gamma I_0(t) g'(t), \\
{}_{t_0}^{RL}D_t^{\alpha} I_1(t) &= \frac{-\beta}{N} g'(t) I_1(t) S_0(t), \\
R_1(t) &= \frac{\gamma}{l} g'(t) I_1(t).
\end{aligned} \tag{14.129}$$

We apply the Riemann–Liouville integral on both sides for the first equation to obtain

$$S_0(t) - S_0(t_0) = \frac{1}{\Gamma(\alpha)} \int_{t_0}^{t} (t-\tau)^{\alpha-1} f_{S_0}(\tau, I_0, S_0) g'(\tau) d\tau. \tag{14.130}$$

Other can be derived in a similar way.

At $t = t_{n+1}$, we have

$$\begin{aligned}
S_0(t_{n+1}) &= \frac{1}{\Gamma(\alpha)} \int_{t_0}^{t_{n+1}} (t_{n+1}-\tau)^{\alpha-1} f_{S_0}(\tau, I_0, S_0) g'(\tau) d\tau, \\
&= \frac{1}{\Gamma(\alpha)} \sum_{j=0}^{n} \int_{t_j}^{t_{j+1}} (t_{n+1}-\tau)^{\alpha-1} f_{S_0}(\tau, I_0, S_0) g'(\tau) d\tau, \\
&\simeq \frac{1}{\Gamma(\alpha)} \sum_{j=0}^{n} \int_{t_j}^{t_{j+1}} (t_{n+1}-\tau)^{\alpha-1} f_{S_0}(\tau, I_0, S_0) \frac{g(t_{j+1}) - g(t_j)}{h} d\tau.
\end{aligned} \tag{14.131}$$

Within $\left[t_j, t_{j+1}\right]$, we have the following approximation:

$$f_{S_0}(\tau, I_0, S_0) \simeq \frac{f_{S_0}(t_{j+1}, I_0^{j+1}, S_0^{j+1}) + f_{S_0}(t_j, I_0^j, S_0^j)}{2}. \tag{14.132}$$

Then we have

$$S_0(t_{n+1}) \simeq \frac{1}{\Gamma(\alpha)} \sum_{j=0}^{n} \frac{g(t_{j+1}) - g(t_j)}{2h} \left(f_{S_0}(t_{j+1}, I_0^{j+1}, S_0^{j+1}) + f_{S_0}(t_j, I_0^j, S_0^j) \right) \times \frac{h^\alpha}{\alpha} \{(n-j+1)^\alpha - (n-j)^\alpha\}. \tag{14.133}$$

Noting that when $n + 1$, we have implicit scheme, to solve this, we do this

$$\begin{aligned} S_0^{n+1} = & \frac{h^{\alpha-1}}{2\Gamma(\alpha+1)} \sum_{j=0}^{n-1} g(t_{j+1}) - g(t_j) \\ & \left(f_{S_0}(t_{j+1}, I_0^{j+1}, S_0^{j+1}) + f_{S_0}(t_j, I_0^j, S_0^j) \right) \\ & \times \left\{ (n-j+1)^\alpha - (n-j)^\alpha \right\} \\ & + \frac{h^{\alpha-1}}{2\Gamma(\alpha+1)} (g(t_{n+1}) - g(t_n)) \left(f_{S_0}(t_{n+1}, \overline{I}_0^{n+1}, \overline{S}_0^{n+1}) + f_{S_0}(t_n, I_0^n, S_0^n) \right), \end{aligned} \tag{14.134}$$

$$\overline{S}_0^{n+1} = \frac{h^{\alpha-1}}{2\Gamma(\alpha+1)} \sum_{j=0}^{n} g(t_{j+1}) - g(t_j) f_{S_0}(t_j, I_0^j, S_0^j) \left\{ (n-j+1)^\alpha - (n-j)^\alpha \right\}, \tag{14.135}$$

$$\begin{aligned} I_0^{n+1} = & \frac{h^{\alpha-1}}{2\Gamma(\alpha+1)} \sum_{j=0}^{n-1} g(t_{j+1}) - g(t_j) \\ & \left(f_{I_0}(t_{j+1}, I_0^{j+1}, S_0^{j+1}) + f_{I_0}(t_j, I_0^j, S_0^j) \right) \\ & \times \left\{ (n-j+1)^\alpha - (n-j)^\alpha \right\} \\ & + \frac{h^{\alpha-1}}{2\Gamma(\alpha+1)} (g(t_{n+1}) - g(t_n)) \left(f_{I_0}(t_{n+1}, \overline{I}_0^{n+1}, \overline{S}_0^{n+1}) + f_{I_0}(t_n, I_0^n, S_0^n) \right), \end{aligned} \tag{14.136}$$

$$\overline{I}_0^{n+1} = \frac{h^{\alpha-1}}{2\Gamma(\alpha+1)} \sum_{j=0}^{n} g(t_{j+1}) - g(t_j) f_{I_0}(t_j, I_0^j, S_0^j) \left\{ (n-j+1)^\alpha - (n-j)^\alpha \right\}, \tag{14.137}$$

$$\begin{aligned} R_0^{n+1} = & \frac{h^{\alpha-1}}{2\Gamma(\alpha+1)} \sum_{j=0}^{n-1} g(t_{j+1}) - g(t_j) \left(f_{R_0}(t_{j+1}, I_0^{j+1}) + f_{R_0}(t_j, I_0^j) \right) \\ & \times \{(n-j+1)^\alpha - (n-j)^\alpha\} \\ & + \frac{h^{\alpha-1}}{2\Gamma(\alpha+1)} (g(t_{n+1}) - g(t_n)) \left(f_{R_0}(t_{n+1}, \overline{I}_0^{n+1}) + f_{R_0}(t_n, I_0^n) \right), \end{aligned} \tag{14.138}$$

$$\overline{R}_0^{n+1} = \frac{h^{\alpha-1}}{2\Gamma(\alpha+1)} \sum_{j=0}^{n} g(t_{j+1}) - g(t_j) f_{R_0}(t_j, I_0^j) \{(n-j+1)^\alpha - (n-j)^\alpha\}, \tag{14.139}$$

$$\begin{aligned} I_1^{n+1} &= \frac{h^{\alpha-1}}{2\Gamma(\alpha+1)}\sum_{j=0}^{n-1} g(t_{j+1})-g(t_j) \\ &\left(f_{I_1}(t_{j+1}, I_1^{j+1}, S_0^{j+1}) + f_{I_1}(t_j, I_1^j, S_0^j)\right) \\ &\times\{(n-j+1)^{\alpha}-(n-j)^{\alpha}\} \\ &+\frac{h^{\alpha-1}}{2\Gamma(\alpha+1)}(g(t_{n+1})-g(t_n))\left(f_{I_1}(t_{n+1}, \overline{I}_1^{n+1}, \overline{S}_0^{n+1}) + f_{I_1}(t_n, I_1^n, S_0^n)\right), \end{aligned} \tag{14.140}$$

$$\overline{I}_1^{n+1} = \frac{h^{\alpha-1}}{2\Gamma(\alpha+1)}\sum_{j=0}^{n} g(t_{j+1})-g(t_j)f_{I_1}(t_j, I_1^j, S_0^j)\{(n-j+1)^{\alpha}-(n-j)^{\alpha}\}, \tag{14.141}$$

$$R_1^n(t) = \frac{\gamma}{l}g'(t)I_1^n(t). \tag{14.142}$$

We replace the classical derivative with the Atangana–Baleanu derivative to have

$$\begin{aligned} {}_{t_0}^{ABR}D_g^{\alpha}S_0(t) &= -\frac{\beta}{N}I_0(t)S_0(t), \\ {}_{t_0}^{ABR}D_g^{\alpha}I_0(t) &= \frac{\beta}{N}I_0(t)S_0(t)-\gamma I_0(t), \\ {}_{t_0}^{ABR}D_g^{\alpha}R_0(t) &= \gamma I_0(t), \\ {}_{t_0}^{ABR}D_g^{\alpha}I_1(t) &= \frac{-\beta}{N}I_1(t)S_0(t), \\ R_1(t) &= \frac{\gamma}{l}I_1(t). \end{aligned} \tag{14.143}$$

As we did before, we can transform the system as

$$\begin{aligned} {}_{t_0}^{ABR}D_t^{\alpha}S_0(t) &= -g'(t)\frac{\beta}{N}I_0(t)S_0(t), \\ {}_{t_0}^{ABR}D_t^{\alpha}I_0(t) &= g'(t)\frac{\beta}{N}I_0(t)S_0(t)-\gamma g'(t)I_0(t), \\ {}_{t_0}^{ABR}D_t^{\alpha}R_0(t) &= \gamma I_0(t)g'(t), \\ {}_{t_0}^{ABR}D_t^{\alpha}I_1(t) &= \frac{-\beta}{N}g'(t)I_1(t)S_0(t), \\ R_1(t) &= \frac{\gamma}{l}g'(t)I_1(t). \end{aligned} \tag{14.144}$$

We use the Atangana–Baleanu integral on both sides for the first equation to produce

$$S_0(t)-S_0(t_0) = (1-\alpha)g'(t)f_{S_0}(t, I_0(t), S_0(t)) \tag{14.145}$$

$$+\frac{\alpha}{\Gamma(\alpha)}\int_{t_0}^{t}(t-\tau)^{\alpha-1}f_{S_0}(\tau, I_0(\tau), S_0(\tau))g'(\tau)d\tau.$$

The rest can be derived in a similar way.

At $t=t_{n+1}$, we have

$$\begin{aligned} S_0(t_{n+1}) = & S_0(t_0) + (1-\alpha)g'(t_{n+1})f_{S_0}(t_{n+1}, I_0(t_{n+1}), S_0(t_{n+1})) \\ & +\frac{\alpha}{\Gamma(\alpha)}\int_{t_0}^{t_{n+1}}(t_{n+1}-\tau)^{\alpha-1}f_{S_0}(\tau, I_0(\tau), S_0(\tau))g'(\tau)d\tau. \end{aligned} \tag{14.146}$$

On the integral side, we apply the same routine we did on the Riemann–Liouville case to obtain

$$\begin{aligned} S_{n+1} = & S_0(t_0) + (1-\alpha)g'(t_{n+1})f_{S_0}(t_{n+1}, I_0(t_{n+1}), S_0(t_{n+1})) \\ & +\frac{\alpha}{\Gamma(\alpha)}\sum_{j=0}^{n}\int_{t_j}^{t_{j+1}}(t_{n+1}-\tau)^{\alpha-1}f_{S_0}(\tau, I_0(\tau), S_0(\tau))g'(\tau)d\tau, \\ \simeq & S_0(t_0) + (1-\alpha)\frac{g(t_{n+1})-g(t_n)}{h}f_{S_0}(t_{n+1}, \overline{I}_0^{n+1}, \overline{S}_0^{n+1}) \\ & +\frac{\alpha h^{\alpha-1}}{2\Gamma(\alpha+1)}\sum_{j=0}^{n-1}(g(t_{j+1})-g(t_j))\left(f_{S_0}(t_{j+1}, I_0^{j+1}, S_0^{j+1}) + f_{S_0}(t_j, I_0^j, S_0^j)\right) \\ & \times\left\{(n-j+1)^{\alpha}-(n-j)^{\alpha}\right\}. \end{aligned} \tag{14.147}$$

Here within $\left[t_j, t_{j+1}\right]$, we have the following approximation:

$$f_{S_0}(\tau, I_0, S_0) \simeq \frac{f_{S_0}(t_{j+1}, I_0^{j+1}, S_0^{j+1}) + f_{S_0}(t_j, I_0^j, S_0^j)}{2}, \tag{14.148}$$

where

$$\begin{aligned} \overline{S}_0^{n+1} = & S_0(t_0) + (1-\alpha)f_{S_0}(t_n, I_0^n, S_0^n) \\ & +\frac{h^{\alpha-1}}{\Gamma(\alpha)}\sum_{j=0}^{n}\left(g(t_{j+1})-g(t_j)\right)f_{S_0}(t_j, I_0^j, S_0^j) \\ & \times\{(n-j+1)^{\alpha}-(n-j)^{\alpha}\}. \end{aligned} \tag{14.149}$$

After applying the same routine, we got other solutions as below

$$
\begin{aligned}
I_{n+1} = \; & I_0(t_0) + (1-\alpha)\frac{g(t_{n+1}) - g(t_n)}{h} f_{I_0}(t_{n+1}, \overline{I}_0^{n+1}, \overline{S}_0^{n+1}) \\
& + \frac{\alpha h^{\alpha-1}}{2\Gamma(\alpha+1)} \sum_{j=0}^{n-1} (g(t_{j+1}) - g(t_j)) \left(f_{I_0}(t_{j+1}, I_0^{j+1}, S_0^{j+1}) + f_{I_0}(t_j, I_0^j, S_0^j) \right) \\
& \times \left\{ (n-j+1)^\alpha - (n-j)^\alpha \right\},
\end{aligned} \tag{14.150}
$$

where

$$
\begin{aligned}
\overline{I}_0^{n+1} = \; & I_0(t_0) + (1-\alpha) f_{I_0}(t_n, I_0^n, S_0^n) \\
& + \frac{h^{\alpha-1}}{\Gamma(\alpha)} \sum_{j=0}^{n} \left(g(t_{j+1}) - g(t_j) \right) f_{I_0}(t_j, I_0^j, S_0^j) \\
& \times \{ (n-j+1)^\alpha - (n-j)^\alpha \}.
\end{aligned} \tag{14.151}
$$

$$
\begin{aligned}
R_{n+1} = \; & R_0(t_0) + (1-\alpha)\frac{g(t_{n+1}) - g(t_n)}{h} f_{R_0}(t_{n+1}, \overline{I}_0^{n+1}) \\
& + \frac{\alpha h^{\alpha-1}}{2\Gamma(\alpha+1)} \sum_{j=0}^{n-1} (g(t_{j+1}) - g(t_j)) \left(f_{R_0}(t_{j+1}, I_0^{j+1}) + f_{I_0}(t_j, I_0^j, S_0^j) \right) \\
& \times \left\{ (n-j+1)^\alpha - (n-j)^\alpha \right\},
\end{aligned} \tag{14.152}
$$

where

$$
\begin{aligned}
\overline{R}_0^{n+1} = \; & R_0(t_0) + (1-\alpha) f_{R_0}(t_n, I_0^n) \\
& + \frac{h^{\alpha-1}}{\Gamma(\alpha)} \sum_{j=0}^{n} g(t_{j+1}) - g(t_j) f_{R_0}(t_j, I_0^j) \\
& \times \{ (n-j+1)^\alpha - (n-j)^\alpha \}.
\end{aligned} \tag{14.153}
$$

$$
\begin{aligned}
I_1^{n+1} = \; & I_1(t_0) + (1-\alpha)\frac{g(t_{n+1}) - g(t_n)}{h} f_{I_1}(t_{n+1}, \overline{I}_1^{n+1}, \overline{S}_0^{n+1}) \\
& + \frac{\alpha h^{\alpha-1}}{2\Gamma(\alpha+1)} \sum_{j=0}^{n-1} (g(t_{j+1}) - g(t_j)) \left(f_{I_1}(t_{j+1}, I_1^{j+1}, S_0^{j+1}) + f_{I_1}(t_j, I_1^j, S_0^j) \right) \\
& \times \left\{ (n-j+1)^\alpha - (n-j)^\alpha \right\},
\end{aligned} \tag{14.154}
$$

where

$$
\begin{aligned}
\overline{I}_1^{n+1} = \; & I_1(t_0) + (1-\alpha) f_{I_1}(t_n, I_1^n, S_0^n) \\
& + \frac{h^{\alpha-1}}{\Gamma(\alpha)} \sum_{j=0}^{n} \left(g(t_{j+1}) - g(t_j) \right) f_{I_1}(t_j, I_1^j, S_0^j) \\
& \times \{ (n-j+1)^\alpha - (n-j)^\alpha \}.
\end{aligned} \tag{14.155}
$$

14.5 Numerical Simulation for SIR Model

In this section, we consider the following model:

$$\begin{aligned}
\frac{dS(t)}{dt} &= \alpha t^{\alpha-1}\left(-\beta I(t)\frac{S(t)}{N}\right), \\
\frac{dI(t)}{dt} &= \alpha t^{\alpha-1}\left(\beta I(t)\frac{S(t)}{N} - \gamma I(t)\right), \\
\frac{dR(t)}{dt} &= \alpha t^{\alpha-1}\left(\gamma I(t)\right),
\end{aligned} \tag{14.156}$$

where initial conditions are given as $S(0) = 997, I(0) = 3, R(0) = 0, N(0) = 1000$. Also the parameters are chosen as $\beta = 0.12,\ \gamma = 0.04,\ \alpha = 0.1$. The numerical simulations are given for different values of fractional order as presented in Figs. 14.1, 14.2, 14.3, 14.4, 14.5, 14.6, 14.7, 14.8, and 14.9.

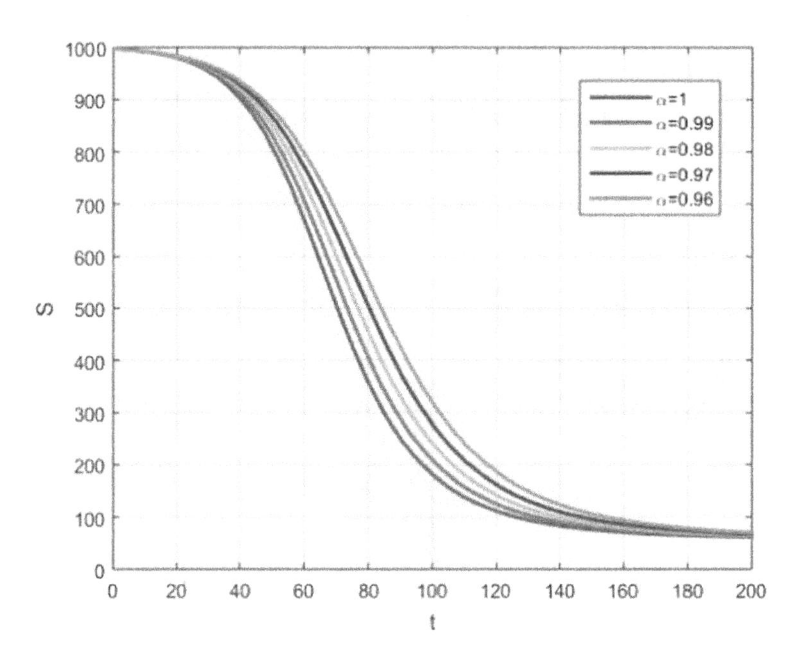

Fig. 14.1 Dynamics of susceptible population when varying alpha

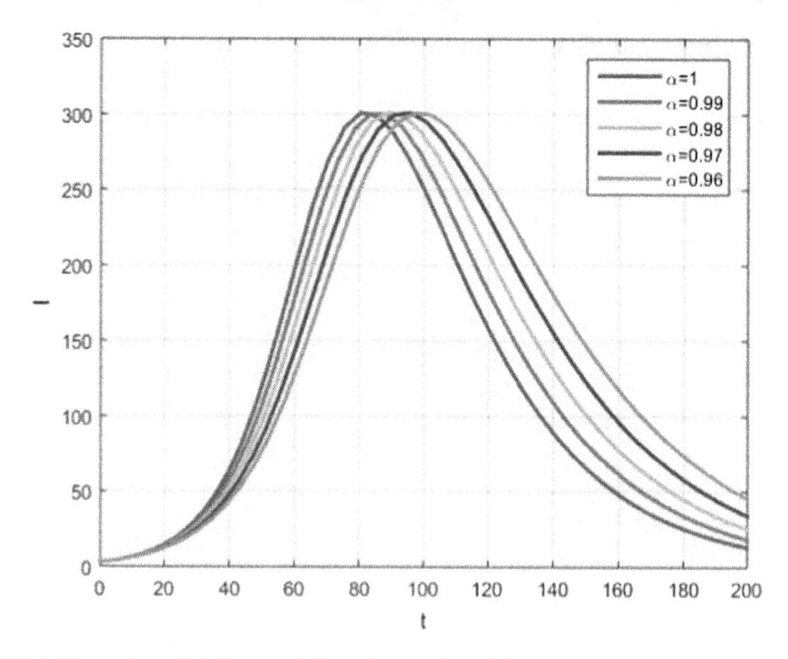

Fig. 14.2 Dynamics of infected population when varying alpha

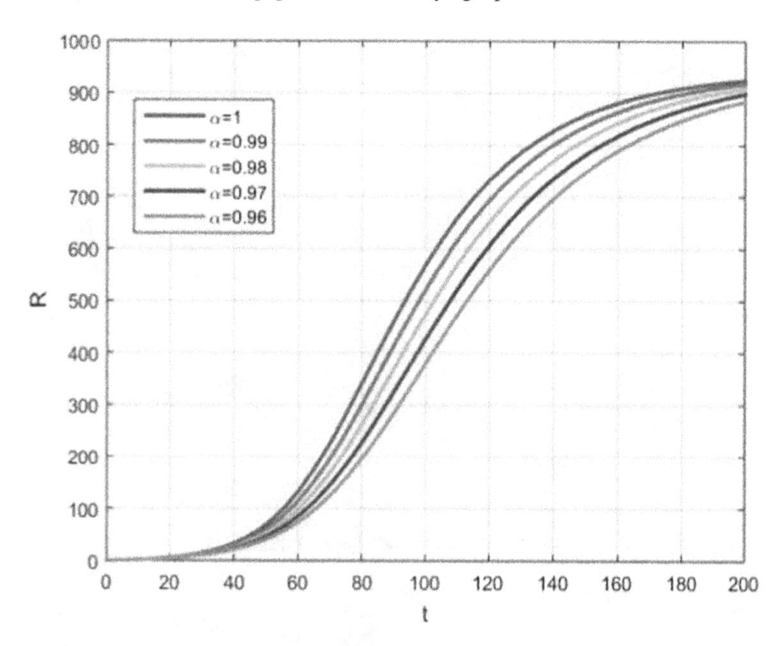

Fig. 14.3 Dynamics of recovered population when varying alpha

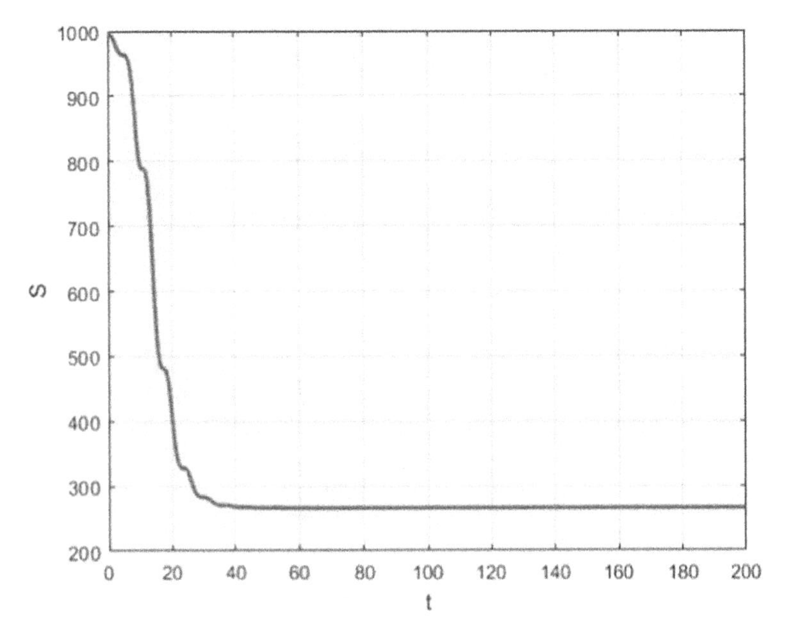

Fig. 14.4 Dynamics of susceptible population when the function g is $sin(t) + t$

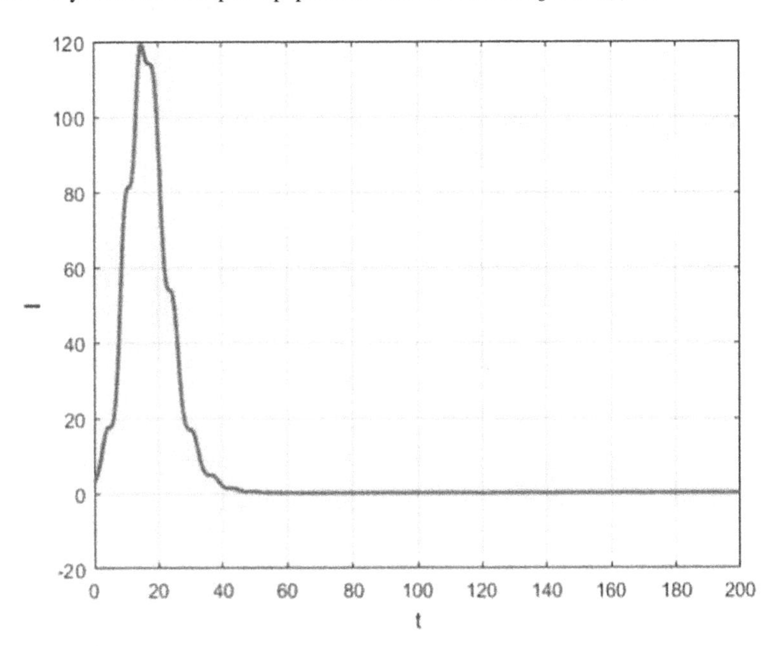

Fig. 14.5 Dynamics of infected population when the function g is $sin(t) + t$

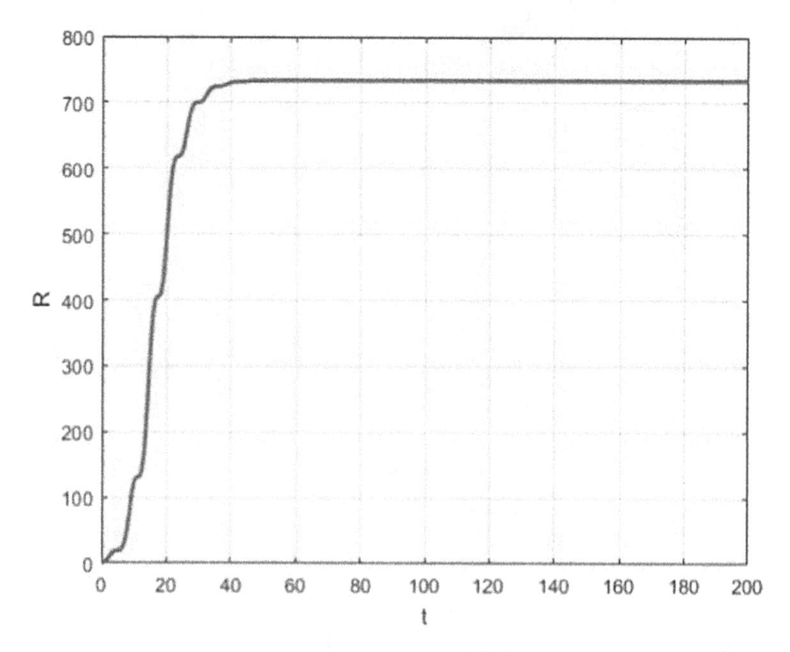

Fig. 14.6 Dynamics of recovered population when the function g is $sin(t) + t$

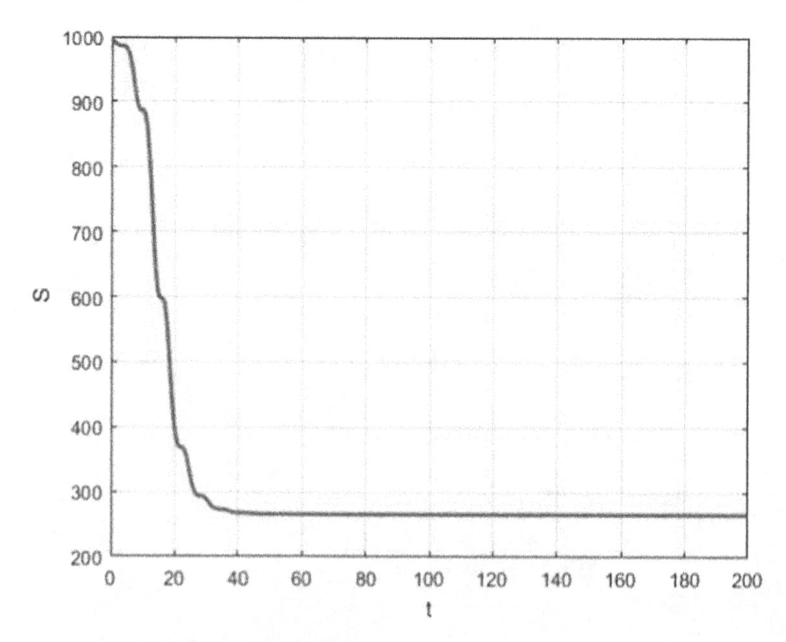

Fig. 14.7 Dynamics of susceptible population when the function g is $\cos(t) + t$

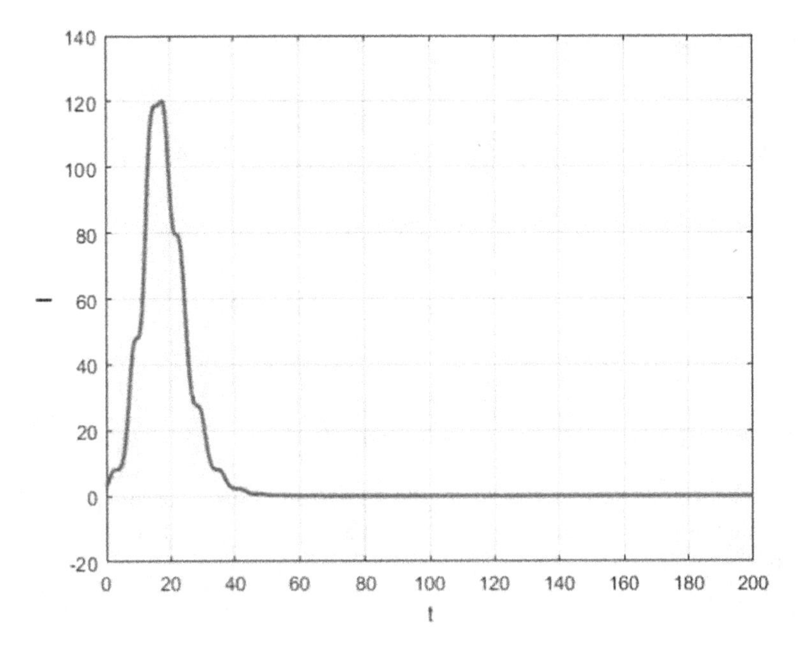

Fig. 14.8 Dynamics of infected population when the function g is $\cos + t$

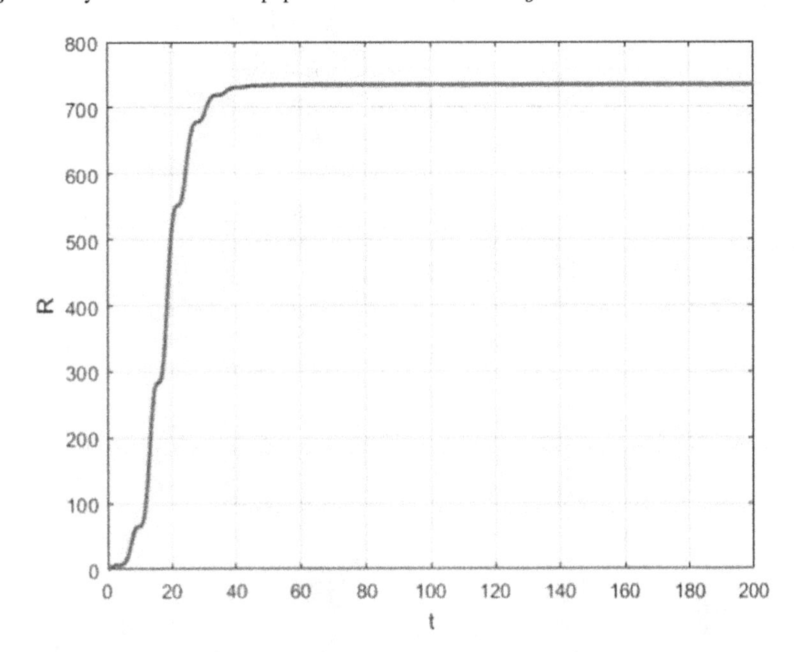

Fig. 14.9 Dynamics of recovered population when the function g is $\cos + t$

14.6 Application to Chaos

In this section, we consider the Rayleigh–Duffing equation. We shall recall that this model provides a description of a spring containing nonlinear properties

$$\begin{cases} \frac{dx}{dt} = y \\ \frac{dy}{dt} = \varepsilon\left(1 - y^2\right) y - x^3 + B\cos \upsilon t \end{cases}. \tag{14.157}$$

We provide for the sake of readers some analysis of the model. When $B = 0$, we have that the system is reduced

$$\begin{cases} \dot{x} = y \\ \dot{y} = \varepsilon\left(1 - y^2\right) y - x^3 \end{cases} \tag{14.158}$$

which is the well-known Vander Pol oscillator. (0, 0) is the equilibrium point which is unstable however is stable limit cycle when $\varepsilon > 0$. We can put the system in the form of

$$\begin{cases} \dot{x} = f_1(x, y, t) \\ \dot{y} = f_2(x, y, t) \end{cases}. \tag{14.159}$$

$$\begin{array}{lll} \frac{\partial f_1}{\partial x} = 0, & \frac{\partial^2 f_1}{\partial x^2} = 0, & \frac{\partial f_1}{\partial y} = 1, \\ \frac{\partial^2 f_1}{\partial y^2} = 0, & \frac{\partial^2 f_1}{\partial x \partial y} = 0, & \frac{\partial^2 f_1}{\partial y \partial x} = 0, \\ \frac{\partial f_2}{\partial x} = -3x^2, & \frac{\partial^2 f_2}{\partial x^2} = -6x, & \frac{\partial f_2}{\partial y} = \varepsilon - 3\varepsilon y^2, \\ \frac{\partial^2 f_2}{\partial y^2} = -6\varepsilon y, & \frac{\partial^2 f_2}{\partial y \partial x} = 0, & \frac{\partial^2 f_2}{\partial x \partial y} = 0. \end{array} \tag{14.160}$$

We have that all f_1 and f_2 are $C^2[0, T]$; therefore since f_1 and f_2 are also Lipschitz with respect to the x and y respectively, the following iterations

$$\begin{cases} x_{n+1}(t) = x(0) + \int_0^t g'(\tau) f_1(\tau, x_n(\tau), y_n(\tau))\, d\tau, \\ y_{n+1}(t) = y(0) + \int_0^t g'(\tau) f_2(\tau, x_n(\tau), y_n(\tau))\, d\tau, \end{cases} \tag{14.161}$$

$$\begin{cases} x_{n+1}(t) = \int_0^t g'(\tau) \frac{(t-\tau)^{\alpha-1}}{\Gamma(\alpha)} f_1(\tau, x_n(\tau), y_n(\tau))\, d\tau, \\ y_{n+1}(t) = \int_0^t g'(\tau) \frac{(t-\tau)^{\alpha-1}}{\Gamma(\alpha)} f_2(\tau, x_n(\tau), y_n(\tau))\, d\tau, \end{cases} \tag{14.162}$$

$$\begin{cases} x_{n+1}(t) = x(0) + (1-\alpha) f_1(t, x_n(t), y_n(t)) g'(t) + \alpha \int_0^t g'(\tau) f_1(\tau, x_n(\tau), y_n(\tau)) d\tau, \\ y_{n+1}(t) = y(0) + (1-\alpha) f_2(t, x_n(t), y_n(t)) g'(t) + \alpha \int_0^t g'(\tau) f_2(\tau, x_n(\tau), y_n(\tau)) d\tau, \end{cases} \tag{14.163}$$

and

$$\begin{cases} x_{n+1}(t) = (1-\alpha) f_1(t, x_n(t), y_n(t)) g'(t) \\ +\frac{\alpha}{\Gamma(\alpha)} \int_0^t (t-\tau)^{\alpha-1} g'(\tau) f_1(\tau, x_n(\tau), y_n(\tau)) d\tau, \\ y_{n+1}(t) = (1-\alpha) f_2(t, x_n(t), y_n(t)) g'(t) \\ +\frac{\alpha}{\Gamma(\alpha)} \int_0^t (t-\tau)^{\alpha-1} g'(\tau) f_2(\tau, x_n(\tau), y_n(\tau)) d\tau \end{cases} \tag{14.164}$$

provide existence and uniqueness of the following equations respectively:

$$\begin{cases} D_g x(t) = f_1(t, x(t), y(t)), \\ D_g y(t) = f_2(t, x(t), y(t)), \end{cases} \tag{14.165}$$

$$\begin{cases} {}_t^{RL} D_g^{\alpha} x(t) = f_1(t, x(t), y(t)), \\ {}_t^{RL} D_g^{\alpha} y(t) = f_2(t, x(t), y(t)), \end{cases}$$

$$\begin{cases} {}_t^{CF} D_g^{\alpha} x(t) = f_1(t, x(t), y(t)), \\ {}_t^{CF} D_g^{\alpha} y(t) = f_2(t, x(t), y(t)), \end{cases}$$

and

$$\begin{cases} {}_{t_0}^{ABR} D_g^{\alpha} x(t) = f_1(t, x(t), y(t)), \\ {}_{t_0}^{ABR} D_g^{\alpha} y(t) = f_2(t, x(t), y(t)). \end{cases} \tag{14.166}$$

In the following interval $[0, \gamma_1]$, $[0, \gamma_2]$, $[0, \gamma_3]$, and $[0, \gamma_4]$ where (γ_i) are defined as in previous chapter. We shall now provide other analysis. Let us reformulate the system for the variational equation, where the variable computer variables' differential equations and initial conditions:

$$\begin{cases} x = \lambda_1(t, x_0, y_0, \varepsilon), \; x_1 \text{ (variable)}, \\ y = \lambda_2(t, x_0, y_0, \varepsilon), \; y_1 \text{ (variable)}. \end{cases} \tag{14.167}$$

The associate differential equations are

$$\begin{cases} \dot{x}_1(t) = x_2(t), \\ \dot{x}_2(t) = \varepsilon\left(1 - x_2^2\right) x_2 - x_1^3 + B\cos \upsilon t, \\ x_1(0) = x_0, \\ x_2(0) = y_0. \end{cases} \tag{14.168}$$

But

$$\begin{aligned} &\frac{\partial \lambda_1}{\partial x_0} \quad x_3 \,(\text{variable})\,, \\ &\frac{\partial \lambda_2}{\partial x_0} \quad x_4 \,(\text{variable})\,, \\ &\frac{\partial \lambda_1}{\partial y_0} \quad x_5 \,(\text{variable})\,, \\ &\frac{\partial \lambda_2}{\partial y_0} \quad x_6 \,(\text{variable})\,. \end{aligned} \tag{14.169}$$

For the above, we have the following differential equations:

$$\begin{cases} \dot{x}_3\,(t) = x_4, \\ \dot{x}_4\,(t) = -3x_1^2 x_3 + \varepsilon\left(1 - 3x_2^2\right) x_4, \\ \dot{x}_5\,(t) = x_6, \\ \dot{x}_6\,(t) = -3x_1^2 x_5 + \varepsilon\left(1 - 3x_2^2\right) x_6, \\ x_3\,(0) = 1, \\ x_4\,(0) = 0, \\ x_5\,(0) = 0, \\ x_6\,(0) = 1. \end{cases} \tag{14.170}$$

But

$$\begin{aligned} &\frac{\partial \lambda_1}{\partial B} \quad x_7 \,(\text{variable})\,, \\ &\frac{\partial \lambda_2}{\partial B} \quad x_8 \,(\text{variable})\,. \end{aligned} \tag{14.171}$$

The differential equations associated with the above variable are given as

$$\begin{cases} \dot{x}_7\,(t) = x_8, \\ \dot{x}_8\,(t) = -3x_1^2 x_7 + \varepsilon\left(1 - 3x_2^2\right) x_8 + \cos vt, \\ x_7\,(0) = 0, \\ x_8\,(0) = 0. \end{cases} \tag{14.172}$$

But

$$\begin{aligned} &\frac{\partial^2 \lambda_1}{\partial x_0^2} \quad x_9 \,(\text{variable})\,, \\ &\frac{\partial^2 \lambda_2}{\partial x_0^2} \quad x_{10} \,(\text{variable})\,, \end{aligned} \tag{14.173}$$

$$\frac{\partial^2 \lambda_1}{\partial y_0^2} \quad x_{11}\,(\text{variable}),$$
$$\frac{\partial^2 \lambda_2}{\partial y_0^2} \quad x_{12}\,(\text{variable}).$$

The above provides the following differential equations:

$$\begin{cases} \dot{x}_9\,(t) = x_{10}, \\ \dot{x}_{10}\,(t) = -3x_1^2 x_9 + \varepsilon\left(1 - 3x_2^2\right) x_{10} - 6x_1 x_3^2 - 6\varepsilon x_2 x_{10}^2 \\ \dot{x}_{11}\,(t) = x_{12}, \\ \dot{x}_{12}\,(t) = -3x_1^2 x_{11} + \varepsilon\left(1 - 3x_2^2\right) x_{12} - 6x_1 x_3 x_5 - 3\varepsilon x_2 x_4 x_6 \\ \dot{x}_{13}\,(t) = x_{14} \\ \dot{x}_{14}\,(t) = -3x_1^2 x_{13} + \varepsilon\left(1 - 3x_2^2\right) x_{14} - 6x_1 x_5^2 - 6\varepsilon x_2 x_{14}^2 \\ x_9\,(0) = 0,\ x_{10}\,(0) = 0,\ x_{11}\,(0) = 0,\ x_{12}\,(0) = 0, \\ x_{13}\,(0) = 0,\ x_{14}\,(0) = 0. \end{cases} \tag{14.174}$$

Finally

$$\frac{\partial^2 \lambda_1}{\partial x_0 \partial B} \quad x_{15}\,(\text{variable}), \tag{14.175}$$
$$\frac{\partial^2 \lambda_2}{\partial x_0 \partial B} \quad x_{16}\,(\text{variable}),$$
$$\frac{\partial^2 \lambda_1}{\partial y_0 \partial B} \quad x_{17}\,(\text{variable}),$$
$$\frac{\partial^2 \lambda_2}{\partial y_0 \partial B} \quad x_{18}\,(\text{variable}).$$

The above provides the following differential equations

$$\begin{cases} \dot{x}_{15}\,(t) = x_{16}, \\ \dot{x}_{16}\,(t) = -3x_1^2 x_{15} + \varepsilon\left(1 - 3x_2^2\right) x_{16} - 6x_1 x_3 x_7 - 6\varepsilon x_2 x_4 x_8 \\ \dot{x}_{17}\,(t) = x_{18}, \\ \dot{x}_{18}\,(t) = -3x_1^2 x_{17} + \varepsilon\left(1 - 3x_2^2\right) x_{18} - 6x_1 x_5 x_7 - 6\varepsilon x_2 x_6 x_8, \\ x_{15}\,(0) = 0,\ x_{16}\,(0) = 0,\ x_{17}\,(0) = 0,\ x_{18}\,(0) = 0. \end{cases} \tag{14.176}$$

We note that the above analysis was performed using the following variational equation for a general two-dimensional non-autonomous system:

$$\begin{cases} \dot{x}\,(t) = f_1\left(t, x, y, \overline{\lambda}\right), \\ \dot{y}\,(t) = f_2\left(t, x, y, \overline{\lambda}\right), \end{cases} \tag{14.177}$$

where $\overline{\lambda}$ is a parameter. We note the solution of the above is possessed via (x_0, y_0) when $t = 0$, therefore can be used for the description of

$$\begin{cases} x(t) = \lambda_1\left(t, x_0, y_0, \overline{\lambda}\right), \\ y(t) = \lambda_2\left(t, x_0, y_0, \overline{\lambda}\right). \end{cases} \tag{14.178}$$

In this case

$$\begin{cases} \dot{\lambda}_1(t) = f_1\left(t, \lambda_1, \lambda_2, \overline{\lambda}\right) \\ \dot{\lambda}_2(t) = f_2\left(t, \lambda_1, \lambda_2, \overline{\lambda}\right) \\ \lambda_1\left(0, x_0, y_0, \overline{\lambda}\right) = x_0, \\ \lambda_2\left(0, x_0, y_0, \overline{\lambda}\right) = y_0, \end{cases} \tag{14.179}$$

where the first variational equation

$$\frac{d}{dt}\begin{pmatrix} \frac{\partial\lambda_1}{\partial x_0} \\ \frac{\partial\lambda_2}{\partial x_0} \end{pmatrix} = \begin{pmatrix} \frac{\partial f_1}{\partial x} & \frac{\partial f_1}{\partial y} \\ \frac{\partial f_2}{\partial x} & \frac{\partial f_2}{\partial y} \end{pmatrix}\begin{pmatrix} \frac{\partial\lambda_1}{\partial x_0} \\ \frac{\partial\lambda_2}{\partial x_0} \end{pmatrix}, \begin{pmatrix} \frac{\partial\lambda_1}{\partial x_0} \\ \frac{\partial\lambda_2}{\partial x_0} \end{pmatrix}|_{t=0} = \begin{pmatrix} 1 \\ 0 \end{pmatrix}, \tag{14.180}$$

$$\frac{d}{dt}\begin{pmatrix} \frac{\partial\lambda_1}{\partial y_0} \\ \frac{\partial\lambda_2}{\partial y_0} \end{pmatrix} = \begin{pmatrix} \frac{\partial f_1}{\partial x} & \frac{\partial f_1}{\partial y} \\ \frac{\partial f_2}{\partial x} & \frac{\partial f_2}{\partial y} \end{pmatrix}\begin{pmatrix} \frac{\partial\lambda_1}{\partial y_0} \\ \frac{\partial\lambda_2}{\partial y_0} \end{pmatrix}, \begin{pmatrix} \frac{\partial\lambda_1}{\partial y_0} \\ \frac{\partial\lambda_2}{\partial y_0} \end{pmatrix}|_{t=0} = \begin{pmatrix} 0 \\ 1 \end{pmatrix},$$

$$\frac{d}{dt}\begin{pmatrix} \frac{\partial\lambda_1}{\partial\overline{\lambda}} \\ \frac{\partial\lambda_2}{\partial\overline{\lambda}} \end{pmatrix} = \begin{pmatrix} \frac{\partial f_1}{\partial x} & \frac{\partial f_1}{\partial y} \\ \frac{\partial f_2}{\partial x} & \frac{\partial f_2}{\partial y} \end{pmatrix}\begin{pmatrix} \frac{\partial\lambda_1}{\partial\overline{\lambda}} \\ \frac{\partial\lambda_2}{\partial\overline{\lambda}} \end{pmatrix} + \begin{pmatrix} \frac{\partial f_1}{\partial\overline{\lambda}} \\ \frac{\partial f_2}{\partial\overline{\lambda}} \end{pmatrix}, \begin{pmatrix} \frac{\partial\lambda_1}{\partial\overline{\lambda}} \\ \frac{\partial\lambda_2}{\partial\overline{\lambda}} \end{pmatrix}|_{t=0} = \begin{pmatrix} 0 \\ 0 \end{pmatrix}.$$

For the second variational equation

$$\frac{d}{dt}\begin{pmatrix} \frac{\partial^2\lambda_1}{\partial x_0^2} \\ \frac{\partial^2\lambda_2}{\partial x_0^2} \end{pmatrix} = D_f\begin{pmatrix} \frac{\partial^2\lambda_1}{\partial x_0^2} \\ \frac{\partial^2\lambda_2}{\partial x_0^2} \end{pmatrix} + \left(\frac{\partial}{\partial x_0}D_f\right)\begin{pmatrix} \frac{\partial\lambda_1}{\partial x_0} \\ \frac{\partial\lambda_2}{\partial x_0} \end{pmatrix}, \tag{14.181}$$

$$\frac{d}{dt}\begin{pmatrix} \frac{\partial^2\lambda_1}{\partial x_0\partial y_0} \\ \frac{\partial^2\lambda_2}{\partial x_0\partial y_0} \end{pmatrix} = D_f\begin{pmatrix} \frac{\partial^2\lambda_1}{\partial x_0\partial y_0} \\ \frac{\partial^2\lambda_2}{\partial x_0\partial y_0} \end{pmatrix} + \left(\frac{\partial}{\partial x_0}D_f\right)\begin{pmatrix} \frac{\partial\lambda_1}{\partial y_0} \\ \frac{\partial\lambda_2}{\partial y_0} \end{pmatrix},$$

$$\frac{d}{dt}\begin{pmatrix} \frac{\partial^2\lambda_1}{\partial x_0\partial\overline{\lambda}} \\ \frac{\partial^2\lambda_2}{\partial x_0\partial\overline{\lambda}} \end{pmatrix} = D_f\begin{pmatrix} \frac{\partial^2\lambda_1}{\partial x_0\partial\overline{\lambda}} \\ \frac{\partial^2\lambda_2}{\partial x_0\partial\overline{\lambda}} \end{pmatrix} + \left(\frac{\partial}{\partial x_0}D_f\right)\begin{pmatrix} \frac{\partial\lambda_1}{\partial\overline{\lambda}} \\ \frac{\partial\lambda_2}{\partial\overline{\lambda}} \end{pmatrix} + \left(\frac{\partial}{\partial x_0}D_{\overline{\lambda}f}\right),$$

$$\frac{d}{dt}\begin{pmatrix} \frac{\partial^2\lambda_1}{\partial y_0\partial\overline{\lambda}} \\ \frac{\partial^2\lambda_2}{\partial y_0\partial\overline{\lambda}} \end{pmatrix} = D_f\begin{pmatrix} \frac{\partial^2\lambda_1}{\partial y_0\partial\overline{\lambda}} \\ \frac{\partial^2\lambda_2}{\partial y_0\partial\overline{\lambda}} \end{pmatrix} + \left(\frac{\partial}{\partial y_0}D_f\right)\begin{pmatrix} \frac{\partial\lambda_1}{\partial\overline{\lambda}} \\ \frac{\partial\lambda_2}{\partial\overline{\lambda}} \end{pmatrix} + \left(\frac{\partial}{\partial y_0}D_{\overline{\lambda}f}\right),$$

where

$$\left(\frac{\partial}{\partial x_0}D_f\right) = \begin{pmatrix} \frac{\partial^2 f_1}{\partial x^2}\frac{\partial \lambda_1}{\partial x_0} + \frac{\partial^2 f_1}{\partial y \partial x}\frac{\partial \lambda_2}{\partial x_0} & \frac{\partial^2 f_1 \partial \lambda_1}{\partial x \partial y \partial x_0} + \frac{\partial^2 f_1}{\partial y^2}\frac{\partial \lambda_2}{\partial x} \\ \frac{\partial^2 f_2}{\partial x^2}\frac{\partial \lambda_1}{\partial x_0} + \frac{\partial^2 f_2}{\partial y \partial x}\frac{\partial \lambda_2}{\partial x_0} & \frac{\partial^2 f_2 \partial \lambda_1}{\partial x \partial y \partial x_0} + \frac{\partial^2 f_2}{\partial y^2}\frac{\partial \lambda_2}{\partial x} \end{pmatrix}, \tag{14.182}$$

$$\left(\frac{\partial}{\partial x_0}D_{\lambda f}\right) = \begin{pmatrix} \frac{\partial^2 f_1}{\partial x \partial \overline{\overline{\lambda}}}\frac{\partial \lambda_1}{\partial x_0} & \frac{\partial^2 f_1}{\partial y \partial \overline{\overline{\lambda}}}\frac{\partial \lambda_2}{\partial x_0} \\ \frac{\partial^2 f_2}{\partial x \partial \overline{\overline{\lambda}}}\frac{\partial \lambda_1}{\partial x_0} & \frac{\partial^2 f_2}{\partial y \partial \overline{\overline{\lambda}}}\frac{\partial \lambda_2}{\partial x_0} \end{pmatrix}.$$

14.7 Numerical Solutions

In this section, we shall present the numerical solution of the model. We consider again the model

$$\begin{cases} D_g x(t) = f_1(t, x, y), \\ D_g y(t) = f_2(t, x, y), \\ \quad x(0) = x_0, \\ \quad y(0) = y_0. \end{cases} \tag{14.183}$$

When $g(t) = t$, we have that

$$\begin{cases} x(t) = x(0) + \int\limits_0^t f_1(\tau, x, y)\, d\tau, \\ y(t) = y(0) + \int\limits_0^t f_2(\tau, x, y)\, d\tau. \end{cases} \tag{14.184}$$

We can solve this system, using the Midpoint or Heun's method. By using Heun's method, we get

$$\begin{cases} \widetilde{x}_{i+1} = x_i + hf_1(t_i, x_i, y_i), \\ \widetilde{y}_{i+1} = y_i + hf_2(t_i, x_i, y_i), \end{cases} \tag{14.185}$$

$$\begin{cases} \widetilde{x}_{i+1} = x_i + \frac{h}{2}\left[f_1(t_i, x_i, y_i) + f_1(t_i, \widetilde{x}_{i+1}, \widetilde{y}_{i+1})\right], \\ \widetilde{y}_{i+1} = y_i + \frac{h}{2}\left[f_2(t_i, x_i, y_i) + f_2(t_i, \widetilde{x}_{i+1}, \widetilde{y}_{i+1})\right], \end{cases}$$

where $g(t) \neq t$; we have that

$$\begin{cases} x(t) = x(0) + \int\limits_0^t f_1(\tau, x, y)\, g'(\tau)\, d\tau, \\ y(t) = y(0) + \int\limits_0^t f_2(\tau, x, y)\, g'(\tau)\, d\tau. \end{cases} \tag{14.186}$$

Using Heun's method we get

$$\begin{cases} \widetilde{x}_{i+1} = x_i + \left(g\left(t_{i+1}\right) - g\left(t_i\right)\right) f_1\left(t_i, x_i, y_i\right), \\ \widetilde{y}_{i+1} = y_i + \left(g\left(t_{i+1}\right) - g\left(t_i\right)\right) f_2\left(t_i, x_i, y_i\right), \end{cases} \tag{14.187}$$

$$\begin{cases} x_{i+1} = x_i + \frac{(g(t_{i+1})-g(t_i))}{2}\left[f_1\left(t_i, x_i, y_i\right) - f_1\left(t_{i+1}, \widetilde{x}_{i+1}, \widetilde{y}_{i+1}\right)\right], \\ y_{i+1} = y_i + \frac{(g(t_{i+1})-g(t_i))}{2}\left[f_2\left(t_i, x_i, y_i\right) - f_2\left(t_{i+1}, \widetilde{x}_{i+1}, \widetilde{y}_{i+1}\right)\right]. \end{cases}$$

The rest can be obtained similarly; however we show the case with Caputo

$$\begin{cases} \widetilde{x}_{i+1} = \frac{h^{\alpha-1}}{2\Gamma(\alpha+1)} \sum_{j=1}^{i-1} \left\{ \begin{array}{l} \left(g\left(t_{j+1}\right) - g\left(t_j\right)\right) f_1\left(t_{j+1}, x_{j+1}, y_{j+1}\right) \\ + \left(g\left(t_j\right) - g\left(t_{j-1}\right)\right) f_1\left(t_j, x_j, y_j\right) \end{array} \right\} \left((i-j+1)^\alpha - (i-j)^\alpha\right), \\ \widetilde{y}_{i+1} = \frac{h^{\alpha-1}}{2\Gamma(\alpha+1)} \sum_{j=1}^{i-1} \left\{ \begin{array}{l} \left(g\left(t_{j+1}\right) - g\left(t_j\right)\right) f_2\left(t_{j+1}, x_{j+1}, y_{j+1}\right) \\ + \left(g\left(t_j\right) - g\left(t_{j-1}\right)\right) f_2\left(t_j, x_j, y_j\right) \end{array} \right\} \left((i-j+1)^\alpha - (i-j)^\alpha\right) \end{cases} \tag{14.188}$$

$$\begin{cases} x_{i+1} = \frac{h^{\alpha-1}}{2\Gamma(\alpha+1)} \sum_{j=1}^{i-1} \left\{ \begin{array}{l} \left(g\left(t_{j+1}\right) - g\left(t_j\right)\right) f_1\left(t_{j+1}, x_{j+1}, y_{j+1}\right) \\ + \left(g\left(t_j\right) - g\left(t_{j-1}\right)\right) f_1\left(t_j, x_j, y_j\right) \end{array} \right\} \left((i-j+1)^\alpha - (i-j)^\alpha\right) \\ \quad + \frac{h^{\alpha-1}}{2\Gamma(\alpha+1)} \left\{\left(g\left(t_{i+1}\right) - g\left(t_i\right)\right) f_1\left(t_{i+1}, \widetilde{x}_{i+1}, \widetilde{y}_{i+1}\right) + \left(g\left(t_i\right) - g\left(t_{i-1}\right)\right) f_1\left(t_i, x_i, y_i\right)\right\}, \\ y_{i+1} = \frac{h^{\alpha-1}}{2\Gamma(\alpha+1)} \sum_{j=1}^{i-1} \left\{ \begin{array}{l} \left(g\left(t_{j+1}\right) - g\left(t_j\right)\right) f_2\left(t_{j+1}, x_{j+1}, y_{j+1}\right) \\ + \left(g\left(t_j\right) - g\left(t_{j-1}\right)\right) f_2\left(t_j, x_j, y_j\right) \end{array} \right\} \left((i-j+1)^\alpha - (i-j)^\alpha\right) \\ \quad + \frac{h^{\alpha-1}}{2\Gamma(\alpha+1)} \left\{\left(g\left(t_{i+1}\right) - g\left(t_i\right)\right) f_2\left(t_{i+1}, \widetilde{x}_{i+1}, \widetilde{y}_{i+1}\right) + \left(g\left(t_i\right) - g\left(t_{i-1}\right)\right) f_2\left(t_i, x_i, y_i\right)\right\}. \end{cases} \tag{14.189}$$

Let us consider the Rayleigh–Duffing equation again which is given below

$$\begin{cases} \frac{dx}{dt} = y \\ \frac{dy}{dt} = \varepsilon\left(1 - y^2\right) y - x^3 + B\cos vt \end{cases}. \tag{14.190}$$

The numerical simulations are presented in Figs. 14.10, 14.11, 14.12, and 14.13 for classical derivative below where initial conditions are given as $x(0) = 0.12$, $y(0) = 1.1$. Also the parameters are chosen as $\varepsilon = 0.02$, $B = 2$, $v = 1$, $alpha = 0.9$. The numerical simulations are presented in Figs. 14.14, 14.15, 14.16, and 14.17 for classical derivative below where initial conditions are given as $x(0) = 0.12$, $y(0) = 1.1$. Also the parameters are chosen as $\varepsilon = 0.25$, $B = 2.5$, $v = 1.5$, $alpha = 0.75$. The numerical simulations are presented in Figs. 14.18, 14.19, 14.20, and 14.21 for Atangana–Baleanu derivative below where initial conditions are given as $x(0) = 0.1$, $y(0) = 0.8$. Also the parameters are chosen as $\varepsilon = 0.01$, $B = 1.5$, $v = 1$, $alpha = 0.86$.

The numerical simulations are presented in Figs. 14.22, 14.23, 14.24, and 14.25 for Atangana–Baleanu derivative below where initial conditions are given as $x(0) = 0.1$, $y(0) = 0.8$. Also the parameters are chosen as $\varepsilon = 0.02$, $B = 2$, $v = 1$, $alpha = 0.70$.

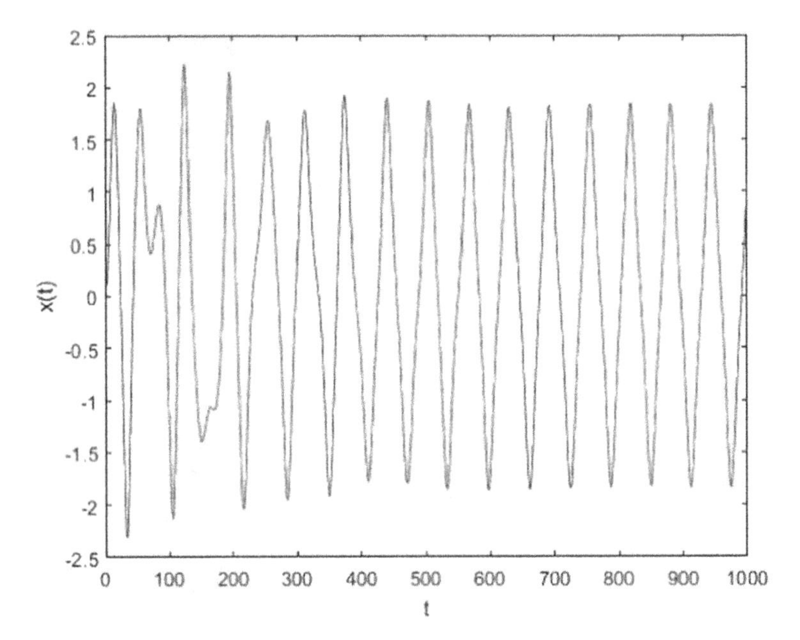

Fig. 14.10 Numerical simulation for $x(t)$

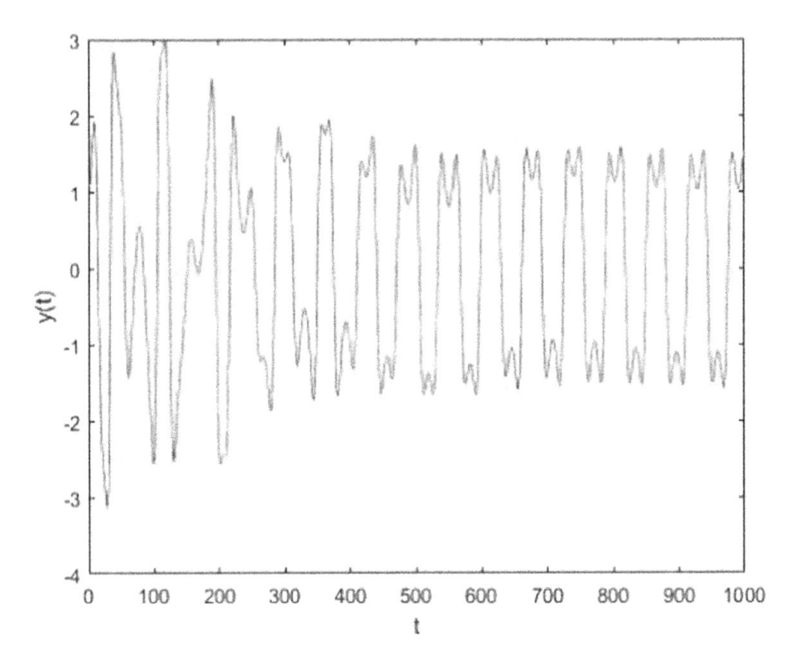

Fig. 14.11 Numerical simulation for $y(t)$

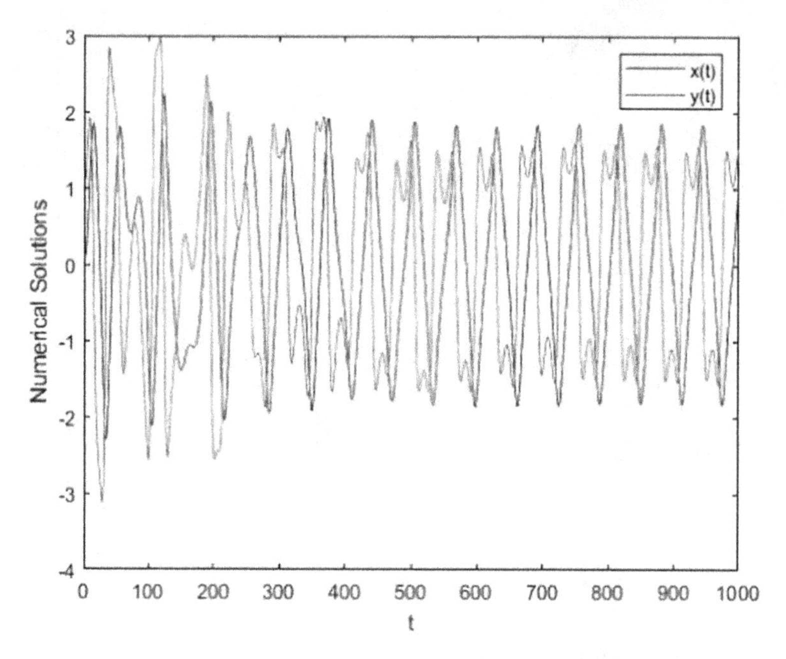

Fig. 14.12 Numerical simulation for $x(t)$ and $y(t)$

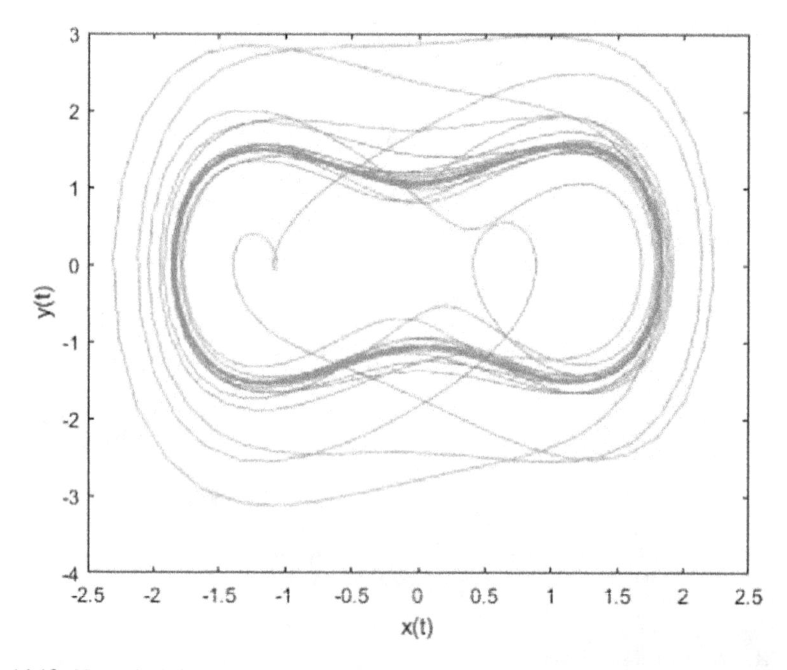

Fig. 14.13 Numerical simulation for chaos

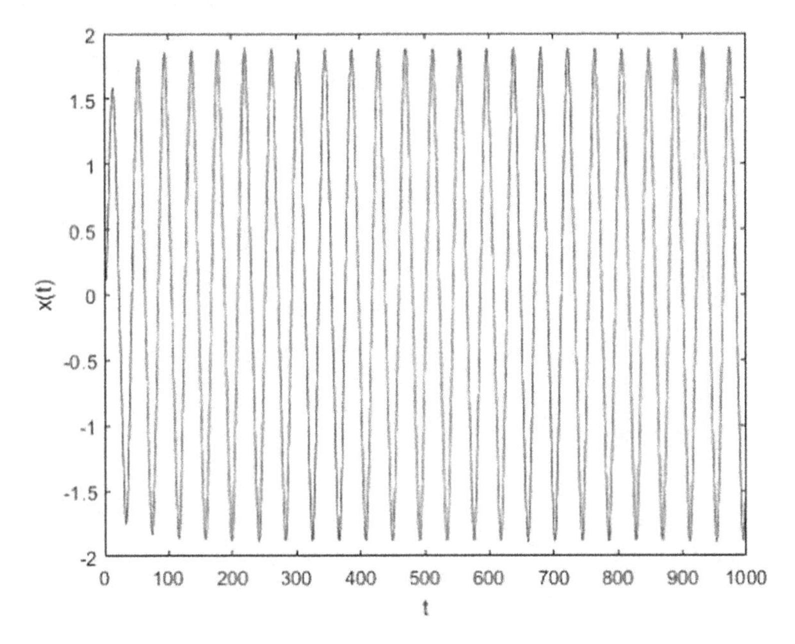

Fig. 14.14 Numerical simulation for $x(t)$

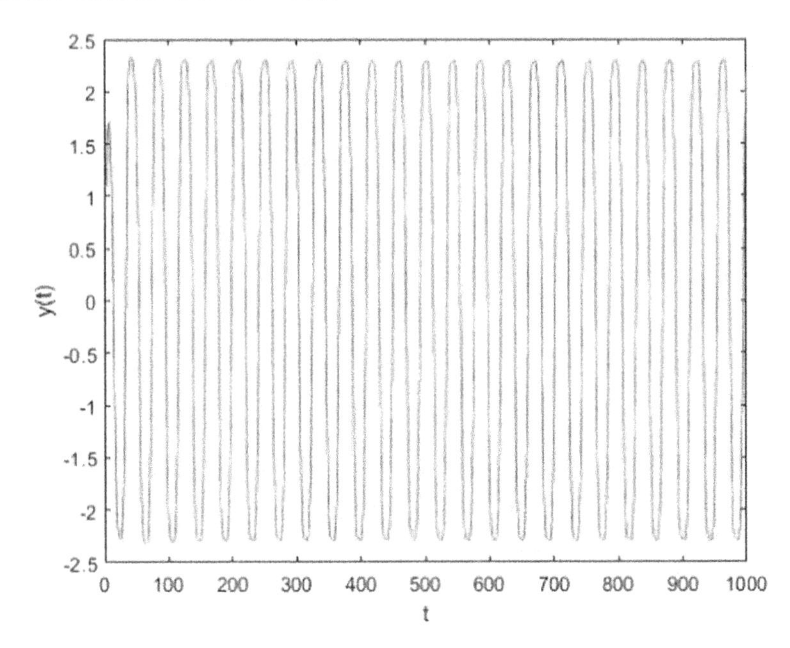

Fig. 14.15 Numerical simulation for $y(t)$

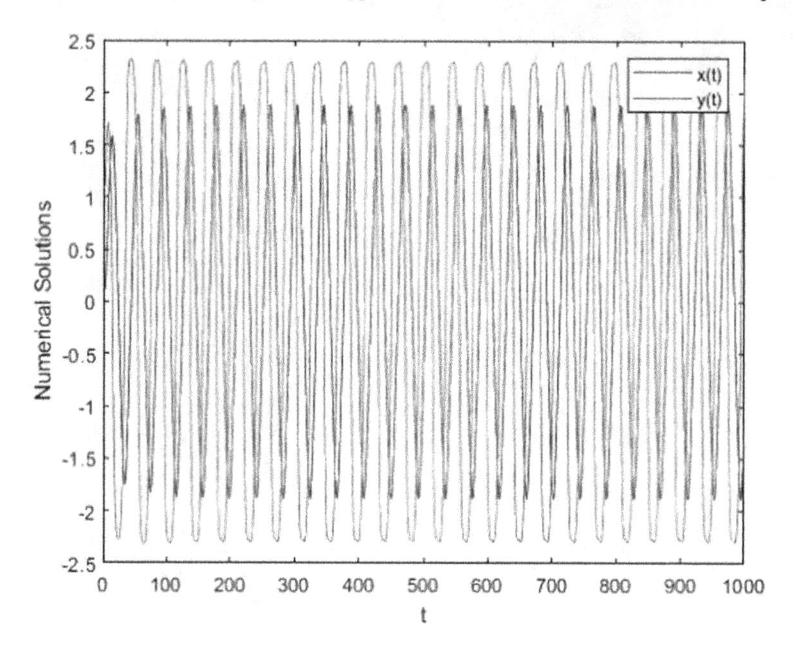

Fig. 14.16 Numerical simulation for $x(t)$ and $y(t)$

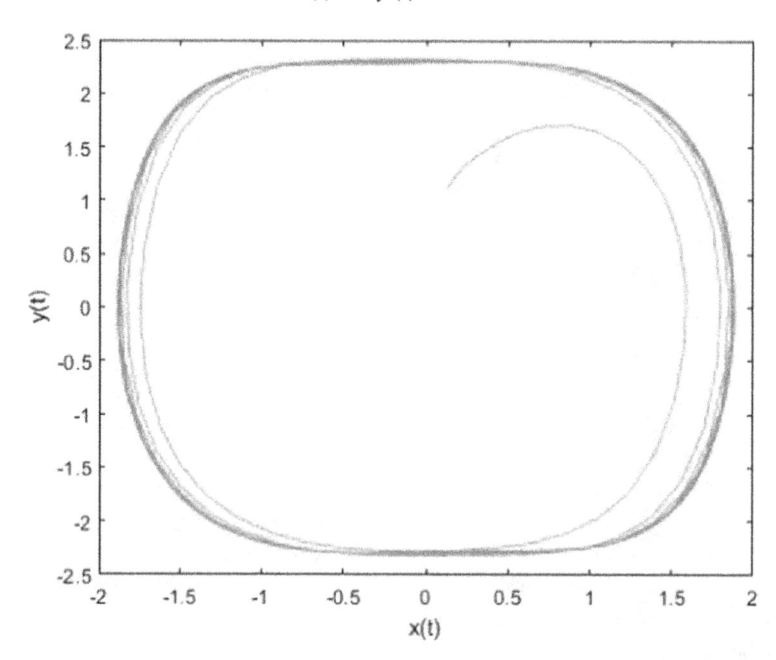

Fig. 14.17 Numerical simulation for chaos

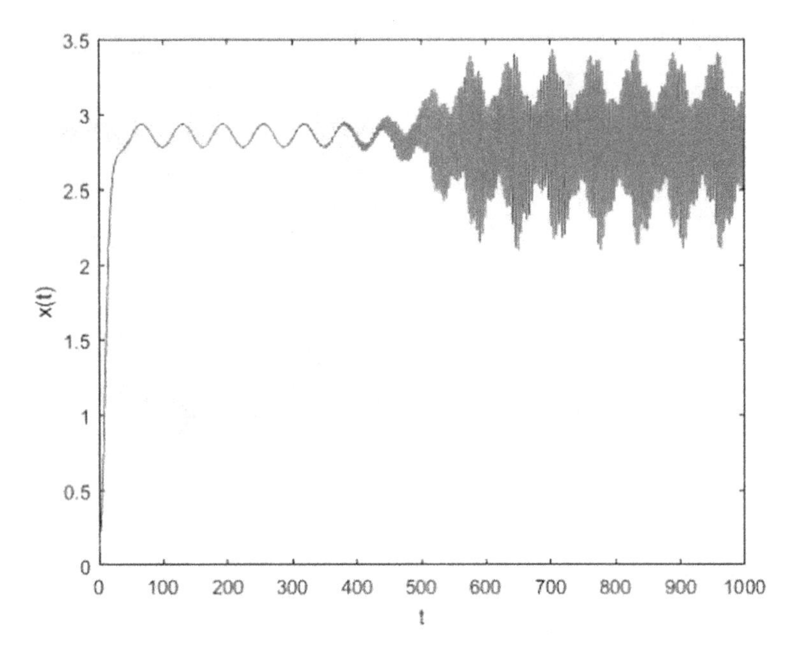

Fig. 14.18 Numerical simulation for $x(t)$

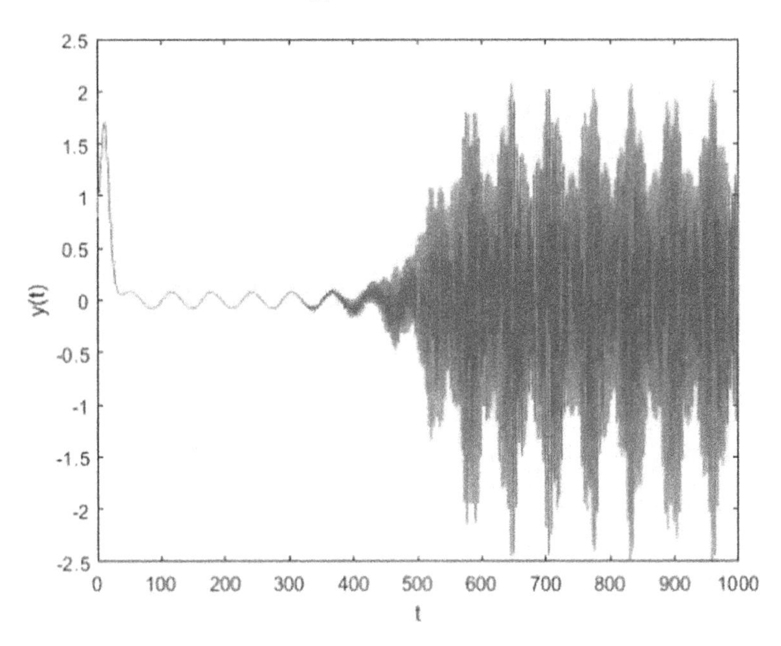

Fig. 14.19 Numerical simulation for $y(t)$

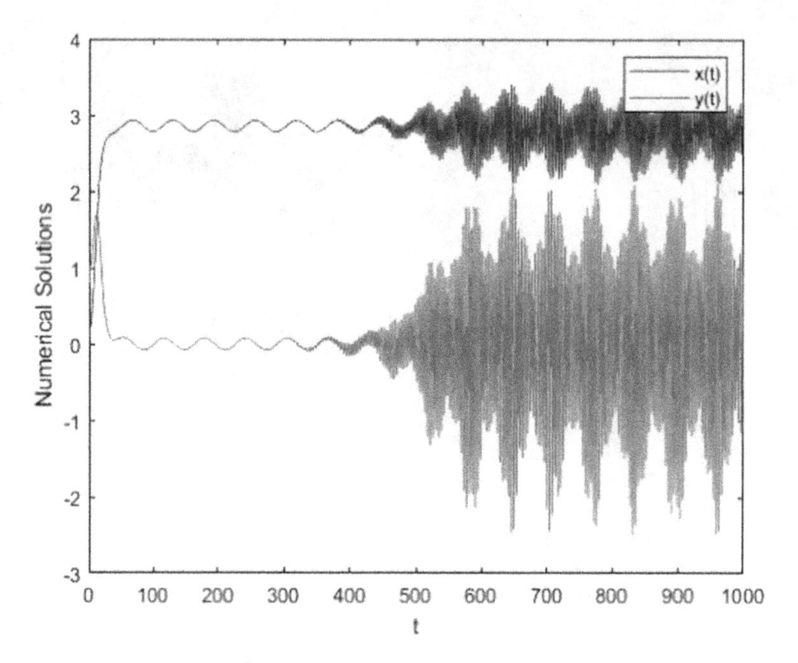

Fig. 14.20 Numerical simulation for $x(t)$ and $y(t)$

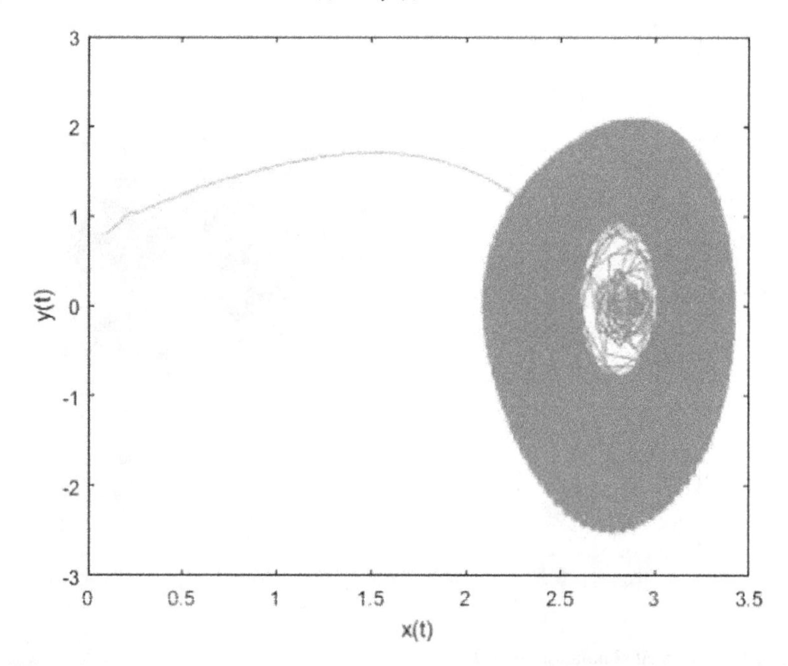

Fig. 14.21 Numerical simulation for chaos

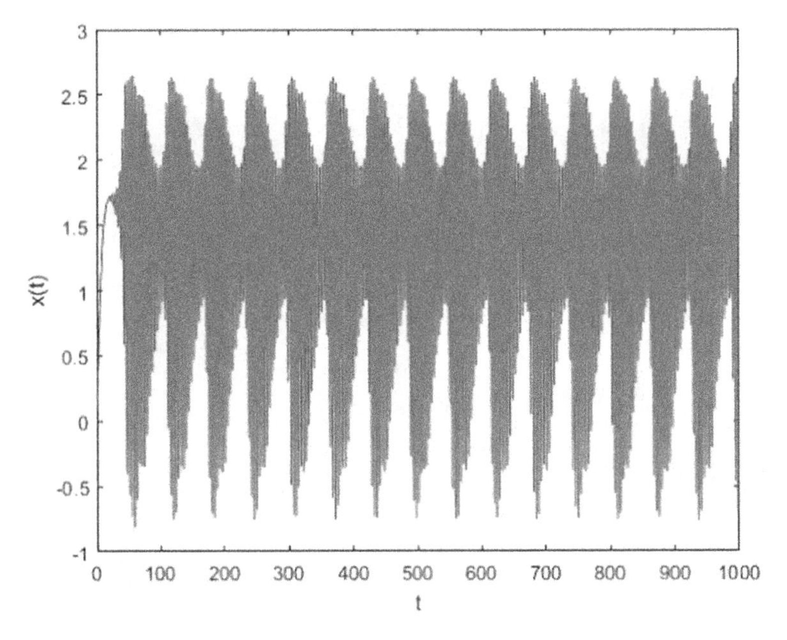

Fig. 14.22 Numerical simulation for $x(t)$

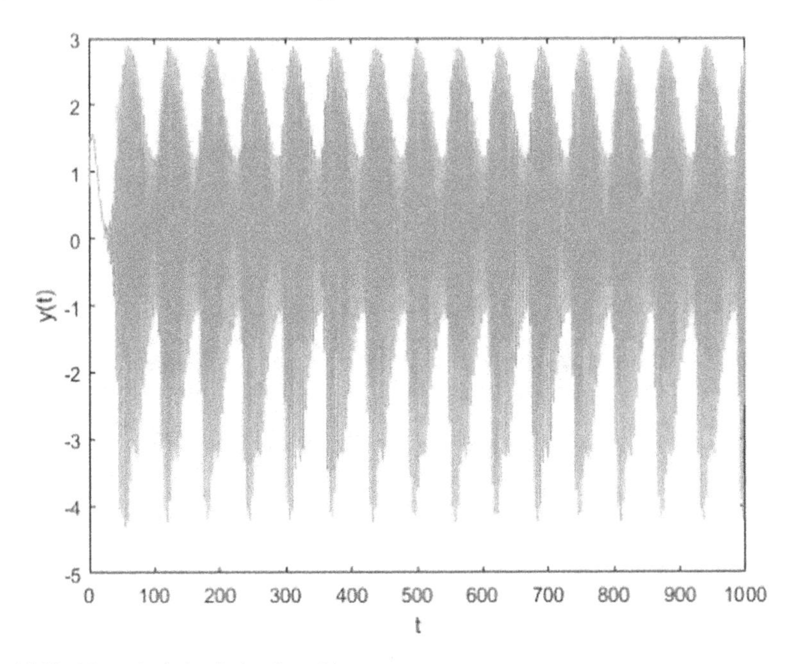

Fig. 14.23 Numerical simulation for $y(t)$

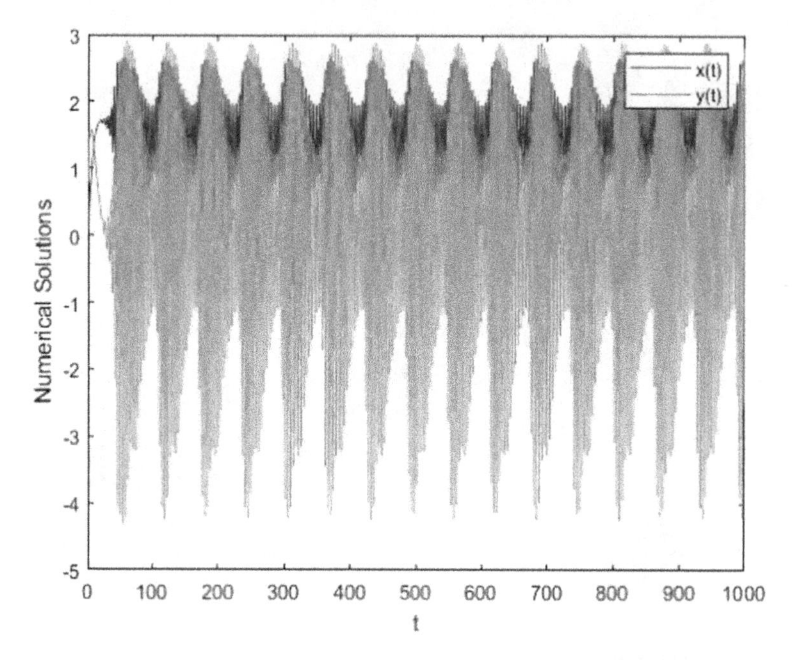

Fig. 14.24 Numerical simulation for $x(t)$ and $y(t)$

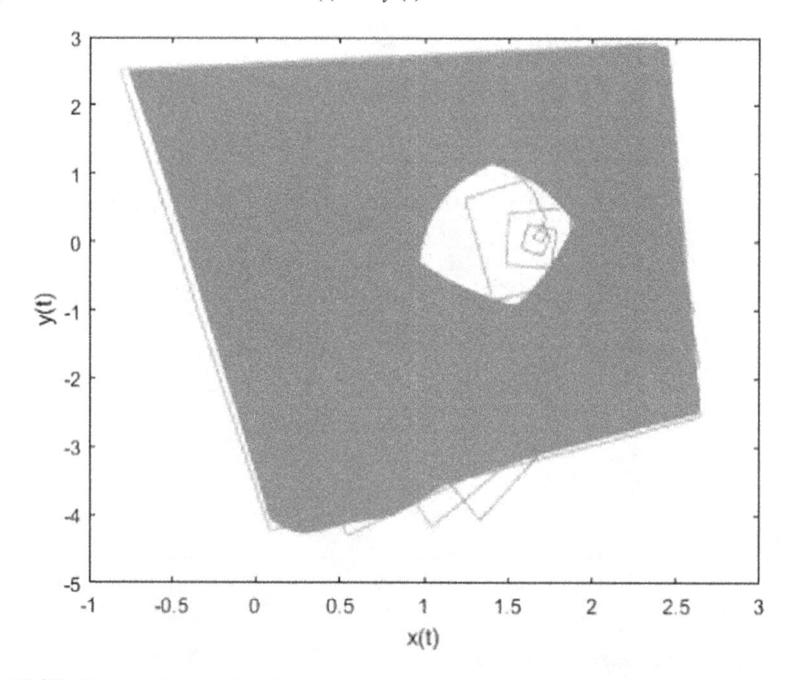

Fig. 14.25 Numerical simulation for chaos

14.8 Partial Differential with Fractional Global Differential Operators

The aim of this section is not to develop any theoretical result on this topic rather present some applications of these concepts to some partial differential equations. However at least in general we shall provide some definitions of these concepts to the framework of the partial differential operator. Let $g(x, t)$ be a continuous function on $[0, X] \times [0, T]$. Let us assume that $g(x, t)$ is differentiable with respect to x and t on $[0, X] \times [0, T]$, such that $\forall t \in [0, T]$, $\frac{\partial g(x,t)}{\partial t} \neq 0$ and $\forall x \in [0, X]$, $\frac{\partial g(x,t)}{\partial x} \neq 0$:

$$\begin{aligned}\frac{\partial f(x,t)}{\partial_t g(x,t)} &= \lim_{\bar{h}\to 0} \frac{f(x,t+h) - f(x,t)}{g(x,t+h) - g(x,t)}, \\ \frac{\partial f(x,t)}{\partial_x g(x,t)} &= \lim_{\bar{h}\to 0} \frac{f(x+h,t) - f(x,t)}{g(x+h,t) - g(x,t)}.\end{aligned} \tag{14.191}$$

Now having both f and g partial differentiable, we get

$$\begin{aligned}\frac{\partial f(x,t)}{\partial_t g(x,t)} &= \lambda(x,t), \\ \frac{\partial f(x,t)}{\partial t} &= \frac{\partial g(x,t)}{\partial t}\lambda(x,t),\end{aligned} \tag{14.192}$$

therefore

$$f(x,t) = f(x,0) + \int_0^t \frac{\partial g(x,\tau)}{\partial \tau}\lambda(x,\tau)\, d\tau. \tag{14.193}$$

Therefore, we have the following general partial differential equation:

$$\begin{cases} {}_0D_{g_t(x,t)}u(x,t) = f(x,t,u(x,t)), \\ u(x,t) = l(x). \end{cases} \tag{14.194}$$

If $u(x, t)$ and $g(x, t)$ are differentiable with respect to t with $\frac{\partial g(x,t)}{\partial t} \neq 0$ everywhere, we get

$$D_t u(x,t) = \frac{\partial g(x,t)}{\partial t} f(x,t,u(x,t)). \tag{14.195}$$

Later, we have

$$u(x,t) = u(x,t_0) + \int_{t_0}^t \frac{\partial g(x,\tau)}{\partial \tau} f(x,\tau,u(x,\tau))\, d\tau. \tag{14.196}$$

At the point (x_i, t_n) we have

$$u\left(x_i, t_n\right) = u(x_i, t_0) + \int_{t_0}^{t_n} \frac{\partial g(x_i, \tau)}{\partial \tau} f\left(x_i, \tau, u(x_i, \tau)\right) d\tau. \tag{14.197}$$

By using the simple Euler approximation, we get

$$\begin{aligned} u_i^{n+1} &= u_i^0 + \int_{t_0}^{t_{n+1}} \frac{\partial g(x_i, \tau)}{\partial \tau} f\left(x_i, \tau, u(x_i, \tau)\right) d\tau, \\ u_i^n &= u_i^0 + \int_{t_0}^{t_n} \frac{\partial g(x_i, \tau)}{\partial \tau} f\left(x_i, \tau, u(x_i, \tau)\right) d\tau, \end{aligned} \tag{14.198}$$

$$\begin{aligned} u_i^{n+1} &= u_i^n + \frac{g\left(x_i, t_{n+1}\right) - g\left(x_i, t_n\right)}{\Delta t} f\left(x_i, t_n, u_i^n\right) \Delta t, \\ u_i^{n+1} &= u_i^n + \left(g_i^{n+1} - g_i^n\right) f\left(x_i, t_n, u_i^n\right). \end{aligned} \tag{14.199}$$

The function $f\left(x_i, t_n, u_i^n\right)$ can be discretized accordingly. Alternatively, we can have different approaches to descritize the above. We shall consider for illustration, the heat equation that will be investigated. We consider the following modified heat equation:

$$\begin{cases} \partial_{g(t)} u(x,t) = \alpha u_{xx}(x,t), \\ u(x,0) = f(x), \\ u(0,t) = 0 = u\left(L,t\right), \forall t > 0,\ 0 \le x \le L,\ t \ge 0. \end{cases} \tag{14.200}$$

We shall first show the uniqueness of the solution. To do this, we consider v and $\overline{v}$ as solutions and put $z = v - \overline{v}$. We define

$$\Lambda = \{(x,t) \,|0 \le x \le X,\ 0 \le t \le T .\} \tag{14.201}$$

Λ is the frontier. We have that the function $g(t)$ is positive and differentiable with $g'(t) \neq 0\ \forall t > 0$. The modified heat equation can now be transformed to

$$\begin{cases} \partial_t u(x,t) = -g(t) u_{xx}(x,t) \text{ in } \Lambda, \\ u(x,t) = 0 \text{ in } \Lambda. \end{cases} \tag{14.202}$$

Therefore

$$\begin{cases} \partial_t z(t) = -g'(t) \Delta z(t) \text{ in } \Lambda, \\ u(0,t) = 0, \\ u(x,0) = f(x). \end{cases} \tag{14.203}$$

By fixing t, we have that $g'(t)$ is fixed and positive. The energy function is given as

$$E(t) = \int_{\Lambda} z(x,t)^2 dx, \tag{14.204}$$
$$E'(t) = \int_{\Lambda} 2z(z_t)dx = \int_{\Lambda} 2zg'(t)\Delta z,$$
$$= -2g'(t)\int_{\Lambda} |Dz|^2 \leq 0.$$

We have therefore obtained $E'(t) \leq 0$ so we can conclude that such energy has a fading process; therefore

$$0 \leq E(t) \leq E(0) = \int_{\Lambda} z(x,0)^2 dx = \int 0 = 0, \tag{14.205}$$
$$E(t) = \int z^2 = 0 \rightarrow z = 0, \text{ so } v - \overline{v} = 0$$

and hence

$$v(x,t) = \overline{v}(x,t), \; \forall (x,t). \tag{14.206}$$

We can now solve our equation using the method of separation of variables

$$u(x,t) = u_1(x)u_2(t), \tag{14.207}$$
$$\frac{u_2'(t)}{\alpha g'(t)u_2(t)} = \frac{u_1''(x)}{u(x)}.$$

Since the above is proportional, we shall have

$$u_2'(t) = -\gamma\alpha g'(t)u_2(t), \tag{14.208}$$
$$u_1''(x) = -\gamma u_1(x).$$

Having $\gamma > 0$, then $\exists\, A, B, C$ such that

$$u_2(t) = A\exp\left[-\gamma\alpha\left(g(t) - g(t_0)\right)\right], \tag{14.209}$$
$$u_1(t) = B\sin\left(\sqrt{\gamma}x\right) + C\cos\left(\sqrt{\gamma}x\right),$$
$$\sqrt{\gamma} = n\frac{\pi}{L}.$$

Therefore, we have

$$u(x,t)=\sum_{n=1}^{\infty}D_n\sin\left(\frac{n\pi x}{L}\right)\exp\left[-\frac{n^2\pi^2}{L^2}\gamma\left(g(t)-g\left(t_0\right)\right)\right] \tag{14.210}$$

where

$$D_n=\frac{2}{L}\int_0^L f(x)\sin\left(\frac{n\pi x}{L}\right)dx. \tag{14.211}$$

We can now consider the following modified equation:

$${}_0^{RL}D_t^{\alpha}u(x,t)=g'(t)\gamma\partial_{xx}u(x,t). \tag{14.212}$$

Applying the Riemann–Liouville integral yields

$$u(x,t)=u(x,0)=\frac{1}{\Gamma(\alpha)}\int_0^t g'(\tau)\gamma\partial_{xx}u(x,\tau)\left(t-\tau\right)^{\alpha-1}d\tau. \tag{14.213}$$

We can solve the above using numerical scheme

$$u_i^{n+1}=u_i^0+\frac{1}{\Gamma(\alpha)}\sum_{j=0}^{n}\int_{t_j}^{t_{j+1}}\frac{g(t_{j+1})-g(t_j)}{\Delta t}\gamma\partial_{xx}u(x_i,\tau)\left(t_{n+1}-\tau\right)^{\alpha-1}d\tau, \tag{14.214}$$

$$u_i^{n+1}=\frac{(\Delta t)^{\alpha-1}}{\Gamma(\alpha+1)}\sum_{j=0}^{n}g(t_{j+1})-g(t_j)\gamma\left[\frac{u_{i+1}^j-2u_i^j+u_{i-1}^j}{\Delta x^2}\right]\delta_{n,j}^{\alpha}$$

where

$$\delta_{n,j}^{\alpha}=(n-j+1)^{\alpha}-(n-j)^{\alpha}. \tag{14.215}$$

We consider now the modified model with the ABR case,

$${}_0^{ABR}D_t^{\alpha}u(x,t)=g'(t)\gamma\partial_{xx}u(x,t). \tag{14.216}$$

The modified model is reformulated to

$$u(x,t)=(1-\alpha)g'(t)\gamma\partial_{xx}u(x,t)+\frac{\alpha\gamma}{\Gamma(\alpha)}\int_0^t g'(\tau)\partial_{xx}u(x,\tau)\left(t-\tau\right)^{\alpha-1}d\tau. \tag{14.217}$$

At (x_i, t_{n+1}) we have

$$u_i^{n+1} = (1-\alpha)\left[\frac{g(t_{n+1})-g(t_n)}{\Delta t}\gamma\right]\left[\frac{u_{i+1}^{n+1}-2u_i^{n+1}+u_{i-1}^{n+1}}{\Delta x^2}\right] \tag{14.218}$$
$$+\frac{\alpha\gamma\,(\Delta t)^{\alpha-1}}{\Gamma\,(\alpha+1)}\sum_{j=0}^{n}\big(g(t_{j+1})-g(t_j)\big)\left[\frac{u_{i+1}^{j}-2u_i^{j}+u_{i-1}^{j}}{\Delta x^2}\right]\delta_{n,j}^{\alpha}.$$

References

1. Simonovits, A.: Mathematical Methods in Dynamical Economics. MacMillan Press (2000)
2. Solow, R.M.: Contribution to the theory of economic growth. Q. J. Econ. **70**, 65 (1956)
3. Anderson, R.M., May, R.M.: Population biology of infectious diseases, part I. Nature **280**, 361–367 (1979)
4. Tornatore, E., Buccellato, S.M., Vetro, P.: Stability of a stochastic SIR system. Phys. A **354**, 111–126 (2005)
5. Strutt, J.M.: Sci. Papers. Macmillan and Co., London (1943)
6. Cao, J., Ma, C., Xie, H., Jiang, Z.: Nonlinear dynamics of Duffing system with fractional order damping. J. Comput. Nonlinear Dyn. **1**(4), 2–6 (2010)
7. Crank, J., Nicolson, P.: A practical method for numerical evaluation of solution of partial, differential equations of the heat conduction type. Proc. Camb. Phil. Soc. **6**, 50–67 (1996)
8. Lazaridis, T., Karplus, M.: Effective energy functions for protein structure prediction. Curr. Opin. Struct. Biol. **10**(2), 139–145 (2000)

Index

A. Atangana and İ. Koca, *Fractional Differential and Integral Operators with Respect to a Function*, Industrial and Applied Mathematics,
https://doi.org/10.1007/978-981-97-9951-0

www.ingramcontent.com/pod-product-compliance
Lightning Source LLC
Chambersburg PA
CBHW061430230225
22421CB00001B/20

* 9 7 8 9 8 1 9 7 9 9 5 0 3 *